Lecture Notes on Data Engineering and Communications Technologies 202

Series Editor

Fatos Xhafa, *Technical University of Catalonia, Barcelona, Spain*

The aim of the book series is to present cutting edge engineering approaches to data technologies and communications. It will publish latest advances on the engineering task of building and deploying distributed, scalable and reliable data infrastructures and communication systems.

The series will have a prominent applied focus on data technologies and communications with aim to promote the bridging from fundamental research on data science and networking to data engineering and communications that lead to industry products, business knowledge and standardisation.

Indexed by SCOPUS, INSPEC, EI Compendex.

All books published in the series are submitted for consideration in Web of Science.

Leonard Barolli
Editor

Advanced Information Networking and Applications

Proceedings of the 38th International Conference on Advanced Information Networking and Applications (AINA-2024), Volume 4

Editor
Leonard Barolli
Department of Information and Communication
Engineering
Fukuoka Institute of Technology
Fukuoka, Japan

ISSN 2367-4512 ISSN 2367-4520 (electronic)
Lecture Notes on Data Engineering and Communications Technologies
ISBN 978-3-031-57915-8 ISBN 978-3-031-57916-5 (eBook)
https://doi.org/10.1007/978-3-031-57916-5

This Springer imprint is published by the registered company Springer Nature Switzerland AG
The registered company address is: Gewerbestrasse 11, 6330 Cham, Switzerland

Paper in this product is recyclable.

Welcome Message from AINA-2024 Organizers

Welcome to the 38th International Conference on Advanced Information Networking and Applications (AINA-2024). On behalf of AINA-2024 Organizing Committee, we would like to express to all participants our cordial welcome and high respect.

AINA is an International Forum, where scientists and researchers from academia and industry working in various scientific and technical areas of networking and distributed computing systems can demonstrate new ideas and solutions in distributed computing systems. AINA is a very open society and is always welcoming international volunteers from any country and any area in the world.

AINA International Conference is a forum for sharing ideas and research work in the emerging areas of information networking and their applications. The area of advanced networking has grown very rapidly and the applications have experienced an explosive growth, especially in the area of pervasive and mobile applications, wireless sensor and ad-hoc networks, vehicular networks, multimedia computing, social networking, semantic collaborative systems, as well as IoT, big data, cloud computing, artificial intelligence, and machine learning. This advanced networking revolution is transforming the way people live, work, and interact with each other and is impacting the way business, education, entertainment, and health care are operating. The papers included in the proceedings cover theory, design and application of computer networks, distributed computing, and information systems.

Each year AINA receives a lot of paper submissions from all around the world. It has maintained high-quality accepted papers and is aspiring to be one of the main international conferences on the information networking in the world.

We are very proud and honored to have two distinguished keynote talks by Prof. Fatos Xhafa, Technical University of Catalonia, Spain, and Dr. Juggapong Natwichai, Chiang Mai University, Thailand, who will present their recent work and will give new insights and ideas to the conference participants.

An international conference of this size requires the support and help of many people. A lot of people have helped and worked hard to produce a successful AINA-2024 technical program and conference proceedings. First, we would like to thank all authors for submitting their papers. We are indebted to Program Track Co-chairs, Program Committee Members and Reviewers, who carried out the most difficult work of carefully evaluating the submitted papers.

We would like to thank AINA-2024 General Co-chairs, PC Co-chairs, Workshops Organizers for their great efforts to make AINA-2024 a very successful event. We have special thanks to the Finance Chair and Web Administrator Co-chairs.

We do hope that you will enjoy the conference proceedings and readings.

AINA-2024 Organizing Committee

Honorary Chair

Makoto Takizawa	Hosei University, Japan

General Co-chairs

Minoru Uehara	Toyo University, Japan
Euripides G. M. Petrakis	Technical University of Crete (TUC), Greece
Isaac Woungang	Toronto Metropolitan University, Canada

Program Committee Co-chairs

Tomoya Enokido	Rissho University, Japan
Mario A. R. Dantas	Federal University of Juiz de Fora, Brazil
Leonardo Mostarda	University of Perugia, Italy

International Journals Special Issues Co-chairs

Fatos Xhafa	Technical University of Catalonia, Spain
David Taniar	Monash University, Australia
Farookh Hussain	University of Technology Sydney, Australia

Award Co-chairs

Arjan Durresi	Indiana University Purdue University in Indianapolis (IUPUI), USA
Fang-Yie Leu	Tunghai University, Taiwan
Marek Ogiela	AGH University of Science and Technology, Poland
Kin Fun Li	University of Victoria, Canada

Publicity Co-chairs

Markus Aleksy	ABB Corporate Research Center, Germany
Flora Amato	University of Naples "Federico II", Italy
Lidia Ogiela	AGH University of Science and Technology, Poland
Hsing-Chung Chen	Asia University, Taiwan

International Liaison Co-chairs

Wenny Rahayu	La Trobe University, Australia
Nadeem Javaid	COMSATS University Islamabad, Pakistan
Beniamino Di Martino	University of Campania "Luigi Vanvitelli", Italy

Local Arrangement Co-chairs

Keita Matsuo	Fukuoka Institute of Technology, Japan
Tomoyuki Ishida	Fukuoka Institute of Technology, Japan

Finance Chair

Makoto Ikeda	Fukuoka Institute of Technology, Japan

Web Co-chairs

Phudit Ampririt	Fukuoka Institute of Technology, Japan
Ermioni Qafzezi	Fukuoka Institute of Technology, Japan
Shunya Higashi	Fukuoka Institute of Technology, Japan

Steering Committee Chair

Leonard Barolli	Fukuoka Institute of Technology, Japan

Tracks Co-chairs and Program Committee Members

1. Network Architectures, Protocols and Algorithms

Track Co-chairs

Spyropoulos Thrasyvoulos	Technical University of Crete (TUC), Greece
Shigetomo Kimura	University of Tsukuba, Japan
Darshika Perera	University of Colorado at Colorado Springs, USA

TPC Members

Thomas Dreibholz	Simula Metropolitan Center for Digital Engineering, Norway
Angelos Antonopoulos	Nearby Computing SL, Spain
Hatim Chergui	i2CAT Foundation, Spain
Bhed Bahadur Bista	Iwate Prefectural University, Japan
Chotipat Pornavalai	King Mongkut's Institute of Technology Ladkrabang, Thailand
Kenichi Matsui	NTT Network Innovation Center, Japan
Sho Tsugawa	University of Tsukuba, Japan
Satoshi Ohzahata	University of Electro-Communications, Japan
Haytham El Miligi	Thompson Rivers University, Canada
Watheq El-Kharashi	Ain Shams University, Egypt
Ehsan Atoofian	Lakehead University, Canada
Fayez Gebali	University of Victoria, Canada
Kin Fun Li	University of Victoria, Canada
Luis Blanco	CTTC, Spain

2. Next Generation Mobile and Wireless Networks

Track Co-chairs

Purav Shah	School of Science and Technology, Middlesex University, UK
Enver Ever	Middle East Technical University, Northern Cyprus
Evjola Spaho	Polytechnic University of Tirana, Albania

TPC Members

Burak Kizilkaya	Glasgow University, UK
Muhammad Toaha	Middle East Technical University, Turkey
Ramona Trestian	Middlesex University, UK
Andrea Marotta	University of L'Aquila, Italy
Adnan Yazici	Nazarbayev University, Kazakhstan
Orhan Gemikonakli	Final International University, Cyprus
Hrishikesh Venkataraman	Indian Institute of Information Technology, Sri City, India
Zhengjia Xu	Cranfield University, UK
Mohsen Hejazi	University of Kashan, Iran
Sabyasachi Mukhopadhyay	IIT Kharagpur, India
Ali Khoshkholghi	Middlesex University, UK
Admir Barolli	Aleksander Moisiu University of Durres, Albania
Makoto Ikeda	Fukuoka Institute of Technology, Japan
Yi Liu	Oita National College of Technology, Japan
Testuya Oda	Okayama University of Science, Japan
Ermioni Qafzezi	Fukuoka Institute of Technology, Japan

3. Multimedia Networking and Applications

Track Co-chairs

Markus Aleksy	ABB Corporate Research Center, Germany
Francesco Orciuoli	University of Salerno, Italy
Tomoyuki Ishida	Fukuoka Institute of Technology, Japan

TPC Members

Hadil Abukwaik	ABB Corporate Research Center, Germany
Thomas Preuss	Brandenburg University of Applied Sciences, Germany
Peter M. Rost	Karlsruhe Institute of Technology (KIT), Germany
Lukasz Wisniewski	inIT, Germany
Angelo Gaeta	University of Salerno, Italy
Angela Peduto	University of Salerno, Italy
Antonella Pascuzzo	University of Salerno, Italy
Roberto Abbruzzese	University of Salerno, Italy
Tetsuro Ogi	Keio University, Japan

Yasuo Ebara	Osaka Electro-Communication University, Japan
Hideo Miyachi	Tokyo City University, Japan
Kaoru Sugita	Fukuoka Institute of Technology, Japan

4. Pervasive and Ubiquitous Computing

Track Co-chairs

Vamsi Paruchuri	University of Central Arkansas, USA
Hsing-Chung Chen	Asia University, Taiwan
Shinji Sakamoto	Kanazawa Institute of Technology, Japan

TPC Members

Sriram Chellappan	University of South Florida, USA
Yu Sun	University of Central Arkansas, USA
Qiang Duan	Penn State University, USA
Han-Chieh Wei	Dallas Baptist University, USA
Ahmad Alsharif	University of Alabama, USA
Vijayasarathi Balasubramanian	Microsoft, USA
Shyi-Shiun Kuo	Nan Kai University of Technology, Taiwan
Karisma Trinanda Putra	Universitas Muhammadiyah Yogyakarta, Indonesia
Cahya Damarjati	Universitas Muhammadiyah Yogyakarta, Indonesia
Agung Mulyo Widodo	Universitas Esa Unggul Jakarta, Indonesia
Bambang Irawan	Universitas Esa Unggul Jakarta, Indonesia
Eko Prasetyo	Universitas Muhammadiyah Yogyakarta, Indonesia
Sunardi S. T.	Universitas Muhammadiyah Yogyakarta, Indonesia
Andika Wisnujati	Universitas Muhammadiyah Yogyakarta, Indonesia
Makoto Ikeda	Fukuoka Institute of Technology, Japan
Tetsuya Oda	Okayama University of Science, Japan
Evjola Spaho	Polytechnic University of Tirana, Albania
Tetsuya Shigeyasu	Hiroshima Prefectural University, Japan
Keita Matsuo	Fukuoka Institute of Technology, Japan
Admir Barolli	Aleksander Moisiu University of Durres, Albania

5. Web-Based Systems and Content Distribution

Track Co-chairs

Chrisa Tsinaraki	Technical University of Crete (TUC), Greece
Yusuke Gotoh	Okayama University, Japan
Santi Caballe	Open University of Catalonia, Spain

TPC Members

Nikos Bikakis	Hellenic Mediterranean University, Greece
Ioannis Stavrakantonakis	Ververica GmbH, Germany
Sven Schade	European Commission, Joint Research Center, Italy
Christos Papatheodorou	National and Kapodistrian University of Athens, Greece
Sarantos Kapidakis	University of West Attica, Greece
Manato Fujimoto	Osaka Metropolitan University, Japan
Kiki Adhinugraha	La Trobe University, Australia
Tomoki Yoshihisa	Shiga University, Japan
Jordi Conesa	Open University of Catalonia, Spain
Thanasis Daradoumis	Open University of Catalonia, Spain
Nicola Capuano	University of Basilicata, Italy
Victor Ströele	Federal University of Juiz de Fora, Brazil

6. Distributed Ledger Technologies and Distributed-Parallel Computing

Track Co-chairs

Alfredo Navarra	University of Perugia, Italy
Naohiro Hayashibara	Kyoto Sangyo University, Japan

TPC Members

Serafino Cicerone	University of L'Aquila, Italy
Ralf Klasing	LaBRI Bordeaux, France
Giuseppe Prencipe	University of Pisa, Italy
Roberto Tonelli	University of Cagliari, Italy
Farhan Ullah	Northwestern Polytechnical University, China

Leonardo Mostarda	University of Perugia, Italy
Qiong Huang	South China Agricultural University, China
Tomoya Enokido	Rissho University, Japan
Minoru Uehara	Toyo University, Japan
Lucian Prodan	Polytechnic University of Timisoara, Romania
Md. Abdur Razzaque	University of Dhaka, Bangladesh

7. Data Mining, Big Data Analytics and Social Networks

Track Co-chairs

Pavel Krömer	Technical University of Ostrava, Czech Republic
Alex Thomo	University of Victoria, Canada
Eric Pardede	La Trobe University, Australia

TPC Members

Sebastián Basterrech	Technical University of Denmark, Denmark
Tibebe Beshah	University of Addis Ababa, Ethiopia
Nashwa El-Bendary	Arab Academy for Science, Egypt
Petr Musilek	University of Alberta, Canada
Varun Ojha	Newcastle University, UK
Alvaro Parres	ITESO, Mexico
Nizar Rokbani	ISSAT-University of Sousse, Tunisia
Farshid Hajati	Victoria University, Australia
Ji Zhang	University of Southern Queensland, Australia
Salimur Choudhury	Lakehead University, Canada
Carson Leung	University of Manitoba, Canada
Syed Mahbub	La Trobe University, Australia
Osama Mahdi	Melbourne Institute of Technology, Australia
Choiru Zain	La Trobe University, Australia
Rajalakshmi Rajasekaran	La Trobe University, Australia
Nawfal Ali	Monash University, Australia

8. Internet of Things and Cyber-Physical Systems

Track Co-chairs

Tomoki Yoshihisa	Shiga University, Japan
Winston Seah	Victoria University of Wellington, New Zealand
Luciana Pereira Oliveira	Instituto Federal da Paraiba (IFPB), Brazil

TPC Members

Akihiro Fujimoto	Wakayama University, Japan
Akimitsu Kanzaki	Shimane University, Japan
Kazuya Tsukamoto	Kyushu Institute of Technology, Japan
Lei Shu	Nanjing Agricultural University, China
Naoyuki Morimoto	Mie University, Japan
Teruhiro Mizumoto	Chiba Institute of Technology, Japan
Tomoya Kawakami	Fukui University, Japan
Adrian Pekar	Budapest University of Technology and Economics, Hungary
Alvin Valera	Victoria University of Wellington, New Zealand
Chidchanok Choksuchat	Prince of Songkla University, Thailand
Jyoti Sahni	Victoria University of Wellington, New Zealand
Murugaraj Odiathevar	Sungkyunkwan University, South Korea
Normalia Samian	Universiti Putra Malaysia, Malaysia
Qing Gu	University of Science and Technology Beijing, China
Tao Zheng	Beijing Jiaotong University, China
Wenbin Pei	Dalian University of Technology, China
William Liu	Unitec, New Zealand
Wuyungerile Li	Inner Mongolia University, China
Peng Huang	Sichuan Agricultural University, PR China
Ruan Delgado Gomes	Instituto Federal da Paraiba (IFPB), Brazil
Glauco Estacio Goncalves	Universidade Federal do Pará (UFPA), Brazil
Eduardo Luzeiro Feitosa	Universidade Federal do Amazonas (UFAM), Brazil
Paulo Ribeiro Lins Júnior	Instituto Federal da Paraiba (IFPB), Brazil

9. Intelligent Computing and Machine Learning

Track Co-chairs

Takahiro Uchiya	Nagoya Institute of Technology, Japan
Flavius Frasincar	Erasmus University Rotterdam, The Netherlands
Miltos Alamaniotis	University of Texas at San Antonio, USA

TPC Members

Kazuto Sasai	Ibaraki University, Japan
Shigeru Fujita	Chiba Institute of Technology, Japan
Yuki Kaeri	Mejiro University, Japan
Jolanta Mizera-Pietraszko	Military University of Land Forces, Poland
Ashwin Ittoo	University of Liège, Belgium
Marco Brambilla	Politecnico di Milano, Italy
Alfredo Cuzzocrea	University of Calabria, Italy
Le Minh Nguyen	JAIST, Japan
Akiko Aizawa	National Institute of Informatics, Japan
Natthawut Kertkeidkachorn	JAIST, Japan
Georgios Karagiannis	Durham University, UK
Leonidas Akritidis	International Hellenic University, Greece
Athanasios Fevgas	University of Thessaly, Greece
Yota Tsompanopoulou	University of Thessaly, Greece
Yuvaraj Munian	Texas A&M-San Antonio, USA

10. Cloud and Services Computing

Track Co-chairs

Salvatore Venticinque	University of Campania "Luigi Vanvitelli", Italy
Shigenari Nakamura	Tokyo Denki University, Japan
Sajal Mukhopadhyay	National Institute of Technology, Durgapur, India

TPC Members

Giancarlo Fortino	University of Calabria, Italy
Massimiliano Rak	University of Campania "Luigi Vanvitelli", Italy
Jason J. Jung	Chung-Ang University, Korea

Dimosthenis Kyriazis	University of Piraeus, Greece
Geir Horn	University of Oslo, Norway
Dario Branco	University of Campania "Luigi Vanvitelli", Italy
Dilawaer Duolikun	Cognizant Technology Solutions, Hungary
Naohiro Hayashibara	Kyoto Sangyo University, Japan
Tomoya Enokido	Rissho University, Japan
Sujoy Saha	NIT Durgapur, India
Animesh Dutta	NIT Durgapur, India
Pramod Mane	IIM Rohtak, India
Nanda Dulal Jana	NIT Durgapur, India
Banhi Sanyal	NIT Kurukshetra, India

11. Security, Privacy and Trust Computing

Track Co-chairs

Ioannidis Sotirios	Technical University of Crete (TUC), Greece
Michail Alexiou	Georgia Institute of Technology, USA
Hiroaki Kikuchi	Meiji University, Japan

TPC Members

George Vasiliadis	Hellenic Mediterranean University, Greece
Antreas Dionysiou	University of Cyprus, Cyprus
Apostolos Fouranaris	Athena Research Center, Greece
Panagiotis Ilia	Technical University of Crete, Greece
George Portokalidis	IMDEA, Spain
Nikolaos Gkorgkolis	University of Crete, Greece
Zeezoo Ryu	Georgia Institute of Technology, USA
Muhammad Faraz Karim	Georgia Institute of Technology, USA
Yunjie Deng	Georgia Institute of Technology, USA
Anna Raymaker	Georgia Institute of Technology, USA
Takamichi Saito	Meiji University, Japan
Kazumasa Omote	University of Tsukuba, Japan
Masakatsu Nishigaki	Shizuoka University, Japan
Mamoru Mimura	National Defense Academy of Japan, Japan
Chun-I Fan	National Sun Yat-sen University, Taiwan
Aida Ben Chehida Douss	National School of Engineers of Tunis, ENIT Tunis, Tunisia
Davinder Kaur	IUPUI, USA

12. Software-Defined Networking and Network Virtualization

Track Co-chairs

Flavio de Oliveira Silva	Federal University of Uberlândia, Brazil
Ashutosh Bhatia	Birla Institute of Technology and Science, Pilani, India

TPC Members

Rui Luís Andrade Aguiar	Universidade de Aveiro (UA), Portugal
Ivan Vidal	Universidad Carlos III de Madrid, Spain
Eduardo Coelho Cerqueira	Federal University of Pará (UFPA), Brazil
Christos Tranoris	University of Patras (UoP), Greece
Juliano Araújo Wickboldt	Federal University of Rio Grande do Sul (UFRGS), Brazil
Haribabu K.	BITS Pilani, India
Virendra Shekhavat	BITS Pilani, India
Makoto Ikeda	Fukuoka Institute of Technology, Japan
Farookh Hussain	University of Technology Sydney, Australia
Keita Matsuo	Fukuoka Institute of Technology, Japan

AINA-2024 Reviewers

Admir Barolli
Aida ben Chehida Douss
Akimitsu Kanzaki
Alba Amato
Alberto Postiglione
Alex Thomo
Alfredo Navarra
Amani Shatnawi
Anas AlSobeh
Andrea Marotta
Angela Peduto
Anne Kayem
Antreas Dionysiou
Arjan Durresi
Ashutosh Bhatia
Beniamino Di Martino
Bhed Bista
Burak Kizilkaya
Carson Leung
Chidchanok Choksuchat
Christos Tranoris
Chung-Ming Huang
Dario Branco
David Taniar
Elinda Mece
Enver Ever
Eric Pardede
Euripides Petrakis
Evjola Spaho
Fabrizio Messina
Feilong Tang
Flavio Silva
Francesco Orciuoli
George Portokalidis

Giancarlo Fortino
Giorgos Vasiliadis
Glauco Gonçalves
Hatim Chergui
Hiroaki Kikuchi
Hiroki Sakaji
Hiroshi Maeda
Hiroyuki Fujioka
Hyunhee Park
Isaac Woungang
Jana Nowaková
Jolanta Mizera-Pietraszko
Junichi Honda
Jyoti Sahni
Kazunori Uchida
Keita Matsuo
Kenichi Matsui
Kiki Adhinugraha
Kin Fun Li
Kiyotaka Fujisaki
Leonard Barolli
Leonardo Mostarda
Leonidas Akritidis
Lidia Ogiela
Lisandro Granville
Lucian Prodan
Luciana Oliveira
Mahmoud Elkhodr
Makoto Ikeda
Mamoru Mimura
Manato Fujimoto
Marco Antonio To
Marek Ogiela
Masaki Kohana
Minoru Uehara
Muhammad Karim
Muhammad Toaha Raza Khan
Murugaraj Odiathevar
Nadeem Javaid
Naohiro Hayashibara
Nobuo Funabiki
Nour El Madhoun
Omar Darwish
Panagiotis Ilia
Petr Musilek
Philip Moore
Purav Shah
R. Madhusudhan
Raffaele Guarasci
Ralf Klasing
Roberto Tonelli
Ronald Petrlic
Sabyasachi Mukhopadhyay
Sajal Mukhopadhyay
Salvatore D'Angelo
Salvatore D'Angelo
Salvatore D'Angelo
Salvatore Venticinque
Santi Caballé
Satoshi Ohzahata
Serafino Cicerone
Shigenari Nakamura
Shinji Sakamoto
Sho Tsugawa
Sriram Chellappan
Stephane Maag
Takayuki Kushida
Tetsuya Oda
Thomas Dreibholz
Tomoki Yoshihisa
Tomoya Enokido
Tomoya Kawakami
Tomoyuki Ishida
Vamsi Paruchuri
Victor Ströele
Vikram Singh
Wei Lu
Wenny Rahayu
Winston Seah
Yong Zheng
Yoshitaka Shibata
Yusuke Gotoh
Yuvaraj Munian
Zeezoo Ryu
Zhengjia Xu

AINA-2024 Keynote Talks

Agile Edge: Harnessing the Power of the Intelligent Edge by Agile Optimization

Fatos Xhafa

Technical University of Barcelona, Barcelona, Spain

Abstract. The digital cloud ecosystem comprises various degrees of computing granularity from large cloud servers and data centers to IoT devices, leading to the cloud-to-thing continuum computing paradigm. In this context, the intelligent edge aims at placing intelligence to the end devices, at the edges of the Internet. The premise is that collective intelligence from the IoT data deluge can be achieved and used at the edges of the Internet, offloading the computation burden from the cloud systems and leveraging real-time intelligence. This, however, comes with the challenges of processing and analyzing the IoT data streams in real time. In this talk, we will address how agile optimization can be useful for harnessing the power of the intelligent edge. Agile optimization is a powerful and promising solution, which differently from traditional optimization methods, is able to find optimized and scalable solutions under real-time requirements. We will bring real-life problems and case studies from Smart City Open Data Repositories to illustrate the approach. Finally, we will discuss the research challenges and emerging vision on the agile intelligent edge.

Challenges in Entity Matching in AI Era

Juggapong Natwichai

Chiang Mai University, Chiang Mai, Thailand

Abstract. Entity matching (EM) is to identify and link entities originating from various sources that correspond to identical real-world entities, thereby constituting a foundational component within the realm of data integration. For example, in order to counter-fraud detection, the datasets from sellers, financial services providers, or even IT infrastructure service providers might be in need for data integration, and hence, the EM is highly important here. This matching process is also recognized for its pivotal role in data augmenting to improve the precision and dependability of subsequent tasks within the domain of data analytics. Traditionally, the EM procedure composes of two integral phases, namely blocking and matching. The blocking phase associates with the generation of candidate pairs and could affect the size and complexity of the data. Meanwhile, the matching phase will need to trade-off between the accuracy and the efficiency. In this talk, the challenges of both components are thoroughly explored, particularly with the aid of AI techniques. In addition, the preliminary experiment results to explore some important factors which affect the performance will be presented.

Contents

The Deployment of E-Learning Application as a Web Service in a Cloud Broker Architecture

Rihem Zorgati[1](✉), Hamdi Hassen[2], and Khlil Ahmad Alsulbi[3]

[1] Business Computing Department, Higher Institute of Management, University of Sousse, Sousse, Tunisia
zorgatirihem@gmail.com
[2] Mir@cl Lab, FSEGS University of Sfax, Sfax, Tunisia
[3] Computers Department, College of Engineering and Computers in Al-Qunfodah Umm Al-Qura University, Makkah, Saudi Arabia
kasulbi@uqu.edu.sa

Abstract. In the wake of the COVID-19 outbreak, university operations were suspended, and online learning therefore was the best option to reduce the spread of the pandemic. To ensure performant, robust and accurate E-learning applications, intelligent algorithms such as Deep Learning (DL) and Machine Learning (ML) are essential, demanding significant resources. Research suggests distributing E-learning applications across Grid Computing, Peer-to-Peer (P2P) networks and Cloud Computing (CC) environments. Exam scenarios, requiring high availability, highlight the inadequacy of a single cloud. Moreover, diverse Internet of Things (IoT) devices used by E-learning users necessitate an adaptable infrastructure. The multicloud or Cloud Broker (CB) architecture is our suggested approach for the deployment of E-learning application to optimize the experience, overcoming time constraints and the challenge of managing numerous accounts. Experimental results validate that multicloud or CB architecture is an effective infrastructure for the development of potent E-learning tools and for boosting performance.

Keywords: E-Learning · Covid-19 · Distributed Computing · Cloud Computing · Cloud Broker · IoT · Real Time

1 Introduction

The conventional university, as an establishment providing courses on-site, must recognize the opportunities and risks offered by new technologies, to maintain its prestigious position [1]. E-learning is one of the most popular academic topics today and is playing a significant part in the contemporary educational environment [2]. In the realm of E-learning, educators engage with a variety of IoT devices, such as tablets, smartphones and other connected gadgets.

The spread of COVID-19 has caused international activities, especially educational ones to switch to online learning using existing educational platforms [3].

L. Barolli (Ed.): AINA 2024, LNDECT 202, pp. 1–12, 2024.
https://doi.org/10.1007/978-3-031-57916-5_1

There are E-learning platforms that are deployed on the cloud, but there are challenges in terms of compliance, security, cost and rigidity [4]. E-learning applications that are high-performing, reliable and precise rely on intelligent algorithms like DL, ML, K-Nearest Neighbors (KNN) and Support Vector Machine (SVM) which require significant bandwidth and computational power.

Multicloud refers to the use of separate cloud platforms, each with a user interface that can connect to various management and implementation domains. CB acts as an intermediary between customers and providers [5]. For thus, we recommend utilizing a distributed platform like CC; however, it may not always be the optimal choice. To overcome the limitations of CC, we propose usage of CB to resolve technical and financial problems.

The remainder of this paper is structured as follows: in Sect. 2, we clarify concepts related to E-learning, CC and multicloud environments and cloud providers. In Sect. 3, we present a summary of the related works. In Sect. 4, we discuss the problem statement and the proposed approach in Sect. 5. In Sect. 6, we outline the experimental study and in the conclusion, we summarize the findings of this research and outline areas for future exploration.

2 E-Learning, Cloud Computing and Multicloud Environments: A General Overview

2.1 E-learning

E-learning, encompassing online and distance learning, has become widely recognized as an optimal solution for curbing the spread of epidemics. Its ease of management allows students to rapidly access instructors and course materials [3].

With the emergence of the IoT environment, E-learning based on this technology poses new challenge. E-learning in an IoT environment refers to the use of IoT technology to enhance and facilitate online education and training [6].

2.2 Cloud Computing Environment

The emergence of CC was driven by advancements in various technologies, including hardware, internet, distributed computing and systems management [7].

A cloud is a parallel and distributed system comprising a network of interconnected and virtualized computers. These computers are dynamically provisioned and presented as one or more cohesive computing resources [4].

2.3 Multicloud Environment

Multicloud, or Cloud of Clouds, describes CC systems where applications are segmented and distributed across a network of multiple clouds. It involves utilizing independent cloud platforms unified under a single interface that can link to a variety of administrative and operational domains [8].

A CB manages and coordinates cloud services and contracts between cloud users and cloud service providers [5].

2.4 Cloud and Cloud Broker Providers

A cloud provider service, offered by companies handling large-scale internet applications, consists of hardware and software solutions [9]. Key players like Amazon Web Services (AWS) [10], Google Cloud Platform (GCP) [11] and Microsoft Windows Azure [12] not only offer individual services but also function in both single and multicloud capacities.

3 Related Works

This section is dedicated to various approaches in the literature, where numerous methodologies have been suggested for E-learning applications.

Table 1. State of the Art

Authors	Proposed Approaches	Algorithms/Methods
Rezvan et al. (2023) [13]	Personalized Learning in Virtual Learning Environments Using Students' Behavior Analysis	Light Gradient Boosting Machine (LGBM)
Fatima et al. (2023) [14]	Integration of evolutionary algorithm in an agent-oriented approach for an adaptive E-learning	Genetic algorithm
Mir et al. (2022) [15]	Artificial Intelligence (AI)-Based Personalized E-Learning Systems	Deep learning recommendation
Yuhui et al. (2022) [16]	Learning Style Integrated Deep Reinforcement Learning (DRL) Framework for Programming Problem Recommendation in Online Judge System	DRL Bidirectional Gated Recurrent Units (Bi-GRUs)
Bens et al. (2022) [17]	AI-Based Learning Style Prediction in Online Learning for Primary Education	Collaborative Filtering (CF)

Table 1 provides an insightful overview of various proposed approaches that showcase diverse strategies for enhanced learning experiences, paving the way for advancements in digital education.

4 Problem Statement

The global impact of COVID-19 has necessitated the adoption of E-learning as a vital solution for uninterrupted education. This transition to E-learning demands significant bandwidth, robust computing resources and a high level of availability. The utilization of intelligent algorithms, including DL and ML plays a role in ensuring optimal performance, reliability and accuracy of E-learning applications. These algorithms are known for their capability to process vast amounts of data, but they also impose significant demands on bandwidth and computational power, which need to be considered.

In the field of E-learning, various research studies propose the adoption of distributed architectures such as Grid Computing, P2P networks and CC environments. These architectures offer the potential for scalability, resource sharing and enhanced performance by leveraging multiple computing resources. Notably, certain research findings caution against a blind reliance on distributed architectures and CC, particularly in critical situations like exams, where the availability and reliability of the E-learning application are of paramount importance. While distributed architectures can provide flexibility and scalability, they may also introduce potential challenges in terms of ensuring consistent and reliable application access.

E-learning faces significant challenges in universities across various nations which are primarily associated with user adoption and utilization. Consequently, numerous issues have been identified and categorized into four main areas for in-depth analysis, as illustrated in Fig. 1. Users of E-learning applications can connect from various IoT devices that differ in terms of storage capacity, computing power, operating systems and more. Consequently, managing these diverse devices poses a significant challenge.

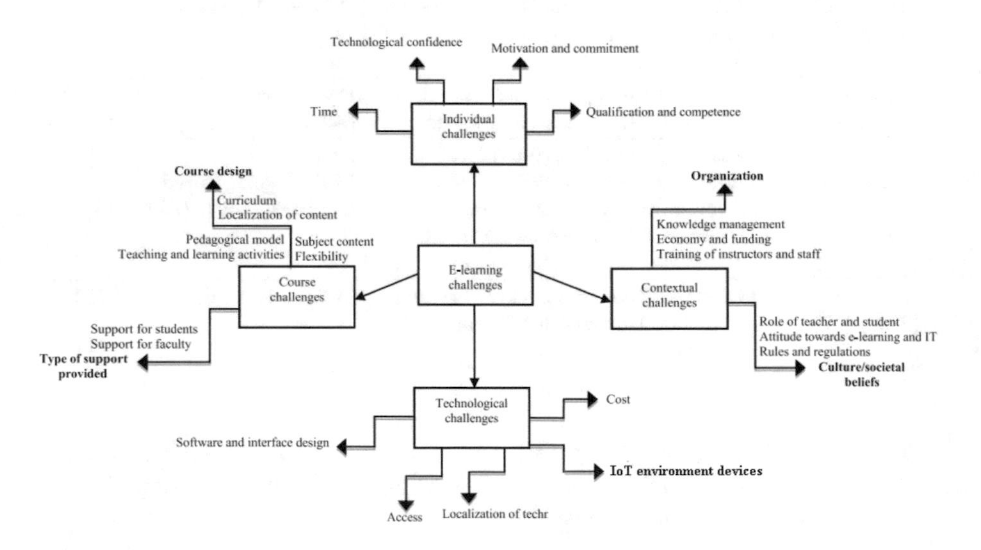

Fig. 1. E-Learning Challenges [18]

Thirty challenges were identified and classified into four major categories: individual (involving instructors and students), course-related, contextual and technological challenges.

4.1 E-learning Technical Obstacles

Information and Communication Technology (ICT) infrastructure E-learning faces hurdles due to underfunded classrooms and energy shortages in rural areas, impacting ICT infrastructure. Technical support is lacking, with inadequate facilities and staff for tasks like installation, operation, maintenance and security, exacerbated by slow Internet speeds and heavy traffic during online learning [19].

4.2 E-learning Based on IoT Devices Obstacles

The efficiency of IoT in education enables diverse models of learning, providing students with access to unlimited resources beyond the classroom. The overarching goal is to create a modern and efficient knowledge acquisition environment tailored to student needs. While IoT offers numerous advantages in E-learning, challenges include communication issues between devices, device capacity, memory constraints, computing power, and bandwidth limitations [19].

4.3 E-learning Financial Obstacles

E-learning, often more costly than traditional education, requires significant financial investment, particularly during COVID-19, leading to higher development costs [19].

5 Proposed Approach

The pivotal inquiry revolves around the efficacy of E-learning approaches in addressing challenges and ushering in significant advancements for students in both their university learning experiences and future employment prospects. Our proposed solution involves deploying our E-learning application on a cloud platform, utilizing it as both a platform and infrastructure for storing educational materials. Additionally, we advocate for leveraging the CB to enhance our E-learning solution.

E-learning has proven invaluable in times of crises, such as wars and pandemics like COVID-19, with substantial advancements over the past three decades. Key to our approach is the incorporation of online examinations, enabling precise assessment but presenting unique challenges in planning and execution. To overcome these challenges, we propose a multicloud environment, utilizing the strengths of multiple cloud vendors to ensure scalability, data security and resource availability during online tests.

This strategic choice aims to enhance the overall reliability, flexibility and accessibility of our E-learning system in the context of IoT-driven educational environments, providing a dynamic platform that meets the evolving needs of students and educators in a digital learning landscape. Furthermore, the success of our E-learning application is deeply rooted in its integration of intelligent algorithms including DL and ML, Play a crucial role in forming the adaptive characteristics of our platform.

By simplifying management and providing a centralized for provisioning, monitoring and billing, the CB streamlines the deployment process and maximizes the benefits derived from utilizing different Cloud Service Providers as shown in Fig. 2. Cloud Services Brokers (CSBs) play a critical role as intermediaries between users and providers like AWS, GCP and Azure, streamlining processes. CSBs simplify multicloud complexities, offering unified interfaces, management and added services like monitoring and cost optimization.

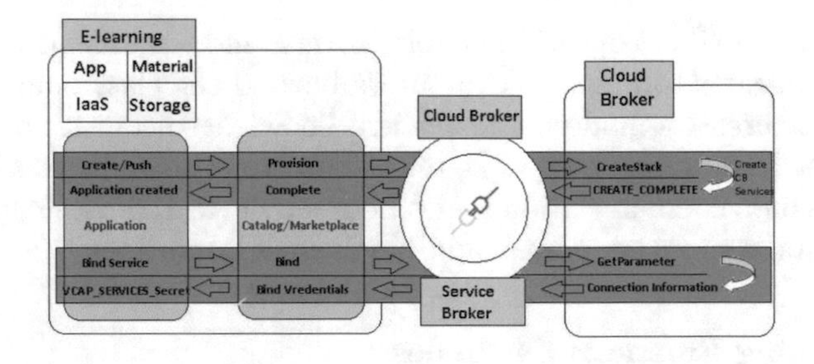

Fig. 2. Cloud Service Brokerage [20]

CSBs maximize cloud benefits for multicloud operations, reducing complexity and boosting productivity while ensuring control and security, aligning with the enhanced capabilities brought by IoT systems in the realm of E-learning.

5.1 E-learning Application in a Cloud Environment

To improve our E-learning solution, we propose leveraging the capabilities of the CC environment. By tapping into the CC framework, we aim to optimize different aspects of our E-learning application, specifically capitalizing on the advantages offered by Infrastructure as a Service (IaaS), Platform as a Service (PaaS) and Software as a Service (SaaS) as shown in Fig. 3.

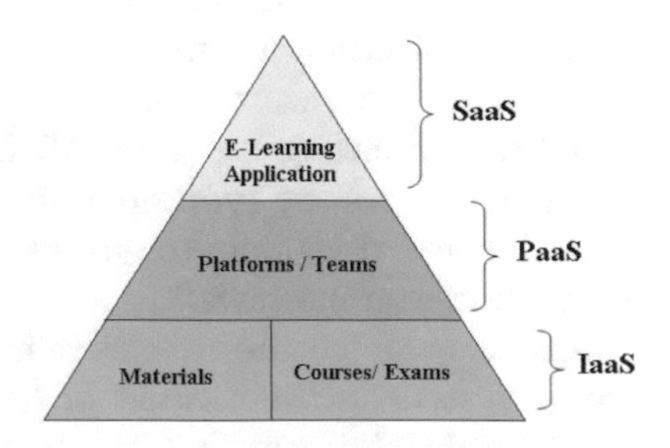

Fig. 3. E-Learning Based on Cloud Environment

We utilize IaaS for convenient access to educational content, leverage PaaS for efficient storage and organization and exemplify SaaS for a scalable and user-friendly E-learning application accessible on-demand. The integrated use of IaaS, PaaS and SaaS within the CC framework enhances the overall performance, accessibility and adaptability of our E-learning system in meeting the dynamic needs of modern digital education.

5.2 E-learning Application in a Multicloud Environment

To enhance our E-learning solution, we propose harnessing the capabilities of the CB environment. The CB operates as a mediator, employing algorithms for negotiation and collaboration with various cloud providers. The CB framework operates by intelligently distributing workloads, optimizing resource allocation and ensuring scalability as shown in Fig. 4.

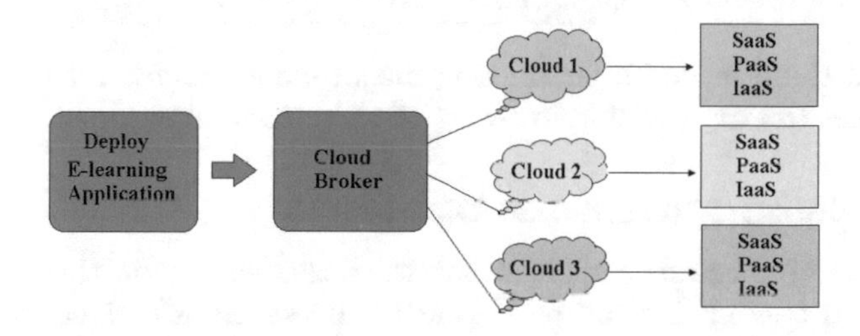

Fig. 4. E-learning Based on Multicloud Environment

CB essentially acts as a coordinator, orchestrating the combined power of IaaS, PaaS and SaaS to enhance the overall performance and flexibility of our E-learning platform. This approach provides flexibility in deployment, allowing us to adapt to changing requirements and optimize resource utilization. The benefits lie in the efficient use of resources, improved scalability and the ability to capitalize on the unique features and strengths offered by a multicloud environment, ultimately enhancing the overall reliability and performance of our E-learning application.

6 Experimental Study

6.1 Datasets

The proposed approach underwent evaluation with a manually selected random database, including video and PDF courses, and exams of varying sizes. The assessment involved a corpus of 20 videos (2 MB to 2 GB) and PDFs (600 KB to 8 MB), considering diverse devices like computers and smartphones for a comprehensive evaluation across platforms.

6.2 Experimental Settings

We deploy our application on AWS, GCP and Azure to compile a thorough of executions. Numerous companies, such as Flexera [21], SpotCloud [22] and BlueWolf [23], offer brokering solutions for the current cloud technologies. Table 2 is the key driving factor for our application's infrastructure.

Table 2. The Ranking of Leading Cloud and Cloud Broker Providers [24]

Standards	Rank-1	Rank-2	Rank-3
Market Entry Date	AWS	GCP	Azure
Availability Regions	AWS	Azure	GCP
Market Share	AWS	Azure	GCP
Growth Rate	GCP	Azure	AWS
Services Offered	AWS	Azure	GCP
Quantity of Data Centers	AWS	Azure	GCP

Based on the provided table ranking major cloud vendors, it is evident that AWS occupies the top position in terms of all criteria mentioned.

6.3 Experimental Results and Discussions

Various trials of E-learning applications were carried out on three single-cloud platforms and one multi-cloud platform. To assess the effectiveness of the proposed solution and illustrate the appeal of developing an E-learning application using CB architecture, the following metrics will be taken into account:

Multi-user Sign-in. According to our test, the average loading time with number of users equals to 3 is presented in Fig. 5:

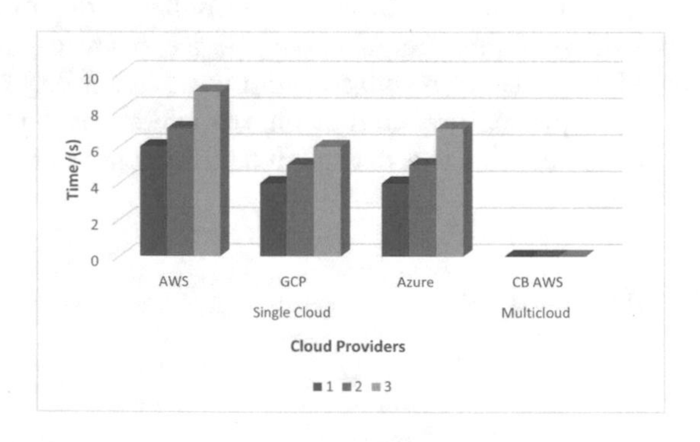

Fig. 5. Multi-User Sign-In

For instance, with one user, AWS takes about 6 s, while GCP and Azure are slightly faster at around 4 s. AWS broker offers nearly instant loading times.

Execution of Video Courses. The Fig. 6 demonstrates the average time for video execution based on file size:

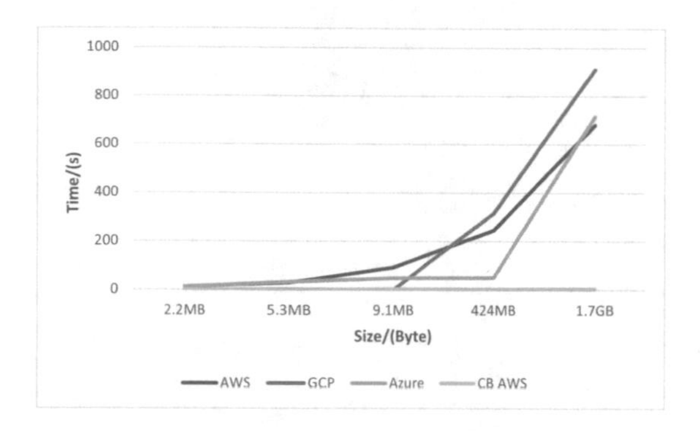

Fig. 6. Execution of Video Courses

For a significantly larger 424-Megabyte video, AWS takes around 4 min, GCP about 5 min, Azure 50 s and AWS broker merely 3 s to load.

Uploading of Video Courses. Fig. 7 illustrates a demonstration of the average time needed to start the uploading video depending on its size:

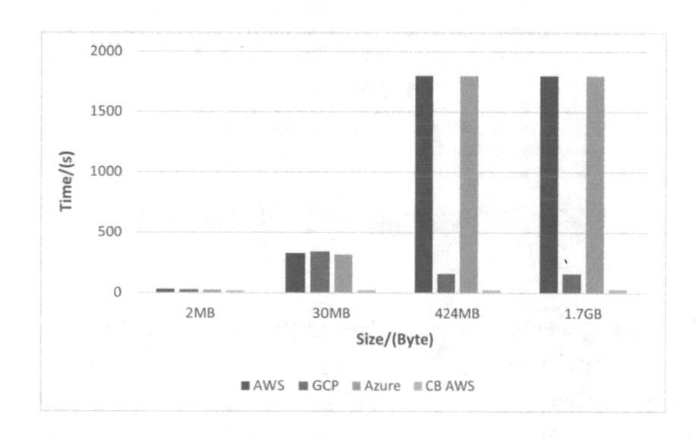

Fig. 7. Uploading of Video Courses

For larger videos like 424 megabytes and 1.7 gigabytes, they take 30 min to load on AWS and Azure, around 2 min on GCP and only 26 s on AWS broker.

Practice the Exam. In the application, especially during exams, the average time for both single cloud and multicloud scenarios is shown in Fig. 8:

In a single cloud setup, task execution typically takes 1 to 3 s.

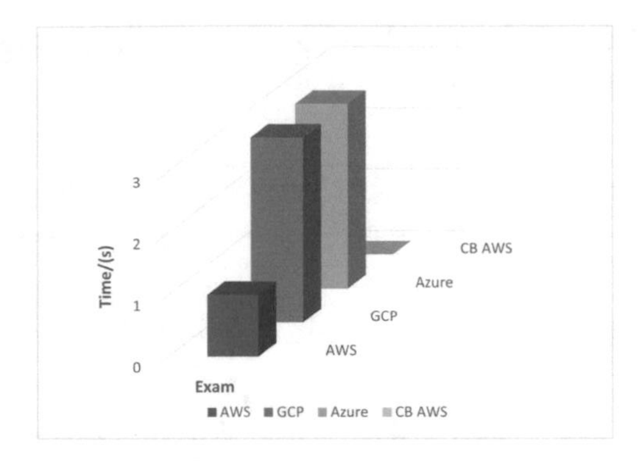

Fig. 8. Practice the Exam

Uploading of PDF Courses. The Fig. 9 shows the average startup time for executing PDF files based on their respective sizes:

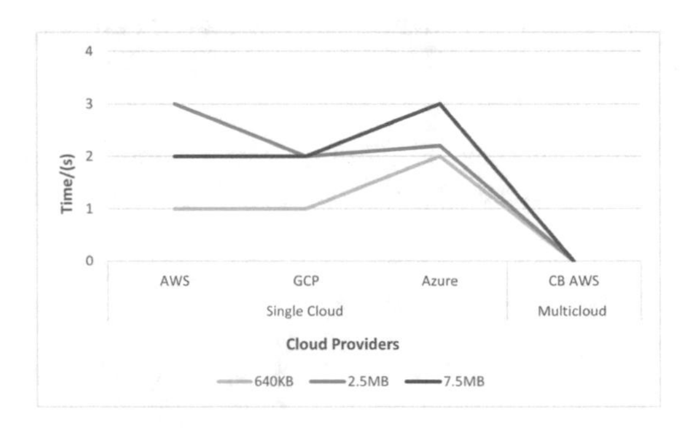

Fig. 9. Uploading of PDF Courses

For a larger 7.5 megabytes PDF, it takes about 2 s to load on AWS and GCP, 3 s on Azure and instantly on AWS broker.

Based on the experiments conducted involving the loading times of various elements, it can be firmly concluded that AWS CB emerges as the superior provider among the options assessed. However, it should be noted that sometimes AWS exhibits times superior to 0 s and we performed calculations in seconds rather than milliseconds.

7 Conclusion

In our work, we propose a new robust E-learning application based on CB architecture. The experimental results support the utilization of a CB architecture to establish a multicloud environment for E-learning applications, significantly enhancing power and reliability. The following benefits are provided by using AWS for our application: cloud interoperability, application portability, elasticity and cost savings. We intend to launch the E-learning application as a web service as additional explorations of the suggested strategy because it is more difficult and less enjoyable than it should be.

While the research has yielded promising results, there is still room for improvement. Future plans include enhancing the functionality of the E-learning application and exploring the utilization of other multicloud platforms to expand the scope of the research. Additionally, there is a goal to deploy it as a service called "E-learning as a Service (E-LeaaS)" in the multicloud environment. This would allow users to easily access the application and its features without the need for complex setup or installation processes.

References

1. Abdeldjalil, L., Hakim, B., Sofiane, M.: Fault tolerance in distributed systems: a survey. In: 3rd International Conference, on Pattern Analysis and Intelligent Systems (PAIS). IEEE (2018)
2. Tunisia Conference: Monastir University Digital Festival (2022). https://fb.watch/bpBZK3dZpt/. Accessed 31 Dec 2023
3. Abdelsalam, M., Ebitisam, K., Shadi, A., Hasan, R., Hadeel, A.: The COVID 19 pandemic and E learning: challenges and opportunities from the perspective of students and instructors. Journal of Computing in Higher Education. Springer Journal (2021). https://doi.org/10.1007/s12528-021-09274-2
4. Mohammad, M., Ali, A., Azlin, N., Imad, F.: Disaster Recovery in Single-Cloud and Multi-Cloud Environments: Issues and Challenges. IEEE (2017)
5. Jing, M., Kenli, L., Senior, M., Zhao, T., Qian, L., Keqin, L.: Profit Maximization for Cloud Brokers in Cloud Computing, 1045-9219 IEEE (2018). https://doi.org/10.1109/tpds.2018.2851246
6. Zaid, A., Tanupriya, C., Subhash, C., Gaurav R.: Internet of Things and its applications in E-learning. In: 3rd IEEE International Conference on Computational Intelligence and Communication Technology, IEEE-CICT (2017)
7. Ali, S.: Internet Computing: Principles of Distributed Systems and Emerging Internet-Based Technologies. Springer Nature Switzerland AG, pp. 195–226 (2020)
8. Leonard, B., Makoto, T., Fatos, X., Tomoya, E.: Web, Artificial Intelligence and Network Applications. Springer Nature Switzerland AG, WAINA, AISC 927, pp. 1055–1068 (2019). https://doi.org/10.1007/978-3-030-15035-8_103
9. Rawan, A., Ali, E., Fahed, J.: A comparative review of high-performance computing major cloud service providers. In: 9th International Conference on Information and Communication Systems (ICICS) (2018)
10. Amazon Web Service. https://aws.amazon.com/. Accessed 31 Dec 2023
11. Google Cloud Platform. https://cloud.google.com/. Accessed 31 Dec 2023
12. Microsoft Windows Azure. https://azure.microsoft.com/. Accessed 31 Dec 2023
13. Rezvan, N., Houshang, D.: Personalized learning in virtual learning environments using students' behavior analysis. Educ. Sci. (2023). https://doi.org/10.3390/educsci13050457
14. Fatima, Z.L., Otman, A.: Integration of evolutionary algorithm in an agent-oriented approach for an adaptive E-learning. Int. J. Electr. Comput. Eng. (IJECE) **13**(2),

1964–1978 (2023). ISSN: 2088-8708. https://doi.org/10.11591/ijece.v13i2.pp1964-1978
15. Mir, M., Yamna, A., Jawwad, A.S., Fahad, S., Mariam, U.: AI-Based Personalized E-Learning Systems: Issues, Challenges, and Solutions. National Center of Artificial Intelligence (NCAI), Pakistan, vol. 10. IEEE (2022). https://doi.org/10.1109/access.2022.3193938
16. Yuhui, X., Qin, N., Shuang, L., Yifei, M., Yangze, Y., Yujia, H.: Learning style integrated deep reinforcement learning framework for programming problem recommendation in online judge system. Int. J. Comput. Intell. Syst. (2022). https://doi.org/10.1007/s44196-022-00176-4
17. Bens, P., Teddy, S., Tjeng, W.C., Digdo, S., Andri, A.: AI-Based Learning Style Prediction in Online Learning for Primary Education. IEEE, vol. 10 (2022). https://doi.org/10.1109/access.2022.3160177
18. Hanan, A., Hosam, A., Samar, G.: How Course, Contextual and Technological Challenges Are Associated With Instructors' Individual Challenges to Successfully Implement E-Learning: A Developing Country Perspective. IEEE 2169-3536, vol. 7 (2019). https://doi.org/10.1109/access.2019.2910148
19. Mohammed, A., Ahmad, K., Ahmad, A.: Exploring the critical challenges and factors influencing the E-learning system usage during COVID-19 pandemic. Educ. Inf. Technol. **25**, 5261–5280 (2020). https://doi.org/10.1007/s10639-020-10219-y
20. Amazon Web Services Service Broker. https://aws.amazon.com/partners/servicebroker/. Accessed 31 Dec 2023
21. Flexera. https://www.flexera.com/. Accessed 31 Dec 2023
22. SpotCloud. https://www.spotcloud.io/. Accessed 31 Dec 2023
23. BlueWolf. https://bluewolftravel.com/. Accessed 31 Dec 2023
24. Manish, S., R.C., Tripathi: Cloud Computing: Comparison and Analysis of Cloud Service Providers-AWS, Microsoft and Google. In: 9th International Conference on System Modeling & Advancement in Research Trends, 4th-5th, Proceedings of the SMART-2020, IEEE Conference ID: 50582, ISBN: 978-1-7281-8908-6, India (2020). https://doi.org/10.1109/smart50582.2020.9337100

CarbonApp: Blockchain Enabled Carbon Offset Project Management

Yining Hu[1(✉)], Alistair McFarlane[2], and Farookh Hussain[1]

[1] University of Technology Sydney, 81 Broadway, Ultimo, NSW 2007, Australia
{yining.hu,farookh.hussain}@uts.edu.au
[2] Ground Floor, 84 Greenhill Road, Wayville, SA 5034, Australia

Abstract. Carbon credits are permits to emit Green House Gases (GHGs) issued by standards bodies. They are generated from carbon offset projects, and purchased by companies that pollute. Due to the lack of transparency in carbon accounting, the fragmented data collection and crediting systems, and the complexity in claiming carbon credits, there have been many criticisms around the integrity of carbon credit schemes. The double-counting of carbon credits, i.e., issuing the same carbon credits to more than one entity, cannot be effectively prevented in existing systems. Therefore, many organizations are reluctant to participate in offset projects. Blockchain has been used in a wide range of supply-chain scenarios for its immutability, decentralisation and security, and has been considered in designing carbon credit trading platforms. However, its potential in offset project management has not been fully explored. In this paper, we propose a novel blockchain based solution–the *CarbonApp*–consisting of a set of smart contracts deployed on a private Polygon blockchain to enable the monitoring and management of offset projects.

1 Introduction

Green house gases (GHGs) originating from human activities are a major contributor to climate change. The Kyoto Protocol [6] and the Paris Agreement [2] both set legally binding targets for individual countries to limit their emissions. To meet these targets, governments around the world are establishing more stringent regulations on Environmental, Social, Governmental (ESG) reporting. In Australia, with the government's legislated commitment to a 43% reduction in emissions from 2005 levels by 2030 and achieving net zero emissions by 2050, companies are challenged to embed climate and sustainability into strategic decision making. As a result, *carbon crediting schemes* are increasingly used to quantify and control emissions. Carbon credits are issued to projects that reduce emissions, and purchased by companies that pollute. One carbon credit represents one ton of CO_2 or its equivalent. Mandatory carbon markets, such as the Emission Reduction Fund (ERF) administered by the Clean Energy Regulator (CER) in Australia, enforces emission reduction targets for participating companies. Voluntary carbon markets regulated by international standards bodies, such as the Verra

L. Barolli (Ed.): AINA 2024, LNDECT 202, pp. 13–25, 2024.
https://doi.org/10.1007/978-3-031-57916-5_2

Verified Carbon Standard (VCS)[1] and United Nations Framework Convention on Climate Change (UNFCCC),[2] allow trading carbon credits like commodities.

To receive carbon credits, an organization first registers an offset project that applies an approved methodology. The methodology should have not been used in previous years, or the *baseline* scenario to provide "additionality". Then, an external validation/verification body validates the baseline data and the project data before the standards body issues carbon credits. The data collection is challenging, as many of these offset projects are carried out on farms where no systematic and reliable data management tools are in place. This makes carbon accounting difficult and can lead to *over-crediting*. In Australia, it has been reported that some offset projects do not achieve the environmental benefits as they claim [11]. Verra was also reported to have miscalculated the carbon credit worth of its Deforestation and Forest Degradation (REDD+) projects [7,9]. Despite the justification provided by CER and Verra, it is undeniable that there can be major disagreements when evaluating an offset project's real impact. Moreover, *double-counting*, which refers to the duplicated issuance of carbon credits for the same offset project to different entities, cannot be effectively prevented. As a result, many organizations are reluctant to participate in offset projects.

For offset projects to achieve their environmental goals, a more transparent system that enables trusted data sharing is needed. Blockchain, being a decentralised, immutable ledger, can potentially be leveraged to manage offset projects and enable more accurate evaluation of their impact [5]. Many existing studies focus on using blockchain for carbon credit trading and ownership management, such as the Carbon Credit Ecosystem proposed by Saraji and Borowczak [16], and the trading platform developed by Patel et al. [13]. However, these studies do not address the provenance of carbon credits or the monitoring of offset projects. In this paper, we present the *CarbonApp*, a blockchain based platform to manage offset projects, consisting of a set of Solidity [8] smart contracts and corresponding RESTful APIs. The CarbonApp was developed primarily for a CH_4 reduction project conducted on a dairy farm in South Australia. The CarbonApp is deployed on a private Polygon [1] blockchain network.

The rest of this paper is organized as follow. Section 2 provides background on Verra carbon schemes, the Polygon blockchain and tokenization. Section 3 explains the carbon reduction methodology used in this project, the practical set-up on farm and data collection. Section 4 and Sect. 5 then demonstrate the design, implementation and evaluation of the CarbonApp. Section 6 surveys the related work. Finally Sect. 7 concludes the paper and discusses future improvements.

2 Background

2.1 Verra Carbon Credits

Verra is a US based standards body that recognizes various emission reduction methodologies such as regenerative farming, enteric CH_4 reduction in ruminants, etc. Our

[1] https://verra.org/programs/verified-carbon-standard/.
[2] https://unfccc.int/.

project followed the methodology VM0041, which evaluates the enteric CH_4 emission reduction in ruminants through the use of an approved feed additive [18]. Upon registration, the project proponent provides details of the participating farms, herds and locations. A Verra project registration can be started by an "aggregator" who works with individual farms. This type of project is known as a "grouped project". It allows new instances to be added without going through the full validation process, which greatly reduces administrative cost and time. In a dairy supply chain, the "aggregator" could be a dairy producer who contracts multiple suppliers. Inspection performed by independent verification bodies or auditors is required before Verra can issue carbon credits. The price of a carbon credit can be affected by the location of the project, the issuing body, and its provenance and verification. Providing traceability information can increase the price of a carbon credit [14].

To establish the baseline emission, the project proponent provides data in the past three years falling in one of the three categories: 1) direct measurement of enteric CH_4 emissions; 2) daily average gross energy intake of the herds, derived from the Dry Matter Intakes (DMIs) and the number of days the animals spent on farm, or 3) number of days the animals spent on farm, multiplied by national or regional emission factors. To monitor the project emissions, the project proponent can: 1) apply the emission reduction factor claimed by the producer of the feed additive, or 2) directly measurement enteric CH_4 emissions.

2.2 The Polygon Blockchain and Tokenization

Blockchain is a peer-to-peer (P2P) system initially designed to support financial transactions [12]. With the development of Ethereum [8], blockchains were extended with smart contracts capabilities to handle non-financial transactions with complex transactional terms [8]. To improve transaction throughput and reduce transaction fees on Ethereum, many scaling solutions were proposed. One of the most successful platforms is Polygon, a *layer 2* solution that can process up to 65,000 transactions per second (compared to 30 transactions per second on Ethereum). Similar to Ethereum, Polygon also supports smart contracts in the Solidity language [8].

Tokenization is a process that uses tokens to represent physical assets or create their *digital twins* on-chain. Tokens can be fungible like coins, or non-fungible if the item they represent is unique. The ERC-1155 Multi Token Standard[3] is compatible with fungible tokens, non-fungible tokens and semi-fungible tokens.

3 Project Set-Up and Data Collection

3.1 Practical Set-Up on Farm

We used a commercially available product, SEAFEED^{TM}–an asparagopsis oil in canola extract–in the dairy cows' regular total mixed ration (TMR) when they were milked. Asparagopsis is an edible, safe type of red seaweed rich in bromoform, an organic compound that can effectively interfere with the digestion process to reduce CH_4 production in ruminants [10, 15].

The cows were milked twice a day, once in the morning between 5am-8am, and once in the afternoon between 2pm-5pm in a herringbone cowshed as shown in Fig. 1. When a cow enters the cowshed, the RFID tag attached to its ear is scanned and recorded in the herd management system DelPro.[3] DelPro also records the milking duration of and the yield of each cow. To measure CH_4 emission, we also installed CH_4 sensors at each cow space just above the feed trough. The CH_4 concentration was measured in parts-per-million (ppm).[4] The CH_4 measurement system cannot measure the exact amount of CH_4 emitted by a cow, however, as the measurement is performed at the same time everyday, it can detect the absolute changes in CH_4 level. The measurement is the most reliable when the cows eat from the troughs, with their heads down. This usually lasts for about 10min from the time they enter the shed.

Fig. 1. Cows being milked in a herringbone shed.

The project was conducted on 40 cows feeding on SEAFEED^{TM}mixed TMR as the trial herd, and 40 cows of similar health conditions feeding only on TMR as the control herd, from March 20 to May 30, a total of 72 days. We started by feeding 170 ml of SEAFEED^{TM}per cow per day (*Phase 1*), then reduced to 140 ml SEAFEED^{TM}per cow per day on April 18 (*Phase* 2), and reduced again to 110 ml per cow per day on May 16 (*Phase 3*). SEAFEED^{TM}was distributed equally between morning and afternoon milkings. Towards the end of *Phase 2*, the control herd transformed from feeding TMR to pasture.

[3] https://www.delaval.com/en-au/discover-our-farm-solutions/farm-management/delaval-delpro/.

[4] A CH_4 concentration of 2 ppm means that 2 out of every 1 million air molecules are CH_4.

3.2 Data Collection

As our CH_4 measurement system does not provide direct and complete measurements, we used *Option 3* for baseline calculation, and *Option 1* for project monitoring (cf. Sect. 2). More specifically, we obtained the herd size and the number of days animals spent on farm from 2020 to 2022 to establish the baseline, and manually recorded the DMIs of each herd in each milking session. To enable project monitoring in real time, and inform the relevant parties if CH_4 reduction has continuously occurred as expected, we use the CH_4 ppm data collected by our CH_4 measurement system as an indicator. Due to the lack of system integration, we downloaded the CH_4 measurements from the CH_4 sensor provider's database, and used a script to map them to the corresponding cows. The raw dataset is available on Github.[5]

3.3 Analysis

The CH_4 sensors collect CH_4 ppm data of each cow once every minute. Hence, we have 10 data points per cow per milking session. We denote the CH_4 values of the trial herd in the morning and afternoon milking sessions as $T_a = \{t_{aij}\}$ and $T_p = \{t_{p_{ij}}\}$ respectively, where $i \in [1,40] \& i \in \mathbb{Z}$, and $j \in [1,10] \& j \in \mathbb{Z}$. Correspondingly, the CH_4 values of the control herd are represented as $C_a = \{c_{aij}\}$ and $C_b = \{c_{bij}\}$. We then calculated the daily herd average using Eqs. 1, 2.

$$\mu_T = (\sum_{i=1}^{40}(\sum_{j=1}^{10} t_{aij}/10)/40 + \sum_{i=1}^{40}(\sum_{j=1}^{10} t_{p_{ij}}/10)/40)/2, (i,j \in \mathbb{Z}) \quad (1)$$

$$\mu_C = (\sum_{i=1}^{40}(\sum_{j=1}^{10} c_{aij}/10)/40 + \sum_{i=1}^{40}(\sum_{j=1}^{10} c_{p_{ij}}/10)/40)/2, (i,j \in \mathbb{Z}) \quad (2)$$

Figure 2 compares the daily CH_4 average of the trial herd and the control herd over the project period. The missing data points on March 28 and May 3 are due to power failures on farm. At 170 ml dosage level, the trial herd produced less CH_4 than the control herd on most of the days, especially after the first week; at 140 ml, the trial herd produced more CH_4 than the control herd on some of the days, while the control herd was showing a similar pattern as in *Phase 1*; and at 110 ml, the trial herd produced more CH_4 than the control herd half of the time. Meanwhile, due to the change from TMR to pasture, the emission pattern of the control herd has also changed notably. Therefore, we conclude that the CH_4 sensors are sensitive enough to detect the changes in the feed given to the cows, including SEAFEED^{TM}, TMR and pasture. This means we can rely on these measurements to decide whether a carbon reduction activity has been continuously carried out or not.

[5] https://github.com/Yining-Hu/CarbonApp_Data.

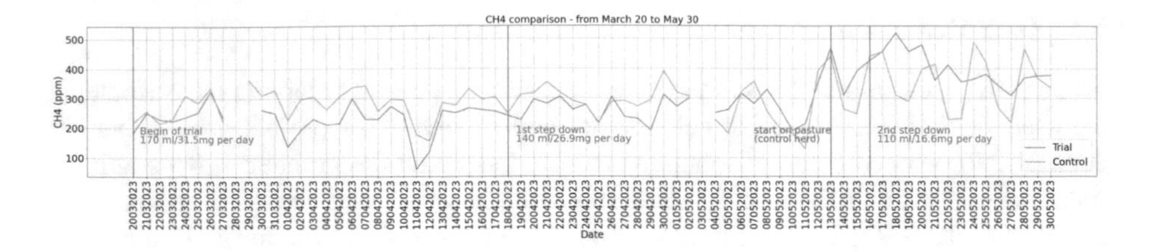

Fig. 2. Average CH_4 (ppm) of trial herd and control herd during the project.

4 CarbonApp Design

4.1 Design Requirements

We aim to design a system primarily for the project aggregator's use, and allow the contracted farmers and auditor to have some visibility into it. The system also accepts data from the CH_4 sensor provider, and the feed additive producer.

To meet the requirements of VM0041 (cf. Sect. 2), our system should store farm data, herd size, DMI, and days the animals spend on farm. To inform users when something affecting the project happens, e.g., a flood that forces farming activities to stop, the system will also store the CH_4 measurements. For the aggregator and farmers to view the status of carbon credit creation, the system should also tokenize carbon credits and represent their issuance and distribution on-chain.

4.2 Architecture

The main functionalities are enabled by a set of smart contracts which we collectively refer to as the *CarbonApp*. Figure 3 illustrates the system architecture showing its connections with other systems through RESTful APIs and a frontend for data that need to be entered manually.

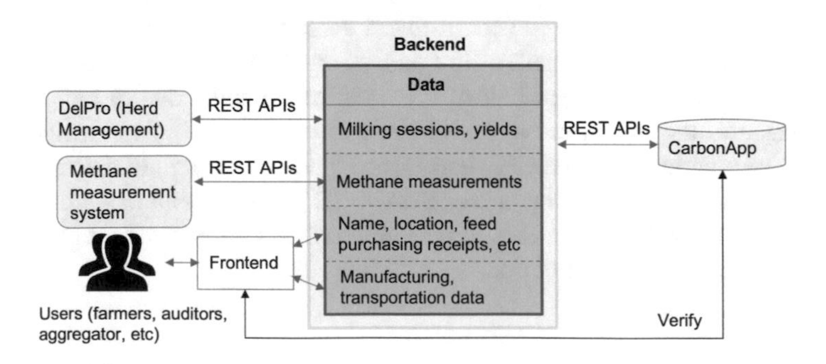

Fig. 3. System architecture.

4.3 Features

The CarbonApp consists of 6 core smart contracts: *FarmRegistry.sol*, *HerdRegistry.sol*, *ProjectRegistry.sol*, *EmissionTracking.sol*, *FeedTracking.sol*, *CarbonToken.sol*. Figure 4 shows how the smart contracts of the CarbonApp relate to each other. More specifically, *ProjectRegistry.sol* handles project registration, taking in information such as the start and end times of the project and the baseline. *FarmRegistry.sol*, *HerdRegistry.sol* handles the registration of farms and herds respectively. *EmissionTracking.sol* and *FeedTracking.sol* track the feed (TMR and SEAFEED^{TM}) and emission data. Due to the expensive and limited storage space on-chain, *EmissionTracking.sol* and *FeedTracking.sol* only require daily herd summaries. For emissions, we use the daily herd average as detailed in Sect. 3. To prevent any changes made to the data after it is recorded on-chain, we also calculate the SHA-256 hash of the dataset and send it to *EmissionTracking.sol* together with the daily herd average. *CarbonToken.sol* manages the issuance and distribution of carbon credits, and allows relevant parties to view the status of carbon credit creation through *carbon tokens*. Carbon tokens are semi-fungible, because the carbon tokens issued for the same project are interchangeable with each other, while the carbon tokens issued for different projects are not. *CarbonToken.sol* also contains functions to verify the generation of carbon tokens. An additional smart contract, *SeafeedRegistry.sol*, is used by the feed additive producer to record the production and transportation details. It can be customised for the selected feed additive.

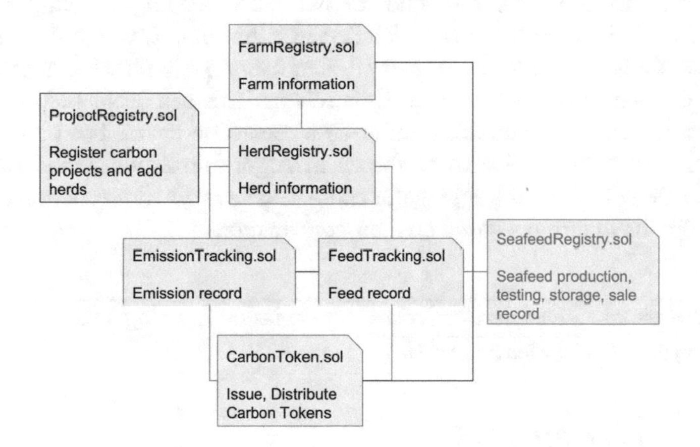

Fig. 4. CarbonApp smart contract connections.

4.4 Workflow

As shown in Fig. 5, upon joining, farmers register farms by providing the location and other information. Next, farmers register herds by providing herd size and other required information. An aggregator initiates a project registration by providing the start and end times of the project and the baseline. After that, the aggregator can add more herds to the project by calling the `addHerdToProject` function. The

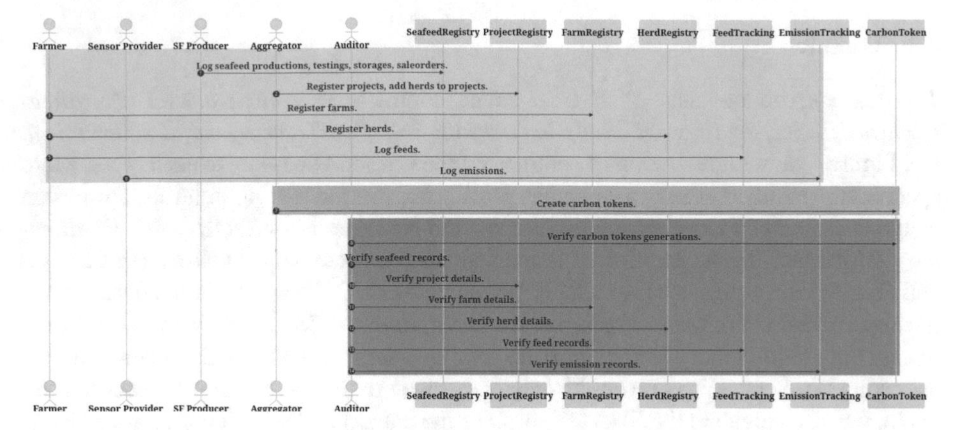

Fig. 5. CarbonApp sequence diagram.

SEAFEED^{TM}producer can log details of production, testing, storage and sale orders on SeafeedRegistry.sol.

During the project, farmers regularly enter the feed data of the herd. The CH_4 sensor provider sends the emission data either manually or automatically. A feed record should be backed by a sale order on SeafeedRegistry.sol, and an emission record can be linked to a feed record occurring on the same day for the same herd. Users can compare the emission values of two herds feeding on different additives using `verifyEmissionValue` as detailed in Algorithm 1, where the variables are explained in Table 1. At a selected interval, i.e., every 3 months, the aggregator calls `issue` on CarbonToken.sol to issue carbon tokens. This can occur before a standards body issues credits as an outlook or after as a record on-chain. Feed records should be supplied when calling `issue` as shown in Algorithm 2, where the variables are explained in Table 1. Then, the aggregator calls the `distribute` function to distribute the carbon credits to farmers based on their contributions.

Algorithm 1. `verifyEmissionValue` on EmissionTracking.sol.

Require: length of IDs_c == length of IDs_t
  $TE_c \leftarrow 0$
  $TE_t \leftarrow 0$
  for i=0,1,2,...,length of IDs_c-1 **do**
    if $IDs_c[i]$ exists & $IDs_t[i]$ exists **then**
      $TE_c \leftarrow TE_c + Es[IDs_c[i]]$
      $TE_t \leftarrow TE_t + Es[IDs_t[i]]$
    end if
  end for
  $result \leftarrow TE_c > TE_t$
  return *result*

Algorithm 2. `issue` on CarbonToken.sol.

Require: caller == admin
Require: the carbon token record $ID_{cbtoken}$ does not exist
Require: the project $ID_{project}$ exists
  Mint *Amount* of tokens and assign them to the admin
  query the start time and end time of the project $ID_{project}$
  for all ID_{feed} in IDs_{feed} **do**
    query the timestamp of feed ID_{feed}
    if the timestamp of feed ID_{feed} falls in the project duration **then**
      change feed status to CLAIMED
    end if
  end for
  create a new carbon token record $ID_{cbtoken}$
  add $ID_{cbtoken}$ to $Alltokens$ array
  change S_{feed} of the $ID_{cbtoken}$ to $true$

Table 1. Input and State variables

Input variable	$ID_{cbtoken}$	ID of the carbon token issuance record
	$Amount$	Amount of carbon tokens
	IDs_{feed}	The feed ids used to back this carbon token record
	$ID_{project}$	ID of the project to which the carbon tokens are issued
	IDs_c	Emission records of the control group
	IDs_t	Emission records of the treatment group
	TE_c	Total emissions of the control group
	TE_t	Total emissions of the treatment group
	Es	All emission records
	$Alltokens$	All carbon token issuance records
State variable	S_{feed}	Claim state of a feed record. Possible states are *CLAIMED*, *UNCLAIMED*

5 Implementation and Evaluation

5.1 Implementation

The CarbonApp is implemented in Solidity 0.8.0 and deployed on a private Polygon blockchain with two nodes–one on a Linux Virtual Private Server (VPS) with two cores, 2GB memory, the other on a Linux machine with four cores, 16GB memory. To enable communications between the CarbonApp and other systems, for each function in the CarbonApp, we create a corresponding route to accept API requests. We also create a `signup` router with the `web3.eth.sendTransaction` and `web3.eth.accounts` APIs for new users to create blockchain accounts and receive initial funding from a *Bank* account. We also implemented a simple frontend in React.[6] Figure 6 demonstrates the landing page for the `carbontoken` router. The table at the top of the page lists all carbon tokens upon loading, and the forms below allow the

[6] https://react.dev/.

aggregator to issue and distribute carbon tokens. The source code of the full stack is available on Github.[7]

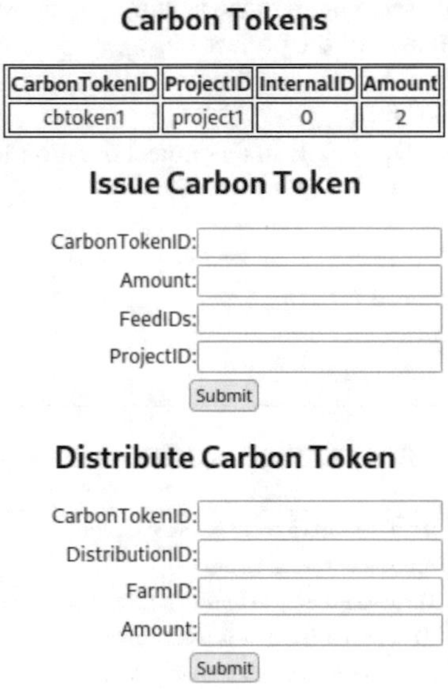

Carbon Tokens

CarbonTokenID	ProjectID	InternalID	Amount
cbtoken1	project1	0	2

Issue Carbon Token

CarbonTokenID:
Amount:
FeedIDs:
ProjectID:
Submit

Distribute Carbon Token

CarbonTokenID:
DistributionID:
FarmID:
Amount:
Submit

Fig. 6. CarbonToken Landing Page

5.2 Evaluation

The blockchain is a decentralised, immutable, secure ledger that can facilitate trusted data sharing among multiple participants. By using a blockchain, our system guarantees the integrity of the data collected on farms, ensures that the aggregator issues and distributes credits equitably among farmers, and the data is comprehensive and straightforward for auditing. Compared to only relying on a standards body like Verra, the blockchain greatly improves the trust and efficiency in project monitoring and carbon crediting. Choosing the Polygon blockchain has allowed us to deploy smart contracts at a low cost. The total gas cost for deploying all 7 smart contracts was 17803140. At the time of writing, the standard gas cost on the public Polygon PoS Chain[8] was 70 gwei, and the MATIC/USD exchange rate was 0.64. This gives an estimated cost of $0.8 USD for deploying all smart contracts, which is a negligible cost for companies interested in implementing the solution.

The CarbonApp design captures the key data points required in VM0041, and uses semi-fungible ERC-1155 tokens to tokenize carbon credits on-chain and incorporates

[7] https://github.com/Yining-Hu/CarbonApp.

[8] https://polygonscan.com/gastracker.

functions to issue and distribute. These allow the aggregator and farmers to track a project's progress and view the carbon credit status. The API routes allow data sharing between the CarbonApp and other systems, while the frontend simplifies manual data collection. The `signup` route enhances the system security by requiring apikeys.

6 Related Work

Blockchain and tokenization have been previously considered in carbon trading to solve the issues around centralization, lack of transparency, and hoarding–when organizations holding carbon credits speculate the market before releasing them [13]. To address these issues, some existing studies proposed blockchain-based carbon trading schemes. Saraji et al. [16] proposed a *Carbon Credit Ecosystem* to handle the tokenization and trading of carbon credits on-chain.

Besides carbon trading, several studies also tackled the measurement, reporting and verification (MRV) of offset projects. In the construction industry, Woo et al. [19] carried out an extensive survey to understand how to design a MRV system that meet the requirements of building energy performance (BEP) audit schemes. They conclude that blockchain can be used to build a reliable, transparent, and affordable MRV system, and can be easily integrated with blockchain-based carbon trading platforms. The authors, however, did not provide an actual design of the system or any implementations.

Another category of studies use blockchain and external sensors for offset project management. FlowerTokens [17] proposed by Terra0 tokenizes plants as ERC-721 tokens[9] and record their conditions on-chain, allowing plants to own and utilize themselves through smart contracts. The project also uses a camera system to monitor the plants 24/7. Despite the intriguing idea of enabling physical assets to utilize themselves, FlowerTokens fell short of its goals as payments in cryptocurrencies are restricted in the global governance framework. Our solution does not rely on cryptocurrencies and we do not intend to bypass centralised authorities to generate carbon credits. Instead, it is intended to assist with existing carbon crediting schemes using blockchain-based data-sharing. Sadawi et al. [4] proposed a framework consisting of a multi-level blockchain network and CO_2 sensors to monitor emissions and trade carbon credits. The framework contains a public blockchain for trading and a private blockchain to record emissions. However, the system design lacks practical considerations as the CO_2 measurement can be challenging, and storing the full emission dataset on-chain will result in excess energy and storage usage. The authors did not propose a technical design of the system. In comparison, our system design and implementation meet all the practical requirements arose from a real-life offset project.

7 Conclusion and Future Work

We have presented a blockchain-based solution to monitor offset projects and manage carbon credits. Our proposal is primarily based on a practical offset project carried out on a South Australian farm, where we used SEAFEED™to reduce dairy cows' CH_4

[9] https://eips.ethereum.org/EIPS/eip-721.

emissions. We started by investigating the data collection requirements of the Verra methodology VM0041. We then presented the CarbonApp design, and implemented it as a set of smart contracts deployed on a private Polygon blockchain and created RESTful APIs and a frontend. To prevent unauthorised accesses, we require users to register and provide an apikey when sending requests. As a result, we have proposed and implemented a unique, low-cost and secure blockchain-based solution for manage carbon offset projects.

In the future, we aim to improve the CarbonApp design for it to be used in a wider range of offset projects. E.g., with a more generic design of the traceability contract, we can easily apply the solution to a different feed additive. The system can be further extended to cover the rest of the agri-food supply chain including the manufacturing and distribution of end products. This will enable us to associate the emission reduction on farm with the end products, allowing producers to make claims such as "carbon-reduced milk", and "environmental friendly products", etc. We will also integrate a SQL database with the CarbonApp to avoid storing the bulk of the raw data on-chain. We will also enable the interoperability between the different systems including the farm management system, the CH_4 measurement system, and the herd testing lab with our CarbonApp to automate the data collection and reduce manual tasks for all users.

Acknowledgements. This project is partially funded by PIRSA AgTech Growth Fund, and Beston Global Food Company, South Adeliade, Adelaide.

References

1. Polygon wiki (2023). https://wiki.polygon.technology/
2. Agreement, P.: Paris agreement. In: report of the conference of the parties to the United Nations framework convention on climate change (21st session, 2015: Paris). Retrived December, vol. 4, p. 2017. HeinOnline (2015)
3. et al., W.R.: Erc-1155: Multi token standard (2018). https://eips.ethereum.org/EIPS/eip-1155
4. Al Sadawi, A., Madani, B., Saboor, S., Ndiaye, M., Abu-Lebdeh, G.: A comprehensive hierarchical blockchain system for carbon emission trading utilizing blockchain of things and smart contract. Technol. Forecasting Soc. Change **173**, 121,124 (2021)
5. Ashley, M.J., Johnson, M.S.: Establishing a secure, transparent, and autonomous blockchain of custody for renewable energy credits and carbon credits. IEEE Eng. Manage. Rev. **46**(4), 100–102 (2018)
6. Breidenich, C., Magraw, D., Rowley, A., Rubin, J.W.: The kyoto protocol to the united nations framework convention on climate change. Am. J. Int. Law **92**(2), 315–331 (1998)
7. Civillini, M.: Verra boss steps down after criticism of its carbon credits (2023). https://www.climatechangenews.com/2023/05/23/verra-boss-steps-down-after-criticism-of-its-carbon-credits/
8. Dannen, C.: Introducing Ethereum and solidity, vol. 1. Springer (2017)
9. Greenfield, P.: Biggest carbon credit certifier to replace its rainforest offsets scheme (2023). https://www.theguardian.com/environment/2023/mar/10/biggest-carbon-credit-certifier-replace-rainforest-offsets-scheme-verra-aoe
10. Kinley, R.D., de Nys, R., Vucko, M.J., Machado, L., Tomkins, N.W.: The red macroalgae asparagopsis taxiformis is a potent natural antimethanogenic that reduces methane production during in vitro fermentation with rumen fluid. Animal Prod. Sci. **56**(3), 282–289 (2016)

11. Morton, A.: Australia's carbon credit scheme 'largely a sham', says whistleblower who tried to rein it in (2022). https://www.theguardian.com/environment/2022/mar/23/australias-carbon-credit-scheme-largely-a-sham-says-whistleblower-who-tried-to-rein-it-in
12. Nakamoto, S.: Bitcoin: A peer-to-peer electronic cash system. Decentralized business review (2008)
13. Patel, D., Britto, B., Sharma, S., Gaikwad, K., Dusing, Y., Gupta, M.: Carbon credits on blockchain. In: 2020 International Conference on Innovative Trends in Information Technology (ICITIIT), pp. 1–5. IEEE (2020)
14. Power, G.V.: Understanding carbon credits and the role of blockchain technology. (2023). https://www.linkedin.com/pulse/understanding-carbon-credits-role-blockchain-technology/
15. Roque, B.M., Brooke, C.G., Ladau, J., Polley, T., Marsh, L.J., Najafi, N., Pandey, P., Singh, L., Kinley, R., Salwen, J.K., et al.: Effect of the macroalgae asparagopsis taxiformis on methane production and rumen microbiome assemblage. Animal Microbiome **1**, 1–14 (2019)
16. Saraji, S., Borowczak, M.: A blockchain-based carbon credit ecosystem. arXiv preprint arXiv:2107.00185 (2021)
17. Seidler, P., Kolling, P., Hampshire, M.: Can an augmented forest own and utilise itself. White Paper (2016)
18. VERRA: Vm0041 methodology for the reduction of enteric methane emissions from ruminants through the use of feed ingredients, v2.0 (2021). https://verra.org/methodologies/revision-to-vm0041-methodology-for-the-reduction-of-enteric-methane-emissions-from-ruminants-through-the-use-of-100-natural-feed-supplement-v1/
19. Woo, J., Fatima, R., Kibert, C.J., Newman, R.E., Tian, Y., Srinivasan, R.S.: Applying blockchain technology for building energy performance measurement, reporting, and verification (mrv) and the carbon credit market: a review of the literature. Build. Environ. **205**, 108,199 (2021)

Investigating the Impact of Congestion Control Algorithms on Edge-Cloud Continuum

Nicolas Keiji Cattani Sakashita, Maurício Aronne Pillon, Charles Christian Miers, and Guilherme Piêgas Koslovski(✉)

Graduate Program in Applied Computing (PPGCAP), Santa Catarina State University (UDESC), Joinville, Santa Catarina, Brazil
nicolas.sakashita@edu.udesc.br,
{mauricio.pillon,charles.miers,guilherme.koslovski}@udesc.br

Abstract. Edge-Cloud Continuum (ECC) is an architecture combining the concepts of Cloud and Edge Computing to improve the performance of distributed services. This combination is necessary to improve Quality-of-Service (QoS) indicators, related to latency, processing time, and energy consumption. Essentially, services are simultaneously hosted by edge and cloud servers, distributing the workload based on user's expectations. Although revolutionary, the ECC architecture is dependent on the interconnection networks, inheriting its research challenges related to the management and shared resources. Even applying optimized provisioning policies (scheduling and allocation), the end-to-end performance of distributed applications remains dependent on Transmission Control Protocol (TCP)'s internal algorithms. We investigate the impact of the congestion control algorithms Cubic, Reno and Bottleneck Bandwidth and Round-trip propagation time (BBR) when used for supporting ECC applications. Our experimental analysis comprises two scenarios: (i) the execution of n-layer application using resources distributed atop edge and cloud providers, and configured with Cubic and Reno algorithms; and (ii) a scenario demonstrating the benefits of using BBR in ECC architectures. In summary, the analyzes demonstrate that the congestion control configuration can improve application performance in ECC.

Keywords: Edge Cloud Continuum · Congestion Control · Reno · Cubic · BBR

1 Introduction

Cloud computing is a paradigm enabling ubiquitous, convenient and on-demand access to interconnected computing resources. These resources constitute a pool of shared and configurable computing elements, including networks, servers, storage, applications, and services that can be swiftly instantiated with minimal management effort or service provider interaction [15]. In general, control, data, and computing systems have been substantially transitioned to the cloud, concentrating the offer of services. Nevertheless, the centralization of services overlooks the opportunity to utilize the computational and storage power of the modern devices largely distributed among end users. The use of these distributed resources disseminated a new computing paradigm known as Edge

L. Barolli (Ed.): AINA 2024, LNDECT 202, pp. 26–37, 2024.
https://doi.org/10.1007/978-3-031-57916-5_3

Computing. It migrates computing applications, data and services away from the centralized nodes of the clouds, relocating them to the network periphery. This paradigm retains the main advantages of the cloud, such as infrastructure support, but places control and decision reliability at the edges, enabling centralized computing applications in humans [7]. The combined execution using edge and cloud computing led to the definition of Edge-Cloud Continuum (ECC) [3]. In this scenario, applications can be entirely or partially executed within any of the mentioned edge and cloud architectures, all aimed at improving the indicators of Quality-of-Service (QoS).

ECC offers opportunities to implement innovative applications, which require efficient resource scheduling and allocation due to strict users' requirements in terms of latency, energy consumption, cost, privacy, among other aspects [14]. Applications benefiting from the ECC architecture include, but are not restricted to, unmanned aerial vehicles, video services, smart cities, smart healthcare, smart manufacturing, and smart homes. Essentially, a distributed application can be spread atop multiple ECC providers. Each provider hosts a subset of the application, which are interconnected using the Internet backbone.

Although revolutionary, the specialized literature on ECC provisioning still lacks information about the impact of the Transmission Control Protocol (TCP) congestion control algorithms on applications within the ECC architecture. In fact, applications share computing and communications resources with traditional applications originating from other scenarios (eventually using different transport-layer protocols), communicating atop Internet backbones with distinct sharing policies. It is evident that there is a management challenge regarding the appropriate configuration for each application, given the distributed nature of ECC and the existence of multiple administrative domains with private and competitive policies. Specifically regarding the TCP congestion control, while some algorithms seek to maximize the overall network throughput (e.g., Reno and Cubic), without optimizing the use of intermediate buffers, other algorithms (e.g., Bottleneck Bandwidth and Round-trip propagation time (BBR)) focus on reducing queuing time, keeping buffers at a minimum occupancy. Such a configuration directly impacts the applications, especially considering that ECC architectures are mostly shared by applications that require QoS requirements (low latency indicators and high throughput) [19].

In this context, this work presents experimental campaigns demonstrating applications' indicators obtained when changing the TCP congestion control algorithm. Reno [9], Cubic [8], and BBR [4] were selected for composing the experimental analysis. Specifically, two representative scenarios are discussed: *(i)* a n-layers distributed application hosted by distinct configurations of ECC providers, and *(ii)* an analysis of the benefits from using BBR in specific links from ECC. The remainder of this work is organized as follows. Section 2 presents the motivation, problem definition, and related work. The experimental protocol is detailed in Sect. 3, while the experimental analysis is discussed in Sect. 4. Section 5 concludes the work and presents perspectives for future research.

2 Motivation and Problem Definition

The challenges in managing cloud computing, mainly the ultra-low latency and high throughput requirements, have motivated the development of new concepts and architectures, such as Fog Computing and Multi-access Edge Computing (MEC) [24]. These architectures proposed to implement computational infrastructures from edge resources to the cloud network, bringing the computational resources closer to the end users. These concepts evolved to contemplate idle servers and devices, adding them to the set [21]. Aiming this goal, technologies such as Software-Defined Networking (SDN) [11], Network Function Virtualization (NFV) [6] and Information Centric Networking (ICN) [1] have emerged as alternatives for implementation. Essentially, the combination of all aforementioned opportunities led to the emergence of a new architecture, allowing the continuous and transparent interconnection of diverse services hosted by multiple providers, termed Edge-Cloud Continuum (ECC) [3]. Notably, fog computing operates by uniting available devices at the edges with resources located in computational clouds, forming a hierarchy of computational capabilities. These capacities can be distributed through access points, routers, and various other devices. In general, ECC can enhance the QoS for applications by distributing computing among resources and adapting the computational scenario to meet the requirements for latency, processing and power consumption.

Data Centers (DCs) alone will not handle the amount of volume and data accumulation of Internet of Things (IoT) devices in the future. Also, high latency is a critical challenge for a number of applications that need end-to-end communication. For this reason, various research initiatives, such as Cloudlets, Central Office Re-architected as a Data center, and HomeCloud, are exploring the integration of NFV and SDN to develop ECC applications and frameworks [18,21]. ECC makes these applications possible by providing the capability to maintain accuracy with minimal latency. However, ECC applications still share the underlying Internet backbone with background traffic. In this sense, congestion control algorithms also impact the quality of applications in the ECC architecture, with specific algorithms designed for stability, high throughput, and fast convergence time, sharing and disputing the available network throughput. This work will focus on studying the standard and most-used congestion control algorithm in Linux devices, Cubic, and comparing it with the former standard, Reno. Additionally, a comparison with BBR, a relatively recent congestion control algorithm known for its speed and convergence time in high-mobility scenarios, will be presented in comparison to Cubic.

2.1 Congestion Control in TCP-Based Networks

Network communication is susceptible to packet loss, often originated by the depletion of intermediary buffers necessitating packet retransmission. The mechanism of congestion control operates monitoring the Congestion Window (CWND), imposing a restriction in the traffic rate in which a TCP sender can send within a network. TCP compels each sender to restrict the network traffic rate for its connection as a function of the network congestion perceived. If the sender perceives low congestion in the path, it will increase the sending rate; otherwise, it will reduce it. Generally, there are two basic models of mechanism for the implementation of the congestion control, end-to-end and network-assisted congestion control. While the network-assisted requires support from forwarding resources, the end-to-end congestion control does not have any kind of explicit support in the network layer, with the TCP being responsible for its implementation. In network-assisted congestion control, components forwarding from the network layer provide specific feedback, i.e., information referring to the state of the congested network. This allows final hosts to make more precise decisions, and is commonly used in controlled administrative domains and local data center networks [2, 16].

The focus of this work is on end-to-end congestion control, specifically Reno, Cubic, and BBR, which represent the algorithms used as a basis for composing the standards of modern operating systems. While Reno and Cubic algorithms exhibit a contrast in stability and high-throughput, making them suitable for different applications, BBR stands out as a prominent algorithm with the potential to complement of even replace Cubic. In this sense, the ECC end-user's applications must define which objective they want to pursue in end-to-end congestion control algorithms: low use of intermediate buffers, resulting in reduced latency; or high throughput for heavy flows, occupying buffers and eventually increasing data loss.

2.2 Related Work

Among research related to the quantification and analysis of the impact of congestion control algorithms on the ECC, [23] presents the optimization of TCP Cubic for IoT, aiming to enhance QoS. In turn, for 5G mobile communications, [13] introduces a new challenge in the mechanisms of implementation of congestion control algorithms, due to the execution in a user-dense environment and with high demands for services and network traffic. In addition, [17, 20] explores the comparison between TCP Cubic, Reno and Data Center TCP (DCTCP) [2] in an SDN centralized environment. The results shows that TCP Reno has reduced performance in a TCP SDN environment with high-latency networks. The choice of the congestion control algorithms used is a local decision, taking into account the contrast in throughput and stability between them [10]. That is why it is important to highlight that algorithms have distinct objectives (e.g., fairness, total use of resources, buffer sharing, latency). This fact will affect the throughput bottleneck of communications and, consequently, will influence the QoS. Conceptually, Cubic exhibits an aggressive growth in the window size, a factor that results in a higher data transmission rate compared to Reno. However, Reno is more stable, making it a better choice for applications that require a constant, stable connection without variance.

Finally, BBR is also receiving attention and relevance, because it has been widely deployed by Google on Linux servers. A comparison of default TCP congestion control algorithm Cubic and BBR in commercial satellite network is also published [5]. It is mentioned that both Cubic and BBR have similar median throughput, however BBR reaches link maximum capacity more often than Cubic.

3 Experimental Scenarios and Protocol

The experimental campaign comprises two scenarios executed atop two distinct SDN-based infrastructures. Although SDN is not a prerequisite for ECC applications, its centralized management facilitates the collection and analysis of results.

***n*-Layer Application**. We employed a *n*-layer architecture[1] which divides the application into logical layers and physical tiers. These layers serve to separate responsibilities and manage dependencies. Originally, *n*-layer architectures were hosted on Infrastructure-as-a-Service (IaaS) providers, with each physical layer executing on a separate group of virtual machines. Currently, an application does not need to be purely IaaS, being able to distribute the components among distinct edge and cloud providers. Figure 1 illustrates the *n*-layer architecture showcasing different placements in cloud and edge providers. A client is submitting requests to an application composed of multiple services: web application firewall (WAF), web servers, messaging queuing system, caches, middle and remote external services, and finally the data source. For investigating the impact of congestion control algorithms on such application, we positioned the composing services atop three experimental configurations, represented by ellipses in blue, yellow, and red colors. Each configuration uses multiple routers for determining the edge and cloud administrative domains, as well as for composing network links with distinct configurations in terms of latency and available bandwidth. In each experiment, the client will establish a connection to the database, crossing all intermediate paths, denoting the end-to-end TCP flows under investigation, that face network congestion. In Experiment 1, represented by blue ellipses, the client will connect directly with the cloud. In Experiment 2, represented by yellow ellipses, an edge (Edge 1) is present between the client and the cloud. In Experiment 3, represented by red ellipses, two edges (Edge 1 and Edge 2) are presented between the client and the cloud. The values of latency and throughput for the cloud and the edges are based on the fact that the cloud is physically more distant, resulting in higher latency and lower throughput compared to the edges, which are closer to the user.

The main traffic between client and the database crosses the WAF and web services, latter directed to the messaging layer, middle tier 2 and finally the data source, returning by cache, middle tier 1 until the client. Specifically, in Experiment 2 and 3, the messaging layer and middle tier 2 represent the transition between the edge and the cloud. For investigating the network congestion, a background traffic was generated using User Datagram Protocol (UDP) connections in this critical region. In Experiment 1 and 2, the maximum throughput in the link is 100 Mbps. Therefore, the UDP throughput was individually restricted to 25 Mbps, 50 Mbps, and 75 Mbps, or deactivated. These settings are represented in the charts (Sect. 4) by the subtitles 25%, 50%, 75% and 0%. In Experiment 3, to maintain proportionality in relation to the maximum throughput in the network (100Mbps) and knowing that the throughput in this link is 250 Mbps, the UDP throughput was increased to 175 Mbps, 200 Mbps, 225 Mbps, or deactivated. This adjustment was made to create a bottleneck proportional to the other scenarios of

[1] Available at: https://learn.microsoft.com/en-us/azure/architecture/guide/architecture-styles/n-tier.

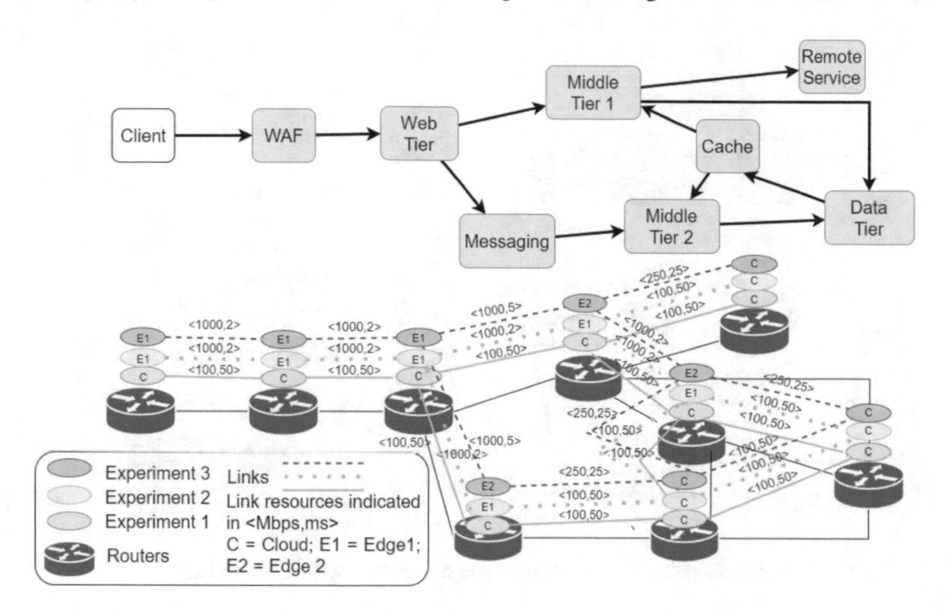

Fig. 1. The *n*-layer architecture used for hosting the experimental scenario.

75 Mbps, 50 Mbps, 25 Mbps, or total. Each experiment was executed with both algorithms running simultaneously for 120 s, and the results are discussed in Sect. 4.1.

BBR as an Opportunity in ECC. BBR is a hybrid TCP congestion controls algorithm that combines loss-based and delay-based approaches. In a congested network or with high link utilization, hybrid algorithms tend to use a delay-based congestion control approach, while in a low link and high-speed network, they tend to employ a more aggressive approach of loss-based algorithms. Its main feature is to overcome the problems faced by loss-based algorithms in scenarios with small buffers, such as investigated in ECC [10]. BBR periodically measures the Round-Trip Time (RTT) and the delivery rate of the network, allowing it to adjust the sending rate accordingly [13]. As an example, the high-throughput application of 5G in high-density and high-frequency urban scenarios may lead to several distortions and long delays in long-distance connections. Some congestion control algorithms, such as BBR, can be deployed and show significant performance improvements in networks with large delays. BBR also outperforms other algorithms when dealing with high-mobility connections to remote or edge servers. It demonstrates fast convergence times in these scenarios, which is an important indicator of adaptability and implies better utilization of its available capacity [22].

Figure 2 summarizes the four experiments composed for investigating the applicability of BBR in ECC scenarios: client-edge, client-cloud, client-edge-cloud without data output to end-user (termed *edge cloud w/o* in experimental analysis), and client-edge-cloud with data output (termed *edge cloud w* in experimental analysis), encompassing distinct network architectures commonly employed in contemporary applications. *Client-Edge-Cloud w* and *w/o* (in red and green colors) represents the ECC architecture, allocating a percentage of the data volume for simultaneous transmission to

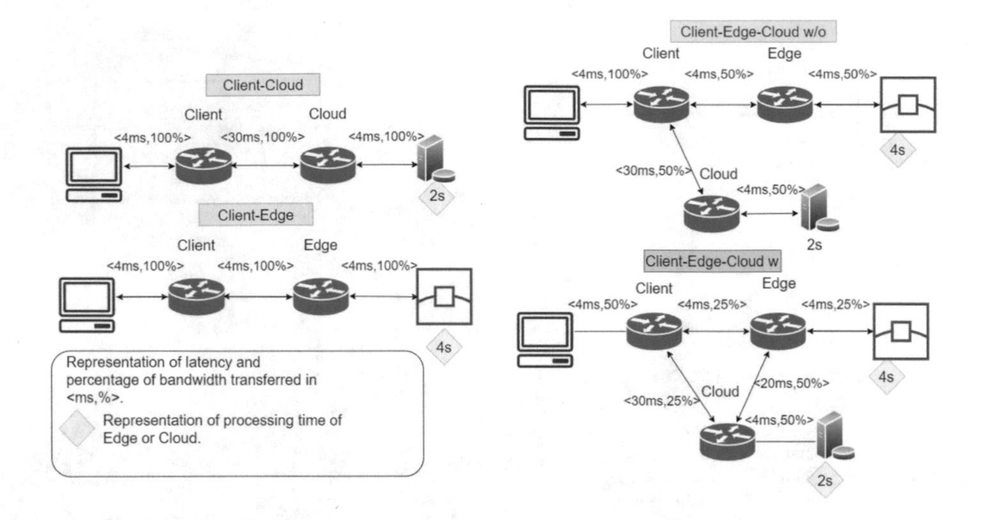

Fig. 2. Experimental scenarios to analyze the use of BBR in ECC.

both the cloud and the edge. In client-edge and client-cloud experiments, 100% of the data volume will be transferred in a single link, while in *Client-Edge-Cloud w* and *w/o* each link will transfer 50% or 25% of the total data volume. The *Client-Edge-Cloud w* scenario involves three links, where 50% of the total data volume is concurrently transferred between the edge and cloud, and 25% is distributed among the links connecting the client and edge, and client and cloud. The transfer of smaller data volumes will impact the QoS of the network, as well as the efficiency of the congestion control algorithms Cubic and BBR. Essentially, the experiment reflects the distribution of data volume in different locations, showcasing the collaborative efforts of edge and cloud to enhance network QoS indicators, realized with a simpler ECC application than the previous n-layer architecture. It is worthwhile to mention that this experiment introduced a representation of processing time between network transfer, useful for increasing variation in network data flows. Moreover, the joint discussion on processing and communication time will highlight how important is the selection of appropriated congestion control algorithms in ECC-based scenarios (results are discussed in Sect. 4.2). In each test case, 16 clients are connecting simultaneously, each requiring a data transfer volume randomly selected between 1 and 100 MB. The ECC application processes the received data and returns, if need, an output of the same volume to the client. The transmitted data undergoes processing at each node before being returned to the client. The data source is distributed among edge and cloud providers, and depending on the category, the data transferred on certain links is represented in percentages, as presented in Fig. 2. Finally, we compared three combinations of congestion control protocols: BBR, Cubic, and a mix of BBR and Cubic.

4 Experimental Analysis

The experimental scenarios were implemented using the Mininet network emulator [12] in conjunction with the Ryu controller[2] hosted by a virtual machine (VM) flavor of 4GB RAM, 4 CPUs, and GNU/Linux Ubuntu 20.04. For representing each link, subnets and their forwarding rules were configured. To generate the data flow, the iPerf3 tool was employed, and the selection of the congestion control algorithm was done directly in the tool. The results for both experimental scenarios are summarized in Fig. 3, and individually discussed in the following sections. The label Scenario 1 represents the experiments with the n-layer application, while Scenario 2 denotes the BBR-based experimental campaign.

We selected two metrics for discussing the experimental analysis: total bandwidth and Flow-Completion Time (FCT). The total bandwidth gives insights on the data sent and received for each connection, as well as the network sharing policies implemented by the congestion control algorithms. In turn, the FCT summarizes the performance in an application-level, accounting all transfer time, even the retransmissions. The data were collected using the iPerf3 tool and then presented as boxplots and Cumulative Distribution Functions (CDFs).

4.1 Results for the Experiment with a n-layer application

Figures 3(a)–3(c) summarize the results of the experiments varying the congestion control for a n-layer application. In Experiment 1, Cubic exhibits a slightly higher throughput in all tests. In Experiment 2, as background traffic increases, Reno's throughput drops significantly while Cubic maintains a satisfactory amount. It is also observed that both algorithms exhibit a more stable throughput at high background traffic rates, with Reno being more stable than Cubic. In Experiment 3, Cubic dominates with higher throughput compared to Reno, which keeps stable and low throughput results. As detailed in Fig. 1, ellipses in blue, red, and yellow represent the placement of the edges in each experiment's respective layers in the n-layer architecture. Experiment 1 lacks any edges, with a latency of 50 ms and a throughput of 100 Mbps in each link. In contrast, Experiments 2 and 3 feature edges with higher throughput (250–1000 Mbps) and lower latency (2–5 ms). The edges are present in all layers except middlet tier 2, remote service, and data tier.

The impact of having or not having edges is evident in Figs. 3(a), 3(b), 3(c). Comparing Experiment 1 Fig. 3(a) to Figs. 3(b) and 3(c), it is evident that the influence of the edges brought greater throughput to Cubic, which stood out in these experiments in relation to Reno. There were only two cases where Cubic and Reno had a similar throughput rate: in Experiment 1 or when there was no UDP background traffic. Mainly, Cubic dominated in all edge experiments with background traffic illustrated in Figs. 3(b) and 3(c). This result highlights the advantages of using Cubic for various network traffic scenarios and complex network architectures. Finally, the results demonstrate that understanding the characteristics of Cubic, Reno, and congestion control algorithms is crucial for achieving better quality and development in ECC applications.

[2] Ryu Controller. Available at: https://ryu-sdn.org.

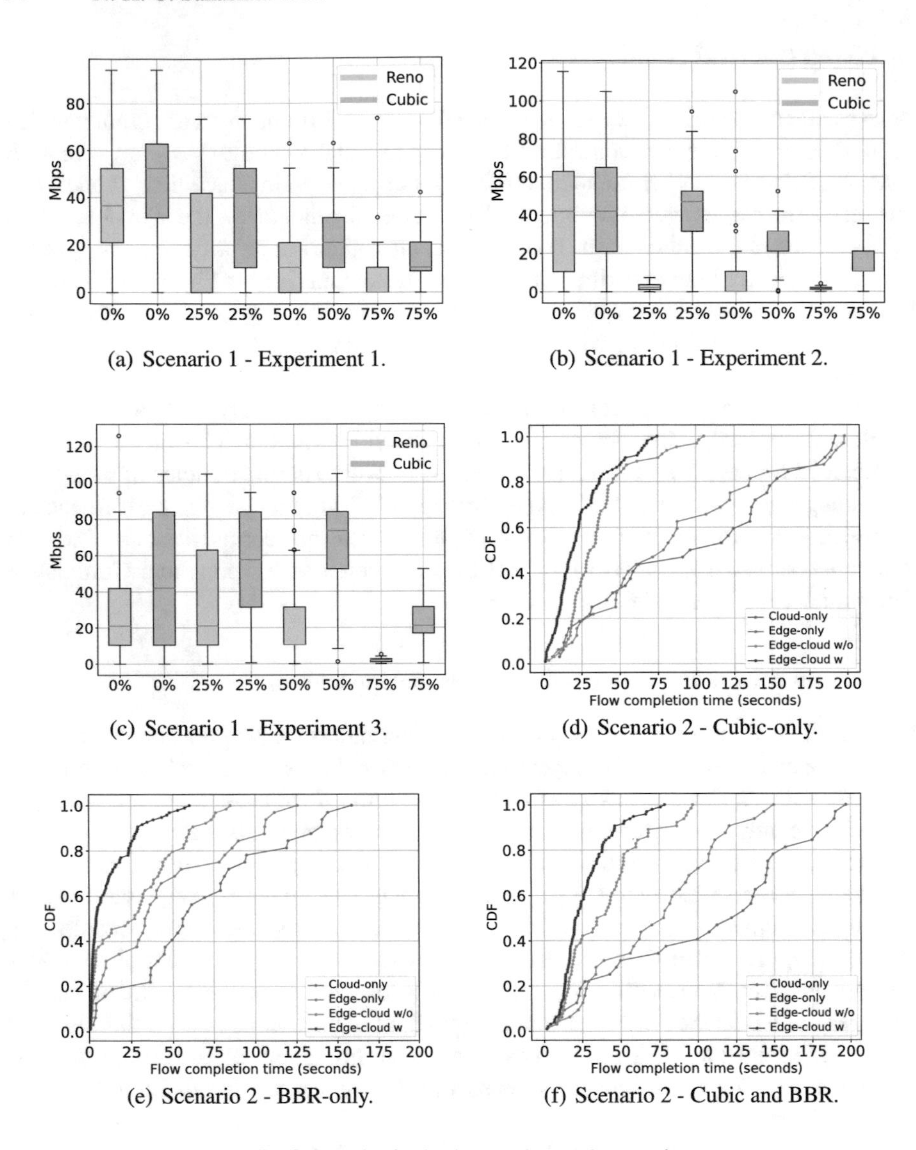

(a) Scenario 1 - Experiment 1.

(b) Scenario 1 - Experiment 2.

(c) Scenario 1 - Experiment 3.

(d) Scenario 2 - Cubic-only.

(e) Scenario 2 - BBR-only.

(f) Scenario 2 - Cubic and BBR.

Fig. 3. Results for both experimental scenarios.

4.2 Using BBR to Improve the Performance of ECC-Based Applications

This scenario represents the dichotomy between processing and communication time in the positioning of ECC applications, with an emphasis on algorithms for congestion control. As depicted by Fig. 2, cloud servers can process faster while edge resources are limited. However, the network QoS values indicate that edge services can be accessed more efficiently in terms of latency and bandwidth. In this sense, Figs. 3(d)–3(f) showcase CDFs illustrating the Flow-Completion Time (FCT) results for the categories tested in the experiment: (a) representing cubic-only; (b) BBR only; and (c) a mix of

Cubic and BBR. As depicted in the CDFs of Figs. 3(d)–3(f), for all categories tested in the experiment *Edge-Cloud w* got the fastest FCT, followed by *Edge-Cloud w/o*. This result is due to the smaller data volumes transferred in each of the links, such as 50% and 25% of the total data, as illustrated in Fig. 2. For Cubic only (Fig. 3(d)), *Edge-Cloud w* contemplates 25 s for around 70% of the cases. In comparison, *Edge-Cloud w/o*, which was similarly fast for 25 s, contemplates only about 40% of the cases. For edge or cloud only within the same time frame, it corresponds to 20%. The tests with BBR, as shown in Fig. 3(e), were quite impressive, showcasing faster results for all categories of proposed congestion control algorithms. Twenty seconds contemplates about 80% of the cases for *Edge-Cloud w*, nearly 50% of the cases for the same time frame in *Edge-Cloud w/o*, and from 20 to almost 40% for the cases in cloud and edge only. It is also remarkable that BBR demonstrated to be significantly more effective in scenarios with large throughput values. For instance, around 20% of edge-cloud connections achieve completion times of nearly 1 s. Furthermore, over 60% of edge-cloud connections in *Edge-Cloud w* complete within 10 s. Lastly, for the proposed mix of Cubic and BBR (Fig. 3(f)), results present similar data to Cubic-only (Fig. 3(d)).

As expected, the category edge and cloud only took a longer time to complete its data transfers. While cloud-only requires data transfer crossing links with higher latency values, edge-only requires more processing time to accomplish a given task, as demonstrated by Fig. 2. Furthermore, the links in the edge and cloud only categories have only one link transferring 100% of the data volume, which also caused a higher time in the FCT. These overall results were expected due to the increased number of nodes processing the entire bandwidth in divided parts, resulting in a faster completion flow time and leveraging the advantages of BBR. The results in this work highlight the superiority of BBR compared to Cubic, particularly in scenarios with small bandwidths and flow completion time within an ECC architecture.

5 Conclusions

The rise of new communication technologies, mobile devices, IoT, and new architectures is driven by the need to address various demands, including computational processing, data storage, bandwidth, and latency. One of them is ECC, which brings the processing of these devices closer to the users, an approach to meet their required demands. Although innovative, ECC-based applications are distributed across the network, and consequently inherit the management challenges. Specifically, this work investigated the impact of selecting congestion control algorithms, in terms of network throughput and FCT. The experimental campaign investigated two representative scenarios: the execution of a n-layer application positioned atop edge and cloud providers, and the use of BBR congestion control algorithm for improving the overall performance of ECC-based applications. The TCP traffic was decomposed into steps to represent the processing of complex data flows. The results highlighted that the selection of an appropriated network congestion control is crucial to take advantage of the benefits offered by ECC. As future work, large-scale applications will be investigated, as well as the impact of user mobility.

Acknowledgements. This work was funding by the National Council for Scientific and Technological Development (CNPq, grant 311245/2021-8), Santa Catarina State Research and Innovation Support Foundation (FAPESC), Santa Catarina State University (UDESC), and developed at Laboratory of Parallel and Distributed Processing (LabP2D).

References

1. Ahlgren, B., Dannewitz, C., Imbrenda, C., Kutscher, D., Ohlman, B.: A survey of information-centric networking. IEEE Commun. Mag. **50**(7), 26–36 (2012)
2. Alizadeh, M., et al.: Data center tcp (dctcp). In: Proceedings of the ACM SIGCOMM 2010 Conference, pp. 63–74 (2010)
3. Bittencourt, L., Immich, R., Sakellariou, R., Fonseca, N., Madeira, E., Curado, M., Villas, L., DaSilva, L., Lee, C., Rana, O.: The internet of things, fog and cloud continuum: integration and challenges. Internet Things **3–4**, 134–155 (2018)
4. Cardwell, N., Cheng, Y., Gunn, C.S., Yeganeh, S.H., Jacobson, V.: Bbr: congestion-based congestion control. Commun. ACM **60**(2), 58–66 (2017)
5. Claypool, S., Chung, J., Claypool, M.: Measurements comparing tcp cubic and tcp bbr over a satellite network. In: 2021 IEEE 18th Annual Consumer Communications & Networking Conference (CCNC), pp. 1–4. IEEE (2021)
6. ETSI, N.F.V.: Network functions virtualisation (nfv). Management and Orchestration **1**, V1 (2014)
7. Garcia Lopez, P., et al.: Edge-centric computing: vision and challenges. SIGCOMM Comput. Commun. Rev. **45**(5), 37–42 (2015)
8. Ha, S., Rhee, I., Xu, L.: Cubic: a new tcp-friendly high-speed tcp variant. SIGOPS Oper. Syst. Rev. **42**(5), 64–74 (2008)
9. Jacobson, V.: Congestion avoidance and control. ACM SIGCOMM Comput. Commun. Rev. **18**(4), 314–329 (1988)
10. Jain, V.K., Mazumdar, A.P., Faruki, P., Govil, M.C.: Congestion control in internet of things: Classification, challenges, and future directions. Sustainable Comput. Inform. Syst. **35**, 100,678 (2022)
11. Kreutz, D., Ramos, F.M., Verissimo, P.E., Rothenberg, C.E., Azodolmolky, S., Uhlig, S.: Software-defined networking: a comprehensive survey. Proc. IEEE **103**(1), 14–76 (2014)
12. Lantz, B., Heller, B., McKeown, N.: A network in a laptop: rapid prototyping for software-defined networks. In: Proceedings of the 9th ACM SIGCOMM Workshop on Hot Topics in Networks, Hotnets-IX. Association for Computing Machinery, New York (2010)
13. Lorincz, J., Klarin, Z., Ožegović, J.: A comprehensive overview of tcp congestion control in 5g networks: research challenges and future perspectives. Sensors **21**(13), 4510 (2021)
14. Luo, Q., Hu, S., Li, C., Li, G., Shi, W.: Resource scheduling in edge computing: a survey. CoRR abs/2108.08059 (2021)
15. Mell, P., Grance, T.: The nist definition of cloud computing (2011)
16. Moro, V., Pillon, M.A., Miers, C.C., Koslovski, G.P.: Analysis of congestion control virtualization on execution of hadoop mapreduce application. In: 2018 Symposium on High Performance Computing Systems (WSCAD), pp. 93–93 (2018)
17. da Silva de Oliveira, F., Pillon, M.A., Miers, C.C., Koslovski, G.P.: Identifying network congestion on sdn-based data centers with supervised classification. In: Barolli, L. (ed.) Advanced Information Networking and Applications, pp. 222–234. Springer, Cham (2023)
18. Pan, J., McElhannon, J.: Future edge cloud and edge computing for internet of things applications. IEEE Internet Things J. **5**(1), 439–449 (2017)
19. Pham, Q.V., et al.: A survey of multi-access edge computing in 5g and beyond: Fundamentals, technology integration, and state-of-the-art. IEEE Access **8**, 116,974–117,017 (2020)

20. Roberts, J., Skandalakis, J., Foard, R., Choi, J.: A comparison of sdn based tcp congestion control with tcp reno and cubic. Technical Report (2016)
21. Rodrigues, D.O., de Souza, A.M., Braun, T., Maia, G., Loureiro, A.A., Villas, L.A.: Service provisioning in edge-cloud continuum: emerging applications for mobile devices. J. Internet Serv. Appl. **14**(1), 47–83 (2023)
22. Sandoval, J.I., Céspedes, S.: Performance evaluation of congestion control over b5g/6g fluctuating scenarios. In: Proceedings of the Int'l ACM Symposium on Design and Analysis of Intelligent Vehicular Networks and Applications, pp. 85–92 (2023)
23. Verma, L.P., Kumar, M.: An iot based congestion control algorithm. Internet Things **9**, 100, 157 (2020)
24. Yousefpour, A., et al.: All one needs to know about fog computing and related edge computing paradigms: a complete survey. J. Syst. Architect. **98**, 289–330 (2019)

Design and Implementation of a Fuzzy-Based System for Assessment of Relational Trust

Shunya Higashi[1(✉)], Phudit Ampririt[1], Ermioni Qafzezi[2], Makoto Ikeda[2], Keita Matsuo[2], and Leonard Barolli[2]

[1] Graduate School of Engineering, Fukuoka Institute of Technology, 3-30-1 Wajiro-Higashi, Higashi-Ku, Fukuoka 811-0295, Japan
{mgm23108,bd21201}@bene.fit.ac.jp

[2] Department of Information and Communication Engineering, Fukuoka Institute of Technology, 3-30-1 Wajiro-Higashi, Higashi-Ku, Fukuoka 811-0295, Japan
qafzezi@bene.fit.ac.jp, makoto.ikd@acm.org, {kt-matsuo,barolli}@fit.ac.jp

Abstract. Trust serves as a cornerstone for decision-making process across diverse contexts, prompting extensive exploration in various research domains. Among the pivotal dimensions of trust, the relational trust has a particular significance. This paper introduces a fuzzy-based system for evaluating relational trust considering three key parameters: Influence (If), Importance (Ip) and Similarity (Sm). The proposed system is evaluated through computer simulations. The simulation results show a positive correlation between If, Ip and Sm parameters with the Relational Trust (RT). So, the increase of these parameters results in the increase of RT.

1 Introduction

Our daily live is becoming increasingly entwined with the digital world, whether through social media platforms, online services, or immersive virtual experiences. This digital realm shapes our interactions, provides unparalleled access to information, and even influences our identities. However, there is paradox in this relationship. While the technology offers us the convenience and connectivity, concerns about trust remain.

Recent reports, such as Edelman Trust Barometer [1], reveal a rapid decline in trust of technology entities and institutions. These concerns of trust are propelled by escalating anxieties surrounding misinformation, breaches of data privacy, and the opaque nature of algorithms that often govern our digital experiences [2]. However, despite these concerns, individuals express a strong desire for the benefits of digital progress, exemplified by the widespread adoption of online healthcare services [3]. This highlights the critical need to address the gap between innovation and trust-building measures.

In order to deal with this challenge, promising initiatives are emerging. Decentralization efforts, driven by technologies such as blockchain, offer glimpses

L. Barolli (Ed.): AINA 2024, LNDECT 202, pp. 38–47, 2024.
https://doi.org/10.1007/978-3-031-57916-5_4

of a future where users have greater control and security over their data and online experiences [4]. Similarly, advances in Artificial Intelligence (AI), particularly in the domain of explainable AI, aim to demystify algorithms and foster increased public trust [5]. However, finding a balance between innovation and ethical considerations, and ensuring responsible AI development, requires ongoing collaboration and dialogue among diverse stakeholders, including researchers, policymakers, and the public [6].

To deal with this issues, in this paper, we proposes a fuzzy-based system for evaluating relational trust. This system considers three parameters: Influence (If), Importance (Ip) and Similarity (Sm). By evaluating relational trust through these parameters, we aim to provide a foundation for building more trustworthy systems. Our proposed system is evaluated through computer simulations. The simulation results show a positive correlation between If, Ip and Sm parameters with the Relational Trust (RT). So, the increase of these parameters results in the increase of RT.

The rest of the paper is constructed into the following sections. In Sect. 2 is presented an overview of trust computing. In Sect. 3, we present Fuzzy Logic (FL). In Sect. 4, we describe the proposed Fuzzy-based system. In Sect. 5, we discuss the simulation results. In Sect. 6, we give conclusions and future work.

2 Trust Computing

In the ever-evolving digital landscape, the concept of trust computing plays a pivotal role in shaping the dynamics of user interactions, data transactions, and overall digital experiences. Trust, in this context, extends beyond mere reliability and security by encompassing a multidimensional framework that involves aspects of transparency, accountability and user confidence [7].

Accountability is another key dimension, holding digital entities responsible for their actions. Trustworthy systems are built on accountability foundation, ensuring that any issue in security, privacy, or ethical considerations are addressed promptly, and the responsible parties are held accountable. User confidence is the most important goal of trust computing, which involves instilling a sense of assurance and reliability in users, encouraging them to participate in digital interactions without undue hesitation. Building user confidence requires a delicate balance of technological robustness and user-centric design.

As our reliance on digital platforms intensifies, trust computing emerges as a critical factor influencing user behavior, business transactions, and societal trust in digital systems. Understanding and implementing trust computing principles not only enhance the overall user experience but also contribute to the establishment of a robust and resilient digital ecosystem [8].

A trust framework model is shown in Fig. 1 [9]. It present the relationship between the trustor and the trustee. The trustor takes actions by creating a trust object for the trustee considering the experience. Then, the trustee by decrypting the trust object decides whether to trust or not relying on its own experience and trust environment. This process can be risky because of the interference from distorted messages or misunderstood actions.

In order to evaluate the trust, the trust computing considers many parameters. The trust assessment consists of individual trust and relational trust as shown in Fig. 2. The individual trust is assessed by considering the individual trustee characteristics and is evaluated from logical thinking and evidence such as feelings and propensities. While, the relational trust assessment derives the trust considering the relationship between the trustor and trustee [10].

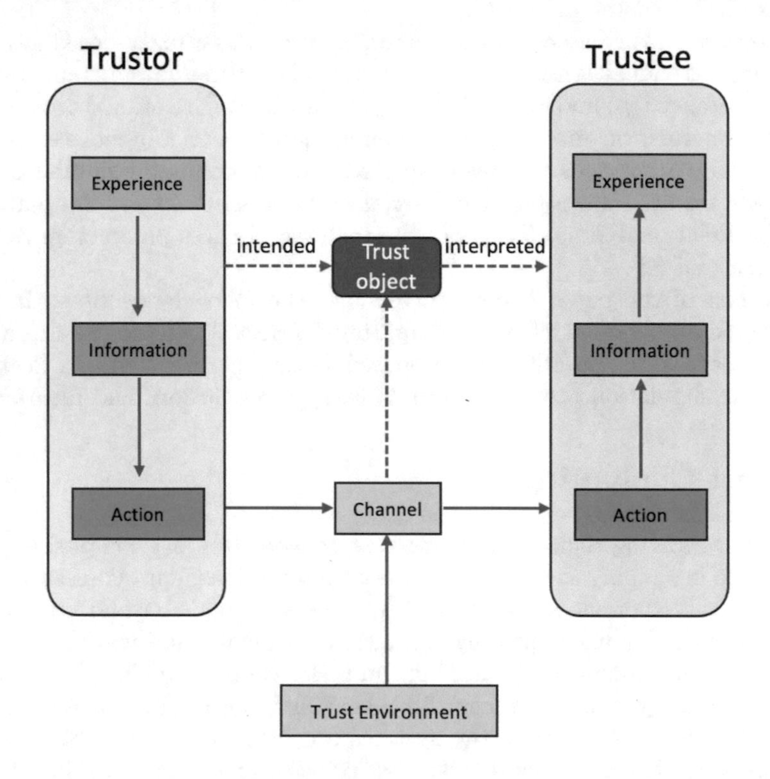

Fig. 1. Trust framework model.

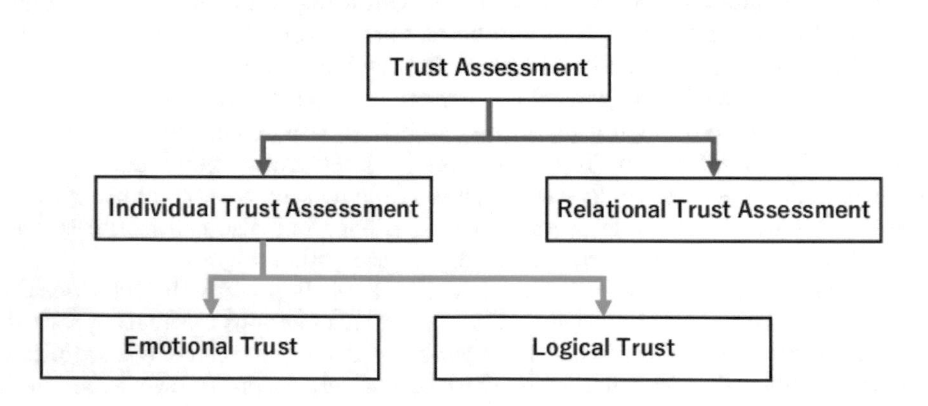

Fig. 2. Trust assessment process.

3 Fuzzy Logic Overview

The FL is a form of logic that deals with continuous values, providing flexibility and adaptability in situations where traditional strict logic may face challenges, particularly within the context of reliability computing. In the framework of digital trust computing, the FL plays a crucial role as a powerful tool for addressing uncertainty and ambiguity. When applied in reliability modeling, the FL allows the consideration of vague conditions and probabilistic elements, contributing to the effective incorporation of uncertain elements and enhancing overall reliability.

In the proposed Fuzzy-based system, the FL Controller (FLC) has a central role. By using fuzzy control rules and input parameters, the controller infers and generates output. This mechanism can address uncertain conditions and intricate relationships providing a reliable evaluation.

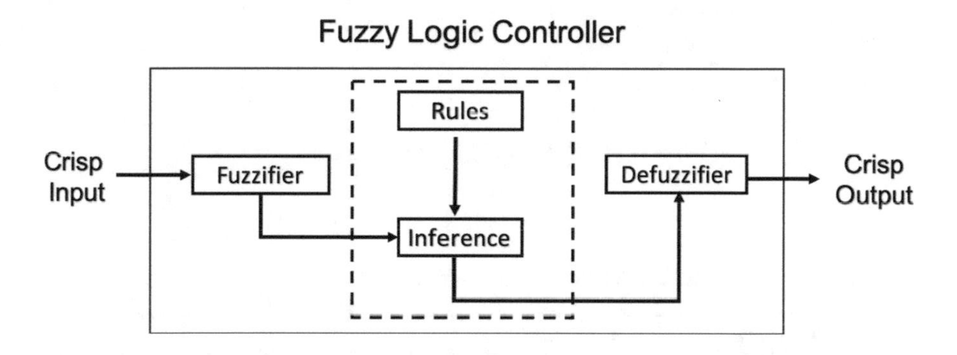

Fig. 3. FLC structure.

The FLC structure consists of four components: Fuzzifier, Inference Engine, Fuzzy Rule Base and Defuzzifier as shown in Fig. 3. The FLC mechanism uses the fuzzy sets and control rules for the transformation from input to output [11,12]. This flexible approach accommodates specific parameters and conditions, enhancing the adaptability and effectiveness of trustworthiness evaluations within the proposed Fuzzy-based system.

4 Proposed Fuzzy-Based System

In this section, we present the proposed Fuzzy-based System for Assessment of Relational Trust (FSART). By employing FL, the FSART evaluates the relational trust based on three input parameters: Influence (If), Importance (Ip) and Similarity (Sm), while the output parameter is Relational Trust (RT). The system structure is depicted in Fig. 4.

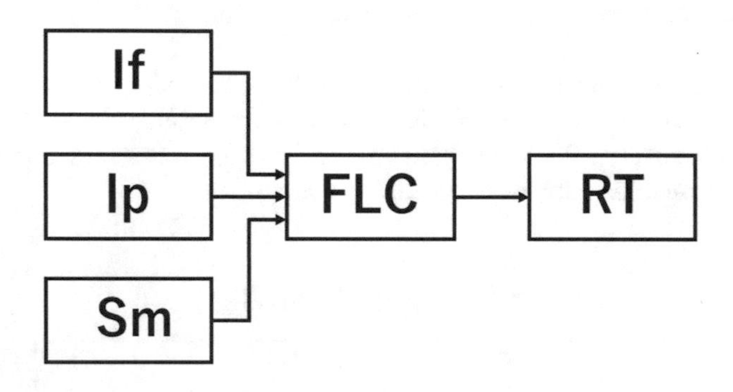

Fig. 4. Proposed system structure.

Table 1. Parameter and their term sets.

Parameters	Term set
Influence (If)	Small (S), Medium (M), Large (L)
Importance (Ip)	Low (Lo), Medium (Me), High (Hi)
Similarity (Sm)	Very low (Vl), Low (Lw), Medium (Md), High (Hg), Very high (Vh),
Relational Trust (RT)	RT1, RT2, RT3, RT4, RT5, RT6, RT7

Membership functions are illustrated in Fig. 5, providing a visual representation of their linguistic values and degrees. Table 1 shows the term sets associated with each parameter presenting their linguistic values and ranges.

The Fuzzy Rule Base (FRB) that controls the decision-making process within FSART is shown in Table 2. The FRB has 45 rules with control rules in the form of "IF condition THEN control action".

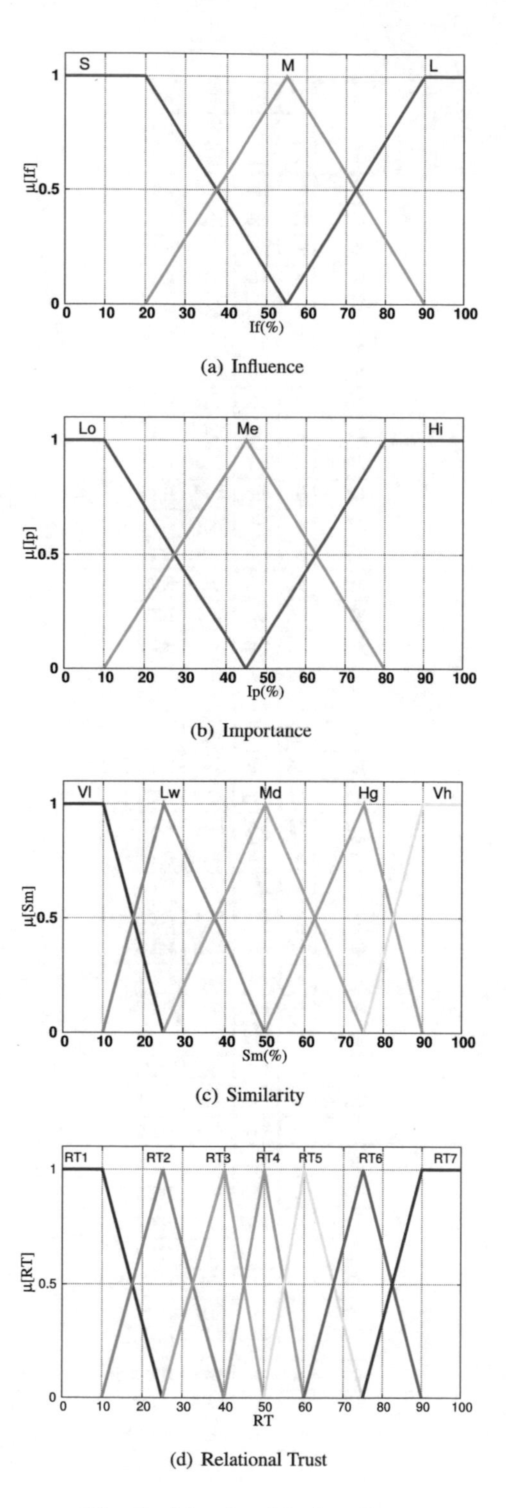

(a) Influence

(b) Importance

(c) Similarity

(d) Relational Trust

Fig. 5. Membership functions.

Table 2. FRB.

Rule	If	Ip	Sm	RT
1	S	Lo	Vl	RT1
2	S	Lo	Lw	RT1
3	S	Lo	Md	RT1
4	S	Lo	Hg	RT2
5	S	Lo	Vh	RT3
6	S	Me	Vl	RT1
7	S	Me	Lw	RT1
8	S	Me	Md	RT2
9	S	Me	Hg	RT3
10	S	Me	Vh	RT4
11	S	Hi	Vl	RT2
12	S	Hi	Lw	RT3
13	S	Hi	Md	RT4
14	S	Hi	Hg	RT5
15	S	Hi	Vh	RT6
16	M	Lo	Vl	RT1
17	M	Lo	Lw	RT2
18	M	Lo	Md	RT3
19	M	Lo	Hg	RT4
20	M	Lo	Vh	RT5
21	M	Me	Vl	RT2
22	M	Me	Lw	RT3
23	M	Me	Md	RT4
24	M	Me	Hg	RT5
25	M	Me	Vh	RT6
26	M	Hi	Vl	RT4
27	M	Hi	Hg	RT5
28	M	Hi	Md	RT6
29	M	Hi	Hg	RT7
30	L	Hi	Vh	RT7
31	L	Lo	Vl	RT2
32	L	Lo	Lw	RT3
33	L	Lo	Md	RT4
34	L	Lo	Hg	RT5
35	L	Lo	Vh	RT6
36	L	Me	Vl	RT3
37	L	Me	Lw	RT4
38	L	Me	Md	RT5
39	L	Me	Hg	RT6
40	L	Me	Vh	RT7
41	L	Hi	Vl	RT5
42	L	Hi	Lw	RT6
43	L	Hi	Md	RT7
44	L	Hi	Hg	RT7
45	L	Hi	Vh	RT7

5 Simulation Results

In this section, we present the simulation results of the proposed FSART system. The simulation results are depicted in Fig. 6, Fig. 7 and Fig. 8 showing the relationship between RT and Sm for different Ip values considering If as a constant parameter.

In Fig. 6, If is held constant at the value of 0.1. For Ip 0.5, when Sm increases from 0.1 to 0.5 and 0.5 to 0.9, the RT increases by 17% and 28%, respectively. This indicates a positive correlation between similarity and RT. When similarity increases, RT also increases, suggesting that a higher degree of similarity contributes to greater relational trust.

By comparing Fig. 6 with Fig. 7, we investigate the impact of If on RT. The If value is increased from 0.1 to 0.5. We see that the RT value is increased by 22% when Ip is 0.5 and Sm is 0.5.

In Fig. 8, the value of If is 0.9. We see that the RT values are increased much more than the results of Fig. 6 and Fig. 7. For values of Sm more than 0.5, all RT values are higher than 0.5, indicating a trustworthy person or device.

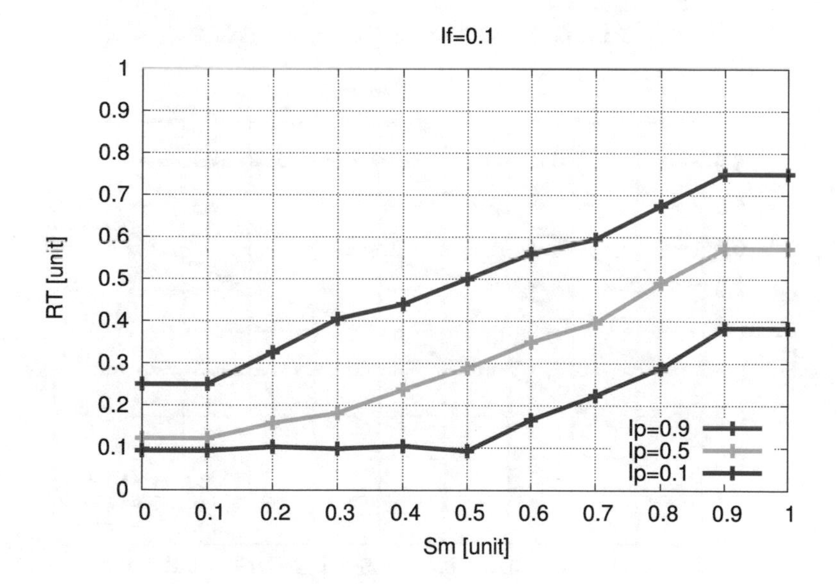

Fig. 6. Simulation results for If = 0.1.

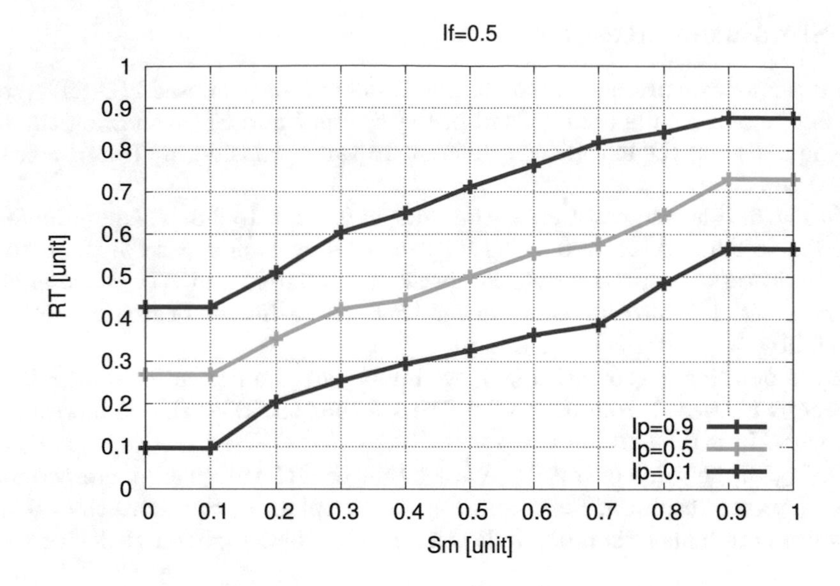

Fig. 7. Simulation results for If = 0.5.

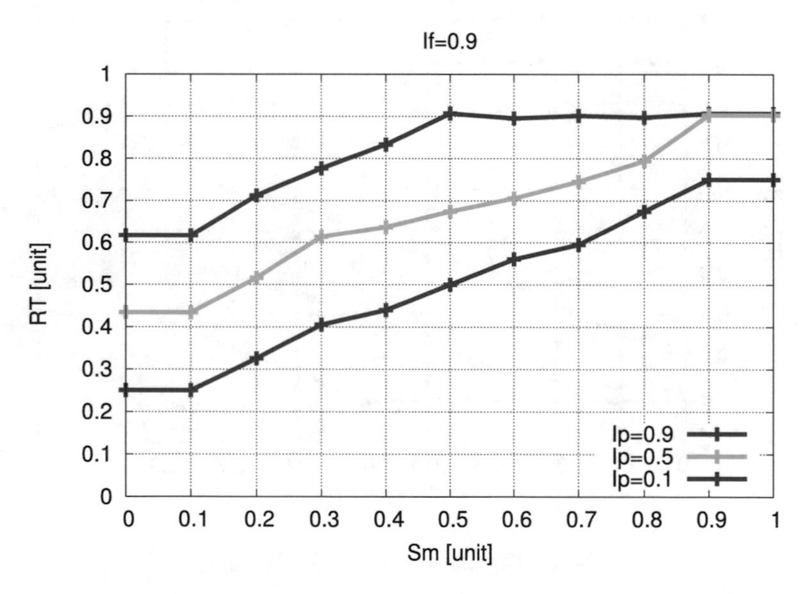

Fig. 8. Simulation results for If = 0.9.

6 Conclusions and Future Work

In this paper, we designed and implemented a Fuzzy-based system for assessment of RT called FSART. We evoluated the FSART system by computer simulations considering the relation of RT with input parameters. The simulation results have shown a positive correlation between If, Ip and Sm parameters with the RT. So, the increase of these parameters resulted in the increase of RT.

In the future, we will consider other parameters and make extensive simulations to evaluate the proposed system.

References

1. Edelman: Edelman trust barometer 2023, January 2023. https://www.edelman.com/trust/2023/trust-barometer. accessed Jan 2024
2. Ziewitz, M.: Governing algorithms. Sci. Technol. Hum. Values **41**(1), 16–30 (2016)
3. Heponiemi, T., Jormanainen, V., Leemann, L., Manderbacka, K., Aalto, A., Hyppnen, H.: Digital divide in perceived benefits of online health care and social welfare services: national cross-sectional survey study. J. Med. Internet Res. **22**(7), e17616, 1–12 (2020)
4. Chen, N., Cho, D.S.-Y.: A blockchain based autonomous decentralized online social network, pp. 186–190 (2021)
5. Adadi, A., Berrada, M.: Peeking inside the black-box: a survey on explainable artificial intelligence (XAI). IEEE Access **6**, 52:138–52:160 (2018)
6. Zhu, L., Xu, X., Lu, Q., Governatori, G., Whittle, J.: AI and ethics - operationalising responsible AI. arXiv, abs/2105.08867 (2021)
7. Frank, R.D., Chen, Z., Crawford, E., Suzuka, K., Yakel, E.: Trust in qualitative data repositories. Proc. Assoc. Inf. Sci. Technol. **54**(1), 102–111 (2017)
8. Jayasinghe, U., Lee, G., Um, T.-W., Shi, Q.: Machine learning based trust computational model for IoT services. IEEE Trans. Sustain. Comput. **4**(1), 39–52 (2019)
9. Schultz, C.D.: A trust framework model for situational contexts. pp. 1–7 (2006)
10. Cho, J.-H., Chan, K., Adali, S.: A survey on trust modeling. ACM Comput. Surv. (CSUR) **48**(2), 1–40 (2015)
11. Lee, C.-C.: Fuzzy logic control systems: fuzzy logic controller - part i. IEEE Trans. Syst. Man Cybern. **20**(2), 404–418 (1990)
12. Mendel, J.: Fuzzy logic systems for engineering: a tutorial. Proc. IEEE **83**(3), 345–377 (1995)

BarongTrace: A Malware Event Log Dataset for Linux

Baskoro Adi Pratomo(✉), Stefanus A. Kosim, Hudan Studiawan, and Angela O. Prabowo

Department of Informatics, Institut Teknologi Sepuluh Nopember, Surabaya, Indonesia
baskoro@if.its.ac.id

Abstract. Previous research that develops a machine learning malware detection model on Linux operating system typically evaluated their approach by using a dataset generated by running a small amount of malware samples. It may cause issues with not having enough data to train and evaluate the model. Thus, the result may not be applicable in the real world. Other issues with the previous research are that they focus on machine-level data (e.g., CPU, memory, and network usage) to identify malware and most of the detection models were trained with Windows-based malware. Therefore, in this research, we generated a dataset by running malicious and benign Linux ELF files on Cuckoo Sandbox. We captured the events generated by running processes with Sysmon for Linux. Sysmon collects various events such as process creation, network connection, file modification, file creation, and DNS queries. The resulting dataset (*BarongTrace*) consists of events that were populated by executing 22,784 ELF files successfully; 10,414 of them are benign and 12,370 of the rest are malicious. The biggest amount of dynamic analysis malware samples. Apart from merely generating a dataset, we also looked at the Sysmon event log generated by samples running on Windows and Linux. We notice several significant differences in the Windows and Linux data. This data discrepancy led to lower malware prediction results as shown in our experiments. When we used the Linux dataset on the pre-trained Windows malware detection machine learning model, the F1 score dropped up to 0.54 points. Therefore, we hope that this finding and our proposed dataset can contribute to the development of a more accurate machine learning-based malware detection model for Linux operating systems.

1 Introduction

Malware is a type of threat to computer systems and is defined as a program that is inserted into a system to disrupt the confidentiality, integrity, or availability of victims' data, applications, or operating systems [14]. More than 450,000 malware is discovered every day. All of which has been investigated and classified based on its characteristics [2]. As malware detection techniques develop, malware evasion techniques also come with more sophisticated anti-detection and anti-analysis capabilities [8].

To cope with those evasion techniques, researchers integrated machine learning (ML) into malware detection systems [17]. Several malware detection methods look at machine-level usage behaviour, e.g., CPU, memory, network, and disk usage, to

L. Barolli (Ed.): AINA 2024, LNDECT 202, pp. 48–60, 2024.
https://doi.org/10.1007/978-3-031-57916-5_5

identify malicious activities running on the system [11]. While this approach can identify which machine running malware, it lacks the ability to find the malicious process that causes it. Therefore, another approach for identifying malware works by analysing the process-level data, such as created processes, created files, updated registry keys, and API calls [10]. Nevertheless, both of these approaches are typically evaluated with publicly known datasets such that the results will be comparable. Currently, the publicly available malware datasets that are widely used are the EMBER [3], MOTIF [7], and SOREL-20m [5] datasets. However, those datasets only contain Windows malware. While Windows malware is more prevalent than its Linux counterparts, many servers use Linux as their operating system. Malware incidents in such computer systems would be more devastating as their users would also be affected.

Another issue with the aforementioned datasets is that they collected data from conducting static analysis on malware samples. While static analysis can give us hints whether a file is malicious, it can be easily avoided [11]. On the other hand, dynamic analysis executes malware samples and analyses its behaviour. We can capture more information with dynamic analysis. The process-level data can only be obtained by using dynamic analysis. Due to the lack of datasets of executed Linux malware, let alone the one containing process-level data, we focus on generating such datasets that contain captured process-level data - the events generated by Sysmon - while executing various Linux malware.

Sysmon [12] is a highly useful and versatile system monitoring tool developed by Microsoft for the Windows operating system but it is now also available for Linux [6]. Designed primarily for enhancing security and providing deep insights into system activities, Sysmon records and logs a wide array of events such as process creations, network connections, file changes, and more. This detailed event logging can be critical in detecting and investigating security incidents, as it allows administrators and security professionals to track potentially malicious behaviour, helping to identify threats and vulnerabilities within a Windows environment. However, Sysmon does not provide analysis results from events and does not protect the system from attackers [12].

We collected the raw data by using Cuckoo Sandbox [1]. Cuckoo Sandbox is a tool that executes malware in a virtualised environment and collects various data during the execution. While Cuckoo can capture various process-level data, such as process creation, created file, modified file, and network connection initiated, Cuckoo does not provide the order of such activities. The order of activities is essential as it shows us what the malware does over time and it can help us to provide early detection, there is no need to wait until the analysis finishes to obtain the result. We then employed Sysmon For Linux [6] and incorporated it into our sandbox to give us the order of activities.

Considering the benefit of the information provided by Sysmon for malware detection and as most malware detection models were developed to detect Windows malware [15], it raises a question as to whether the Windows malware detection models can identify malicious activities in Linux as is. Therefore, our main contribution is that we generated *BarongTrace*, the first and the biggest Linux malware dataset that contains Sysmon events which were generated from executing 22,784 ELF files. It is available at our GitHub repository[1]. The generated dataset was then used to evaluate the existing Windows malware detection models. The evaluation result can provide insights on cross operating system malware detection model.

The rest of the article is structured as follows: Sect. 2 describes existing datasets related to this research. Section 3 explains the methodology to generate the proposed dataset and how we evaluated the generated dataset with the existing malware detection models. Section 4 describes the experimental results and provides insights gathered from the experiments. Finally, conclusions and future works are presented in Sect. 5.

2 Related Work

A malware dataset is crucial for any malware detection research. It is used to evaluate the performance of the detection approach, particularly when the approach employs machine learning. There have been several malware datasets which were publicly released. Several datasets were generated by conducting static analysis on both malware and benign samples [3,5,7]. Some others were generated by executing the samples and capturing machine-level information such as CPU, memory, network, and disk usage [11,13], while our dataset - *BarongTrace* - captured the Sysmon events generated by executing each sample. This section will start by explaining existing malware datasets and then comparing them with our dataset.

EMBER (Anderson & Roth, 2018) is a dataset that collected data from static analysis of portable executable files in Windows [3]. The features were obtained by extracting features from malicious and benign PE files using the LIEF project [16].

SOREL-20M (Harang & Rudd, 2020) is a dataset that covers the shortcomings of the EMBER dataset [5]. This dataset is a large-scale dataset consisting of 20 million files with features and metadata that have been pre-extracted, have labels derived from various sources and additional "tags" as additional targets. This research also produces Python code to interact with features and data, as well as a baseline neural network and decision tree model with the results that have been trained.

MOTIF (Joyce et al., 2021) is another dataset that consists of 3,095 malware samples [7]. The dataset collection time was from January 2016 to January 2021 with labels obtained from threat reports from both samples and industry.

In the study conducted by Rhode et al. [11], a dataset containing both malware and benignware was created. This was achieved by running malicious and benign samples in a virtualized environment using the Cuckoo sandbox. While the samples were executed, they collected data at the machine level, including information on CPU usage, memory usage, network activity, and the number of running processes.

Similarly, in the work of Sihwail et al. [13], a malware dataset was generated using the Cuckoo sandbox. However, in this research, the focus was primarily on capturing the API calls made by the analysis machine during the execution of the samples, rather than collecting data on CPU, memory, and network usage, as seen in the study by Rhode et al. [11].

[1] https://github.com/bazz-066/linux-malware-dataset.

Table 1 shows the comparison of our dataset with the existing ones. *BarongTrace* is the first dataset that was generated using dynamic analysis and contains Linux malware samples and granular process-level data.

Table 1. The comparison of *BarongTrace* with previous malware datasets

Dataset	OS	Static/Dynamic	Machine-level/Process-level Data
EMBER [3]	Windows	Static	N/A
MOTIF [7]	Windows	Static	N/A
SOREL-20m [5]	Windows	Static	N/A
Rhode, et al. [11]	Windows	Dynamic	Machine
Sihwail, et al. [13]	Windows	Dynamic	Machine
BarongTrace	Linux	Dynamic	Process

3 Dataset Generation Methodology

Our methodology to generate the malware dataset starts with installing Cuckoo Sandbox and configuring the firewall and DNS. The malware samples will then be executed on an analysis virtual machine, but before that, there are several requirements that need to be installed and configured in the analysis machine. Once the analysis machines are ready, Cuckoo will automatically fire up each sample on every analysis machine. The report will be collected afterwards including Sysmon events stored in `syslog`. Once the data were collected, we map the event attributes from the Sysmon logs on Linux to the attributes we see on Windows as there are several differences between them. Afterwards, we used the dataset to evaluate detection models which had been trained with data from Windows process execution to see whether the models could perform well in detecting Linux malware.

3.1 Cuckoo Sandbox Installation

Sandboxes are the foundation of dynamic analysis. It is essential to run malware on a sandbox to prevent harm to our physical system. In this research, we used Cuckoo Sandbox 2.0.7. Despite needing the discontinued Python 2, Cuckoo Sandbox is still the best tool for conducting dynamic analysis due to its extensible nature.

We followed the Cuckoo Sandbox documentation [4] to install the Cuckoo host on a physical machine. It is worth noting that Werkzeug [9] has a more recent version than what the documentation suggests and is not compatible with Cuckoo Sandbox 2.0.7. We then installed Werkzeug version 0.16.1.

The firewall and DNS in the Cuckoo host are configured such that the machine can receive all network traffic from the analysis machines. We set the default policy of iptables to ACCEPT, disabled tcpdump, and changed the appointed DNS server to Google Public DNS.

3.2 Preparing the Analysis Machine

The eight analysis machines were created using VirtualBox with Ubuntu 18.04 OS. We also installed Sysmon for Linux and the dependencies required by Cuckoo Sandbox for the analysis machine. The dependencies required by Cuckoo Sandbox for the analysis machine follow the Cuckoo Sandbox documentation [4]. Systemtap installed on the operating system has a version that is not compatible with the documentation and needs to be downgraded to systemtap version 4.3.

Sysmon for Linux configuration was carried out using Sysmon version commit 44441f8 on December 17 2022 [6]. Sysmon for Linux requires eBPF which was issued by Sysinternals. The commit version used for eBPF is c086aca which is compatible with Sysmon commit version 44441f8. The eBPF installation process is carried out using the make and cmake tools. The Sysmon configuration used is the configuration created by Olaf Hartong [6]. We put a script on the analysis machines to pull the event logs generated by Sysmon which was located in `syslog`. The script used `awk` to take the sixth column to the last one from all rows in `syslog` that contain the word "sysmon".

All the analysis machines are connected to a virtual switch and a virtual router set up on the Cuckoo host. The architecture of our sandbox is shown in Fig. 1. The analysis machines will be reset to the clean state when a malware sample has finished execution.

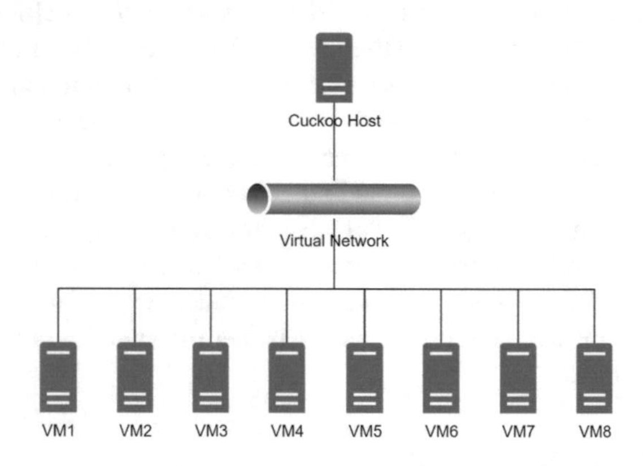

Fig. 1. The architecture of our sandbox that was used to run malicious and benign samples.

3.3 Cuckoo Sandbox Modification

We had to modify the core part of Cuckoo as it had a bug when the analysis machine ran on Linux. Modifications are made to the code in the Cuckoo Sandbox located at `resultserver.py` in the `netlog_sanitize_fname(path)` function. We replaced RESULT_UPLOADABLE with RESULT_DIRECTORIES.

Cuckoo Sandbox collected various information from the analysis machine during malware execution, but the Sysmon event log is not one of them, let alone the event log on Linux. We needed to modify Cuckoo Sandbox such that it would collect the Sysmon event log as described in the previous section. We created a Cuckoo auxiliary module in the `.cuckoo/analyzer/linux/modules/auxiliary` directory. A Cuckoo module has three functions, i.e., start, stop, and _upload_file. The start function will be executed when the analysis process starts running and will return a True value. The stop function will be executed when the analysis process is complete. This function will run a script inside the analysis machine and call the _upload_file function. The _upload_file function will check for the existence of the output file from the script and send the resulting file to the local computer using the NetlogFile module.

3.4 Executing Samples

We collected malware samples from VirusShare; the ZIP file we downloaded was the VirusShare_ELF_20190212 which had been uploaded on February 12th, 2019. The benignware was obtained from default Linux installation by downloading applications with snapd. Benignware samples were collected from files inside the "/bin", "/usr/bin", and "/usr/local/bin" directories. More information on the samples will be explained in Sect. 4.

To execute the samples, we developed a script that looped through a directory containing benign and malicious ELF files. The script sent a sample to the analysis machine every 40 s. Each sample was executed for 400 s in one of the idle analysis machines. The analysis results from Cuckoo Sandbox are then stored in `.cuckoo/storage/analyses/` which then would be curated in the following process.

3.5 Data Processing

We collected the XML files from the Cuckoo Sandbox analysis results and combined every hundred results into a single file. We labelled the resulting XML for benign and malicious according to the filename suffix; 0 means benign and 1 means malicious. We put a"malware.xml" suffix to the XML analysis result files for the malicious one and "benign.xml" for the benign one.

For further processing using the `xmltodict` library, an XML tag is added to the root tag of the file and the '&' character is replaced by '&'. Then, for each Sysmon event in the XML file that is being read, we converted the System and EventData tags into a comma-separated value format such that it is easier to process the data and consumes less storage.

XML files that have been converted to the CSV format are combined into one file for every 20 CSV files that have been converted. An example of the resulting dataset can be seen in Table 2. Some data are truncated for brevity, the full list is available at the dataset Github repository. "Attribute" shows the name of the feature recorded by Sysmon for Linux, some empty-valued attributes are omitted for brevity. The next six columns in Table 2 show an example of the feature value. An empty cell indicates that Sysmon does not record any values for that feature or that the feature does not belong to the associated EventID.

From our initial exploration of the dataset, we found out that Sysmon does not always use the same feature name in Windows and Linux. For some features, they have

Table 2. An example of Sysmon events generated by a sample

Attribute	EventID				
	1 (Process Creation)	3 (Network Connection)	10 (Process Access)	11 (File Create)	23 (File Delete)
Provider_Name	Linux-Sysmon	Linux-Sysmon	Linux-Sysmon	Linux-Sysmon	Linux-Sysmon
Provider_Guid	{ff032593-a8d3-4f13-b0d6-01fc615a0f97}	{ff032593-a8d3-4f13-b0d6-01fc615a0f97}	{ff032593-a8d3-4f13-b0d6-01fc615a0f97}	{ff032593-a8d3-4f13-b0d6-01fc615a0f97}	{ff032593-a8d3-4f13-b0d6-01fc615a0f97}
EventID	1	3	10	11	23
Version	5	5	3	2	5
Level	4	4	4	4	4
Task	1	3	10	11	23
Opcode	0	0	0	0	0
Keywords	0x8000000...	0x8000000...	0x8000000...	0x8000000...	0x8000000...
TimeCreated_SystemTime	2023-04-04 19:01:16.11475	2023-04-04 19:01:16.245641	2023-04-23 08:49:29.139374	2023-04-04 19:10:26.485888	2023-04-04 19:10:26.486064
EventRecordID	45990	46000	47200	46624	46625
Execution_ProcessID	851	851	857	837	837
Execution_ThreadID	851	851	857	837	837
Channel	Linux-Sysmon/ Operational	Linux-Sysmon/ Operational	Linux-Sysmon/ Operational	Linux-Sysmon/ Operational	Linux-Sysmon/ Operational
Computer	Cuckoo	Cuckoo	Cuckoo	Cuckoo	Cuckoo
UtcTime	4/4/2023 7:01:16 PM	4/4/2023 7:01:16 PM	4/4/2023 6:59:41 PM	4/4/2023 7:10:26 PM	4/4/2023 7:10:26 PM
ProcessGuid	{4c908a74-73fc-642c-a8b4-7ae7f6550000}	{4c908a74-73fc-642c-a8b4-7ae7f6550000}	-	{4c908a74-745e-642c-b86d-028ed0550000}	{4c908a74-7622-642c-b86d-028ed0550000}
ProcessId	1476	1476	-	244	1580
Image	/usr/lib/colord/ colord-sane	/usr/lib/colord/ colord-sane	-	/lib/systemd/ systemd-udevd	/lib/systemd/ systemd-udevd
User	colord	colord	-	-	-
label	0	0	1	0	0
TargetFilename	-	-	-	-	/run/udev/data/ c13:33
CreationUtcTime	-	-	-	4/4/2023 7:10:26 PM	-
CommandLine	/usr/lib/colord/ colord-sane	-	-	-	-

a different name in Windows and Linux. Some others only exist either in Windows or in Linux. Therefore, we mapped the dataset such that our existing Windows malware detection model can work with the dataset.

We first look for features that share column names. For each feature in the Barong-Trace dataset that does not have the exact feature name in Windows, we look at features with similar values and names. Table 3 shows the mapping that can be done for each column in the Windows and Linux datasets. The Windows Feature column shows the name of the Sysmon event attribute/feature on Windows. The Sample Value column shows a possible value of the feature. While the Linux Feature column shows the name of the Sysmon event attribute/feature on Linux.

Apart from the differences in the event attributes, several Sysmon event IDs only appeared in our Linux malware dataset, but not in our Windows malware dataset [10]. As we would evaluate the detection model which was trained on the Windows malware dataset, we provide two sets of Sysmon events. The first set contains Sysmon event IDs that also occur in the Windows malware dataset, while the second set contains all events captured by Sysmon for Linux.

3.6 Experiment with Windows Malware Detection Model

At the testing stage, four experiments were carried out, namely trial dataset with EventID adapted to Sysmon Windows using supervised learning, trial dataset with all EventIDs recorded on Sysmon Linux using supervised learning, trial dataset with EventID

Table 3. The mapping of Sysmon event attributes/features from Windows to Linux

Linux Feature	Sample value	Windows Feature	Sample value
Provider_Name	Linux-Sysmon	-	-
Provider_Guid	{ff032593-a8d3-4f13-b0d6-01fc615a0f97}	-	-
EventID	5	EventID	1
Version	3	-	-
Level	4	-	-
Task	5	-	-
Opcode	0	-	-
Keywords	0x8000000000000000	-	-
TimeCreated_SystemTime	2023-04-04T18:59:18.350051000Z	timestamp	0
EventRecordID	46591	-	-
Correlation		-	-
Execution_ProcessID	857	-	-
Execution_ThreadID	857	-	-
Channel	Linux-Sysmon/Operational	-	-
Computer	cuckoo	host_name	WKSTN-5.COFFEE.LOCAL
Security_UserId	0	-	-
RuleName	-	RuleName	Alert, Sysinternals Tool Used
UtcTime	59:18.4	UtcTime	44735.31687
ProcessGuid	{4c908a74-7386-642c-a03f-b066a2550000}	ProcessGuid	{C89119D8-17F1-62B4-FC00-000000001A00}
ProcessId	1536	ProcessId	1320
Image	/usr/bin/truncate	Image	[omitted]\cmd.exe
User	root	User	NT AUTHORITY\SYSTEM
TargetFilename	/run/udev/data/c13:33	TargetFilename	[omitted] Roaming\Microsoft\ctjrba.exe
CreationUtcTime	59:34.0	CreationUtcTime	38474.35269
Hashes	-	-	-
IsExecutable	-	-	-
Archived	-	-	-
FileVersion	-	FileVersion	0
Description	-	Description	Windows Command Processor
Product	-	Product	Microsoft Windows Operating System
Company	-	Company	Microsoft Corporation
OriginalFileName	-	OriginalFileName	0
CommandLine	/usr/lib/colord/colord-sane	StartFunction	EtwpNotificationThread
CurrentDirectory	/tmpQ4P6ce	-	-
LogonGuid	{4c908a74-0000-0000-0000-000000000000}	LogonGuid	{C89119D8-FA34-62B3-E703-000000000000}
LogonId	0	LogonId	0x3e7
TerminalSessionId	4294967295	TerminalSessionId	0
IntegrityLevel	no level	IntegrityLevel	System
ParentProcessGuid	{4c908a74-739e-642c-6892-8b7c60550000}	ParentProcessGuid	{C89119D8-17F1-62B4-FB00-000000001A00}
ParentProcessId	1572	ParentProcessId	772
ParentImage	/bin/dash	-	-
ParentCommandLine	sh	-	-
ParentUser	root	-	-
label	1	label	0

adapted to Sysmon Windows using unsupervised learning, test dataset with all EventIDs recorded on Sysmon Linux using unsupervised learning. This trial was carried out to determine the suitability of the dataset produced in this research with a detection model that had been trained using a dataset from Sysmon Windows.

The supervised learning trial was carried out using a model that had been trained using the Sysmon dataset on Windows using the Random Forest Classifier method. This model is the model with the best results from previous research. The features used when creating the detection model are ParentProcessId, TargetObject, TargetFilename, Description, TargetProcessGuid, ParentProcessGuid, Product, PreviousCreationUtcTime, RuleName, Company, User, TargetImage, EventID, IntegrityLevel, EventType, TerminalSessionId, StartFunction, DestinationPort, DestinationPortName, LogonId. To adapt to the features used, pre-processing is carried out. All columns that have the value "-" are changed to "NaN", then the column that is a feature is immediately taken and the other columns are ignored.

The unsupervised learning trial was carried out using a model that had been trained using the Sysmon dataset on Windows with the Local Outlier Factor, One-Class SVM, and Isolation Forest algorithm models. The features used when creating the detection model are TargetFilename, EventID, TargetProcessGuid, label, User, EventType, TargetImage, PreviousCreationUtcTime, DestinationHostname, Company, Description, Product, IntegrityLevel, CreationUtcTime, StartFunction, ParentProcessGuid, LogonId, ParentProcessId, TerminalSessionId, RuleName, and TargetObject. To adapt the data used to the train results of this model, pre-processing is used. Pre-processing is carried out by changing the label value to -1 and the label value 0 to 1. The malware value is made -1 because it is an outlier.

4 Experimental Results and Analysis

In this section, we will first explain the result of our malware and benignware execution which will be the foundation of our dataset. Then, we explore the results of our evaluation of Windows malware detection models on our Linux dataset.

As previously mentioned, we collected malicious and benign ELF files and executed them in a sandbox. However, not all files were successfully executed. We executed 10,426 malicious and 13,470 benign ELF files, but only 10,414 malicious ELFs that were successfully executed and 12,370 benign ELFs as shown in Fig. 2a and Fig. 2b. Thus, in total, we collected 22,784 XML files containing analysis results from Cuckoo. An example of the XML pulled from Cuckoo is shown in Fig. 3.

Each event has the tag `Event` that contains two child elements, i.e., `System` and `EventData` as in Fig. 3. Information that can be extracted is obtained from the attributes and values of each of those elements. The content of EventData varies depending on the EventID (the type of event) recorded by Sysmon, but the content of System will always be the same, namely Provider, EventID, Version, Level, Task, Opcode, Keywords, TimeCreated, EventRecordID, Correlation, Execution, Channel, Computer, and Security. The total Sysmon events in the dataset amounted to 16,555,075 events after removing duplicates.

The Sysmon events obtained in this study were then transformed into a CSV format through a process involving the extraction of elements and attributes from the source

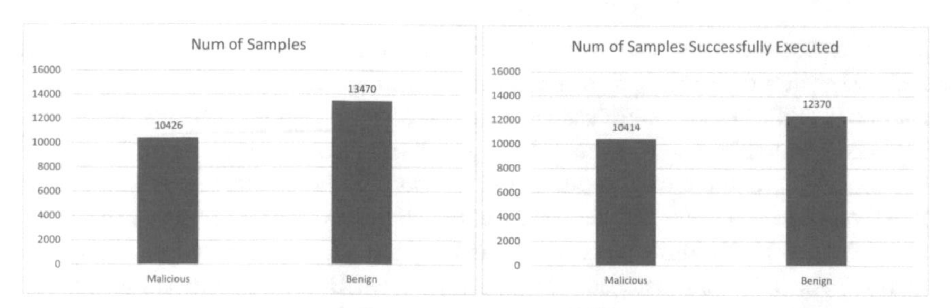

(a) Number of total samples (b) Number of successfully executed samples

Fig. 2. Number of samples.

Fig. 3. An example of Sysmon event data

XML file as explained in Sect. 3. The resulting CSV files are then combined into 124 CSV files containing benign events and 105 CSV files containing malicious events. We then grouped them into 12 final CSV files with labels to make further processing easier. The total number of columns in the CSV file initially varied depending on the `EventID`. To make all Sysmon events have the same number of columens, we put the first 14 columns representing attributes derived from the `System` element, while the remaining columns correspond to attributes sourced from the `EventData` element. Notably, the `EventData` element exhibits adaptability to the EventID, resulting in numerous empty values within the CSV dataset that correlate with the characteristics of the `EventData` element.

Once we had the dataset that contained all Linux events, we needed to map those events such that they could be used to evaluate the Windows malware detection model. As explained in Sect. 3, there are two mapping versions, namely, one encompassing all

EventIDs in Linux (`all-linux-events`) and the other including only EventIDs 1, 3, 5, and 11 (`win-events-only`), as other EventIDs did not exist in the Windows malware dataset [10]. Both versions were used to assess malware detection models that had been trained on Sysmon Windows events. In light of the dataset pertinent to the Windows operating system and the dataset generated as an outcome of this research, a comparative analysis can be conducted between the two datasets, focusing on the features and values recorded by Sysmon. This comparison was undertaken based on the common EventIDs present in both datasets, specifically, EventID 1, 3, 5, and 11. Notably, certain features present in Sysmon Windows, namely "node_id" and "parent_node_id," are absent in Sysmon Linux, as these columns are supplemental in nature and do not constitute inherent components of Sysmon.

EventID 1 corresponds to an event denoting process creation and exhibits a disparity in the number of columns employed between Sysmon versions, featuring 20 columns in Windows and 39 columns in Linux. Through the mapping process, 18 columns from Sysmon Windows align with their counterparts in Sysmon Linux, while 2 columns do not exhibit conformity. The columns unique to Sysmon Windows, "node_id" and "parent_node_id", are not mirrored in Sysmon Linux due to their auxiliary status.

EventID 3 pertains to an event signifying a network connection. We have 15 features related to this EventID in Windows and 33 features in Linux. Mapping this event type reveals that 13 features of this EventID in Sysmon Windows align with their counterparts in Sysmon Linux, while two other features do not display consistency. We omitted those features.

EventID 5 represents an event denoting process termination. All events of this ID also show a difference in the number of features used between Sysmon versions, comprising 10 features in Windows and 22 features in Linux. Upon mapping, it is observed that 8 columns from Sysmon Windows correspond to their counterparts in Sysmon Linux.

EventID 11 embodies an event about file creation. We have 10 features in Windows and 24 features in Linux. Mapping reveals that 8 columns from Sysmon Windows align with their counterparts in Sysmon Linux.

The experiment with malware detection models that were trained on Sysmon events in Windows comprised evaluating four distinct models, specifically the Random Forest Classifier for the supervised approach, the Isolation Forest, One-Class SVM, and Local Outlier Factor for the unsupervised approach. Two dataset versions were employed, one encompassing all EventIDs captured by Sysmon on Linux and the other limited to EventIDs overlapping with those captured by Sysmon on Windows. `win-events-only` consisted of 12,420,236 events, while `all-linux-events` accounted for 16,555,075 events. In the context of supervised learning, a single algorithm was employed, as the machine learning model derived from our preliminary research exclusively relied on the Random Forest Classifier algorithm. Table 4 provides an overview of the methodology employed during the experimentation, showcasing the specific algorithm utilised, the mapping version (`all-linux-events`, `win-events-only`, or CerberusTrace, the Windows malware dataset), and a range of metrics, including F1 Score, Precision, and Recall.

Table 4. The performance of existing malware detection models on our Linux malware dataset

Learning Method	Algorithm	Testing Set	Metric		
			F1 Score	Precision	Recall
Supervised	Random Forest	`all-linux-events`	0,324	0,437	0,257
	Random Forest	`win-events-only`	0,342	0,436	0,282
	Random Forest	CerberusTrace	0.887	0.875	0.899
Unsupervised	Isolation Forest	`all-linux-events`	0,569	0,435	0,821
	Isolation Forest	`win-events-only`	0.606	0.434	1
	Isolation Forest	CerberusTrace	0.696	0.747	0.647
	One-Class SVM	`all-linux-events`	0.608	0.436	1
	One-Class SVM	`win-events-only`	0.606	0.434	1
	One-Class SVM	CerberusTrace	0.945	0.895	1
	Local Outlier Factor	`all-linux-events`	0.608	0.436	1
	Local Outlier Factor	`win-events-only`	0.606	0.434	1
	Local Outlier Factor	CerberusTrace	0.987	0.975	0.999

As shown in Table 4, all algorithms, be it supervised or unsupervised, show decreasing performance when they were evaluated with the Linux malware dataset. One-Class SVM and Local Outlier Factor reached F1-scores of 0.945 and 0.987 respectively when they faced the Windows malware dataset, CerberusTrace. But, their performance dropped by 0.337 and 0.379 respectively. The supervised learning approach, Random Forest, has the largest drop in the F1 score, a difference of 0.545.

There are two take-away messages from this experiment. Firstly, it is clearly shown that malware detection models that were trained on Windows malware will perform poorly in detecting Linux malware. Features in the Linux malware dataset, such as TargetObject, Description, PreviousCreationUtcTime, and RuleName, have many missing values. Secondly, it seems to be consistent with what we know about unsupervised vs supervised approaches, unsupervised approaches still perform better when facing unseen data.

5 Conclusion and Future Work

We have generated *BarongTrace*, a Linux malware dataset containing malware activities in the form of Sysmon events. *BarongTrace*[2] was obtained by executing 22,784 ELF files, both malicious and benign, in a sandbox environment. Our proposed dataset stores the data in CSV format, making it easily accessible by any machine learning library.

We also analysed the Sysmon events generated in Linux and compared it with the Sysmon events generated in Windows. From our analysis, Sysmon for Windows generated 42 features while Sysmon for Linux generated 75 features, more various events

[2] https://github.com/bazz-066/linux-malware-dataset.

were also being captured in Linux. Eight features do not exist in Sysmon for Linux but exist in Sysmon for Windows, while 33 are not generated by Sysmon for Windows but are generated by Sysmon Linux.

From our experiment, the performance of the malware detection model trained with Sysmon events generated by Windows applications dropped. The F1 score of the Random Forest model dropped to 0.342 from 0.886. The Local Outlier Factor's F1 score dropped from 0.974 to 0.606. This experiment shows that a malware detection model developed for an operating system cannot be used in another OS as the data may exhibit different patterns. However, it remains to be seen if the performance will increase if we retrain the model using the new dataset.

To summarise, by creating a dataset that has application behaviour information in the form of Sysmon events, we hope that it can help further research regarding malware detection models, especially for Linux malware.

References

1. Cuckoo sandbox - automated malware analysis. https://cuckoosandbox.org/
2. Malware statistics & trends report — av-test. https://www.av-test.org/en/statistics/malware/
3. Anderson, H.S., Roth, P.: EMBER: an open dataset for training static PE malware machine learning models. CoRR **abs/1804.04637** (2018). http://arxiv.org/abs/1804.04637
4. Cuckoo: Installing the Linux host (2020). https://docs.cuckoosandbox.org/en/latest/installation/guest/linux/
5. Harang, R.E., Rudd, E.M.: SOREL-20M: a large scale benchmark dataset for malicious PE detection. CoRR **abs/2012.07634** (2020). https://arxiv.org/abs/2012.07634
6. Hartong, O.: Sysmon for Linux (2021). https://medium.com/@olafhartong/sysmon-for-linux-57de7ca48575
7. Joyce, R.J., Amlani, D., Nicholas, C., Raff, E.: MOTIF: a large malware reference dataset with ground truth family labels. CoRR abs/2111.15031 (2021). https://arxiv.org/abs/2111.15031
8. Naval, S., Laxmi, V., Rajarajan, M., Gaur, M.S., Conti, M.: Employing program semantics for malware detection. IEEE Trans. Inf. Forensics Secur. **10**(12), 2591–2604 (2015). https://doi.org/10.1109/TIFS.2015.2469253
9. Pallets: Werkzeug (2007). https://werkzeug.palletsprojects.com/en/3.0.x/
10. Pratomo, B.A., Jackson, T., Burnap, P., Hood, A., Anthi, E.: Enhancing enterprise network security: comparing machine-level and process-level analysis for dynamic malware detection (2023). http://arxiv.org/abs/2310.18165
11. Rhode, M., Burnap, P., Jones, K.: Early-stage malware prediction using recurrent neural networks. Comput. Secur. **77**, 578–594 (2018)
12. Russinovich, M., Garnier, T.: Sysmon - windows sysinternals. https://learn.microsoft.com/en-us/sysinternals/downloads/sysmon
13. Sihwail, R., Omar, K., Zainol Ariffin, K.A., Al Afghani, S.: Malware detection approach based on artifacts in memory image and dynamic analysis. Appl. Sci. **9**(18), 3680 (2019)
14. Stallings, W., Brown, L.: Computer Security: Principles and Practice. 4th edn global (2018)
15. Talukder, S., Talukder, Z.: A survey on malware detection and analysis tools. Int. J. Netw. Secur. Appl. **12** (2020). https://doi.org/10.5121/ijnsa.2020.12203
16. Thomas, R.: Lief - library to instrument executable formats (2017). https://lief.quarkslab.com/
17. Ucci, D., Aniello, L., Baldoni, R.: Survey of machine learning techniques for malware analysis. Comput. Secur. **81**, 123–147 (2019)

Specifying SSI over EAP: Towards an Even Better Eduroam in the Future

Ronald Petrlic(✉)

Nuremberg Institute of Technology, 90489 Nuremberg, Germany
ronald.petrlic@th-nuernberg.de

Abstract. The *Extensible Authentication Protocol* (EAP) is a popular protocol for authentication in (wireless) local area networks. The protocol is recommended for use in the 802.1X standard and has, thus, found widespread usage in (enterprise) network environments. One of its advantages is its flexibility: it supports multiple authentication methods.

Self-Sovereign Identity (SSI) is a new approach towards user-centric identity management that promises users full control over their data.

We are the first to propose combining both technologies by specifying *EAP-SSI* as a new EAP method. The integration of SSI to EAP entails several advantages: it eases roaming for users and provides enhanced security and better privacy protection. We will point out the advantages of our approach on the example of *Eduroam*.

Keywords: Authentication · EAP · Self-Sovereign Identity · Eduroam

1 Introduction

The **Extensible Authentication Protocol** (EAP) is a wide-spread, flexible authentication framework, which supports multiple authentication methods. The specific method is negotiated during the authentication. EAP has become the standard for authentication in enterprise (wireless) local area network environments, as it is proposed to be used by the *IEEE 802.1X* standard.

In this paper, we will come up with a new EAP method that allows authentication based on **Self-Sovereign Identity** (SSI). SSI is a new approach to identity management that gives users full control of their digital identities and, thus, their personal data. We will show that integration of SSI in EAP can have tremendous benefits to network roaming infrastructures such as *Eduroam*.

1.1 Background

Extensible Authentication Protocol (EAP)

EAP in Detail: EAP is specified in RFC 3748. EAP typically runs directly over data link layers such as IEEE 802. In EAP terminology, the *authenticator* (in practice: a switch or an access point) is the party that initiates the authentication process. The client device, requiring access to a network is called *peer*[1]. The

[1] In IEEE 802.1X terminology, the peer is called *supplicant*.

L. Barolli (Ed.): AINA 2024, LNDECT 202, pp. 61–73, 2024.
https://doi.org/10.1007/978-3-031-57916-5_6

backend authentication server provides an authentication service to an authenticator. This is the party where the authentication data are stored. The server executes the EAP methods for the authenticator, which provides the advantage that the authenticator does not need to be updated to support new authentication methods—it rather serves as a pass-through agent.

The authenticator starts the authentication process by sending an authentication request to the peer (typically an identity request). The data that is requested is indicated in the type field of the message. The peer responds to the request (typically with its identity in the first step—the identity being transmitted unencrypted) and indicates the type of data in the type field as well. Then the authenticator sends an additional request (typically requiring authentication data) and the peer answers with the requested data. This process is continued until authentication either succeeds and the authenticator answers with an EAP success (code 4) or fails, resulting in an EAP failure (code 3) message.

There is an initial set of authentication methods (like MD5-Challenge, One Time Password, and Generic Token Card) that are already defined as types in the RFC. Other methods have been (and can be) developed. The methods are identified via new types. The method to be used for authentication is negotiated via the type header field.

EAP Methods: A popular authentication method is the *EAP-TLS Authentication Protocol*, specified in RFC 5216. This method employs TLS over EAP, which provides mutual authentication based on certificates, an integrity-protected cipher suite negotiation, and a key derivation for further communication. This method is often used for authentication in the enterprise mode of 802.11 communication. The TLS frames are encapsulated within EAP frames. The derived key material can be used for encryption and authentication of the communication between the peer and the access point. An advantage of the method is that the authentication server (AS) authenticates towards the peer, which is an important requirement in 802.11 environments where the risk of fake access points is inherent. However, the method also entails two disadvantages. First, the transmission of identities (within certificates) is done in clear text, meaning that the identity of a peer is disclosed to eavesdroppers during authentication.[2] Second, the method only supports certificate-based authentication. While this is fine for the authentication of the AS, it also means that the peers need to have a certificate as well, which complicates the setup.

EAP-TTLS (specified in RFC 5281) and *Protected EAP (PEAP)* are two methods that both solve the issues of EAP-TLS. Both methods employ two phases: in the first phase, a TLS connection is initiated between the peer and the AS and the AS authenticates to the peer with a certificate. In the second phase, the client authenticates within the established TLS tunnel. Thus, the peers' identities are only transmitted within the protected TLS channel and the

[2] RFC 4282 specifies a privacy network access identifier with which this problem can be circumvented.

client authentication can be based on other authentication mechanisms than only on certificates as with EAP-TLS.

PEAP is not specified in an RFC but there exist expired Internet drafts. The protocol is recommended for use with WPA2 and WPA3 and is widely used in practice. The most popular peer authentication mechanism is EAP-MSCHAPv2, an old authentication protocol introduced with Microsoft NT 4.0 SP4 and Windows 98, which allows authentication towards an active directory at the AS. Other peer authentication mechanisms are EAP-GTC (Generic Token Card) and EAP-SIM (see related work section).

EAP-TTLS provides more mechanisms for client authentication than PEAP, but PEAP is more widely used in practice due to its support by Microsoft.

EAP in the Bigger Picture: EAP is recommended to be used within the 802.1X standard and is therefore employed in professional wired and wireless networks. To give an overview of how it interplays with 802.1X, we show how it is used in the popular roaming network infrastructure *Eduroam*, which academic institutions around the globe use. Figure 1 shows the process of authentication from a peer at a foreign institution towards its home institution.

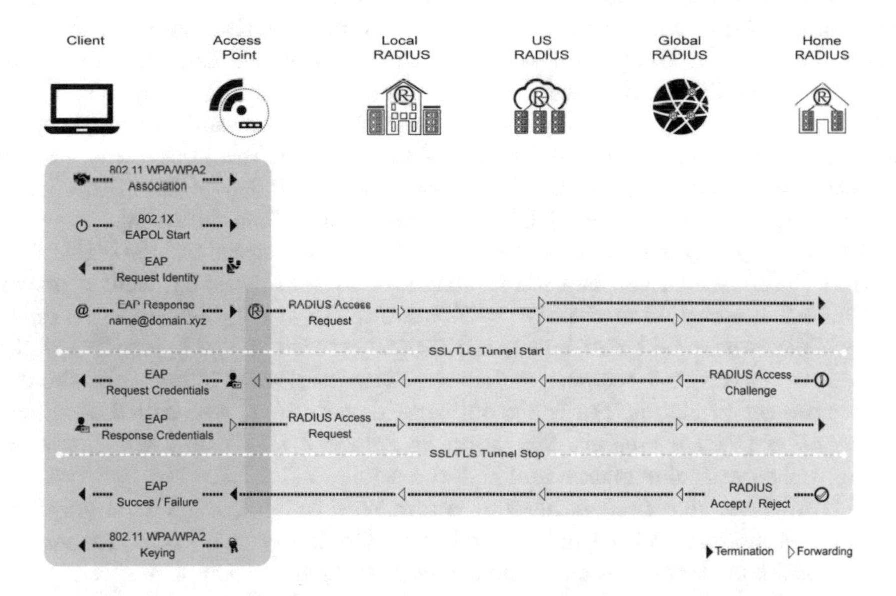

Fig. 1. Eduroam Overview. Source: from Technical Documentation for Administrators from former US Eduroam website (not available any longer).

The initial EAP steps are performed between the peer (client in the picture) and the authenticator (access point in the picture). Then, the authenticator establishes a connection to the peer's home institution (Home RADIUS in the picture) via the RADIUS protocol. The peer is then able to establish a secure

TLS tunnel with the AS of its home institution via PEAP. Within the secure tunnel, the peer authentication takes place via MSCHAPv2. The radius servers in the backend just act as pass-through agents. They encapsulate the EAP frames within RADIUS frames and forward them to the next hop.

Self-Sovereign Identity (SSI). The basis for SSI is a new identifier that does not rely on a central trust authority. The *Decentralized Identifier* (DID), which can be created on one's own, identifies a subject (a person, organization, object, etc.). The DID is managed by a DID controller (in most cases the same person as the DID subject) who can cryptographically prove that he controls the DID [1]. Typically, a DID document, containing data like a public key or information about the initiation of a secure connection (i.e., a URL to the DID subject), is assigned to a DID. The DID and the corresponding DID document are typically stored on a verifiable data registry (VDR)—in most cases on a Blockchain [1–3].

Building upon (layer 1) DIDs, the *DID Communication Protocol* (DIDComm) is located on layer 2. DIDComm enables asynchronous and end-to-end encrypted connections between communication partners. DIDComm can be used independently of the transport mechanism (e.g., HTTP, Bluetooth, email, NFC,...). There are different ways of communication initiation. The sending party might either know the recipient's DID (and can then look up the necessary data on the Blockchain—i.e., the URL of the recipient and its public key) or the sending party receives a communication invitation by scanning a QR code (and thereby retrieving the necessary data). The sending party's agent then encrypts and signs the message and sends it to the recipient—whose agent performs the decryption and signature check (the public key of the sender being retrieved through the sender's DID document on the Blockchain) [3].

Another layer provides trust: On layer 3, we have so-called *Verifiable Credentials* (VCs), which can be used to vouch for certain facts, e.g. that a person is a student at a certain university [4]. VCs are the digital equivalent of ID documents. They are issued by an *issuer* and digitally stored by the *holder* in a wallet and can be shown to a *verifier* on demand. The individual statements about a subject are called claims. For the proof towards a verifier, a so-called *Verifiable Presentation* (VP) is created. VPs support *Selective Disclosure*, which enables sharing only particular claims and not the whole VC. There are two types of VCs: *W3C Verifiable Credentials* and *AnonCreds*. With W3C VCs a standard for the data model of VCs is determined. AnonCreds were developed earlier and have stronger protection of anonymity as a goal. Due to their widespread, they are the de-facto standard for zero-knowledge proof (ZKP) based credentials [3].

The rest of the paper is structured as follows. In Sect. 2 we point out the requirements for our solution before we present the specification of our new EAP-SSI method in Sect. 3. We evaluate our approach in Sect. 4. Finally, we present related work in Sect. 5 before we conclude the paper in Sect. 6.

2 Requirement Analysis

Before we discuss the requirements in detail, we point out why the integration of SSI in EAP is advantageous, i.e., which improvements can be achieved.

Mitigation of the Password Risk: In many cases, peer authentication is based on passwords in practice. With Eduroam, for example, PEAP in combination with MSCHAPv2 is a very popular authentication method chosen by institutions. This is often due to the straightforward integration in Windows network environments. The disadvantage is that weak user passwords can lead to unauthorized network access by attackers. With our approach, we will get rid of peer authentication based on weak user passwords.

No central storage of user keys: As peer authentication is not performed based on user passwords any longer with our approach, there is no need for a central database with user passwords at the AS anymore. This minimizes the risk of an attack and unauthorized network access.

Mitigation of Unknown CAs: One of the problems of Eduroam arises during the first connection to the Eduroam network by a peer: The user is asked to install an unknown certificate issued by its institution to its device. This is due to the Eduroam PKI where certificates are issued by the participating institutions. However, these certificates are not certified by a trusted root CA (in most cases). The AS of an institution thus presents a certificate to the peer (during the first connection) that is unknown to the device and the user needs to install that certificate to proceed. If an attack happens during this step, the user would accept a rogue certificate with which an attacker could cause a lot of harm. With our approach, we are getting rid of traditional CAs and this problem is prevented.

Data Minimization: By using selective disclosure, the user only needs to share the data of a VC that are needed for authentication. It would not even be necessary to reveal the user's identity during network access—it would suffice to prove that the user is a student at some academic institution to get access to an Eduroam-like network, for example. Even if the user's identity is needed for authentication, it is not disclosed to eavesdropping parties as is the case with the use of EAP-TLS, for example.

Single Verification: A user can use a VC for authentication at different institutions. There is no need for a second account, an exchange for a password, or additional verification. The AS gets all the needed data for peer authentication through VCs (or even claims from different VCs in a VP).

Increase of Comfort: The user does not need to remember a strong password (like with PEAP and MSCHAPv2) or retrieve a certificate from his home institution (like with EAP-TLS) with our approach; the user only needs to unlock his wallet and show the claims.

Integration in Existing Infrastructure: Our approach allows both: the existing infrastructure (e.g. Eduroam) may be maintained, as the strength of EAP is its expandability via new EAP methods, but it would also be possible to completely get rid of the backend infrastructure (the RADIUS roaming servers) and just setting up our approach to get the same functionality as with Eduroam.

2.1 Requirements

Functional Requirements

- FR1: The peer can log in to a network with his SSI
- FR2: Claims can be extracted from a VC to enable authentication
- FR3: CAs get replaced by a decentralized solution
- FR4: The login to a network shall be possible without prior registration
- FR5: The AS does not need to store user identities and passwords
- FR6: The user only needs to share those data that are needed for authentication (selective disclosure)
- FR7: The access to the network is still done via standard 802.1X, i.e. the existing infrastructure does not need to be modified

Non-functional Requirements

- NFR1: The wallet with the SSI is on the same device on which the network login shall take place
- NFR2: The approach shall be compatible with networks that do not support EAP-SSI

3 Concept

3.1 Assumptions

To be able to log in to a network, a user needs credentials from a trustworthy issuer. With SSI, a governance framework is used to establish trust [4]. Moreover, we assume that the user has a wallet setup on his device.

3.2 Merging of EAP and SSI Components

Let us start the concept description with a high-level overview, shown in Fig. 2.

As we can see in the overview, the AS is extended by an SSI Agent. This implies a main change in responsibilities at the AS: The AS is not responsible for managing user accounts any longer—instead, the SSI Agent verifies users' SSI credentials and provides the AS with proper identity information.

The peer authenticates towards the AS using our proposed EAP-SSI method. The general setting stays unaltered, though. The peer connects with an authenticator (serving as a pass-through agent) and the authenticator grants network access to the peer after successful authentication and permission by the AS. We omitted the authenticator in the figure for the sake of clarity in this overview.

Fig. 2. Merging of EAP and SSI components.

3.3 New EAP Type

We propose a new EAP method called *EAP-SSI* in this paper. For the participating entities to negotiate the use of this new method, a new EAP type needs to be specified. It is therefore necessary that we standardize our approach in an RFC and retrieve a new EAP type from the *Internet Assigned Numbers Authority* (IANA). EAP types are numbered and in the following protocol description we refer to our new EAP type as "x".

3.4 Protocol in Detail

Now we present the protocol of our new EAP method *EAP-SSI* in detail.

Phase 1: Handshake

1. The peer and the authenticator start the authentication process by initiating the association process via 802.1X (encapsulating the EAP frame within EAPOL as carrier protocol).
2. The authenticator requests the peer's identity with the "Request Identity" EAP message (i.e., EAP message with type of 1).
3. The peer responds with an EAP message (type 1) indicating *anonymous@SSI* as identity; thereby signaling that it supports SSI authentication and requests EAP-SSI as the authentication method.
4. The authenticator forwards the message to the AS (by encapsulating the EAP message within a RADIUS access request message frame).
5. The AS now initiates the EAP-SSI method by generating an EAP Request packet with type "x" (standing for the EAP-SSI method) and the S (Start) bit set. This indicates to the peer that it should initiate a DIDComm connection to the AS. The EAP packet is encapsulated within a RADIUS Access Challenge message and sent to the authenticator, indicating to the authenticator that the AS requests more information to allow access for the peer. By sending this message, the AS confirms that it supports SSI authentication by the peer via EAP-SSI.
6. The authenticator unwraps the RADIUS packet and forwards the EAP packet to the peer.

7. The peer retrieves the EAP packet and initiates a DIDComm connection to the AS (i.e., to the SSI Agent of the AS) by responding with an EAP Response message. During this process, the AS is authenticated using standard DIDComm authentication. This process is done by exchanging EAP packets (encapsulating DIDComm messages) between peer and AS (comparable with a TLS handshake as in EAP-TLS). The result of this process is a secure DIDComm connection (tunnel) between the peer and the AS.

If the AS does not support SSI authentication, it signals the use of an alternative EAP method towards the peer during step 5.

Phase 2: Tunnel

The following messages are exchanged within the secure (i.e., confidential, integrity-protected, and authenticated) tunnel between the peer (the user's wallet) and the SSI Agent of the AS. The exchanged data are encapsulated in sequences of attribute-value pairs, as is common for EAP methods.

1. The SSI Agent requests a proof request with the relevant attributes for authentication from the peer via an EAP Request Credentials message.
2. The wallet shows the user the proof request and the user chooses the proper credentials. After the user's confirmation, a proof presentation is sent to the SSI Agent via an EAP Response Credentials message. All the advantages of SSI, like Selective Disclosure, can be used in this step.
3. The SSI Agent retrieves and verifies the proof.

Phase 3: Termination

1. The DIDComm connection between the peer and the AS is terminated.
2. The AS sends a RADIUS Accept message to the authenticator if the authentication was successful (i.e., the proof verification succeeded); otherwise, it sends a RADIUS Reject message.
3. The authenticator sends an EAP Success message to the peer and grants access to the network if it receives a RADIUS Accept message by the AS; otherwise it sends an EAP Failure message and does not grant access.

Like with other EAP methods (such as EAP-TLS or EAP-TTLS), the AS should distribute keying material to the authenticator that can be used to secure further communication between the authenticator and the peer in the last message. We do not stipulate how the AS chooses the key material at this point—the AS is free to choose any mechanism for key derivation.

Moreover, it is also possible at this point (like with other EAP methods) that the AS provides further authorization information to the authenticator in the last message. The AS can retrieve the proper information for its authorization decision from the attributes provided by the peer during SSI authentication.

4 Evaluation

Our concept fulfills all requirements stated in Sect. 2.

4.1 Security Evaluation

Wireless connections are considerably susceptible to eavesdropping and man-in-the-middle attacks—enabling dictionary attacks against low-entropy passwords, as pointed out in RFC 5281. Moreover, users' anonymity and locational privacy are threatened if protocols (like EAP-TLS) are used in wireless networks where usernames are transmitted in cleartext. RFC 5281 also stresses that the authenticator does not always reside in the administrative domain of the AS (Eduroam is a good example of just that) and, thus, there is potential for eavesdropping attacks in the untrusted network of the authenticator. Last but not least we face the challenge that the peer connects with a potentially malicious authenticator, which is responsible for routing data to the user's home domain (RFC 5281).

Considering all these threats, we need the following guarantees to have strong security during authentication in a wireless environment:

1. The peer shall not reveal its identity during the authentication process in clear text to potential eavesdroppers on the network.
2. The AS needs to authenticate towards the peer.
3. Authentication by the peer shall not be based on passwords.
4. The peer might not even need to provide its real identity to the AS.
5. Replay attacks need to be mitigated.
6. Roaming among access domains with which the user has no relationship and which will have limited capabilities for routing authentication requests shall be supported (as stated in RFC 5281).

We make use of the DIDComm protocol to secure the connection between the user wallet and the SSI Agent. Thus, the whole authentication process is authenticated, integrity-protected and confidential. No identity information is leaked to eavesdroppers. (1.)

The authentication of the AS is done using the DIDComm protocol. The user can choose, based on the (authenticated) identity information provided by the AS, whether he trusts the AS and wants to proceed with the authentication. (2.)

The user uses SSI to authenticate towards the AS, which means that public-key cryptography is used for authentication instead of passwords. (3.) The user can decide which attributes he wants to present during the authentication process via verifiable credentials (4.) and the proof of the credentials is only valid for a single authentication process, thus mitigating replay attacks (5.).

The support for roaming among domains with which the user has no relationship is one of the main advantages of the integration of SSI in EAP. There is no need for prior registration as the user can show all necessary (certified) attributes via verifiable credentials during authentication. The remote AS (which needs to authenticate the peer) does not need to get in touch with the home AS of the peer (like in Eduroam) but has all the information at its disposal to decide for itself whether it grants access to the network or not. (6.)

4.2 Comparison to Other EAP Methods

In comparison with EAP-TLS and PEAP, the most widely used EAP methods in wireless network environments, our EAP-SSI method provides better flexibility and usability as SSI will be used in the future for other purposes anyway and, thus, no further registration and no further credentials are needed. In comparison to EAP-TLS our approach does not leak any identity information to eavesdroppers (as is the case with X.509 user certificates) and in comparison to PEAP our approach provides better security as public-key cryptography is used for peer authentication (instead of a password-based authentication with MSCHAPv2). The problem of unknown server certificates (as is often the case in practice with Eduroam, for example) is also prevented with our approach. As we do not make use of TLS, but of DIDComm instead, there might not be a need for such a regular change of protocol and cipher suites as was the case with TLS in the past years due to found flaws. The AS does not need to store any user credentials with our approach, which decreases the risk of compromise.

4.3 Towards a Better Eduroam

We motivated our new EAP-SSI method by stating that we can have an even better Eduroam network than we have today.

Looking at the Eduroam infrastructure today, as depicted in Fig. 1, we can see that several RADIUS servers are needed at different hierarchy levels to support roaming. The servers therefore need to be made aware of each other (e.g., the national US server needs to know the RADIUS servers of US institutions). With our approach, the Eduroam network would not need all these roaming-supporting RADIUS servers. Each local AS can authenticate a peer by only looking at the provided credentials. The AS does not need to ask the user's home AS, whether the user is a student there, for example, because the role of the user is provided via the verifiable credential. This shows another big advantage of our approach: the user does not even need to provide his identity to the local AS but it suffices that he provides proof that he is a student at a recognized institution. By using SSI, the user is in full control of his identity and can use privacy-preserving mechanisms such as Selective Disclosure for that purpose. Thus, locational privacy is guaranteed for users: neither do institutions learn the identities of users visiting them, nor do home institutions learn which institutions their users are visiting at the moment by looking at incoming authentication requests as would be possible with Eduroam today. As attributes (representing roles of users) can be shared by users towards an AS, these data can also be used for authorization: e.g., a professor connecting to the network might get other access rights than a student visiting from another institution. It is even imaginable to provide strong accounting with our approach as well. Even if our approach would allow to get completely rid of the existing Eduroam infrastructure, it would also be an option that EAP-SSI is supported within the existing environment. The advantage would then still be that SSI (as a potential future standard) can be used for Eduroam authentication, requiring no further credentials to be managed by institutions and users. EAP-SSI would just need to be supported as another EAP method by the AS.

It is well worth noting that not only the Eduroam network can benefit from our approach, but new roaming networks can emerge based on EAP-SSI. For example, a user who has a certain status with a hotel chain or an airline can get a VC from that institution. Whenever the user is visiting a hotel from that chain or flying with that airline, the user can instantly get access to the network by authenticating via EAP-SSI and presenting the proper VC. The same approach could be used by mobile network providers that also operate WiFi access points. A seamless authentication can take place for different devices with EAP-SSI instead of being bound to just one device as with EAP-SIM, for example.

Of course, all the benefits of EAP-SSI cannot only be achieved in wireless networks but with wired networks based on 802.1X as well. A company could grant access to visitors from other institutions based on their credentials presented during the authentication process.

5 Related Work

5.1 SSI Integration for Authentication Protocols

SSI integration has been proposed for other authentication protocols in the past.

In most cases, an integration is only proposed in higher-level protocols. Integration in the OAuth 2.0 protocol is presented by HONG ET AL. [5]. The authors propose to store personal data in smart contracts on the Ethereum blockchain, contradicting privacy protection. GRÜNER ET AL. [6] and LUX ET AL. [7] propose concepts for SSI integration in OpenID and YILDIZ ET AL. [8] proposes to integrate SSI in the SAML protocol. KUPERBERG ET AL. [9] show in their study how SSI can be integrated into conventional software using established protocols. They investigate popular SSI implementations and check whether they work together with typical IAM protocols.

PETRLIC ET AL. [3] propose to integrate SSI into the Kerberos protocol, with the main advantage of getting rid of (insecure) user passwords and allowing cross-realm authentication.

5.2 EAP Methods

There exist several more EAP methods than the most popular ones that we have already discussed in Sect. 1.1.

One method, specified in RFC 4186, is called *EAP-SIM*. This method provides mutual authentication based on SIM cards. This EAP type can be used to authenticate devices that roam between commercial WiFi hotspots and mobile carrier networks for example. The advantage is that there is no need for a pre-established password between the peer and the AS.

Another method is based on the use of Kerberos over EAP. ZRELLI ET AL. [10] were the first to specify Kerberos over EAP. As the authors point out, this method provides the advantage that institutions can re-use the users' credentials for Kerberos for network access authentication instead of needing to manage a different set of credentials such as Unix passwords or public key certificates. Eum et al. [11] propose another concept for use of Kerberos within EAP authentication, called *EAP-Kerberos II*. They propose to download a key from a server instead of deriving it on the peer, which provides better protection of messages and more efficient authentication, as the authors point out.

5.3 Eduroam Problems

Palamà et al. [12] showed in a recent study from 2022 that most pragmatic Eduroam configurations appear to be grossly insecure. They were able to steal Eduroam user credentials of more than a third of participants in a completely passive attack where users had their devices in their pockets.

Brenza et al. [13] found out that a lot of peers lack the necessary X.509 root certificates to authenticate the AS due to erroneous configuration manuals and a lack of knowledge on the user side. The authors come up with an attack to steal authentication data in a real-time MITM attack scenario. More than half of several hundred investigated devices were prone to their attack.

Asokan et al. [14] show an MITM attack on tunneled authentication protocols (such as PEAP and EAP-TTLS) that is possible due to a missing cryptographic binding between the server authentication and the peer authentication.

6 Conclusion and Outlook

We are the first to present an approach to integrate the usage of Self-Sovereign Identity (SSI) in the Extensible Authentication Protocol (EAP). The main advantage of our approach is that we can enable a secure, privacy-friendly, and usable authentication in (wireless) local area networks. We believe that our approach has great potential to be used in practice in the future as SSI usage is pushed by the European Union with new regulations such as eIDAS 2.0.

In future work, we will come up with a prototype implementation of our approach and we will also standardize our approach in an RFC.

References

1. Reed, D., Sabadello, M., Sporny, M., Guy, A.: Decentralized Identifiers (DIDs) v1.0 (2022). https://www.w3.org/TR/2022/RECdid-core-20220719/
2. Preukschat, A., Reed, D.: Self-Sovereign Identity. Manning Publications, Shelter Island (2021)
3. Petrlic, R., Lange, C.: Kerberssize us: providing sovereignty to the people. In: Dolev, S., Schieber, B. (eds.) Stabilization, Safety, and Security of Distributed Systems, pp. 259–273. Springer, Cham (2023)
4. Introduction to Trust Over IP (2021). https://trustoverip.org/wp-content/uploads/Introduction-to-ToIP-V2.0-2021-11-17.pdf

5. Hong, S., Kim, H.: VaultPoint: a blockchain-based SSI model that complies with OAuth 2.0. Electronics **9**(8) (2020). https://www.mdpi.com/2079-9292/9/8/1231
6. Grüner, A., Mühle, A., Meinel, C.: An integration architecture to enable service providers for self-sovereign identity. In: 2019 IEEE 18th International Symposium on Network Computing and Applications (NCA), pp. 1–5 (2019)
7. Lux, Z.A., Thatmann, D., Zickau, S., Beierle, F.: Distributed-ledger-based authentication with decentralized identifiers and verifiable credentials. In: 2020 2nd Conference on Blockchain Research and Applications for Innovative Networks and Services (BRAINS), pp. 71–78. IEEE (2020)
8. Yildiz, H., Ritter, C., Nguyen, L.T., Frech, B., Martinez, M.M., Küpper, A.: Connecting self-sovereign identity with federated and user-centric identities via SAML integration. In: IEEE Symposium on Computers and Communications (ISCC) 2021, pp. 1–7 (2021)
9. Kuperberg, M., Klemens, R.: Integration of self-sovereign identity into conventional software using established IAM protocols: a survey. In: Roßnagel, H., Schunck, C.H., Mödersheim, S. (eds.) Open Identity Summit 2022, pp. 51–62. Gesellschaft für Informatik e.V, Bonn (2022)
10. Zrelli, S., Shinoda, Y.: Specifying kerberos over EAP: towards an integrated network access and Kerberos single sign-on process. In: 21st International Conference on Advanced Information Networking and Applications (AINA '07), pp. 490–497 (2007)
11. Eum, S.-H., Choi, H.-K.: EAP-kerberos II: an adaptation of Kerberos to EAP for mutual authentication. In: 2008 8th International Conference on ITS Telecommunications, pp. 78–83 (2008)
12. Palamà, I., Amici, A., Gringoli, F., Bianchi, G.: "Careful with that Roam, Edu": experimental analysis of Eduroam credential stealing attacks. In: 17th Wireless On-Demand Network Systems and Services Conference (WONS), pp. 1–7 (2022)
13. Brenza, S., Pawlowski, A., Pöpper, C.: A practical investigation of identity theft vulnerabilities in Eduroam. In: Proceedings of the 8th ACM Conference on Security and Privacy in Wireless and Mobile Networks. WiSec '15. Association for Computing Machinery, New York, NY, USA (2015). https://doi.org/10.1145/2766498.2766512
14. Asokan, N., Niemi, V., Nyberg, K.: Man-in-the-middle in tunnelled authentication protocols. In: Christianson, B., Crispo, B., Malcolm, J.A., Roe, M. (eds.) Security Protocols 2003. LNCS, vol. 3364, pp. 28–41. Springer, Heidelberg (2005). https://doi.org/10.1007/11542322_6

FastSGX: A Message-Passing Based Runtime for SGX

Subashiny Tanigassalame[1(✉)], Yohan Pipereau[1], Adam Chader[1], Jana Toljaga[1], and Gaël Thomas[2]

[1] Telecom SudParis - IP, Paris, France
{subashiny.tanigassalame,yohan.pipereau,adam.chader, jana.toljaga}@telecom-sudparis.eu
[2] Inria, Paris, France
gael.thomas@inria.fr

Abstract. Designing an efficient privacy-preserving application with Intel SGX is difficult. The problem comes from the prohibitive cost of switching the processor from the non-secure mode to the secure mode. To avoid this cost, we propose to design an SGX application as a distributed system with worker threads that communicate by exchanging messages. We implemented FastSGX, a runtime that exposes this programming model to the developer, and evaluated it with several data structures. Our evaluation with different workloads shows that the applications designed with FastSGX consistently outperform, and by up to 2.8x, the equivalent applications designed with the software development kit provided by Intel to use SGX.

1 Introduction

Today, citizens deploy their sensitive data in cloud infrastructures. However, a cloud infrastructure is an untrusted system. A cloud infrastructure is shared among many users, which makes it an especially interesting target for an attacker. A cloud provider may also be honest but curious. Protecting user data when it is processed in a cloud infrastructure is thus today paramount to protect the privacy of the citizens.

In order to help users protect their personal data, Intel proposes a Trusted Execution Environment (TEE) named Intel SGX [6]. A TEE is a secure computation mode provided by a processor. A TEE is able to protect a memory zone, which is named an *enclave*, against an attacker, who fully controls the operating system, the hypervisor, and even the hardware. For that, a TEE relies on cryptography to enforce the confidentiality, integrity, and authenticity of an enclave.

Because of the cost of entering or leaving an enclave, designing an efficient application for Intel SGX is difficult. This cost is prohibitive: while a standard call costs only a few cycles, entering or leaving an enclave costs 7000 cycles [18,27,28]. Several research works show that we can avoid this cost by using *switchless calls* [1,24,28,33,34]. A switchless call consists of leveraging *worker threads* in order to avoid switching the processor from the non-secure mode to the secure mode. In detail, each worker thread runs in a *security domain*: either the non-secure domain or in an

L. Barolli (Ed.): AINA 2024, LNDECT 202, pp. 74–85, 2024.
https://doi.org/10.1007/978-3-031-57916-5_7

enclave. In order to perform a call from one domain to another, a worker thread of one domain sends a message to a worker thread in the other. To send this message, the worker thread simply writes a value in a shared memory zone named an activation zone. Transferring the control from one domain to another costs a single cache miss: the cache miss that loads the activation zone from the core of the sender thread in the core of the receiver thread. Since transferring a cache line costs a few hundred cycles instead of several thousand, a switchless call significantly improves performance compared to switching the processor mode.

Whereas switchless call is a well-known technique, using this technique efficiently remains challenging for three reasons.

The first issue comes from the fact that the worker threads are hidden to the developer and configured statically. In detail, the SGX runtime provided by Intel can internally use worker threads to optimize the time to transfer the control from/to an enclave. These workers threads consume CPU resources when they idle because they actively spin on the activation zone. Unfortunately, the developer can only statically configure the number of worker threads, which is inadequate if the workload evolves over time. In such a case, the developer either over-provisions the number of worker threads, which wastes CPU resources when the workload is low, or under-provisions the number of worker threads, which leads to inefficiencies when the workload increases [33].

The second issue comes from the function semantic exposed by the SGX runtime provided by Intel. With a call semantic, the caller is suspended during a call. The caller uselessly wastes a CPU while actively waiting for the termination of the call. The developer can thus not use the wasted CPU resource to execute useful code in parallel.

The third issue comes from the design of the current SGX runtimes, which prevent a direct call from one enclave to another. Because of this design, while a worker thread in an enclave could easily transfer directly the control to a worker thread in another enclave, a worker thread has first to transfer the control back to a thread in non-secure mode in order to transfer the control to another enclave.

In this paper, we propose to use the switchless call technique more efficiently with a new programming model. In detail, we propose to explicitly design an SGX application as a distributed system with worker threads that communicate by exchanging messages. With our programming model, as with switchless calls, each worker thread runs in its security domain. However, the worker threads are made visible to the developer, who can create and destroy worker threads on the fly. The developer can thus adapt the number of worker threads to the workload on the fly. Moreover, instead of exposing a function-call abstraction, we propose to expose a message-passing abstraction. Thanks to this modification, a worker thread can continue its execution while another worker thread proceeds a message, which avoids wasting the CPU of the sender during a call. Finally, by exposing to the developer an interface to send and receive messages, a developer can send a message from any worker threads to any other worker threads, which avoids the need to uselessly transfer the control to a worker thread in the non-secure domain in the case of an inter-enclave call.

We implemented our message-passing programming model in a new SGX runtime named FastSGX. We evaluated FastSGX with two classical data structures: a hashmap and a treemap. We evaluated versions of these data structures with one and two enclaves. Our evaluation with different access patterns shows that:

- The interface of FastSGX, with its 4 main functions, is simple enough to be usable in practice,
- FastSGX can be used to design efficient multi-enclave applications, while current switchless call runtimes become especially inefficient in this case,
- The data structures implemented with FastSGX consistently outperform the equivalent data structures implemented with the Intel development of Intel.

The remainder of the paper is organized as follows: Sect. 2 gives the background, Sect. 3 presents the design of FastSGX, Sect. 4 details our evaluation, Sect. 5 presents related works, and Sect. 6 concludes.

2 Background and Threat Model

In order to use Intel SGX, a developer defines enclaves. An enclave is a contiguous memory zone located inside the virtual address space of a process. Intel SGX protects an enclave by leveraging two processor execution modes: a non-secure mode and a secure mode. When the processor runs in non-secure mode, it prevents any access to the memory of the enclaves. When the processor enters secure mode, it gains access to a single enclave. In secure mode, the processor can access the memory of that enclave, the memory located outside any enclave, but not the memory of the other enclaves.

In order to protect the enclaves, the processor first prevents read and write access to the pages that belong to the enclave. Only preventing read and write access from the processor is not enough if we suppose an attacker that controls the operating system, the hypervisor, or the hardware. Such an attacker can bypass the protection mechanisms of the processor by using direct memory access from the devices. In order to prevent such attacks, the processor encrypts the cache lines before sending them to the main memory, which enforces confidentiality. With Intel SGX v1, the processor additionally enforces authenticity by maintaining a tree of hashes used to detect unintended writes [23].

In the remainder of the paper, we suppose an attacker that fully controls a machine (operating system and hypervisor included). We suppose that the attacker cannot read or write the memory of the enclaves protected by Intel SGX. For that, we suppose that the processor, the FastSGX runtime, and the software development kit provided by Intel to use SGX are correct and do not contain bugs. We do not consider side-channel attacks since Intel SGX does not address this attack vector.

```
1 // initialize the runtime with nenclaves enclaves
2 int initialize(size_t nenclaves, ...);

3 // create a worker thread in the enclave eid
4 int new_worker(pthread_t* tid, size_t eid);

5 // send the value to the enclave eid with the message id mid
6 void send(size_t eid, size_t mid, union value value);

7 // receive a message with the message id mid from eid
8 void recv(size_t eid, size_t mid, union value* value);
```

Fig. 1. Main functions of the FastSGX interface.

3 FastSGX Design

FastSGX is a message-oriented runtime for Intel SGX. It manages a set of enclaves. For each enclave, FastSGX creates a communication channel. Internally, FastSGX implements a communication channel as a lock-free FIFO queue stored in unsafe memory [9]. FastSGX also implements a lock-free memory allocator in order to allocate and free the messages. For that, FastSGX uses a simple lock free stack [9].

Additionally to the enclaves and communication channels, FastSGX manages a set of worker threads. Each worker thread is associated to a single enclave. It receives messages through the communication channel of its enclave. If several worker threads are associated to the same enclave, they share the communication channel. In this case, a message is not broadcasted to the worker threads: each message is only received by a single worker thread.

3.1 Interface

Figure 1 presents the main functions provided by FastSGX. To initialize the runtime, a developer calls the initialize function. This function takes the number of enclaves that have to be created as a parameter and, for each enclave, a path to the binary that has to be loaded in the enclave. The initialize function gives sequentially an enclave identifier for each enclave by starting with the eid 1. After initialization, FastSGX considers that the pseudo-enclave with the eid 0 represents the code and data located in unsafe memory. It also considers that the main thread of a process, i.e., the thread that called initialize, is a worker thread associated to enclave 0.

In order to start executing code in the enclaves, the developer has to call the `new_worker` function. This function creates a new worker thread. The new worker thread enters the enclave eid given as a parameter by executing the function named `start_routine`, which is located in the binary loaded in enclave eid.

As soon as worker threads execute in the enclaves, they can communicate by exchanging messages. For that, FastSGX provides the `send` and `recv` functions. The `send` function takes three parameters: the destination (eid), a message identifier (mid), and a content (value). The message identifier is used to address a message to a specific `recv` function, which is useful to avoid confusing two messages with two different meanings received in parallel. The value exchanged in a message is a union that has the size of a machine word (64 bits on an Intel).

The `recv` function takes the same three parameters. If eid is equal to -1, the function receives messages from any enclave, otherwise, it only receives messages from eid. If mid is equal to -1, the function receives messages with any mid, otherwise, it only receives messages with the mid given as a parameter. The `recv` function blocks while busy waiting until a message that matches eid/mid is received. When such a message arrives in the communication channel of an enclave, the `recv` function removes the message from the channel, fills the value with the content of the message, and returns.

3.2 Hazard Pointers

Since FastSGX implements the communication channels with lock-free queues, it is subject to the ABA problem [17]. This problem appears with lock-free data structures implemented with compare and swaps.

To illustrate, we first give an overview of the algorithm used to implement a lock-free queue. We suppose a queue with two messages A and X. To dequeue A, a receiver thread t_1 loads a pointer to A from the head of the queue and then loads a pointer to X from the `next` field of A. Finally, the receiver thread replaces the head by the pointer to X. Since another receiver thread could dequeue A between the read of the head and its update in t_1, t_1 only updates the head from A to X if the head is still equal to A. For that, t_1 uses an atomic compare and swap instruction, which atomically compares head to A, and, if they are equal, replaces head by X. This simple algorithm is correct, but only if a message is never freed. In FastSGX, a receiver has to free a message when it is consumed in order to avoid a memory leak, which leads to the ABA problem described below.

To illustrate the ABA problem, we also suppose a queue with two messages A and X. The receiver thread t_1 loads A and X, and then, another receiver thread t_2 is scheduled. t_2 dequeues A and X, and frees the messages. Finally, a sender thread t_3 inserts a new message B. At this step, the queue contains a single message: B. When t_1 is re-scheduled, the compare and swap is supposed to fail because B is not the A message. However, this is not necessarily the case: since A is free when t_3 allocates a message, B may be allocated at the location of A. In such a case, the compare and swap of t_1 succeeds since t_1 thinks that A is still the head of the list. t_1 thus install a pointer to X as the head of the queue, which is incorrect because X was freed by t_2.

FastSGX avoids the ABA problem by using hazard pointers [17]. Technically, when a receiver loads the head pointer, it signals to the other threads that freeing the pointed message is unsafe. For that, the receiver thread records the head pointer in an array named the hazard pointer array. Then, to free a message, FastSGX leverages two lists: a purgatory list and a free list. When the application frees a message, FastSGX adds the message in the purgatory list. When an application allocates a message, FastSGX uses the free list. If the free list is empty, FastSGX tries to move a batch of N messages from the purgatory list to the free list. It inspects the purgatory list by starting from the least recently added message, and, if the message is still referenced by the hazard array, FastSGX ignores the message since a receiver may still use the message. Otherwise, FastSGX moves the message to the free list.

4 Evaluation

We evaluate the performance of FastSGX on an Intel i5-9500 CPU 3 GHz with 16 GiB memory. This 6-core CPU ships SGX version 1 with a maximal memory size usable by the enclaves of 93 MiB. The machine runs Linux 5.15.0, glibc 2.31, clang 10.0.0, and Intel SGX SDK 2.19.100.

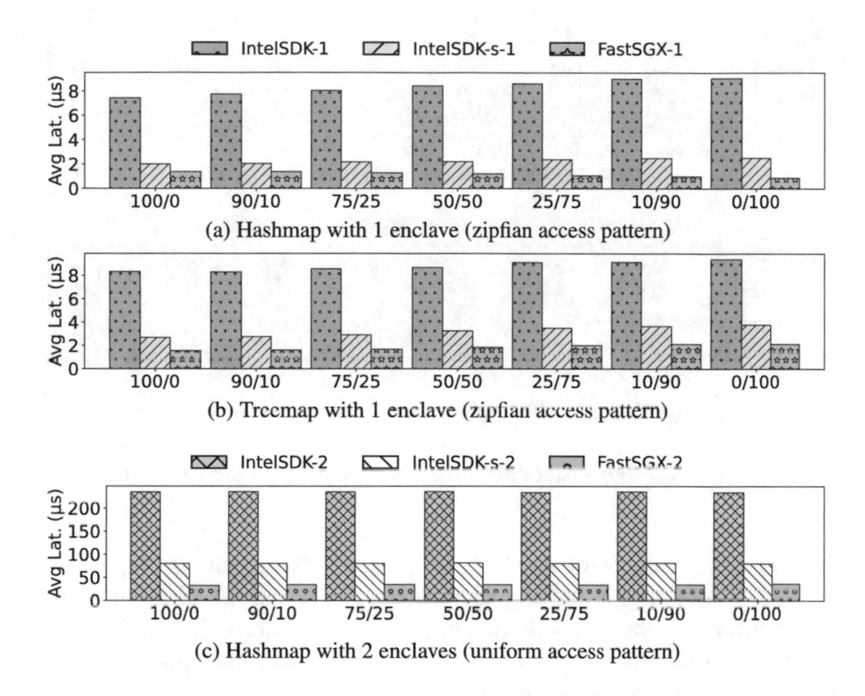

(a) Hashmap with 1 enclave (zipfian access pattern)

(b) Treemap with 1 enclave (zipfian access pattern)

(c) Hashmap with 2 enclaves (uniform access pattern)

Fig. 2. Latency of the data structures. X/Y: get/put ratio.

4.1 Data Structures

We first evaluate two maps that associate keys to values: a hashmap and a treemap. The hashmap uses a separate chaining algorithm: it is designed as an array of linked-lists, in which each linked-list contains the keys that collide. The treemap is implemented as a red-black tree, which ensures that the tree remains balanced.

We evaluate three versions with a single enclave: IntelSDK-1, IntelSDK-s-1 and FastSGX-1. These versions store the whole map in a single enclave. The two IntelSDK versions expose the `put/get` functions to the non-secure domain. IntelSDK-1 switches the processor mode during a call, while IntelSDK-s-1 uses switchless calls. FastSGX-1 is implemented with FastSGX. For a `get`, the worker thread in the non-secure domain sends a message to execute the get in the enclave, and waits for the result. For a `put`, since the result of a `put` is not used, the worker thread in the non-secure domain sends a message to execute `put`, but continues its execution in parallel.

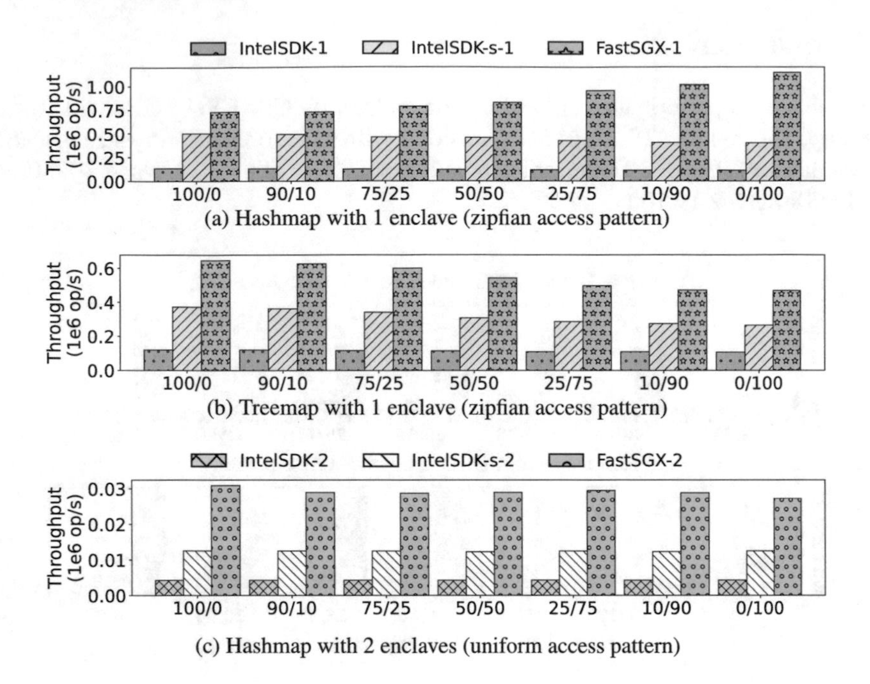

(a) Hashmap with 1 enclave (zipfian access pattern)

(b) Treemap with 1 enclave (zipfian access pattern)

(c) Hashmap with 2 enclaves (uniform access pattern)

Fig. 3. Throughput of the data structures. X/Y: get/put ratio.

We also evaluate three versions with two enclaves. These versions store the keys in one enclave and the values in another, which makes the communication pattern between the enclaves more complex. As with a single enclave, we evaluate (i) IntelSDK-2, which switches the processor mode, (ii) IntelSDK-s-2, which uses switchless calls, and (iii) FastSGX-2, which uses message passing.

Figure 2 reports the latency and Fig. 3 the throughput with different `put/get` ratios. With one color, we execute 100 000 operations in a map with 100 000 keys, and with two colors, 20 000 operations in a map with 20 000 keys. A key is 8-bytes long and a value 1024-byte long.

With one enclave, we observe that, as expected, IntelSDK-s-1 consistently performs better than IntelSDK-1. We also observe that FastSGX-1 is consistently better than IntelSDK-1 and IntelSDK-s-1. The better performance of FastSGX-1 comes from two complementary phenomenons.

First, with a high number of puts, a put is executed in parallel with the load injector in FastSGX-1. This is not the case with IntelSDK-s-1, which suspends the caller during a call. The better parallelism of FastSGX-1 explains why FastSGX-1 is better than IntelSDK-s-1 with a high number of puts (right of the curves).

Second, FastSGX is designed with lock-free data structures while Intel SDK uses a blocking scheme. In detail, with IntelSDK-s-1, a thread takes a spin-lock when it accesses an activation zone. FastSGX-1 does not take a lock since it uses lock-free data structures. Thanks to the use of lock-free data structures, FastSGX-1 is also better than IntelSDK-s-1 with a high number of get operations (left of the curve).

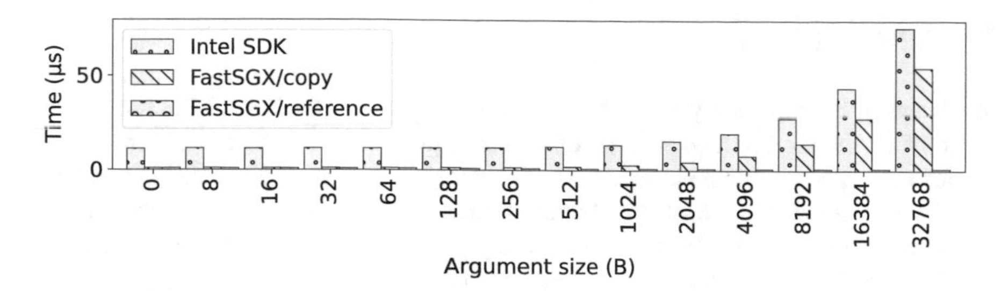

Fig. 4. Ping-pong time with Intel SDK and with FastSGX.

With two enclaves, we observe similar results. With two enclaves, the relative difference between FastSGX and the Intel SDK versions is higher because FastSGX allows the two enclaves to communicate directly. This is not the case with the Intel SDK versions, which pay an additional transfer of control to the non-secure domain for each operation. This result confirms that using an explicit message-passing scheme is important to optimize a multi-enclave application.

4.2 Ping-Pong

In this experiment, we compare the cost of switching the processor with the cost of exchanging messages. For that, we evaluate a ping pong application in three configurations: FastSGX/copy, FastSGX/reference and Intel SDK.

With FastSGX/copy and FastSGX/reference, we run two worker threads: one worker thread in the non-secure domain and one worker thread in an enclave. The non-secure worker thread sends a ping to the enclave worker thread, which answers with a pong message. The ping message contains a pointer to an array of bytes. In FastSGX/copy, the application copies the pointed array in a buffer before replying with a pong. In FastSGX/reference, the application ignores the argument.

With Intel SDK, the non-secure thread of the application calls a ping function provided by the enclave. The non-secure thread of the application is blocked during the execution of the ping function, which is equivalent to waiting for a pong. With Intel SDK, the buffer is copied from the non-secure domain into the enclave.

Figure 4 reports the results of the experiment. We first observe that, when the size of the argument is equal to 0, using message passing instead of mode switching divides the function execution time by 12 (11.2 µs for Intel SDK versus 0.9 µs for FastSGX/copy or FastSGX/nocopy). This result highlights the benefit of using messages instead of switching the processor mode.

We also observe that, when the argument size increases, the cost of copying the buffer becomes larger than the cost of transferring the control. With a buffer of 32 KiB, we observe, however, that using message passing still saves 28% of the time (76.1 µs for Intel SDK versus 54.6 µs for FastSGX/copy). This result shows that, even for an application that protects large user data sets by copying them in an enclave, using message passing remains interesting.

5 Related Works

Many applications directly rely on the Intel SDK to use SGX [3,5,7,11,22,30,35]. Since using the Intel SDK can be complex for legacy applications, several frameworks propose to run a complete application with its dependencies in an enclave [1,2,16,19, 25]. These frameworks are not satisfactory because they lead to a large trusted computing base. Other tools propose to automatically partition an application by starting from variables or functions annotated as sensitive [4,8,10,12–15,20,26,29,31,32]. These tools ease the development while minimizing the trusted computing base. These tools are complementary to FastSGX: they could rely on FastSGX to optimize the time to transfer the control from/to an enclave.

As presented in the introduction, several runtimes rely on switchless calls to avoid the cost of switching the processor from/to secure mode [1,24,28,33,34]. These runtimes hide the worker threads to the developer, which makes the dynamic optimization of the number of worker threads difficult, prevents the execution of code in parallel in the worker threads, and is sub-optimal for a multi-enclave application. With FastSGX, by exposing the worker threads to the developer, and by exposing a message-passing interface, we avoid these three limitations.

EActors [21] proposes to design a SGX application as a set of actors. As in FastSGX, EActors runs worker threads in the enclaves. The worker threads execute eactors, which communicate by exchanging messages. Using EActors requires a whole redesign of an application in order to implement the application with actors. Using FastSGX is more straightforward since only adding calls to send and receive messages is required. Moreover, with FastSGX, the developer explicitly creates on the fly the worker threads, which allows the developer to dynamically adjust the number of worker threads to the workload. This is not the case with EActors. With EActors, the number of worker threads is configured statically, which is inadequate for an application with a dynamic workload because worker threads may become useless in case of a low workload, or saturated in case of a high workload.

6 Conclusion

This paper presents FastSGX, a message-based runtime for SGX. FastSGX relies on the switchless call principle, but avoids the limitations of the current implementations. In detail, FastSGX (i) allows the developer to adjust the number of worker thread to the actual workload on the fly, (ii) allows the developer to execute code in parallel in the sender while a receiver proceeds a message, and (iii) allows the developer to directly transfer the control from an enclave to another. Thanks to these properties, our evaluation with different data structures and workloads shows that FastSGX consistently outperforms the SGX development kit of Intel, and with one or two enclaves. Our evaluation also shows that FastSGX, with its 4 main functions, is simple and usable in practice.

References

1. Arnautov, S., et al.: SCONE: secure linux containers with intel SGX. In: 12th USENIX Symposium on Operating Systems Design and Implementation (OSDI 16)
2. Baumann, A., Peinado, M., Hunt, G.: Shielding applications from an untrusted cloud with haven. In: 11th USENIX Symposium on Operating Systems Design and Implementation (OSDI 14), pp. 267–283, Broomfield, CO, October 2014. USENIX Association
3. Brenner, S., et al.: SecureKeeper: confidential zookeeper using intel SGX. In: Proceedings of the 17th International Middleware Conference, Middleware '16, New York, NY, USA (2016). Association for Computing Machinery
4. Brumley, D., Song, D.: Privtrans: automatically partitioning programs for privilege separation. In: Proceedings of the 13th Conference on USENIX Security Symposium - vol. 13, SSYM'04, p. 5, USA (2004). USENIX Association
5. Chen, L., Li, J., Ma, R., Guan, H., Jacobsen, H.-A.: EnclaveCache: a secure and scalable key-value cache in multi-tenant clouds using intel SGX. In: Proceedings of the 20th International Middleware Conference, Middleware '19, pp. 14–27, New York, NY, USA (2019). Association for Computing Machinery
6. Costan, V., Devadas, S.: Intel SGX explained. IACR Cryptol. ePrint Arch. **2016**, 86 (2016)
7. Decouchant, J., Kozhaya, D., Rahli, V., Yu, J.: DAMYSUS: streamlined BFT consensus leveraging trusted components. In: Proceedings of the Seventeenth European Conference on Computer Systems, EuroSys '22, pp. 1–16, New York, NY, USA (2022). Association for Computing Machinery
8. Ghosn, A., Larus, J.R., Bugnion, E.: Secured routines: language-based construction of trusted execution environments. In: 2019 USENIX Annual Technical Conference (USENIX ATC 19), pp. 571–586, Renton, WA (2019). USENIX Association
9. Herlihy, M., Shavit, N.: The Art of Multiprocessor Programming, 1st edn. Revised Reprint. Morgan Kaufmann Publishers Inc., San Francisco, CA, USA (2012)
10. Jiang, J., et al.: Uranus: simple, efficient SGX programming and its applications. In: Proceedings of the 15th ACM Asia Conference on Computer and Communications Security (ASIACCS 2020), pp. 826–840, Taipei, Taiwan (2020)
11. Kim, T., Park, J., Woo, J., Jeon, S., Huh, J.: ShieldStore: shielded in-memory key-value storage with SGX. In: Proceedings of the Fourteenth EuroSys Conference 2019, EuroSys '19, New York, NY, USA (2019). Association for Computing Machinery
12. Lind, J., et al.: Glamdring: automatic application partitioning for intel SGX. In: 2017 USENIX Annual Technical Conference (USENIX ATC 17), pp. 285–298, Santa Clara, CA (2017). USENIX Association
13. Liu, S., Tan, G., Jaeger, T.: PtrSplit: supporting general pointers in automatic program partitioning. In: Proceedings of the 2017 ACM SIGSAC Conference on Computer and Communications Security, CCS '17, pp. 2359-2371, New York, NY, USA (2017). Association for Computing Machinery
14. Liu, Y., Zhou, T., Chen, K., Chen, H., Xia, Y.: Thwarting memory disclosure with efficient hypervisor-enforced intra-domain isolation. In: Proceedings of the 22nd ACM SIGSAC Conference on Computer and Communications Security, CCS '15, pp. 1607–1619, New York, NY, USA (2015). Association for Computing Machinery
15. Mambretti, A., et al.: Trellis: privilege separation for multi-user applications made easy. In: Monrose, F., Dacier, M., Blanc, G., Garcia-Alfaro, J. (eds.) RAID 2016. LNCS, vol. 9854, pp. 437–456. Springer, Cham (2016). https://doi.org/10.1007/978-3-319-45719-2_20
16. Ménétrey, J., Pasin, M., Felber, P., Schiavoni, V.: Twine: an embedded trusted runtime for web assembly. In: 37th IEEE International Conference on Data Engineering, ICDE 2021, Chania, Greece, April 19-22, 2021, pp. 205–216. IEEE (2021)

17. Michael, M.M.: Hazard pointers: safe memory reclamation for lock-free objects. IEEE Trans. Parallel Distrib. Syst. **15**(6), 491–504 (2004)
18. Orenbach, M., Lifshits, P., Minkin, M., Silberstein, M.: Eleos: Exitless OS services for SGX enclaves. In: Proceedings of the Twelfth European Conference on Computer Systems, EuroSys '17, pp. 238–253 (2017)
19. Priebe, C., et al.: SGX-LKL: securing the host OS interface for trusted execution. arXiv preprint arXiv:1908.11143 (2019)
20. Rubinov, K., Rosculete, L., Mitra, T., Roychoudhury, A.: Automated partitioning of android applications for trusted execution environments. In: Proceedings of the 38th International Conference on Software Engineering, ICSE '16, pp. 923–934, New York, NY, USA (2016). Association for Computing Machinery
21. Sartakov, V.A., Brenner, S., Ben Mokhtar, S., Bouchenak, S., Thomas, G., Kapitza, R.: Eactors: fast and flexible trusted computing using SGX. In: Proceedings of the 19th International Middleware Conference, Middleware '18, pp. 187-200, New York, NY, USA (2018). Association for Computing Machinery
22. Schuster, F., et al.: VC3: trustworthy data analytics in the cloud using SGX. In: 2015 IEEE Symposium on Security and Privacy (SSP 15), pp. 38–54, San Jose, CA, USA. IEEE (2015)
23. Taassori, M., Shafiee, A., Balasubramonian, R.: Vault: reducing paging overheads in SGX with efficient integrity verification structures. In: Proceedings of the Twenty-Third International Conference on Architectural Support for Programming Languages and Operating Systems, ASPLOS '18, pp. 665-678, New York, NY, USA (2018). Association for Computing Machinery
24. Tian, H., et al.: Switchless calls made practical in intel SGX. In: Proceedings of the 3rd Workshop on System Software for Trusted Execution, SysTEX '18, pp. 22–27, New York, NY, USA (2018). Association for Computing Machinery
25. Tsai, C.-C., Porter, D.E., Vij, M.: Graphene-SGX: a practical library OS for unmodified applications on SGX. In: 2017 USENIX Annual Technical Conference (USENIX ATC 17), pp. 645–658 (2017)
26. Tsai, C.-C., Son, J., Jain, B., McAvey, J., Popa, R.A., Porter, D.E.: Civet: an efficient Java partitioning framework for hardware enclaves. In: 29th USENIX Security Symposium (USENIX Security 20), pp. 505–522, Online August (2020). USENIX Association
27. Weichbrodt, N., Aublin, P.-L., Kapitza, R.: SGX-PERF: a performance analysis tool for intel SGX enclaves. In: Proceedings of the 19th International Middleware Conference, Middleware '18, pp. 201–213, New York, NY, USA (2018). Association for Computing Machinery
28. Weisse, O., Bertacco, V., Austin, T.: Regaining lost cycles with HotCalls: a fast interface for SGX secure enclaves. In: Proceedings of the 44th Annual International Symposium on Computer Architecture, ISCA '17, pp. 81–93, New York, NY, USA (2017). Association for Computing Machinery
29. Wu, Y., Sun, J., Liu, Y., Song Dong, J.: Automatically partition software into least privilege components using dynamic data dependency analysis. In: Proceedings of the 28th IEEE/ACM International Conference on Automated Software Engineering, ASE '13, pp. 323–333. IEEE Press (2013)
30. Yuhala, P., Felber, P., Schiavoni, V., Tchana, A.: Plinius: secure and persistent machine learning model training. In: 2021 51st Annual IEEE/IFIP International Conference on Dependable Systems and Networks (DSN), pp. 52–62, Los Alamitos, CA, USA (2021). IEEE Computer Society
31. Yuhala, P., et al.: SecV: secure code partitioning via multi-language secure values. In: Proceedings of the 24nd International Middleware Conference, Middleware '23. Association for Computing Machinery (2023)

32. Yuhala, P., et al.: Montsalvat: Intel SGX shielding for GraalVM native images. In: Proceedings of the 22nd International Middleware Conference, Middleware '21, pp. 352–364, New York, NY, USA (2021). Association for Computing Machinery
33. Yuhala, P., Paper, M., Zerbib, T., Felber, P., Schiavoni, V., Tchana, A.: SGX switchless calls made configless. In: 53rd Annual IEEE/IFIP International Conference on Dependable Systems and Network, DSN 2023, Porto, Portugal, June 27-30, 2023, pp. 229–238. IEEE (2023)
34. Yuhala, P., Paper, M., Zerbib, T., Felber, P., Schiavoni, V., Tchana, A.: SGX switchless calls made configless (PER). In: Proceedings of the International Conference on Dependable Systems and Networks, DSN'23. IEEE Computer Society (2023)
35. Zheng, W., Dave, A., Beekman, J.G., Popa, R.A., Gonzalez, J.E., Stoica, I.: Opaque: an oblivious and encrypted distributed analytics platform. In: 14th USENIX Symposium on Networked Systems Design and Implementation (NSDI 17), pp. 283–298, Boston, MA, USA (2017)

Leveraging Chat-Based Large Vision Language Models for Multimodal Out-of-Context Detection

Fatma Shalabi[1,2](✉), Hichem Felouat[1,2], Huy H. Nguyen[2], and Isao Echizen[1,2,3]

[1] The Graduate University for Advanced Studies, SOKENDAI, Hayama, Japan
{fatmafaek,hichemfel,iechizen}@nii.ac.jp
[2] National Institute of Informatics, Tokyo, Japan
nhhuy@nii.ac.jp
[3] The University of Tokyo, Tokyo, Japan

Abstract. Out-of-context (OOC) detection is a challenging task involving identifying images and texts that are irrelevant to the context in which they are presented. Large vision-language models (LVLMs) are effective at various tasks, including image classification and text generation. However, the extent of their proficiency in multimodal OOC detection tasks is unclear. In this paper, we investigate the ability of LVLMs to detect multimodal OOC and show that these models cannot achieve high accuracy on OOC detection tasks without fine-tuning. However, we demonstrate that fine-tuning LVLMs on multimodal OOC datasets can further improve their OOC detection accuracy. To evaluate the performance of LVLMs on OOC detection tasks, we fine-tune MiniGPT-4 on the NewsCLIPpings dataset, a large dataset of multimodal OOC. Our results show that fine-tuning MiniGPT-4 on the NewsCLIPpings dataset significantly improves the OOC detection accuracy in this dataset. This suggests that fine-tuning can significantly improve the performance of LVLMs on OOC detection tasks.

1 Introduction

Since misinformation is growing rapidly in digital communication channels, it refers to false or misleading information that spreads rapidly through these channels, including social media, news articles, and word-of-mouth. It can have harmful consequences, such as leading people to make poor decisions regarding their health, finances, and safety. Multimodal misinformation is a particularly concerning form of misinformation that combines images with text to deceive or mislead people, making it appear more realistic and challenging to detect. Many terms are related to misinformation, such as rumor, fake news, false information, spam, disinformation, and multimodal misinformation [1,2]. Figure 1 illustrates the related terms of misinformation.

One common type of multimodal misinformation is out-of-context (OOC), which involves separating authentic images from their original context and pairing them with misleading texts. This can lead to the loss or change of the intended meaning and can be used to deceive or mislead the audience intentionally. Figure 2 depicts an example

F. Shalabi and H. Felouat—These authors contributed equally.

L. Barolli (Ed.): AINA 2024, LNDECT 202, pp. 86–98, 2024.
https://doi.org/10.1007/978-3-031-57916-5_8

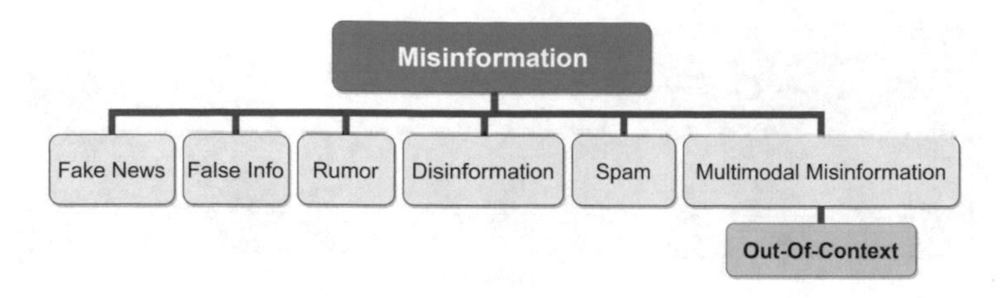

Fig. 1. Key types related to misinformation [2].

of a multimodal OOC concept, where authentic images and captions are mismatched to create a deceptive narrative. This manipulation intentionally alters the original message, potentially misleading or deceiving the audience. To address this problem, researchers are developing systems that can detect OOC information in images and texts. One approach uses pre-trained VLMs, which have demonstrated effectiveness across various tasks. Fine-tuning the parameters of these models enables them to be adapted to specific downstream tasks, such as OOC detection. Fine-tuning LVLMs involves training the models on a labeled OOC dataset and authentic image-text pairs. This process enables the models to learn the nuances of OOC manipulation and identify subtle inconsistencies between images and texts that may indicate OOC content [3,4].

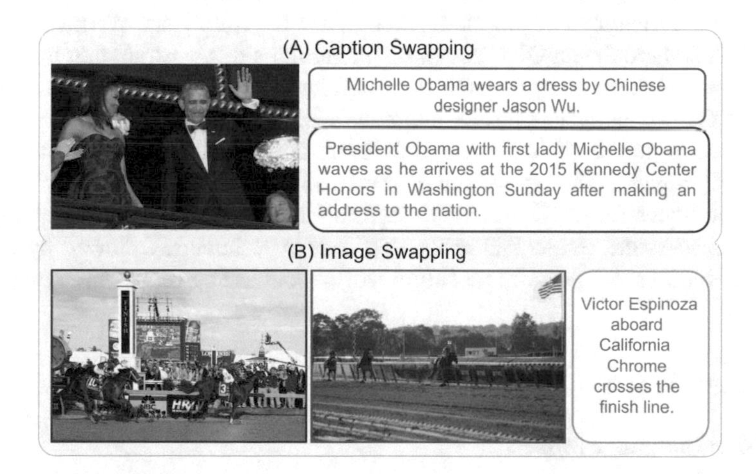

Fig. 2. Examples from our dataset show that OOC content generation arises from swapping authentic images and captions. In example A, the red caption, taken from a different context, is paired with the image. In example B, the green image is the original visual representation of the caption, while the red image is from a different context.

LVLMs are well-suited for OOC detection tasks due to their ability to comprehend the relationship between visual and textual information. This means they can identify anomalies in a scene, even if they are not explicitly mentioned in the text. Also,

Fig. 3. Illustration of our model's ability to detect contextual consistency in an image caption. Green: The image and caption are in context (Match). Red: The image and caption are out of context (Mismatch).

LVLMs can learn from various data sources, enabling them to detect various OOC situations. Moreover, LVLMs can be fine-tuned for specific tasks, making them adaptable to detecting types of anomalies. Inspired by LVLMs' success in computer vision, this study comprehensively evaluated the OOC detection capabilities of MiniGPT-4 on a well-designed dataset. Figure 3 shows the model's effectiveness in pinpointing multimodal anomalies, highlighting its potential for this crucial task.

The key contributions of this study encompass the following: We investigated the ability of LVLMs to detect OOC content. Our findings demonstrate that these models cannot achieve high accuracy on OOC detection tasks. We demonstrate that fine-tuning LVLMs on OOC datasets can further improve their OOC detection accuracy. This suggests that fine-tuning can help these models learn more specific features that are indicative of OOC content. The rest of this paper is structured as follows: Sect. 2 surveys existing techniques for OOC detection. Section 3 then details our proposed approach. Section 4 presents the results and analysis of our experiments, followed by a discussion of the findings and avenues for future work. Concluding remarks are presented in Sect. 5.

2 Related Work

2.1 Vision Language Models

Advancements in machine learning have given rise to powerful techniques for pre-training visual-language representations. The CLIP model [5], a prominent neural network architecture, leverages the inherent semantic connections between language and vision to extract transferable visual representations from natural language descriptions. Trained on a massive dataset of (image, text) pairs, CLIP effectively predicts a given image's most relevant textual description. This capability, similar to the zero-shot learning abilities of GPT-2 [6] and GPT-3 [7], facilitates effective cross-modal understanding.

Leveraging pre-trained language models has demonstrated efficacy in enhancing the comprehension of both visual and textual information using machines. Flamingo [8] further improved this idea by aligning a pre-trained image processor and language model through gated cross-attention. This model learned from billions of image-text pairs and could quickly learn from a small number of examples. BLIP-2 [9] built on this using Flan-T5 [10] and Q-Former to better connect visual features with language.

The release of GPT-4 [11] has brought even better visual understanding and reasoning abilities. Models like ChatGPT, which is a type of language model, have been effective in working together with other specialized models for vision and language tasks. For example, Visual ChatGPT [12] and MM-REACT [13] have shown how ChatGPT can collaborate with different visual models to handle more complex tasks. On the other hand, MiniGPT4 [14] directly connects visual information with the language model to accomplish different tasks involving images and texts without needing external vision models.

2.2 Multimodal Misinformation Detection

Misinformation is a growing threat, as using a combination of text with images is a valid tool for creating realistic-looking misinformation called multimodal misinformation, often leading to real-world consequences like public distrust, political instability, and so on. Researchers have proposed several methods for detecting multimodal misinformation to address this threat. One approach is extracting valuable features from images and texts and then aligning them before feeding them into a classifier. For example, Singhal *et al.* [15] proposed an approach that suppresses information from weaker modalities and extracts relevant information from the strongest modality. They used a gating mechanism to control the flow of information between the text and image modalities and were trained to determine which modality was more informative for a given sample.

Another approach focuses on fusing the multimodal features using a co-attention mechanism. For example, Wu *et al.* [16] proposed the use of stacking multiple co-attention layers to learn the relationships between the text and image features. Jing *et al.* [17] proposed a progressive fusion network (MPFN) for multimodal disinformation detection. The MPFN captures the representational information of each modality at different levels and progressively fuses the information from the same level and different levels using a mixer to establish a strong connection between the modalities. Zhang *et al.* [18] relied on neural-symbolic multimodal learning methods to propose a model that symbolically disassembles the text-modality information to a set of fact queries based on the abstract meaning representation of the caption and then forwards the query-image pairs into a pre-trained LVLM to select the "evidence" that enables them to detect misinformation.

Moholdt *et al.* [19] developed a method to detect OOC by comparing AI-generated images and captions from the COSMOS dataset [20]. They created new datasets using synthetic text-to-image generative models, Stable Diffusion, and DALL-E 2, used object detection and encoding to compute image similarity, and employed cosine similarity for comparison.

2.3 Leveraging Pre-trained Vision Language Models in Multimodal OOC Detection

In recent years, the field of multimodal OOC detection has seen significant advancements, driven in part by the emergence of powerful pre-trained VLMs. Leveraging these models has become a key research focus, as they offer the potential to improve the accuracy and robustness of OOC detection systems. In this section, we review related work that explores the integration of pre-trained VLMs into the context of OOC detection. Luo *et al.* [21] developed a new method for identifying mismatches between images and their corresponding captions. Their method uses the large pre-trained VLM, CLIP, to classify mismatches based on retrieval. They also evaluated both CLIP and VisualBERT [4] on the proposed dataset and achieved classification accuracies of 60.23% and 54.81%, respectively. However, they also found that both machine and human mismatch detection is still limited, suggesting that this task is challenging.

Huang *et al.* [22] used CLIP and VinVL [23] to detect inconsistencies in multimedia content. They encoded each image and its corresponding caption separately and then compared their embeddings. Dissimilar embeddings suggested that the text-image pair was out of context. The authors evaluated their method on several large-scale datasets. However, their method does not address the potential biases in the VLMs used, which may have affected the accuracy of the OOC detection task. Additionally, their method relies on the availability of textual information associated with the image, which may not always be available.

Fatma *et al.* [24] presented a novel methodology for detecting OOC content by employing synthetic data generation. They used two large pre-trained VLMs: BLIP-2 to generate captions describing original images and Stable Diffusion to generate images from original captions. Their approach calculated the similarity between original and generated data using CLIP, Vision transformer (ViT) [25], and Sentence-BERT [26] to assess the coherence of the input image-caption pair. This research validated the efficacy of synthetic data generation in addressing data limitations associated with OOC detection, achieving a classification accuracy of 68.0%.

3 Proposed Approach

We propose a new method for OOC detection that relies on LVLMs, as shown in Fig. 4. Our method includes MiniGPT-4 [14], a new VLM that can generate detailed image descriptions, construct websites from handwritten drafts, and craft stories and poems inspired by images.

The model undergoes a two-stage training process. In the first stage, the model learns the basics of visual and linguistic domains by training on a large collection of raw image-text pairs. It uses a ViT backbone fused with a querying transformer inspired by BLIP-2 to process images. However, this initial training phase produces unnatural language output, leading to the need for a second stage focused on improving natural language generation.

The second stage involves training the model on a curated dataset structured like a conversation, with questions and corresponding answers. A linear projection layer ensures seamless alignment between extracted visual and linguistic information. This layer alone is trained, while the pre-trained vision encoder and language decoder remain frozen. The layer's output feeds into Vicuna [27], a fine-tuned open-source chatbot that generates natural language descriptions based on fine-tuning LLaMA [28]. This stage enhances the association between images and text, producing more reliable and natural language outputs.

3.1 Fine-Tuning Large Language Models

Fine-tuning has been widely used in recent large language model (LLM) studies to align model outputs with specific task or domain requirements [29]. Without fine-tuning, LLMs are susceptible to biases, inaccuracies, and irrelevance to the context of a particular application. Fine-tuning effectively enriches LLMs with domain knowledge, enhances their task-specific capabilities, improves the fairness and reliability of their outputs, and mitigates potential harm caused by hallucinations [29]. However, fine-tuning LLMs incurs significant computational costs and demands substantial domain-specific data, which may not be readily available due to privacy concerns. Hu *et al.* [30] proposed adapting pre-trained LLMs to specific domains by fine-tuning modules with limited trainable parameters to address the computational challenge. To address the data privacy concern, federated learning (FL) emerged as a promising solution where a distributed learning paradigm enables multiple entities to optimize a model collaboratively without directly sharing their data [31].

3.2 OOC Problem Formulation

In general, the formulation of the OOC detection task focuses on the samples that contain both text and image. Let each sample be $X = (x_{img}, x_{txt})$. Denote the ground-truth label as Y, when $Y = 0$, X is a match; otherwise, it is a mismatch. A rich set of features are first extracted from x_{img} and x_{txt}, then fused and projected into a single value of y', i.e., match or mismatch.

$$y' = F_{cls}(F_{Mix}(F_{img}(x_{img}), F_{txt}(x_{txt}))) \tag{1}$$

The procedure is depicted in Eq. 1, where F_{img} and F_{txt} are unimodal feature extractors, F_{Mix} is a feature fusing model, and F_{cls} is the classification head. Instead of applying sophisticated and black-box feature fusing networks, we propose a simple yet effective method that relies on leveraging the VLM.

3.3 Fine-Tuning MiniGPT-4

To fine-tune MiniGPT-4 for the specific task of OOC detection, the first essential requirement pertains to assembling a dataset comprising n (x_{img}, x_{txt}) image-caption pairs. Additionally, these pairs must be labeled Y as either match or mismatch pairs. Subsequently, a transformative step involves restructuring this dataset to match the structure of MiniGPT-4's dataset, on which it has been trained in the second stage. This transformative procedure establishes a harmonious alignment between the task-specific dataset and the existing knowledge within MiniGPT-4, thereby facilitating the subsequent fine-tuning process. Moreover, as an auxiliary step, we need to adjust Vicuna's output to match the actual *True label* format during training. This change enables us to compare Vicuna's output y' (predicted label) with the actual *True label Y*, which aids in calculating the loss and accuracy during each epoch.

3.3.1 Data Preparation In the second stage of MiniGPT-4 training, the authors used a dataset of image-caption pairs. The prompts were used to guide MiniGPT-4 in generating descriptions for the images. The generated descriptions were then evaluated against the original captions to measure accuracy. To improve the accuracy and usability of our multimodal detection model, we fine-tuned MiniGPT-4 on a labeled multimodal dataset. We modified this dataset structure to match the model's input after our modifications, resulting in the following format: *(image, caption, label)*. The model's expected output was a simple *"Yes" or "No"* response. The loss was computed by comparing the model's output with the true label assigned to the pair.

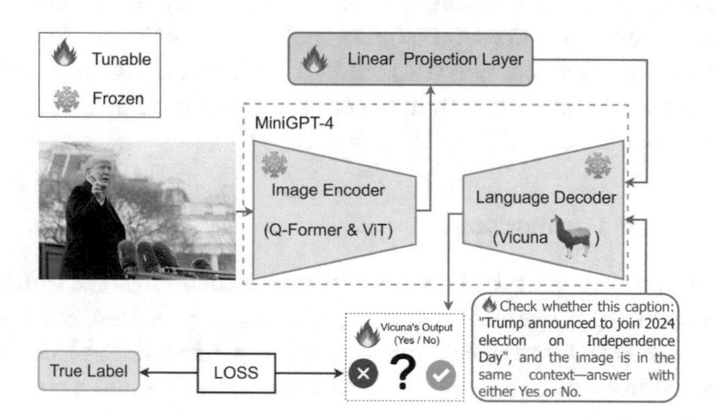

Fig. 4. The workflow of our approach entails exclusively training the linear projection layer to align visual features with the Vicuna and establish weights consistent with our dataset structure. Throughout the training phase, we directed MiniGPT-4 to generate responses in a binary *"Yes" or "No"* format to verify the contextual relevance of the image caption relative to the provided image.

3.3.2 Fine-Tuning MiniGPT-4 for Multimodal OOC Detection The original MiniGPT-4 model is trained on triplets of images, prompts, and corresponding image descriptions. During training, it takes an image *(img)* and a prompt *(p)* as input and attempts to generate a new description d_{pred} that closely matches the original description provided d_{true}. This is achieved by minimizing a loss function that measures the difference between the generated and original descriptions, as shown in Eq. 2 on the left side. When used for inference, the model only requires an image and prompt as input and outputs a predicted description of the image, as shown in Eq. 2 on the right side.

$$M_{train}(img, p, d_{true}) \rightarrow d_{pred} \quad | \quad M_{inf}(img, p) \rightarrow d_{pred} \tag{2}$$

In our approach, we fine-tuned the MiniGPT-4 model for the multimodal OOC detection task. This involved providing both an image *(img)* and prompt *(p)* to the model, where the prompt consisted of a combined question and caption *(f(q, cap))*, a function f used to combine the question and caption, and attempts to predict a label lab_{pred}, as shown in Eq. 3 on the right side. We used the weights from the first training stage to align the model with the specific requirements of our dataset's modified structure. Subsequently, we focused on retraining the second stage using this approach to acquire weights compatible with the modified dataset structure.

$$M'_{train}(img, f(q, cap), lab_{true}) \rightarrow lab_{pred} \quad | \quad M'_{inf}(img, f(q, cap)) \rightarrow lab_{pred} \tag{3}$$

During training (Eq. 3 on the left side), the fine-tuned MiniGPT-4 model takes an image and a combined prompt consisting of a question and a caption. Its objective is to predict a label lab_{pred} matching the true label lab_{true}. Since the nature of this task is a binary classification, we used the CrossEntropy loss function (Eq. 4).

$$L(x, y) = \{l_1, ..., l_n\}^T - l_n = -w_{yn} \times log\left(\frac{exp(x_{n,yn})}{\sum_{c=1}^{2} exp(x_{n,c})}\right) \tag{4}$$

This function guides the fine-tuned MiniGPT-4 towards generating binary responses *("Yes/No" or "Match/Mismatch")* while also adjusting to be compatible with the new input format of the combined prompt (Eq. 5). This approach enabled us to calculate the loss by comparing the model's output with the ground-truth labels. Figure 4 shows a flowchart illustrating these key steps in our approach.

Where x is the input, y is the target, w is the weight, and N is the batch size.

$$L'(x = (img, f(q, cap)), y = lab_{true}) \tag{5}$$

4 Experimental and Results

4.1 Dataset

We used the NewsCLIPpings dataset [21], which is a large dataset of challenging mismatched image-caption pairs constructed based on the VisualNews corpus [32]. It includes news articles from four major news outlets: the BBC, The Guardian, The Washington Post, and USA Today. The dataset was created using various techniques to introduce mismatches between images and captions, such as using semantic similarity between images and captions, matching captions semantically but with different images, associating captions mentioning the same individuals, and identifying captions describing similar scenes. The dataset provides samples representing challenging mismatches between captions and images that can mislead humans. Table 1 presents several statistics for the NewsCLIPpings dataset.

4.2 Experimental Settings

The experiments were conducted on an NVIDIA A100 G80 GPU. The model was trained with a batch size of 4 for a total of 30 epochs. The number of iterations per epoch was equal to the number of examples in the dataset divided by the batch size.

4.3 Results and Discussion

We present our results in Table 1 and compare them with the best results obtained in the NewsCLIPpings paper for each dataset split [21]. Our method outperforms the NewsCLIPpings paper results on every dataset split with a gain of $\geq 8\%$ as shown in Fig. 5, indicating that fine-tuning LVLMs provides fundamental improvements in OOC detection. We further validate our results by comparing them with other methods in the same field that used the Merged/Balanced split of the NewsCLIPpings dataset, excluding methods that rely on additional information from the Internet for classification. As shown in Table 2, our approach is among the best methods for OOC detection.

In our experiments with LVLMs for OOC detection (MiniGPT-4 [14] and IDEFICS[1]), as shown in Fig. 6, we observed the following limitations: LVLMs tend to provide descriptive responses rather than direct answers, making it challenging to evaluate their accuracy. The lack of a clear evaluation metric hinders our ability to determine the accuracy of LVLM-generated descriptions. These limitations suggest that LVLMs may not be the most suitable tool for OOC detection tasks.

However, we used MiniGPT-4 to assess the contextual coherence between image-caption pairs. We prompted the model with questions requiring *("Yes" or "No")* answers concerning the alignment between the image and its caption. Nevertheless, instead of directly providing *Yes/No* responses, the model generated descriptions that incorporated the intended response, as shown in Fig. 6. We developed a post-processing step to evaluate performance, extracting the core *Yes/No* answer from these descriptions. This enabled us to obtain binary classification results, which were used to calculate MiniGPT-4's performance in zero-shot inference. These results are presented in Table 1 and Fig. 5.

Fine-tuning LVLMs for OOC detection enables them to produce direct binary responses *("Yes" or "No")*. This approach facilitates a more straightforward assessment and comparison of their performance compared to relying on descriptive responses. While this approach has shown promise in this field, it falls short in providing explanations for the answers it generates. This is an area we intend to address in future research as we strive to enhance this approach into a framework capable of both classifying and interpreting the responses it provides.

[1] https://huggingface.co/blog/idefics.

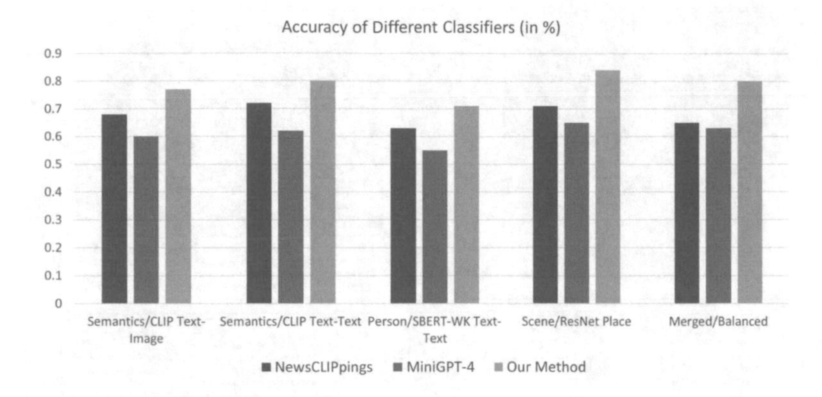

Fig. 5. Our method achieves accuracy gains of $\geq 8\%$ across diverse classification splits of the NewsCLIPpings dataset, compared to NewsCLIPpings and MiniGPT-4 (zero-shot) classifiers.

Table 1. Comparison between our performance results, NewsCLIPpings paper results [21] (Accuracy, Pristine, Falsified, and AUC in %, and MiniGPT-4: Zero-Shot inference of MiniGPT-4).

Split	Train	Val	Test	NewsCLIPpings			MiniGPT-4			Our Method			
				ACC	P	F	**ACC**	P	F	**ACC**	P	F	AUC
a)Semantics/CLIP Text-Image	453,128	47,248	47,288	**0.68**	0.74	0.61	**0.60**	0.59	0.61	**0.77**	0.76	0.79	0.77
b)Semantics/CLIP Text-Text	516,072	53,876	54,164	**0.72**	0.74	0.70	**0.62**	0.60	0.63	**0.80**	0.80	0.81	0.80
c)Person/SBERT-WK Text-Text	17,768	1,756	1,816	**0.63**	0.70	0.57	**0.55**	0.54	0.56	**0.71**	0.66	0.74	0.70
d)Scene/ResNet Place	124,860	13,588	13,636	**0.71**	0.77	0.65	**0.65**	0.63	0.67	**0.84**	0.83	0.85	0.83
Merged/Balanced	71,072	7,024	7,264	**0.65**	0.67	0.64	**0.63**	0.62	0.64	**0.80**	0.78	0.81	0.79

Table 2. Comparison results of our proposed approach with other approaches using the same dataset (The merged/balanced split of the NewsCLIPpings dataset) (Accuracy in %). (NN: Neural Network, LLM: Large Language Model, VLM: Vision-Language Model).

Paper	Year	Model	Detector Based on	ACC
Luo *et al.* [21]	2021	CLIP, VisualBERT	VLM	65.9
Huang *et al.*[22]	2022	CLIP, VinVL	VLM	65.2
Sahar *et al.* [33]	2022	ResNet152, CLIP, SBERT	NN, LLM, VLM	66.1
Zhang *et al.* [18]	2023	VisualBERT, CLIP, VinVL,FaceNet+BERT	NN, LLM, VLM	62.8
Fatma *et al.* [24]	2023	CLIP, SBERT, ViT	NN, LLM, VLM	68.8
Ours	2023	MiniGPT-4	VLM	**80.0**

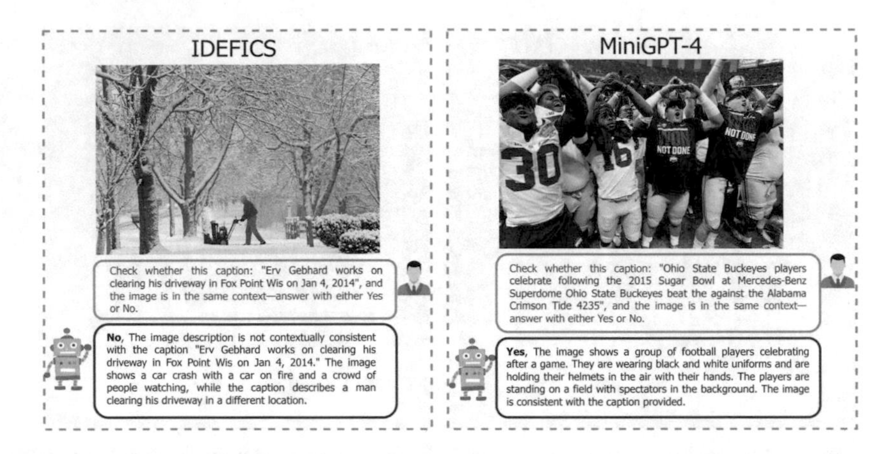

Fig. 6. OOC detection results for LVLMs MiniGPT-4 and IDEFICS without fine-tuning.

5 Conclusion

Large vision-language models (LVLMs) have demonstrated the potential for effective out-of-context (OOC) detection tasks. While initial studies suggest that LVLMs may be unsuitable for OOC detection due to their tendency to provide descriptive responses rather than direct answers, fine-tuning LVLMs on OOC data has shown promising results in improving their OOC detection accuracy. This highlights the importance of fine-tuning LVLMs for specific tasks to enhance their performance. However, further research is needed to address the limitations of LVLMs in providing explanations for their answers. This opacity presents obstacles in comprehending their decision-making process and hinders trust and reliability. To address this gap, we will strive to develop a more robust and interpretable framework for OOC detection by addressing these limitations.

Acknowledgments. This work was partially supported by JSPS KAKENHI Grant JP21H04907, and by JST CREST Grants JPMJCR18A6 and JPMJCR20D3, Japan.

References

1. Lin, X., Liao, X., Xu, T., Pian, W., Wong, K.: Rumor detection with hierarchical recurrent convolutional neural network. In: Natural Language Processing And Chinese Computing: 8th CCF International Conference, NLPCC 2019, Dunhuang, China, October 9-14, 2019, Proceedings, Part II 8, pp. 338–348 (2019)
2. Islam, M., Liu, S., Wang, X., Xu, G.: Deep learning for misinformation detection on online social networks: a survey and new perspectives. Soc. Netw. Anal. Min. **10**, 1–20 (2020)
3. Su, W., et al.: Vl-BERT: pre-training of generic visual-linguistic representations. ArXiv Preprint ArXiv:1908.08530 (2019)
4. Li, L., Yatskar, M., Yin, D., Hsieh, C., Chang, K.: VisualBERT: a simple and performant baseline for vision and language. ArXiv Preprint ArXiv:1908.03557 (2019)

5. Radford, A., et al.: Learning transferable visual models from natural language supervision. Int. Conf. Mach. Learn., 8748–8763 (2021)
6. Radford, A., et al.: Language models are unsupervised multitask learners. OpenAI Blog. **1**, 9 (2019)
7. Brown, T., et al.: Language models are few-shot learners. Adv. Neural. Inf. Process. Syst. **33**, 1877–1901 (2020)
8. Alayrac, J., et al.: Flamingo: a visual language model for few-shot learning. Adv. Neural. Inf. Process. Syst. **35**, 23716–23736 (2022)
9. Li, J., Li, D., Savarese, S., Hoi, S.: BLIP-2: bootstrapping language-image pre-training with frozen image encoders and large language models. ArXiv Preprint ArXiv:2301.12597 (2023)
10. Chung, H., et al.: Scaling instruction-finetuned language models. ArXiv Preprint ArXiv:2210.11416 (2022)
11. OpenAI GPT-4 Technical Report (2023)
12. Wu, C., Yin, S., Qi, W., Wang, X., Tang, Z., Duan, N.: Visual ChatGPT: talking, drawing and editing with visual foundation models. ArXiv Preprint ArXiv:2303.04671 (2023)
13. Yang, Z., et al.: MM-ReAct: prompting ChatGPT for multimodal reasoning and action. ArXiv Preprint ArXiv:2303.11381 (2023)
14. Zhu, D., Chen, J., Shen, X., Li, X., Elhoseiny, M.: MiniGPT-4: enhancing vision-language understanding with advanced large language models. ArXiv Preprint ArXiv:2304.10592 (2023)
15. Singhal, S., Pandey, T., Mrig, S., Shah, R., Kumaraguru, P.: Leveraging intra and inter modality relationship for multimodal fake news detection. Companion Proc. Web Conf. **2022**, 726–734 (2022)
16. Wu, Y., Zhan, P., Zhang, Y., Wang, L., Xu, Z.: Multimodal fusion with co-attention networks for fake news detection. Find. Assoc. Comput. Linguist.: ACL-IJCNLP **2021**, 2560–2569 (2021)
17. Jing, J., Wu, H., Sun, J., Fang, X., Zhang, H.: Multimodal fake news detection via progressive fusion networks. Inf. Process. Manage. **60**, 103120 (2023)
18. Zhang, Y., Trinh, L., Cao, D., Cui, Z., Liu, Y.: Detecting out-of-context multimodal misinformation with interpretable neural-symbolic model. ArXiv Preprint ArXiv:2304.07633 (2023)
19. Moholdt, E., Khan, S., Dang-Nguyen, D.: Detecting out-of-context image-caption pairs in news: a counter-intuitive method. ArXiv Preprint ArXiv:2308.16611 (2023)
20. Aneja, S., Bregler, C., Niessner, M.: COSMOS: catching out-of-context image misuse using self-supervised learning. Proc. AAAI Conf. Artif. Intell. **37**, 14084–14092 (2023)
21. Luo, G., Darrell, T., Rohrbach, A.: NewsCLIPpings: automatic generation of out-of-context multimodal media. EMNLP, pp. 6801–6817 (2021)
22. Huang, M., Jia, S., Chang, M., Lyu, S.: Text-image de-contextualization detection using vision-language models. ICASSP, pp. 8967–8971 (2022)
23. Zhang, P., et al.: VinVL: revisiting visual representations in vision-language models. CVPR, pp. 5579–5588 (2021)
24. Shalabi, F., Nguyen, H., Felouat, H., Chang, C., Echizen, I.: Image-text out-of-context detection using synthetic multimodal misinformation. In: 2023 Asia-Pacific Signal And Information Processing Association Annual Summit and Conference (APSIPA ASC) (2023)
25. Dosovitskiy, A., et al.: An image is worth 16×16 words: transformers for image recognition at scale. ICLR (2021)
26. Reimers, N., Gurevych, I.: Sentence Embeddings using Siamese BERT-Networks. EMNLP, Sentence-BERT (2019)
27. Chiang, W., et al.: Vicuna: an open-source chatbot impressing GPT-4 with 90% ChatGPT quality. See https://vicuna.Lmsys.Org (2023). Accessed 14 Apr 2023
28. Touvron, H., et al.: LLaMA: open and efficient foundation language models. ArXiv Preprint ArXiv:2302.13971 (2023)

29. Ji, Z., et al.: Survey of hallucination in natural language generation. ACM Comput. Surv. **55**, 1–38 (2023)
30. Hu, E., et al.: LoRA: low-rank adaptation of large language models. ArXiv Preprint ArXiv:2106.09685 (2021)
31. Kuang, W., et al.: FederatedScope-LLM: a comprehensive package for fine-tuning large language models in federated learning. ArXiv Preprint ArXiv:2309.00363 (2023)
32. Liu, F., Wang, Y., Wang, T., Ordonez, V.: Visual news: benchmark and challenges in news image captioning. In: EMNLP, pp. 6761–6771 (2021)
33. Abdelnabi, S., Hasan, R., Fritz, M.: Open-Domain, Content-based, Multi-modal Fact-checking of Out-of-Context Images via Online Resources. In: CVPR, pp. 14940–14949 (2022)

Enhancing the 5G-AKA Protocol with Post-quantum Digital Signature Method

Gabriel Rossi Figlarz(✉) and Fabiano Passuelo Hessel

Pontifícia Universidade Católica do Rio Grande do Sul-PUCRS, Porto Alegre, Brazil
gabriel.figlarz@edu.pucrs.br, fabiano.hessel@pucrs.br

Abstract. Data communicated in 5G is crucial to the operation of many areas in our society which are under constant cyber-attacks. The Authenticated Key Agreement (5G-AKA) is a standardized protocol to ensure secure communication between devices and the network. However, the current state of cryptography that is used to provide security in the 5G-AKA will be surpassed by algorithms performed in Quantum Computers. The advance in Quantum Computing implies that its capacity to solve complex problems can be used to perform attacks in current cryptographic systems. Thus, the security in 5G communication needs to be revisited and adapted for the post-quantum era. This work proposes the implementation of a post-quantum algorithm in the 5G-AKA protocol. The Dilithium Crystals digital signature algorithm in the protocol's Initiation Phase to ensure a secure communication between an user equipment and a serving network. The algorithm was successfully simulated providing a post-quantum communication. However, the time of execution can be improved to offer a scalable and universal solution in a post-quantum communication era.

Keywords: post-quantum cryptography · 5G-AKA · Dilithium Crystals

1 Introduction

Mobile communication technology allows a device to connect wirelessly to an antenna and subsequently to a service provider [17]. The 3rd Generation Partnership Project (3GPP), responsible for standardization of mobile telephony technologies sets the Authenticated Key Agreement (5G-AKA) protocol for secure communication between mobile subscribers and the service providers [16].

The 5G technology are responsible for millions of connected devices per square kilometer in many different areas [1]. Hence, 5G networks rely on robust authentication protocols as the 5G-AKA [16]. The 5G-AKA uses elliptical cryptography and challenge-response mechanisms to authenticate communication between devices and networks, ensuring the security and privacy of communication [11]. The protocol's security is menaced by the rise of Quantum Computers (QC) capacity to solve complex problems as the ones that 5G-AKA relies on.

L. Barolli (Ed.): AINA 2024, LNDECT 202, pp. 99–110, 2024.
https://doi.org/10.1007/978-3-031-57916-5_9

QC uses properties of quantum mechanics to perform computations and solve problems beyond the capacity of current computation [25]. Quantum algorithms have already been implemented in areas like finance, healthcare, manufacturing, cybersecurity, and blockchain being used from drug discovery to optimization. This represents an opportunity to explore a groundbreaking area that promises revolutionary solutions. According to a survey by [25], companies like Google, Alibaba, Amazon, Microsoft, and, IBM have been building quantum computers and tools to explore the possibilities that QC offers.

The potential of Quantum Computing makes it possible for current cryptographic algorithms to be broken in a reduced computing time, raising concerns for the National Security Agency (NSA). For this reason, the NSA recommends increasing the security of cryptographic protocols, proposing intermediate solutions for a transition phase between current protocols and quantum cryptographic protocols know as post-quantum solutions [6].

Adaptation of current security methods is an urgent need in view of the vulnerability that current cryptographic systems have against attacks from quantum machines. In this work, we propose an improvement in the 5G-AKA protocol to support the Post-Quantum (PQ) digital signature Dilithium Crystals algorithm.

2 Authenticated Key Agreement 5G-AKA

2.1 Overview

The 5G-AKA protocol architecture consists of three components: User Equipment (UEs), Home Networks (HNs), and Serving Networks (SNs) [5]. UEs are usually smartphones or IoT devices that contain a USIM card. HNs contain the subscriber's database, making them in charge of authentication. SNs control the base station, as an antenna communicating with the UEs through a wireless channel [18]. These components are present in the protocol's phases that guarantee the communication. In the following sessions, the 5G-AKA Initiation Phase and, the Challenge-Response Phase will be described.

2.2 Initiation Phase

The UE needs to establish a secure channel with the SN after authenticating itself to the HN, as described by Borgaonkar *(et al.)* [5]. The UE has a Subscription Permanent Identifier (SUPI), an embedded parameter in the devices that identifies them uniquely. However, to avoid sending the SUPI as plain text, a mechanism to conceal SUPI is applied. This is necessary to protect the communication against reply attacks in case of SUPI's interception. The Elliptic Curves Integrated Encryption Scheme (ECIES) is used for that. This encryption is a hybrid scheme composed of Key Encapsulation Mechanism (KEM) and Data Encapsulation Mechanism (DEM).

The first step of the KEM schema contains an algorithm called Encap and is described by the standard ISO [2]. The Encap outputs a cipher text C_0 and a secret key k_s as Fig. 1a shows. Subsequently, the algorithm SEnc from the DEM portion of the ECIES, described by Shoup [22], uses k_s to encrypt the SUPI to a Subscriber Concealed Identifier (SUCI).

Subsequently, as Fig. 1b represents, the SN sends the SUCI to the HN and the Decap algorithm from the ISO standard KEM derives k_s [2]. Next, the algorithm SDec from DEM decrypts the SUCI. This allows the HN to retrieve the key K and the sequential number SQN_{HN} from its memory [18]. The key K is a long term secret key unique for each subscriber. The SQN_{HN} is incremented in every message and contains meta-data that represents its freshness and will be used to generate the challenge to authenticate the communication [26].

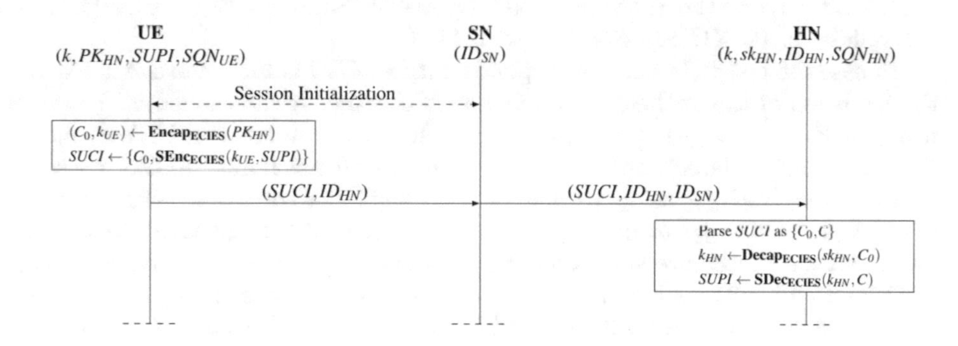

Fig. 1. Initiation Phase

2.3 Challenge-Response Phase

The first step of the challenge-response phase is the authentication vector (AV) generation. It is done in the HN supported by a series of cryptographic functions f_1, f_2, f_3, f_4, f_5, f_1* and f_5* functions described by Fodor (*et al.*) [12]. As Algorithm 1 shows, RAND is a random 128-bit vector that is used to compute together with the key K, the result AK. Subsequently, the MAC identifier is computed with f_1 considering key K and SQN_{HN} retrieved from the Initiation Phase, and RAND. The AUTN vector considers the multiplication of AK with SQN and the MAC identifier number. To compute XRES, the key K and the random vector RAND are plugged into f_2. To generate the Security Anchor Function (k_{seaf}), a KeyDerivation function described by 3GPP TS 33.501 report [3] is applied with K, RAND, SQN_{HN} and the SN Identification (ID_{SN}). Finally, the AV is generated with all its components and sent to the SN. This process is represented in Fig. 2a.

After receiving the AV, the SN stores it and sends RAND and AUTN to the UE that authenticates the message using the Algorithm 2. First, the UE calculates AK with the key K and RAND using f_5 such as the HN in Algorithm 1.

Algorithm 1. Authentication Vector Generation

1: RAND = $(0, 1)^{128}$
2: AK = f_5(K, RAND)
3: MAC = $f_1(K, , \text{RAND})$
4: AUTN = (AK * SQN_{HN}, MAC)
5: XRES = $f_2(K, \text{RAND})$
6: k_{seaf} = KeyDerivation(K, RAND, SQN_{HN}, ID_{SN}
7: AV = (RAND, AUTN, XRES, k_{seaf})

Since AUTN contains the concealed SQN_{HN} with MAC and, AK was calculated, the UE can retrieve the SQN_{HN} by parsing AUTN. Subsequently, the UE verifies the validity of RAND and SQN_{HN} with MAC.

In case the test fails, the UE responds with a MAC failure message $\perp$. Next, the freshness of the authentication vector is checked. If the sequential number from the HN (SQN_{HN}) is smaller than the sequential number in UE's memory (SQN_{UE}) plus a determined δ, the UE responds with a failure AUTS message that conceals SQN_{UE} using f_5. The next step is followed by a synchronization of $SQN_{UE} = SQN_{HN}$ to update the terms to consider the message's freshness. When all the checks are successful, k_{seaf} is derived with K, RAND, SQN_{UE}, and the SN's identification (ID_{SN}) to calculate the response RES. Finally, the UE returns the response with RES and k_{seaf} to the SN. This step of the process where the checks are made and the RES is obtained is shown in Fig. 2b.

Algorithm 2. Authentication in the UE

1: AK = f_5(K, RAND)
2: Parse AUTN as AK * SQN_{HN}, MAC
3: Check if $f_1(K, \text{RAND}, SQN_{HN})$ = MAC, if false, return $\perp$
4: Check $SQN_{UE} < SQN_{HN} < SQN_{UE} + \delta$
5: If the check does not pass: MAC* = $f_1*(K, \text{RAND}, SQN_{UE})$
Returns AUTS = $f_5(K, \text{RAND})$ * SQN_{UE}, MAC* $SQN_{UE} = SQN_{HN}$
6: k_{seaf} = KeyDerivation(K, RAND, SQN_{UE}, ID_{SN})
7: RES = $f_2(K, \text{RAND})$
8: Return(k_{seaf}, RES)

As Fig. 2c shows, after receiving the RES from SN, the HN authenticates it comparing it with a stored XRES calculated by Algorithm 1. If the results are valid, the SUPI is sent to the SN and the communication is authenticated completing the 5G AKA protocol successfully.

2.4 Vulnerability

Pierre-Alain *(et al.)* [13] states that 5G-AKA has vulnerabilities that compromise user's privacy and therefore, require attention and improvement. Among them,

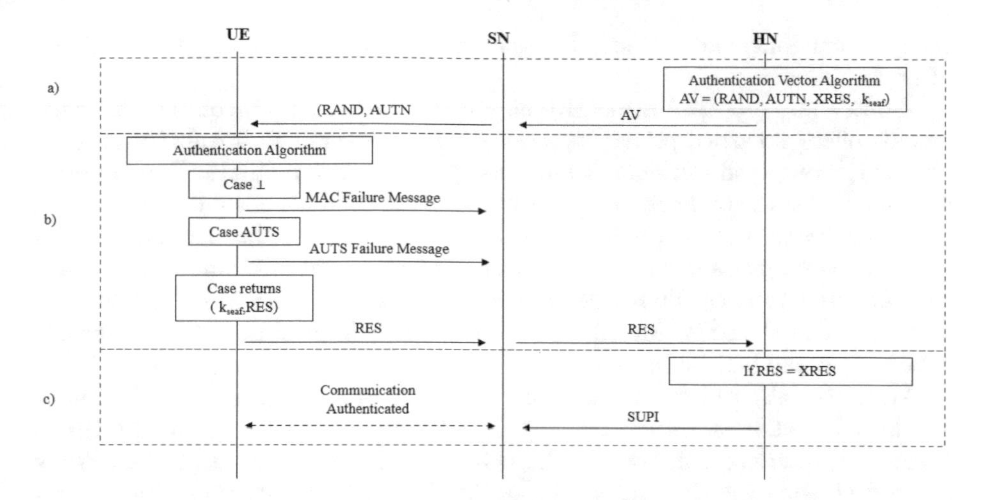

Fig. 2. Challenge-Response Phase

Zhang *(et al.)* [27] presents the possibility of a client leaving a corrupted network area and the attacker impersonating it and replaying messages. Additionally, old authentication vectors can be intercepted and replayed, generating access to the network to unauthorized parties [4].

The key K shared between the Universal Subscriber Identify Module (USIM) and the issuing network is the foundation of all 5G security [20]. It means that if K is intercepted, communication privacy is compromised. The 5G AKA protocol provides limited protection to K even against classic computer attacks. Once a malicious attacker obtains K, they can impersonate any user deriving session's information from the key [9].

Considering that the protocol security is based on particular secret key, attacks using quantum algorithms menace 5G communication. Grover's algorithm, which provides a square root speed up to oracle-based search problems and can reduce N's effective level of security up to 64-bit [14,24]. The decrease in the security level to a 64-bit key makes it possible to use search algorithms to discover K. This means that the current security level that K provides is not effective against quantum-attacks.

This results in a threat to secret-key algorithms and to the 5G-AKA protocol. Considering the possibilities that quantum computers brings, a reevaluation of cellular networks shows necessary. Therefore, implementing a post-quantum solution to improve the 5G-AKA protocol becomes essential to ensure privacy in 5G communication.

3 Related Works

To the best of our knowledge, there are few works study its application to secure communication in 5G network. This is because Quantum Computing is a relatively recent topic and it is still being developed and explored its full potential of applications.

Koutos [17] proposed a security enhanced 5G-AKA protocol. The author's version offers enhanced privacy against active attackers and linkability attacks. One of the proposed solutions is hide the type of message failure. This resource avoids attackers to understand which phase of the protocol has failed, increasing the difficulty of a second attempt to hack the channel. The approach avoids replay attacks, guaranteeing the freshness of the communication. Nevertheless, the proposed protocol does not offer a post-quantum approach to secure its cryptographic elements. Meaning that if the key K is obtained by malicious attackers, it can be broken by Groover's algorithm.

Wang (et al.) [26] states that the 5G AKA protocol presents vulnerabilities against active attackers, putting its privacy at risk. The authors propose a privacy-preserving solution for the 5G-AKA protocol. The study introduces a protocol using the Key Encapsulation Mechanisms (KEM) algorithm to protect the communication against attacks. The algorithm is used encrypt the keys in the initiation phase increasing the difficulty to be broken by an attacker. It was stated that the proposed protocol increases privacy guarantee without extra bandwidth costs satisfactorily. Despite the successful results, in case of the keys interception, the authentication of the UE can be wrongly attributed to the attacker.

Marchsreiter and Sepúlveda [19] acknowledge the treat of quantum-attack to Transport Layer security (TLS) protocol. Hence, post-quantum cryptography algorithms were implemented hybridly with traditional ones in the TLS protocol. The experimentation was applied in a PQC-only approach where PQC fully replaces the protocol's current cryptography and a PQC enhanced with hybrid cryptography that combines classic and PQC. The PQC algorithms presented acceptable results regarding performance and key sizes. However, despite PQC performed better than classic cryptography in speed, it has disadvantages with bandwidth constrained devices. The work highlights the need to study and develop post-quantum solutions and integrate it to existing communication protocols.

A post-quantum 5G AKA prototype was proposed by Damir and Niemi [8]. The authors proposed a copy of the original 5G-AKA with the addition of a post-quantum asymmetric KEM algorithm. The PQC algorithms act on SUPI's encryption in the UE and decryption in the HN. The authors implemented post-quantum algorithms and measured the communication cost in bytes and execution time. The results demonstrate that PQC outperforms classical ECIES algorithms in execution time but not in key size metric. Therefore, the study concluded that classical ECIES offer better storage and communication cost comparing with PQC. The unsatisfactory performance results from the study open discussion to different approaches to bring post-quantum solutions to improve the 5G-AKA protocol against quantum-attacks.

4 Dilithium Crystals

Dilithium Crystals is a digital signature cryptography method that relies its security on the hardness of finding short vectors in lattices [10]. The approach is divided into a key generation, signing procedure, and the verification phase. The key generation phase represented in Algorithm 3 describes how to obtain the public and secret keys [21]. Firstly, the algorithm generates a matrix A with k and l dimensions. In the next step, s and e, which are sampled random key vectors are generated. Finally, b is computed resulting in the public key pk and secret key s as output.

Algorithm 3. Dilithium Crystals Key Generation

1: **Input:** none
2: Generate $A \in R_q^{kxl}$
3: Samples $s \in R_q^l$
4: Samples $e \in R_q^q$
5: Calculate $b = As + e$
6: **Output:** public key $pk = (A, b)$, secret key s

The signing process presented in Algorithm 4 is probabilistic. In the algorithm's step 2, a random vector $y \in R_q^l$ is sampled. In the following step, a given Ay vector of polynomials is rounded and stored as w. In step 4, c is formed by hashing the message m and w. The hash function H maps an input with coefficients in $\{-1, 0, 1\}$. Since z depends on s, it can lead to security vulnerability [21]. Thus, Dilithium uses a technique called the rejection sampling approach to remove the statistical dependencies between z and the secret key s. This means that in case z is rejected, the algorithm starts from the beginning.

Algorithm 4. Dilithium Crystals Signing Process

1: **Input:** public key $pk = Ab$, secret key $s, message m \in \{0, 1\}$ Until z is valid:
2: Sample $y \in R_q^l$
3: Calculate $w = Ay$
4: Calculate $c = H((w), M) \in B_{60}$
5: Calculate $z = y + cs$
6: Output: signature $\sigma = (z, c)$

To validate the signature, the verification method is described in Algorithm 5. The recovered w' is used to recalculate c'. Subsequently, c' is compared with c. In the case which c is valid, the communication between both parties is valid and secure, otherwise considered invalid.

Algorithm 5. Dilithium Verification

1: **Input:** public key $pk = Ab$, secret key $s, messagem \in \{0,1\}$, signature $\sigma = (z, c)$
2: Calculate $w' = round(Az - bc)$
3: Calculate $c' = H(m||w')$
4: Output: valid if $c = c'$, else invalid

5 Proposal

As mentioned in Sect. 2.4, the key K is the core of the 5G-AKA security and if leaked, can compromise the validity of the communication network. Dilithium Crystals ensures that the communication between the UE and the SN is valid, avoiding sharing the key K with malicious interceptors and eavesdroppers. The interception of the key K allows replay attacks to attempt creating unauthentic connections to other UEs. Therefore, the proposal of this work is to enhance the security of 5G AKA protocol introducing the post-quantum digital signature algorithm Dilithium Crystals in the initiation phase.

Our proposal is represented in Fig. 3. The Dilithium Crystals algorithm is applied on the 5G-AKA protocol's initiation phase. The protocol follows the standard procedure firstly, where the UE initiates the session with the SN. Subsequently, the Dilithium Crystals algorithm is initiated as Fig. 3a shows. The UE follows Algorithm 3 from Sect. 4 to generate the keys followed by Algorithm 4 to sign the validation message.

Next, the UE transmits the hashed message and the signature with the private key to the SN that verifies it with Algorithm 5. If the signature is validated, the communication is verified and secure to proceed ensuring that the UE and SN are authentic. With valid communication, the UE generates and communicates the SUCI to the SN. The role of the HN's is represented in Fig. 3b and follows the original 5G-AKA protocol. The signature process using the Dilithium Crystals algorithm guarantees that the communication is being established with an authentic UE.

In our proposal, the connection between the UE and the SN has its security enhanced against quantum attacks using the Dilithium Crystals algorithm. This enhancement allows only verified UEs to establish a connection regardless if the key k was intercepted. Besides ensuring communication privacy, restricting the connection attempts to verified connections only avoids network overload by unauthorized UEs trying to connect to the HN.

6 Study Case and Results

We applied the Dilithium Crystals algorithm in python to simulate the communication between an UE and the SN. The network is composed by a thread application that allows one party to establish connection to another. Besides that, it is possible to connect multiple entities to the server, emulating the communication of multiple IoT devices to a SN.

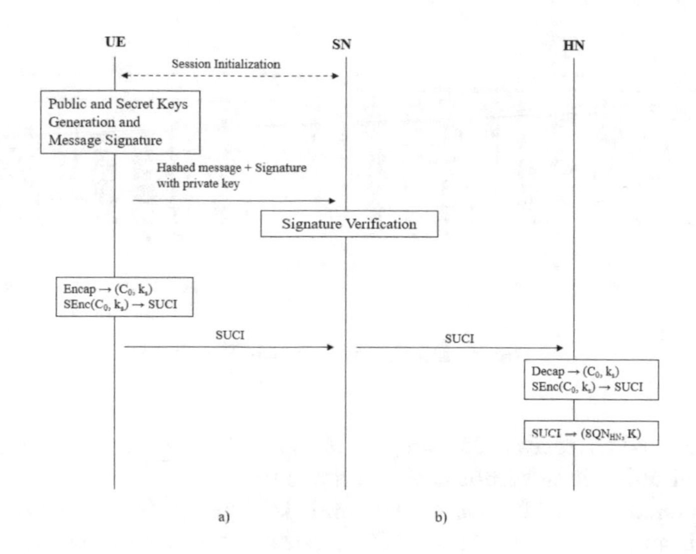

Fig. 3. Proposal Initiation Phase

According to the 5G-AKA protocol, the public and secret keys are generated with Algorithm 3 on the client side of the communication, representing the UE. Additionally, the message is signed by Algorithm 4 and send to the server, which represents the SN in the connection. The SN verifies the signature with Algorithm 5 and concludes the communication as successful, if validated with $c = c'$.

We simulated the communication between the UE and the SN iteratively for 10 times and with different message sizes of prime numbers N. According to Hoffstein [15], the higher the N, higher is the polynomial order and therefore, more complex the calculations are.

The simulation of the Dilithium Crystals algorithm between the multiple UEs and the SN in the Initiation Phase of the 5G-AKA protocol is a representation of security enhancement in communication. The post-quantum algorithm provides a quantum-proof authentication methodology that ensures that the message is being transmitted from a legit UE. The implementation was applied in a 4 CPU cores and 16 GB of memory workstation. We consider this environment adequate, since it is a realistic resource setup when it comes to mobile devices communicating to a network server.

The time of execution and memory usage in both UE and SN channels were averaged and the results found in the simulation are displayed in Figs. 4 a) and b) respectively.

According to Fig. 4 a), the time of execution in the UE increases with size of the message N finding a plateau close to 0.17 s. Nevertheless, in the SN, the time of execution escalates more abruptly with the message size, stabilizing around 0.17 s as well. From N of 433 on, the time of execution for the UE and SN are

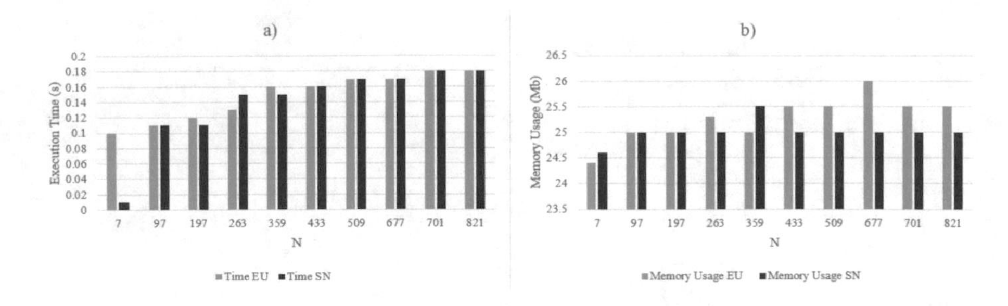

Fig. 4. Dilithium Crystals Results

not impacted. This reveals that for this use case, higher levels of security for the polynomial order do not impact the performance.

The memory usage from both UE and SN, showed in Fig. 4 b), performs a more uniform result in this study. The results of the simulation do not present outliers for the EU and SN. The figures remained in an interval between 24.40 and 26.00 Mb across the considered N values.

Even though the memory usage presented stable results, when it comes to 5G communication, the speed to establish connection is imperative. Due the 5G communication's nature, it is necessary to provide a secure authentication in a fast and reliable fashion. The resulted numbers, when compared to a mean of 0.82 ms and 0.16 ms for message signing and verification from Sikeridis (et al.) [23] or the 10ms from Marchsreiter (et al.) [19] still are not in a satisfactory standard.

Nevertheless, a post-quantum algorithm was successfully implemented in the 5G AKA protocol. By applying the Dilithium Crystals digital-signature algorithm, we ensured that the communication between a legitimate UE and the HN is established avoiding attacks from quantum-algorithms. This complements the issue identified by Damir's work [7], providing protection against an attacker that intercepted K. The enhancement proposed in this work allows only UE that passes the signature verification with the SN to advance in the protocol's Initiation Phase.

7 Conclusion and Future Works

This work proposed an enhancement in the 5G-AKA protocol's Initiation Phase We integrated the Dilithium Crystals algorithm to protect the communication against quantum-attacks. Given that the key K can be intercepted from the devices and used to attempt connection from a deceitful UE to the SN with the aid of a quantum algorithm like Groover's, a quantum-resistant solution becomes necessary. The main goal is to ensure that the connection established is between an authentic UE and the SN. This enhancement secures the communication against reply attacks driven by intercepted K keys from UEs.

We proved that the implementation of a post-quantum algorithm in the 5G-AKA protocol is possible, particularly in the Initiation Phase. Additionally, we tested the algorithm with different message sizes and assessed the performance. Despite the successful implementation of the algorithm, the results for time of execution are still not suitable for a mobile 5G case scenario when compared to the literature, as mentioned in Session Sect. 6.

For this work, we consider the first step implementing the Dilithium Crystals algorithm in the 5G-AKA protocol as a preliminary achievement. Therefore, the methodology can be improved to provide a post-quantum security level resource to the protocol in a more adequate execution time for 5G communication. Subsequently, the Key Encapsulating Mechanism (KEM) algorithm can be applied in the 5G AKA's challenge-response phase. Using KEM instead of the current elliptic curve cryptographic algorithms provides resistance against Shor's algorithm. If performed with the resources of a quantum computer, Shor's algorithm is capable of breaking consolidated algorithms in polynomial time [11].

As the time that quantum-computers leave the theoretical field and become gradually part of reality, post-quantum solutions become more urgent to be developed. Adapting the security in communication protocols as the 5G-AKA is essential and compelling due to its relevance to all its areas of acting.

References

1. 5g: Everything you need to know. https://www.lifewire.com/5g-wireless-4155905. Accessed 01 Jun 2023
2. Iso/iec 18033-2: Information technology - security techniques - encryption algorithms - part 2: Asymetric ciphers. ISO/IEC International Standards (2006)
3. 33.501, T.: Security architecture and procedures for 5g system (release 16). technical specification, 3rd generation partnership project (3gpp). Technical Specification Group Services and System Aspects (SA3) (2020)
4. Arapinis, M., et al.: New privacy issues in mobile telephony: fix and verification. In: Proceedings of the 2012 ACM Conference on Computer and Communications Security, pp. 205–216 (2012)
5. Borgaonkar, R., Hirschi, L., Park, S., Shaik, A.: New privacy threat on 3g, 4g, and upcoming 5g aka protocols. Cryptology ePrint Archive (2018)
6. Boyle, M.: Information assurance standards: a cornerstone for cyber defense. J. Inf. Warfare **13**(2), 8–18 (2014)
7. Damir, M.T., Meskanen, T., Ramezanian, S., Niemi, V.: A beyond-5g authentication and key agreement protocol. In: Network and System Security: 16th International Conference, NSS 2022, Denarau Island, Fiji, December 9–12, 2022, Proceedings, pp. 249–264. Springer (2022). https://doi.org/10.1007/978-3-031-23020-2_14
8. Damir, M.T., Meskanen, T., Ramezanian, S., Niemi, V.: On post-quantum perfect forward secrecy in 6g. arXiv preprint arXiv:2207.06144 (2022)
9. Dehnel-Wild, M., Cremers, C.: Security vulnerability in 5g-aka draft. Department of Computer Science, University of Oxford, Technical report, pp. 14–37 (2018)
10. Ducas, L., et al.: Crystals-dilithium: a lattice-based digital signature scheme. IACR Transactions on Cryptographic Hardware and Embedded Systems, pp. 238–268 (2018)

11. Fernández-Caramés, T.M.: From pre-quantum to post-quantum IoT security: a survey on quantum-resistant cryptosystems for the internet of things. IEEE Internet Things J. **7**(7), 6457–6480 (2019)
12. Fodor, G., Do, H., Ashraf, S.A., Blasco, R., Sun, W., Belleschi, M., Hu, L.: Supporting enhanced vehicle-to-everything services by lte release 15 systems. IEEE Commun. Stand. Mag. **3**(1), 26–33 (2019)
13. Fouque, P.A., Onete, C., Richard, B.: Achieving better privacy for the 3gpp aka protocol. Cryptology ePrint Archive (2016)
14. Grover, L.K.: Quantum mechanics helps in searching for a needle in a haystack. Phys. Rev. Lett. **79**(2), 325 (1997)
15. Hoffstein, J., Pipher, J., Silverman, J.H., Silverman, J.H.: An introduction to mathematical cryptography, vol. 1. Springer (2008)
16. Khan, H., Martin, K.M.: On the efficacy of new privacy attacks against 5g aka. In: ICETE (2), pp. 431–438 (2019)
17. Koutsos, A.: The 5g-aka authentication protocol privacy (technical report). arXiv preprint arXiv:1811.06922 (2018)
18. Koutsos, A.: The 5g-aka authentication protocol privacy. In: 2019 IEEE European Symposium on Security and Privacy (EuroS&P), pp. 464–479. IEEE (2019)
19. Marchsreiter, D., Sepúlveda, J.: Hybrid post-quantum enhanced tls 1.3 on embedded devices. In: 2022 25th Euromicro Conference on Digital System Design (DSD), pp. 905–912. IEEE (2022)
20. Mitchell, C.J.: The impact of quantum computing on real-world security: a 5g case study. Comput. Secur. **93**, 101825 (2020)
21. Richter, M., Bertram, M., Seidensticker, J., Tschache, A.: A mathematical perspective on post-quantum cryptography. Mathematics **10**(15), 2579 (2022)
22. Shoup, V.: A proposal for an iso standard for public key encryption. Cryptology ePrint Archive (2001)
23. Sikeridis, D., Kampanakis, P., Devetsikiotis, M.: Post-quantum authentication in tls 1.3: a performance study. Cryptology ePrint Archive (2020)
24. Ulitzsch, V., Seifert, J.P.: Breaking the quadratic barrier: Quantum cryptanalysis of milenage, telecommunications' cryptographic backbone. Cryptology ePrint Archive (2022)
25. Upama, P.B., et al.: Evolution of quantum computing: A systematic survey on the use of quantum computing tools. arXiv preprint arXiv:2204.01856 (2022)
26. Wang, Y., Zhang, Z., Xie, Y.: Privacy-preserving and standard-compatible aka protocol for 5g. In: USENIX Security Symposium, pp. 3595–3612 (2021)
27. Zhang, M., Fang, Y.: Security analysis and enhancements of 3g pp authentication and key agreement protocol. IEEE Trans. Wireless Commun. **4**(2), 734–742 (2005)

SoK: Directions and Issues in Formal Verification of Payment Protocols

Hideki Sakurada[1,2(✉)] and Kouichi Sakurai[2,3]

[1] NTT Communication Science Laboratories, NTT Corporation, 3-1 Morinosato-Wakamiya, Atsugi-shi, Kanagawa 234-0198, Japan

[2] Graduate School of Information Science and Electrical Engineering, Kyushu University, 744 Motooka, Nishi-ku, Fukuoka-shi, Fukuoka 819-0395, Japan

hideki.sakurada@ntt.com

[3] Advanced Telecommunications Research Institute International, Kyoto, Japan

Abstract. Consumers use various payment methods to purchase goods and services from retailers, such as cash, credit cards, debit cards, prepaid cards, and barcodes/two-dimensional codes. In the past, in the case of in-store payments using credit cards, the in-store terminal read the card number from the magnetic strip on the card and sent it with other purchase information to the credit card network. Recently, to prevent counterfeiting, the IC chip on the credit card and the in-store terminal communicates to authenticate each other and process the payment transaction. The medium of communication is not only contact but also contactless ("touch" payment), Moreover, the in-store terminal may process the payment either online or offline and optionally may require the customer to input their PIN. Various protocols and protocol flows are used depending on the medium and how the payment is processed. Credit cards are also used for remotely purchasing goods or services; in this case, other protocols and protocol flows are used. In some such protocols, researchers found serious security flaws that allow a malicious party to fraudulently purchase goods in such a way that is not allowed for legitimate customers. Such flaws must be fixed, but it is hard to fix and deploy protocols after they are widely used. Formal verification is a method to analyze and verify the security of such protocols and to detect flaws before they are widely deployed. In this paper, we will discuss the research trends in formal verification of the security of various cashless payment protocols, as well as future issues.

1 Introduction

Consumers use various payment methods to purchase goods and services from retailers, such as cash, credit cards, debit cards, prepaid cards, and barcodes/two-dimensional codes. In the past, in the case of in-store payments using credit cards, the in-store terminal read the card number from the magnetic strip on the card and sent it with other purchase information to the credit card network. Recently, to prevent counterfeiting, the IC chip on the credit card and the in-store terminal communicates to authenticate each other and process the payment transaction. The medium of communication is not only contact but also contactless ("touch" payment), Moreover, the in-store terminal may process the payment either online or offline and optionally may require the customer to input their PIN. Various protocols and protocol flows are used depending on

L. Barolli (Ed.): AINA 2024, LNDECT 202, pp. 111–119, 2024.
https://doi.org/10.1007/978-3-031-57916-5_10

the medium and how the payment is processed. Credit cards are also used for remotely purchasing goods or services; in this case, other protocols and protocol flows are used. In some such protocols, researchers found serious security flaws that allow a malicious party to fraudulently purchase goods in such a way that is not allowed for legitimate customers. Such flaws must be fixed, but it is hard to fix and deploy protocols after they are widely used. Formal verification is a method to analyze and verify the security of such protocols and to detect flaws before they are widely deployed. In this paper, we will discuss the research trends in formal verification of the security of various cashless payment protocols, as well as future issues. In particular, we first view various protocols for credit card payments and some efforts that verify some of these protocols. The credit card payment systems are described in detail in the reference document [21], which is also used as a reference in this paper. This paper deals with retail payment protocols and does not deal with payments between financial institutions and companies. A short version of this paper written in Japanese has been presented at [27], and this paper is an extended version.

2 Overview of Credit Card Payments

This section provides an overview of the credit card payment mechanism. Prepaid card payments (prepaid payments) and debit card payments (immediate payments) are also processed using the same mechanism as credit cards if the cards are issued under international credit card brands (Visa, MasterCard, etc.). The difference between credit card payment and debit card payment is basically that payment can be settled in advance, immediately, or later. The debit card payment system (J-Debit) using bank cash cards, which is used in Japan, is based on a unique system and is not mentioned here.

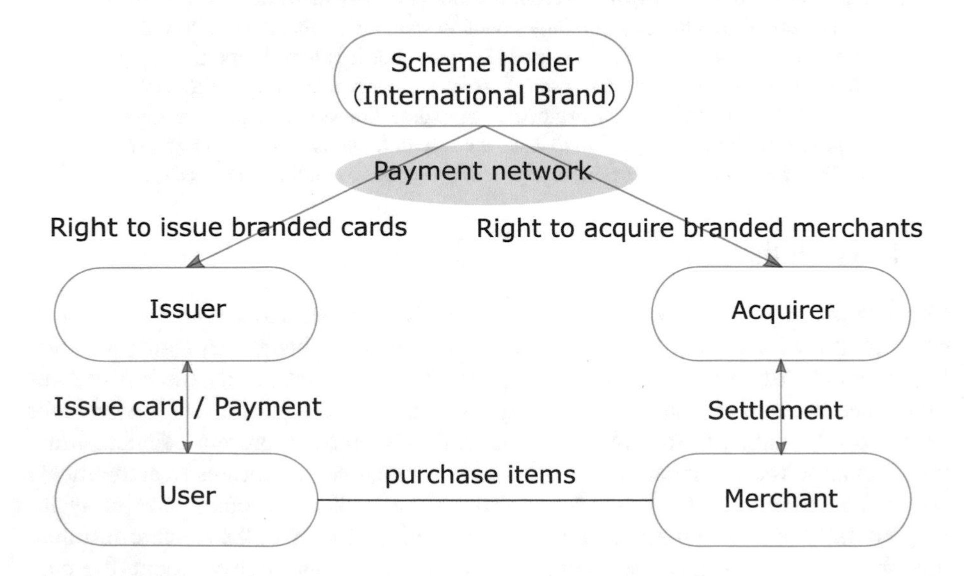

Fig. 1. Credit card payment network (based on [21])

An overview of the credit card payment is shown in Fig. 1. Before using a credit card, the consumer (user) must have the credit card issued by an issuer who is given the right to issue the branded card from an international brand (scheme holder). Before processing payments, a retailer (merchant) must make a contract with a merchant acquiring company (acquirer) and install terminals and other equipment for payment at their stores. Card-based payment can be broadly classified into two types: payment by presenting the card at the store and payment by presenting the card information via the Internet or other means of communication.

2.1 Card-Present Payments

In a payment at a merchant's store, the user presents their card to the merchant. The merchant inserts the card into the terminal and obtains the information necessary for payment through the terminal. The merchant uses the information to request the acquirer to authorize the payment and the settlement. If the payment is authorized, the payment is settled from the user's bank account to that of the merchant through the credit card payment network.

In the past, the card number (primary account number, PAN) was read from the magnetic stripe on the card. Since the contents of the magnetic stripe can be easily read, a malicious employee of the merchant can read the magnetic strip using a small device and forge another credit card indistinguishable from legitimate cards by terminals. This type of theft is called skimming. For this reason, nowadays, an IC chip is embedded in a card and communicates with the terminal to authenticate each other and optionally verify the user by asking them to input the card's PIN code. The protocol for the communication is defined as the EMV specifications [17], the industry standard for credit card payments. There are two types of communication between the IC chip and the terminal: contact communication through electrodes on the IC chip and contactless communication such as NFC. In addition, there are two types of transactions: online and offline transactions. In an online transaction, the merchant's terminal asks the issuer to authorize the transaction while communicating with the IC chip. In an offline transaction, it asks for authorization after the communication.

2.2 Card-Not-Present Payments

In a payment via telephone, facsimile, or the Internet, necessary information is sent without physically presenting the card. These are called card-not-present (CNP) payments. The most basic method is for the user to transmit the card number (PAN) to the merchant, but it is possible to purchase goods using a fraudulently obtained PAN through skimming or other means. Therefore, countermeasures are being taken to prevent unauthorized use.

Security Codes

The simplest and most widely used countermeasure is the security code, a three- or four-digit number on the back of the card. Security codes were introduced by major credit card brands around 1997. During a payment, the user sends the security code to the merchant. The merchant sends it to the acquirer. The acquirer verifies that it is the correct security code associated with the card. Since it is not recorded on the magnetic stripe, it is not stolen by skimming. Of course, if the merchant illegally stores it during the transaction, it can be used for fraudulent purposes.

Secure Electronic Transaction (SET)
One drastic countermeasure is the use of communication protocols designed for card payment. As an initial effort, the SET protocol was developed mainly by VISA and MasterCard. In this protocol, the user and merchant each digitally sign the transaction information and send it to the acquirer (more precisely, a payment gateway operated by the acquirer). The user encrypts the PAN using public-key cryptography so that only the acquirer knows the PAN. The acquirer first verifies the signatures from both parties, confirms that the signed transaction's contents match, and then processes the transaction through the credit card payment network. This protocol enables the authentication of transactions by digital signatures, but this requires the user to be issued a key for digital signatures and to use dedicated software to sign the transaction. This is thought to have hindered its widespread use.

Visa 3-D Secure (1.0)
In order to be accepted by more users, Visa proposed another protocol that does not require the installation of software, Visa 3-D Secure [7]. In Visa 3-D Secure, users are supposed to purchase products using web browsers. After a user enters their PAN of a card on a merchant's web page, the web page forwards the user's browser to the web page of an access control server operated by the issuer. On the web page, the access control server first authenticates the user by the password and confirms the payment with the user. The "3-D" in 3-D Secure refers to three domains: the issuer domain, consisting of the user and issuer; the acquirer domain, consisting of the merchant and acquirer; and the interoperable domain, which connects the issuer and acquirer domains. Payment services using 3-D Secure are called by credit card companies (mainly international brand companies) with their own service names, such as "Verified by VISA" or "MasterCard ID Check." Since 3-D Secure requires users to enter their passwords for all transactions, users may give up on a purchase if they cannot remember their passwords. This is commonly referred to as "cart abandonment" and is thought to be a major cause of opportunity loss for merchants.

EMV 3-D Secure (2.0)
In order to reduce opportunity loss due to "cart abandonment," credit card companies developed another protocol called the EMV 3-D Secure, aiming to reduce checkout friction while preventing fraud. VISA's 3-D Secure is often called 3-D Secure Version 1.0, and EMV 3-D Secure is called 3-D Secure Version 2.0. In Version 2.0, the risk of a transaction is assessed based on the user's device information and payment details. If the risk is low, the user is not asked to enter their password. In addition, to enhance usability, users are allowed to use smartphones, laptop PCs, and tablet PCs as user devices, and various user authentication methods are supported, including passwords, one-time passwords, and biometrics. Furthermore, in 2023, as a new method for confirming payments with users, the use of SPC (Secure Payment Confirmation), which was developed by W3C, is incorporated into the protocol core functional specification [18]. SPC is a Web API for the authentication during payment transactions. Since the

EMV 3-D Secure can be implemented as a payment system in various ways, The security standards, PCI 3DS, for the core components of the 3-D Secure implementation are defined by PCI, an association founded by the credit card industry. PCI also makes some security standards for credit cards, such as the Payment Card Industry Data Security Standard (PCI DSS), to protect user information. International credit card brands (VISA, MasterCard, JCB, and AMEX) have terminated the use of Version 1.0 and are transitioning to Version 2.0 as of October 2022. In this way, the credit card industry is working to prevent fraud and improve user experience in payment transactions in CNP payments.

3 Tokenization of Card Numbers

Tokenization is a method to strengthen the protection of the card number (PAN) and to prevent fraudulent use, regardless of whether or not the card is presented. In tokenization, a token is issued as an identifier tied to the card and used instead of the PAN for payments. Uses of the token are limited in terms of amount, payee, expiration date, etc., to limit unauthorized uses. Furthermore, when repeated payments are required, for example, for annual sports club membership fees, the merchant can store tokens instead of PANs, eliminating the need to store PANs. In such a case, tokenization reduces the cost of securely holding PANs. The PCI DSS Tokenization Guideline [28] is the security guideline for tokens. There are three primary forms of tokens, as described below.

Payment tokens (EMV tokens)
Payment tokens are specified by EMVCo [16]. These tokens are issued by token service providers, entities registered with EMVCo, and, during transactions to be processed in credit card networks, replaced (de-tokenized) with the PANin credit card networks. They are 13–19 digit numbers in the same format as PANs. This allows tokens to be processed in the same manner as conventional PANs. Examples of the use of payment tokens include Apple Pay and Google Pay, systems for making card payments via smartphones.

Acquirer/merchant tokens
An acquirer or merchant token is issued by an acquirer or merchant and de-tokenized prior to processing a transaction on a credit card payment network. As part of its Apple Pay service, Apple has developed a merchant token for its merchants. [6] for merchants.

Issuer tokens
An issuer token is issued by a credit card issuer. It is also called a virtual card number. It has the same format as a PAN and is detokenized by the issuer. Therefore it is processed by merchants and acquirers without being distinguished from PANs.

4 Formal Verification of Credit Card Paymets

4.1 Card-Present Payments

De Ruiter and Poll [25] verified contact-based transactions using ProVerif [4], a major tool for automatically analyzing security protocols. In particular, three card authentication methods by merchants and online and offline transactions were formulated as ProVerif scripts to verify the security of the following two security requirements.

- Correctness of card authentication. That is, if the card is authenticated by the terminal using one of the methods, the transaction is actually being conducted using that card number.
- Correctness of user verification. That is, if the terminal verifies the user with a PIN code, the information about the result of the user verification (i.e., success or failure) is consistent between the terminal and the card.
- Correctness of transaction authentication. That is, if the terminal accepted the transaction, the card also completed the transaction with the same content.

As a result of the formal verification, they successfully found the attack previously found by Murdoch et al. [22]. In this attack, by intervening in the communication between the IC chip and the merchant terminal, the attacker can make the merchant accept a transaction without entering the correct PIN code, while the card believes that the terminal does not support PIN verification.

Basin et al. [9] extracted a formal model of the contact and contactless payment protocols from the over 2000 pages of the EMV specifications and attempted to verify its security using Tamarin [5], a protocol verification tool, and found a novel man-in-the-middle attack on a Visa contactless protocol, in which PIN verification is bypassed, as well as a few attacks, including the one found by Murdoch et al. In subsequent work, they discovered an attack in which the terminal misidentified a non-Visa brand card as a Visa brand card and showed that the former attack on Visa brand cards is possible with other brands of cards [8]. Moreover, they also found another novel attack in which user verification is bypassed by exploiting a flaw in the processing when a merchant terminal fails to authenticate an offline transaction. They proposed a fixed protocol and verified the security using Tamarin [10].

4.2 Card-Not-Present Payments

Bolignano [14] formulated a simplified version of SET and verified its security. In addition to the confidentiality of PAN and other properties, they verified that when a payment gateway accepts an authorization request for a transaction, there is always an authorization request from the corresponding merchant. The protocol and security formulations and verifications were performed on a general-purpose theorem-proving tool, Coq [2].

Sakurada [26] simplified the SET payment protocol, leaving out the parts essentially related to payment authorization, and verified the confidentiality of PANs of the protocol. The protocol was formulated and verified using Paulson's inductive method on the general-purpose theorem proving tool Isabelle/HOL [3].

In parallel with Sakurada's work, Bella et al. [12] first verified the security of the user registration protocol of SET, which is required to be executed before using the payment protocol of SET. In [11, 13], they also verified the security of the payment protocols and several properties of payment authorization, including the property that when a payment gateway authorizes a payment authorization request message, there is indeed a merchant and a user who request the authorization.

Pasupathinathan et al. [23] analyzed Visa 3-D Secure as well as MasterCard Secure Code, which is a card-not-present secure payment protocol developed by MasterCard.

They analyzed the secrecy of various data, including users' passwords, and the authenticity of transactions, including the property that a user and an access control server agree on the payment information exchanged in any successfully terminated transaction. As a result of the analysis, they found that a malicious merchant can forward a user to a malicious access control server and obtain the user's password, assuming that the user cannot verify certificates of access control servers. They used Casper [1,24] and FDR2 [19] for the analysis, which are tools for translating security protocols into processes in the CSP language [20] and analyzing CSP processes.

Watanabe and Yoneyama [29] verified the challenge flows, which involve user verification, of EMV 3-D Secure using ProVerif. They verified the secrecy of challenge data, such as passwords, and the authenticity of users, i.e., the property that if an access control server verifies a user for a transaction, there is a user that sends the challenge data for the transaction. They verified both the browser-based and app-based challenge flows. Since their verification assumed that users always use or browse correct apps and access control servers, it may be interesting to see whether the attack found in [23] is possible without this assumption. Since their verification focused on user verification, they did not verify the agreement of the payment information between users and merchants.

Recently, application software executed on smartphones as well as web browsers have been used for card-not-present payments and allows for various user verification methods, such as biometrics and multifactor authentication. It causes EMV 3-D Secure to contain various protocol flows and requires more effort for verification. Moreover, modern web browsers give web developers more programmability and have complex security mechanisms. This may allow a malicious merchant to fool users through a browser-based interface and cause an attack similar to that in [23]. Investigating payment protocols in such circumstances may be interesting, as in the work by Do et al. [15].

5 Conclusion

This paper describes the mechanism of so-called cashless payment, focusing on credit cards, and formal verification of the protocols used and their security. In addition to those described in this paper, there are other cashless payment methods, such as bar codes, two-dimensional codes, and electronic money (e.g., those using IC chips and networks).

Although it was beyond the scope of this paper, the development and introduction of new payment protocols are motivated not only by the spread of cashless payments and the user environment, such as user terminals (smartphones and web browsers,) but also by legal considerations. Specifically, the European Payment Services Command (PSD2) requires strong user authentication for remote payments, and EMV 3-D Secure has been adopted to meet this standard.

As described above, payments are made using a greater variety of means than in the past, and the security requirements are increasing. It is important to rigorously analyze and verify the degree to which payment protocols meet security requirements, and formal verification is considered to be a powerful tool for this purpose.

References

1. Casper: a compiler for the analysis of security protocols. https://www.cs.ox.ac.uk/gavin.lowe/Security/Casper/
2. The Coq proof assistant. https://coq.inria.fr/
3. Isabelle. https://isabelle.in.tum.de/
4. ProVerif: cryptographic protocol verifier in the formal model. https://bblanche.gitlabpages.inria.fr/proverif/
5. Tamarin prover. https://tamarin-prover.com/
6. Apple Inc.: Introducing apple pay merchant tokens. https://developer.apple.com/apple-pay/merchant-tokens/
7. Association, V.I.S.: 3-D Secure introduction, version 1.0.2 (2002)
8. Basin, D., Sasse, R., Toro-Pozo, J.: Card brand mixup attack: bypassing the PIN in non-Visa cards by using them for visa transactions. In: 30th USENIX Security Symposium (USENIX Security 21), pp. 179–194. USENIX Association (2021). https://www.usenix.org/conference/usenixsecurity21/presentation/basin
9. Basin, D., Sasse, R., Toro-Pozo, J.: The EMV standard: break, fix, verify. In: 2021 IEEE Symposium on Security and Privacy (SP), pp. 1766–1781. ieeexplore.ieee.org (2021). https://doi.org/10.1109/SP40001.2021.00037
10. Basin, D., Schaller, P., Toro-Pozo, J.: Inducing authentication failures to bypass credit card PINs. In: 32nd USENIX Security Symposium (USENIX Security 23), pp. 3065–3079. USENIX Association, Anaheim, CA (2023). https://www.usenix.org/conference/usenixsecurity23/presentation/basin
11. Bella, G., Massacci, F., Paulson, L.C.: Verifying the SET purchase protocols. J. Autom. Reasoning **36**(1), 5–37 (2006). https://doi.org/10.1007/s10817-005-9018-6
12. Bella, G., Massacci, F., Paulson, L.C., Tramontano, P.: Formal verification of cardholder registration in SET. In: Sixth European Symposium on Research in Computer Security: ESORICS 2000 (2000)
13. Bella, G., Paulson, L.C.: Verifying set purchase protocols. Report 524, Computer Laboratory, University of Cambridge (2001)
14. Bolignano, D.: Towards the formal verification of electronic commerce protocols. In: 10th IEEE Computer Security Foundations Workshop, pp. 133–146 (1997)
15. Do, Q.H., Hosseyni, P., Küsters, R., Schmitz, G., Wenzler, N., Würtele, T.: A formal security analysis of the W3C web payment APIs: attacks and verification. In: 2022 IEEE Symposium on Security and Privacy (SP), pp. 215–234 (2022). https://doi.org/10.1109/SP46214.2022.9833681
16. EMVCo, LLC: EMV payment tokenisation. https://www.emvco.com/emv-technologies/payment-tokenisation/
17. EMVCo, LLC: EMV specifications and associated bulletins. https://www.emvco.com/specifications/
18. EMVCo, LLC: EMV 3-D Secure protocol and core functions specification (ver. 2.3.1.1). https://www.emvco.com/specifications/?post_id=90911 (2023)
19. Formal Systems Ltd: Fdr2 user manual (1998)
20. Hoare, C.A.R.: Communicating sequential processes. Commun. ACM **21**(8), 666–677 (1978). https://doi.org/10.1145/359576.359585
21. Miyai, M.: Essence of Payment service and Cashless society (in Japanese). Kinzai Institute for Financial Affairs, Inc. (2020)
22. Murdoch, S.J., Drimer, S., Anderson, R., Bond, M.: Chip and pin is broken. In: 2010 IEEE Symposium on Security and Privacy, pp. 433–446 (2010). https://doi.org/10.1109/SP.2010.33

23. Pasupathinathan, V., Pieprzyk, J., Wang, H., Cho, J.Y.: Formal analysis of card-based payment systems in mobile devices. In: Proceedings of the 2006 Australasian workshops on Grid computing and e-research - Volume 54, ACSW Frontiers '06, pp. 213–220. Australian Computer Society, Inc., AUS (2006). https://dl.acm.org/doi/10.5555/1151828.1151853
24. Roscoe, A.W.: Modelling and verifying key-exchange protocols using CSP and FDR. In: 8th IEEE Computer Security Foundations Workshop, pp. 98–107 (1995)
25. de Ruiter, J., Poll, E.: Formal analysis of the EMV protocol suite. In: Mödersheim, S., Palamidessi, C. (eds.) TOSCA 2011. LNCS, vol. 6993, pp. 113–129. Springer, Heidelberg (2012). https://doi.org/10.1007/978-3-642-27375-9_7
26. Sakurada, H.: Verification of secrecy of the set payment protocol (in Japanese). IPSJ J. **44**(8), 2106–2116 (2003)
27. Sakurada, H., Sakurai, K.: Directions and issues in formal verification of payment protocols (in Japanese). In: 2024 Symposium on Cryptography and Information Security (SCIS2024) (2024)
28. Scoping SIG, Tokenization Taskforce, PCI Security Standards Council: information supplement: PCI DSS tokenization guidelines. https://listings.pcisecuritystandards.org/documents/Tokenization_Guidelines_Info_Supplement.pdf (2011)
29. Watanabe, K., Yoneyama, K.: Formal verification of challenge flow in EMV 3-D Secure (in Japanese). In: 2024 Symposium on Cryptography and Information Security (SCIS2024) (2024)

PCPR: Plaintext Compression and Plaintext Reconstruction for Reducing Memory Consumption on Homomorphically Encrypted CNN

Takuya Suzuki(✉) and Hayato Yamana

Waseda University, Tokyo, Japan
{t-suzuki,yamana}@yama.info.waseda.ac.jp

Abstract. In the big data era, data privacy is a concern for everybody. Adopting homomorphic encryption is a promising way of preserving data privacy; however, it consumes large memory. Previously proposed lazy encoding encapsulates a vector into one data on demand, decreasing memory consumption. However, it results in 2.10–2.49× application latency increase in our experiment, compared to without lazy encoding. This paper proposes a novel technique called plaintext compression and plaintext reconstruction (PCPR), a lightweight pre-encoding and on-demand processing, which achieves almost the same memory consumption decrease as lazy encoding with a shorter latency. Our ideas are 1) dividing data into masks and corresponding scalars and 2) using lightweight operations instead of encoding. Experimental results show that PCPR achieves 1.16–2.03× shorter latency with 0.07–0.15 GiB larger memory consumption than lazy encoding, reducing memory consumption by 16.97–68.17 GiB compared to a method without lazy encoding and PCPR.

1 Introduction

Recently, big data have been used in many applications, such as image recognition and large language models. However, a data owner is concerned about data privacy; for example, an input image in image recognition can include sensitive information.

Homomorphic encryption (HE) is a solution for preserving data privacy. HE enables the evaluation of the addition and multiplication of ciphertexts without decryption. Thus, no sensitive or confidential information in the data is leaked from an HE application because the ciphertexts can be decrypted only by a party with a corresponding secret key. However, although HE has been improved since Gentry's proposal in 2009 [1], it still has several disadvantages. One of the disadvantages is large memory consumption, such as tens or hundreds of GB. One reason for the large memory consumption is that the size of encoded data, called "plaintext" in this paper, is larger than the original data. In HE, data must be encoded before encryption or evaluation of an HE operation with the data. Thus, optimization for plaintexts or encoding is one solution to the large memory consumption problem.

L. Barolli (Ed.): AINA 2024, LNDECT 202, pp. 120–132, 2024.
https://doi.org/10.1007/978-3-031-57916-5_11

Previously proposed lazy encoding [2], an HE-specific technique, enables on-demand data encoding to reduce memory consumption; that is, only the necessary data are encoded just before their evaluation with a homomorphic operation. Thus, we can avoid keeping all plaintexts not used now; that is, memory consumption can be reduced. However, lazy encoding increases application latency because the latency of encoding is several times higher than that of homomorphic addition or multiplication. Thus, a memory consumption reduction technique with a short latency increase is needed.

This study proposes a novel technique called plaintext compression and plaintext reconstruction (PCPR) to reduce memory consumption of plaintexts, which achieves a shorter latency than lazy encoding. The plaintext compression encodes data into plaintexts to reduce the memory consumption as a preprocessing. The plaintext reconstruction is on-demand processing during application execution and evaluates lightweight operations to obtain plaintexts used for homomorphic operations with a short latency. Instead of lazy encoding, the plaintext reconstruction generates the plaintexts using outputs of the plaintext compression. The following is an overview of PCPR processes and reasons for the small memory consumption and short latency.

1. The plaintext compression divides n_{pt} original plaintexts, i.e., given plaintexts, into n_{mask} mask plaintexts and $n_{pt} \times n_{mask}$ coefficient plaintexts, where n_{mask} is smaller than n_{pt}; a mask plaintext is an encoded vector, and a coefficient plaintext is an encoded scalar. For example, $[1, 4, 4, 4]$, $[3, 2, 2, 2]$, and $[5, 7, 7, 7]$ (three originals) are divided into $[1, 0, 0, 0]$ and $[0, 1, 1, 1]$ (two masks) and 1, 4, 3, 2, 5, and 7 (3×2 coefficients). The data size of a coefficient plaintext is negligible because it is thousands of times smaller than the data size of the original plaintext by the theoretical optimization [2]. Thus, as n_{pt}/n_{mask} increases, the total memory consumption is significantly reduced and can be almost the same as that of lazy encoding. Note that the optimized plaintext is referred to as *ScalarPlaintext* in this study.
2. The plaintext reconstruction evaluates plaintext-*ScalarPlaintext* multiplications of each pair of mask plaintexts and corresponding coefficient plaintexts and calculates the sum of the products on demand to reconstruct the original plaintext, for example, $[1, 0, 0, 0] \times 1 + [0, 1, 1, 1] \times 4 = [1, 4, 4, 4]$. Because the plaintext-*Scalar Plaintext* multiplication is several times faster than encoding, the plaintext reconstruction can achieve a shorter latency than lazy encoding.

Our contributions are as follows: 1) we propose an automatic plaintext compression algorithm to reduce memory consumption; 2) we implement plaintext- *ScalarPlaintext* multiplication, which is not implemented in present HE libraries such as Microsoft SEAL [3] and OpenFHE[1], and evaluate its effectiveness; and 3) we show the use case, advantages, and disadvantages of the proposed method.

The remainder of this paper is organized as follows. Section 2 discusses related work on memory usage reduction in HE. Section 3 explains the details of the proposed method and theoretical considerations regarding its effectiveness. Section 4 presents the experimental evaluation, followed by Sect. 5, which presents the results and discussion. Finally, Sect. 6 concludes this paper.

[1] https://www.openfhe.org/.

2 Related Work

In 2014, Smart et al. proposed packing [4], which encodes a vector into a single plaintext or ciphertext and enables single-instruction multiple-data (SIMD) computations. An element in a vector is called a slot, and the number of elements in a vector is called the slot count. Packing is frequently used in HE, including the full residue-number-system (RNS) variant of the Cheon-Kim-Kim-Song (CKKS) scheme [5], which evaluates the addition and multiplication of fixed-point complex numbers over ciphertexts. In the CKKS scheme, the slot count is $N/2$, where N is the degree of the polynomial modulus, an HE parameter. Thus, the data size of a plaintext with packing is $N/2$ times smaller than that without packing when all slots are utilized. However, when a large amount of plaintexts are still used in an HE application even though we adopt packing, a large amount of memory is required.

In 2019, Boemer et al. proposed nGraph-HE2 [2], in which lazy encoding, a technique for encoding data into a plaintext on demand for a homomorphic operation, was proposed. Lazy encoding reduces memory consumption compared to encoding data in advance because not all the plaintexts are needed for each execution. Furthermore, Boemer et al. proposed an optimization for a plaintext when all the slots in the plaintext had the same real value. The data size of the optimized plaintext was N times smaller. However, encoding is a time-consuming operation in HE; thus, although lazy encoding reduces memory usage, the execution latency is increased. Furthermore, the plaintext optimization can be adopted in limited cases.

Natarajan and Dai proposed SEAL-Embedded [6] for IoT devices in 2021. One of their targets was to reduce the data size because IoT devices have fewer computational resources than servers. Although the major HE libraries utilize 64-bit integers for implementation, SEAL-Embedded utilizes 32-bit integers when a 30-bit or smaller modulus is used. Thus, the data size of a plaintext or ciphertext is half that of the original one, and memory consumption is reduced. However, whether this optimization can be adopted depends on the HE parameters configured by an application design, such as output precision, and security level.

Although several other studies have reduced memory consumption by preprocessing before encryption to reduce the number of ciphertexts [7, 8] or by improving packing method and reducing the public key size [9], they did not focus on memory consumption of plaintexts. However, the memory consumption by plaintexts is not negligible; that is, we need to decrease the memory consumption by plaintexts.

3 Plaintext Compression and Plaintext Reconstruction (PCPR)

We state the research questions **RQ1** and **RQ2**: **RQ1** is how to reduce memory consumption, and **RQ2** is how to reduce latency increase. The plaintext compression is a solution to **RQ1**, and the plaintext reconstruction is a solution to **RQ2**. The plaintext compression is processed before application execution, whereas the plaintext reconstruction is processed during execution. The plaintext reconstruction consists of homomorphic additions and multiplications, which require a smaller computational cost than encoding. Furthermore, PCPR, which targets the CKKS scheme [5], does not depend on HE parameters.

3.1 Plaintext Compression and Reconstruction

The plaintext compression divides n_{pt} original plaintexts, which are given plaintexts before applying the plaintext compression, into n_{mask} mask plaintexts $mask_k$ and $n_{pt} \times n_{mask}$ coefficient plaintexts $c_{i,k}$, where $c_{i,k}$ corresponds to the k-th mask plaintext for i-th original plaintext. Here, we define $v_i[j]$ as the j-th slot value of i-th original plaintext, $m_k[j]$ as the j-th slot value of the k-th mask plaintext to be encoded into $mask_k$, and $u_{i,k}$ as a scalar to be encoded into $c_{i,k}$. For example, in case where $n_{slot} = 4$, $v_1 = [1, 4, 4, 4]$, $v_2 = [3, 2, 2, 2]$, and $v_3 = [5, 7, 7, 7]$, the plaintext compression generates $m_1 = [1, 0, 0, 0]$, $m_2 = [0, 1, 1, 1]$, $u_{1,1} = 1$, $u_{1,2} = 4$, $u_{2,1} = 3$, and $u_{2,2} = 2$, $u_{3,1} = 5$, and $u_{3,2} = 7$ and encodes them into $mask_k$ and $c_{i,k}$.

Although the mask plaintexts and coefficient plaintexts can be generated manually by considering slot value patterns in the original plaintexts, similar to the generation in CHET [10], we propose an automatic generation algorithm of mask and coefficient plaintexts from the slot values in the n_{pt} original plaintexts, shows in Algorithm 1. In Algorithm 1, encode() is an encoding operation, and $c_{i,k} = \text{encode}(u_{i,k})$ and $mask_k = \text{encode}(m_k)$ are satisfied (line 18). For simplicity, we assume that $m_k[j]$ is either zero or one, and we assume $m_k[j'] = 1$ when $m_k[j] = 1$ and $v_i[j] = v_i[j']$ are satisfied, where $j' \neq j$ (lines 6–12). Under the above assumption, $u_{i,k}$ is $v_i[j]$, where $m_k[j] = 1$ is satisfied when $mask_k$ is a mask for the i-th original plaintext (lines 13–14). Furthermore, $u_{i,k}$ is 0 when $mask_k$ is not a mask for the i-th original plaintext (lines 15–17).

The CKKS scheme uses a scale to encode real or complex numbers into a plaintext. The product of the mask and coefficient plaintext scales should be the same as that of the original plaintext scale. In this study, we configure the mask plaintext and coefficient plaintext scales as square roots of the original plaintext scale.

3.2 Plaintext Reconstruction

The plaintext reconstruction evaluates the plaintext-*ScalarPlaintext* multiplications of each pair of mask plaintexts and corresponding coefficient plaintexts, calculates the sum of the products, and then outputs a reconstructed plaintext with slot values v_i', as shown in (1). Subsequently, instead of a homomorphic operation on an input ciphertext and original plaintext, the homomorphic operation on the input ciphertext and reconstructed plaintext can be evaluated. Note that we can omit the multiplication when $c_{i,k}$ is 0 to reduce the computational cost.

$$v_i'[j] = \sum_{k=1}^{n_{mask}} mask_k[j] \times u_{i,k}. \tag{1}$$

Algorithm 1. Generation of mask and coefficient plaintexts. Inputs are 1) v, slot values in the original plaintexts, where $v_i[j]$ is the j-th slot value of i-th original plaintext with n_{slot} slots, and 2) n_{pt}, the number of original plaintexts. Outputs are mask and coefficient plaintexts.

```
 1: generate(v, n_pt):
 2:   M ← []
 3:   for i = 1 to n_pt:
 4:     T ← {}
 5:     for j = 1 to n_slot
 6:       if (v_i[j] == 0) or (j in T): continue
 7:       X ← double[n_slot]
 8:       for j' = 1 to n_slot:
 9:         X[j'] ← (v_i[j] == v_i[j'] ? 1 : 0)
10:       if X not in M:
11:         M.push_back(X)
12:         m_|M| ← X
13:       for k = 1 to |M|:
14:         if X == m_k: u_i,k ← v_i[j]
15:   for i = 1 to n_pt:
16:     for k = 1 to |M|:
17:       if u_i,k is undefined: u_i,k ← 0
18:   return ({encode(m_k) | 1 ≤ k ≤ |M|},
                {encode(u_i,k) | 1 ≤ i ≤ n_pt, 1 ≤ k ≤ |M|})
```

3.3 Compression Ratio

This section describes the compression obtained using the plaintext compression. First, we define S_{pt} and S_{spt}, the theoretical data sizes of the plaintext and *ScalarPlaintext*, respectively, as in (2) and (3), where n_{moduli} is the moduli count of a plaintext, *ScalarPlaintext*, or ciphertext.

$$S_{pt} = n_{moduli} \times N \times 8[\text{byte}]. \tag{2}$$

$$S_{spt} = n_{moduli} \times 8[\text{byte}]. \tag{3}$$

The outputs of the plaintext compression are n_{mask} mask plaintexts and $n_{pt} \times n_{mask}$ coefficient plaintexts. Thus, the total output data size S_{pc} is calculated using (4). The compression ratio r_{cp} is calculated using (5), where the compression ratio is defined as the ratio of the reduced amount to the original amount.

$$S_{pc} = n_{mask} \times S_{pt} + n_{pt} \times n_{mask} \times S_{pt}. \tag{4}$$

$$r_{cp} = 1 - \frac{S_{pc}}{n_{pt} \times S_{pt}} = 1 - \left(\frac{n_{mask}}{n_{pt}} + \frac{n_{mask}}{N}\right). \tag{5}$$

For example, in the case of $N = 2^{14}$, $n_{pt} = 32$, and $n_{mask} = 2$, r_{cp} is 0.937, which shows the total data size is 15.9× smaller than the original.

3.4 Reconstruction Latency

Latency of Homomorphic Operations. Before estimating the plaintext reconstruction latency, we present the latencies of encoding, homomorphic addition, and homomorphic multiplication in Table 1. Here, pt, ct, spt, add, and mult are defined as plaintext, ciphertext, *ScalarPlaintext*, homomorphic addition, and homomorphic multiplication, respectively. To measure the latencies, we implemented *ScalarPlaintext*, plaintext-plaintext addition, and homomorphic multiplication with *ScalarPlaintext* for Microsoft SEAL version 4.1.1 [3].

Table 1. Latency of primitive homomorphic operations.

#moduli	Latency [us]					
	Encoding	pt-pt add	ct-ct add	pt-spt mult	ct-spt mult	ct-pt mult
1	703.9	34.0	69.4	49.5	99.6	107.3
2	993.3	68.5	140.8	99.4	201.0	215.0
3	1309.2	103.5	211.3	150.3	299.5	324.1
4	1626.3	138.8	281.1	199.5	402.1	431.1
5	2005.8	174.6	350.3	249.0	500.8	540.0
6	2261.5	210.4	420.1	300.2	600.3	645.7
7	2628.1	246.8	495.5	353.7	701.5	753.9

The latencies were measured by single-threaded 1,000 times execution for each operation on Intel® Xeon® Platinum 8280. The following HE parameters were used: N was 2^{14}, the first modulus was 60-bit, the other moduli were 50-bit, and the inputs' scales were 25-bit.

Latency Comparison. Here, we estimate the latency of the plaintext reconstruction and compare it to those of other methods. We consider homomorphic multiplications of n_{ct} pairs of ciphertexts (ct_i) and plaintexts (pt_i) followed by their summation, as shown in (6), in which the plaintext compression is not adopted, and (7) represents lazy encoding enabled (6), where *encode*() and v_i represent an encoding operation and slot values of i-th plaintext. Equation (8) represents a naïve method using plaintext compression. Equation (9) represents the PCPR-enabled computation that adopts plaintext compression and reconstruction. Here, we assume that all $u_{i,k}$ are non-zero values; that is, no homomorphic multiplication with $c_{i,k}$ is omitted. Table 2 lists the homomorphic operation counts in (6), (7), (8), and (9).

$$y = \sum_{i=1}^{n_{ct}} ct_i \times pt_i. \tag{6}$$

$$y = \sum_{i=1}^{n_{ct}} ct_i \times encode(v_i). \tag{7}$$

$$y = \sum_{k=1}^{n_{mask}} mask_k \times \sum_{i=1}^{n_{ct}} ct_i \times c_{i,k}. \tag{8}$$

$$y = \sum_{i=1}^{n_{ct}} ct_i \times \sum_{k=1}^{n_{mask}} mask_k \times c_{i,k}. \tag{9}$$

Table 2. Operation counts in (6), (7), (8), and (9).

Operation	Equation (6)	Equation (7) Lazy encoding	Equation (8)	Equation (9) Proposed method
w/ compression			✓	✓
w/ reconstruction				✓
Encoding	0	n_{ct}	0	0
pt-pt add	0	0	0	$n_{ct} \times (n_{mask} - 1)$
ct-ct add	$n_{ct} - 1$	$n_{ct} - 1$	$n_{mask} \times n_{ct} - 1$	$n_{ct} - 1$
pt-spt mult	0	0	0	$n_{ct} \times n_{mask}$
ct-spt mult	0	0	$n_{mask} \times n_{ct}$	0
ct-pt mult	n_{ct}	n_{ct}	n_{mask}	n_{ct}

We define $cost_{(8)}$ and $cost_{(9)}$ as the computational costs of (8) and (9), respectively. Equation (10) represents the difference between $cost_{(8)}$ and $cost_{(9)}$, where lt_{op} is the latency of operation *op* in Table 2. For simplicity, (11) is approximated (10) by $lt_{\text{(ct-ct add)}} \approx 2 \times lt_{\text{(pt-pt add)}}$ and $lt_{\text{(ct-spt mult)}} \approx 2 \times lt_{\text{(pt-spt mult)}}$.

$$\begin{aligned} cost_{(8)} - cost_{(9)} = n_{ct} \times (n_{mask} - 1) \times \left(lt_{\text{(ct-ct add)}} - lt_{\text{(pt-pt add)}}\right) + \\ n_{mask} \times n_{ct} \times \left(lt_{\text{(ct-spt mult)}} - lt_{\text{(pt-spt mult)}}\right) + (n_{mask} - n_{ct}) \times lt_{\text{(ct-pt mult)}}. \end{aligned} \tag{10}$$

$$\begin{aligned} cost_{(8)} - cost_{(9)} \approx n_{ct} \times (n_{mask} - 1) \times lt_{\text{(pt-pt add)}} + n_{mask} \times n_{ct} \times \\ lt_{\text{(pt-spt mult)}} + (n_{mask} - n_{ct}) \times lt_{\text{(ct-pt mult)}} \end{aligned}. \tag{11}$$

Equation (11) is always positive when n_{ct} is one because of $n_{mask} \geq 1$. When $n_{mask} \geq 2$, (11) is also always positive because of $n_{ct} \times \left(lt_{\text{(pt-pt add)}} + n_{mask} \times lt_{\text{(pt-spt mult)}}\right) > n_{ct} \times lt_{\text{(ct-pt mult)}}$. Thus, PCPR achieves a shorter latency than the naïve method under the above conditions. However, PCPR performs worse than the naïve method when $n_{mask} = 1$. This is because $n_{ct} \times lt_{\text{(pt-spt mult)}} + (1 - n_{ct}) \times lt_{\text{(ct-pt mult)}}$ can be negative, depending on n_{ct}. This is a disadvantage of the plaintext reconstruction.

We compare PCPR with lazy encoding. As shown in Table 2, the larger the number of masks is, the longer the latency is required for the plaintext reconstruction. Since the difference in costs between lazy encoding and PCPR is $n_{ct} \times \left\{lt_{\text{(encode)}} - (n_{mask} - 1) \times lt_{\text{(pt-pt add)}} - n_{mask} \times lt_{\text{(pt-spt mult)}}\right\}$, n_{moduli} changes the difference and $ub(n_{moduli})$, the upper bound of n_{mask}, to ensure that the latency of PCPR is shorter than that of lazy encoding. For example, in the case of $n_{moduli} = 4$, the difference is $n_{ct} \times \{1626.3 - (n_{mask} - 1) \times 138.8 - n_{mask} \times 199.5\}$, and $ub(4)$ is five. Thus, when n_{mask} is larger than $ub(n_{moduli})$, lazy encoding is better than PCPR.

4 Experimental Evaluation

4.1 Target Application

Our experiment uses a convolutional neural network (CNN) inference with the CIFAR-10 dataset [11]. We assume that an input image is a ciphertext and model parameters are plaintexts. We adopt a six-layer CNN model based on ResNet [12] without a shortcut connection, as shown in Fig. 1. The stride of the second convolutional layer was 2×2, and that of the other convolutional layers was 1×1. We skip the softmax function in the last activation layers in the inference with HE because the softmax function incurs a high computational cost in HE [13] and does not affect the output label. The other activation layers evaluate the sigmoid-weighted linear unit [14] using a two-degree polynomial approximation by the least-squares method with the approximation range of $[-8, 8]$ and the Y-interception to zero.

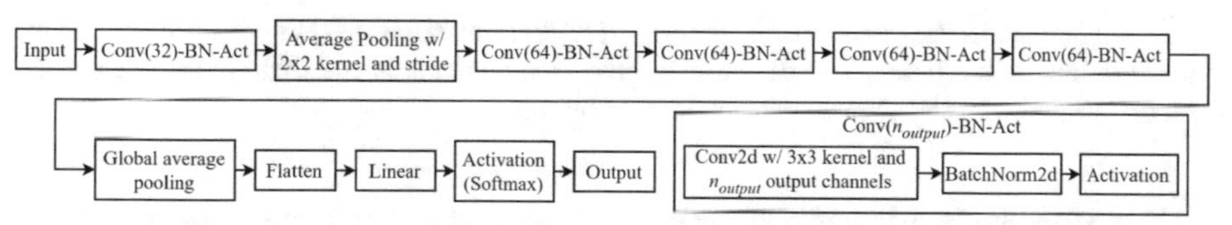

Fig. 1. CNN Model. Conv2d and BatchNorm2d represent a 2-D convolutional layer and a 2-D (channel-wise) batch normalization layer, respectively.

The CNN model was trained using the PyTorch[2] with 256 batches, 1,000 epochs, AdamW as the optimization algorithm, and ConsineLRScheduler[3] in timm (PyTorch Image Models[4]) to change the learning rate.

4.2 Implementation of Convolutional Neural Network with HE

We pack one or more channels of a layer input into a single ciphertext. H and W represent the height and width of the convolutional-layer input, respectively. Further, we introduce the batch size B, a parameter that improves throughput and determines the number of channels in a single ciphertext. Figure 2 shows the packing method for layer inputs with a batch size of two, that is, four channels in a ciphertext, where H' and W' represent the height and width of an input image of the CNN model, respectively; that is, H' and W' are 32 for the CIFAR-10 dataset. We convert an original tensor of $H \times W$ to a converted

[2] https://pytorch.org/.

[3] Parameters of ConsineLRScheduler are as follows: t_initial is 1,000, lr_min and warmup_lr_init are 10^{-4}, warmup_t is 100, and warmup_prefix is true.

[4] https://github.com/huggingface/pytorch-image-models.

tensor of $H' \times W'$, and then we pack the converted tensor into a part of a ciphertext. Intervals of a pair of neighbor valid pixels for a specific batch in the converted tensor are $B \times H'/H - 1$ and $B \times W'/W - 1$ for the height and width axis, respectively.

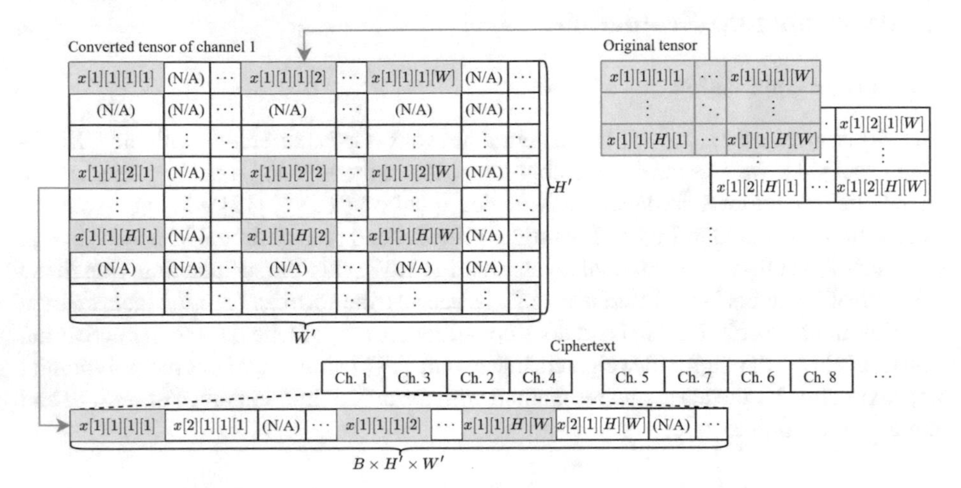

Fig. 2. Ciphertext packing method. $x[b][ch][h][w]$ Represents a pixel of (h, w) in the ch-th channel of the b-th image in a batch. The blue-highlighted area is pixels in the first channel of the first batch ($x[1][1][h][w]$). (N/A) means any value can be stored.

Figure 3 shows how to pack a filter in the convolutional layer into plaintexts. Before convolution, the input ciphertext is rotated to align its slots with the output ciphertext slots to reduce the computation cost. Additionally, the slots that pack the weights in the filter are aligned with slots in the rotated input ciphertext.

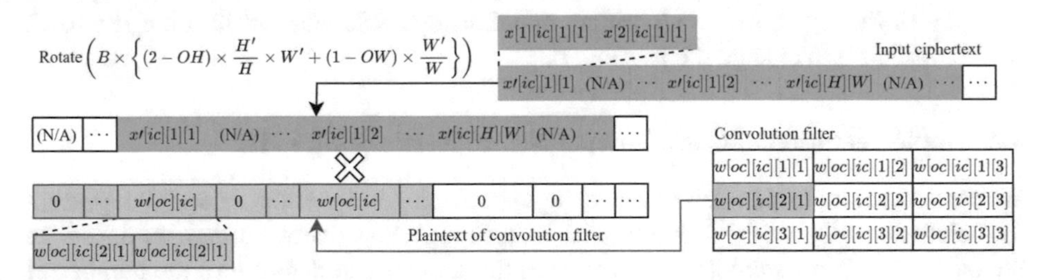

Fig. 3. Packing method for filters in the convolutional layer. oc and ic denote the output and input channels, respectively. "Rotate" is a function to rotate packed values in a ciphertext to the left. (OH, OW) is (2, 2) for the convolution filer because its central position is (2, 2). The green-highlighted area represents the ic-th channel of the input ciphertext. The red-highlighted area represents the target weight ($w[oc][ic][2][1]$) of the filter. The other areas represent other channels or weights.

4.3 Configuration of Homomorphic Encryption Parameters

First, to keep precision, we configured the default and lower-bound scales of all ciphertexts and original plaintexts as 25-bit. We then adopted Avgpool, Activation, and BatchNorm folding [15], but not global average pooling, to reduce the multiplicative depth and total moduli count. Thus, we evaluate rescaling just after each convolutional layer and linear layer. We configured the modulus bit list q_i as $q_0 = 60, q_j = 50$ for $1 \leq j \leq 5$, and $q_6 = 25$, and we set the modulus for key switching to 60-bit. Finally, we configured the polynomial modulus degree N to be 2^{14} and the slot count to be 2^{13} to ensure 128-bit security. Furthermore, we prepare all the required rotation keys, which are needed to evaluate the corresponding rotations, to reduce the inference latency, although the rotation keys increase memory usage.

4.4 Adaption of PCPR

We adopt PCPR for convolutional layers and adopt the plaintext compression for plaintexts every position in convolution filters to reduce the mask count for each original plaintext. Thus, n_{mask} for each original plaintext is $8/B$.

We do not adopt PCPR for the other layers. There are two reasons for this: in each layer except for the convolutional layers, 1) the plaintext count is small, and 2) the compression ratio r_{cp} is small compared to the convolutional layers, that is, the efficiency of PCPR is not high. Moreover, the model parameters in layers except for the convolutional layers are encoded and stored in the memory in advance.

4.5 Evaluation Method

We used a server with two Intel® Xeon® Platinum 8280 and 1.5 TB memory in total. We then used only the NUMA node 1 to remove the effects of multi-CPUs. We used a modified version of Microsoft SEAL, as described in Sect. 3.

We also measured the performance of the following three baselines to evaluate PCPR. In addition, all the plaintexts in the layers, except for the convolutional layers, were encoded in advance.

- *Pre-encoding*: All the plaintexts are stored in memory in advance.
- *Lazy-encoding*: Lazy encoding is adopted for all the convolutional layers.
- *Naïve*: The plaintext compression is adopted; however, the plaintext reconstruction is not adopted.

The evaluation metrics include memory consumption for **RQ1**, latency for **RQ2**, and accuracy. After we execute the inference 11 times with a single thread for each combination of method and batch size, we measure the virtual memory resident set size (VmRSS) as memory usage and calculate the average latency, except for the first execution. Note that the measured memory consumption is at its peak because Microsoft SEAL implements a memory allocator that does not deallocate memory until the end of the execution. Furthermore, we measure the accuracy of PCPR using all the 10,000 test images in the CIFAR-10.

5 Results and Discussion

5.1 Memory Consumption

Table 3 lists memory consumption and latency. *Pre-encoding* consumes the largest memory because it stores all the plaintexts in memory. However, *Lazy-encoding* consumes the smallest memory because it keeps no plaintexts before and after their corresponding homomorphic operations. PCPR and *Naïve* consumed 2.76–7.31× smaller memory than *Pre-encoding* due to the plaintext compression and only 1.007–1.034× (0.07–0.15 GiB) larger memory than *Lazy-encoding*. Thus, PCPR successfully answered **RQ1**.

Table 3. Memory usage and latency.

Method	Memory consumption [GiB]				Latency [sec]			
	$B = 1$	$B = 2$	$B = 4$	$B = 8$	$B = 1$	$B = 2$	$B = 4$	$B = 8$
Pre-encoding	13.21	21.47	40.38	78.97	33.12	45.10	71.42	123.29
Lazy-encoding	4.53	4.35	6.13	10.73	58.75	94.55	167.67	306.55
Naïve	4.83	4.50	6.22	10.81	116.93	115.01	115.82	117.37
PCPR (proposed)	4.79	4.50	6.22	10.80	70.56	81.65	105.86	150.83

5.2 Inference Latency

Table 3 shows that PCPR was 1) 1.16–2.03× faster than *Lazy-encoding* when the batch size was two, four, or eight and 2) 1.09–1.41× faster than *Naïve* when the batch size was one, two, or four. However, in other cases, *Lazy-encoding* or *Naïve* was faster than PCPR. Thus, PCPR answers **RQ2** when n_{mask} is appropriate.

5.3 Discussion

Although PCPR with $B = 8$ achieved 83.47% accuracy, PCPR with other batch sizes degraded the accuracy by a maximum of 0.14% from accuracy without HE (83.47%), which might be a rounding error after splitting the scale during the plaintext compression. One solution is to improve the scale configuration depending on the application and requirements when adopting the plaintext compression; for example, configuring mask plaintexts and coefficient plaintexts with different scales while maintaining the total scale, for example, the scale configuration in CHET [10].

Our plaintext reconstruction has another disadvantage: the latency was longer than that of the plaintext compression when the mask count is one. Thus, to reduce the application latency, the plaintext reconstruction should be adopted when $2 \leq n_{mask} \leq ub(n_{moduli})$ is satisfied.

6 Conclusion

HE is one of the solutions for preserving data privacy; however, large memory consumption is a problem. To solve this, we proposed PCPR, plaintext compression and reconstruction, to reduce the memory consumption with small latency. Our experimental evaluation on CNN determined that PCPR achieved 16.97–68.17 GiB reduction of memory consumption with small latency, which is 1.16–2.03× shorter than the latency of previously proposed *Lazy-encoding* while allowing a negligible increase in memory consumption by 0.7–5.7% compared to *Lazy-encoding*.

Our future studies reduce the precision loss caused by plaintext compression by revising the scale configuration for mask plaintexts and coefficient plaintexts. Moreover, we aim to adopt PCPR for more HE applications.

References

1. Gentry, C.: Fully homomorphic encryption using ideal lattices. In: The 41st annual ACM symposium on Theory of computing (STOC 2009), pp. 169–178. ACM, New York (2009). https://doi.org/10.1145/1536414.1536440
2. Boemer, F., Costache, A., Cammarota, R., Wierzynski, C.: nGraph-HE2: a high-throughput framework for neural network inference on encrypted data. In: Proceedings of the 7th ACM Workshop on Encrypted Computing & Applied Homomorphic Cryptography, pp. 45–56. ACM, New York (2019). https://doi.org/10.1145/3338469.3358944
3. Microsoft SEAL (release 4.1). https://github.com/Microsoft/SEAL. Accessed 21 Apr 2023
4. Smart, N.P., Vercauteren, F.: Fully homomorphic SIMD operations. Des. Codes Cryptogr. **71**, 57–81 (2014). https://doi.org/10.1007/s10623-012-9720-4
5. Cheon, J.H., Han, K., Kim, A., Kim, M., Song, Y.: A full RNS variant of approximate homomorphic encryption. In: Cid, C., Jacobson Jr., M. (eds.) Selected Areas in Cryptography – SAC 2018. SAC 2018. LNCS, vol. 11349, pp. 347–368. Springer, Cham (2019). https://doi.org/10.1007/978-3-030-10970-7_16
6. Natarajan, D., Dai, W.: SEAL-embedded: a homomorphic encryption library for the Internet of Things. IACR Trans. Cryptogr. Hardware Embed. Syst. **2021**(3), 756–779 (2021). https://doi.org/10.46586/tches.v2021.i3.756-779
7. Koseki, R., Ito, A., Ueno, R., Tibouchi, M., Homma, N.: Homomorphic encryption for stochastic computing. J. Cryptogr. Eng. **13**, 251–263 (2023). https://doi.org/10.1007/s13389-022-00299-6
8. Wang, Y., Chen, L., Wu, G., Yu, K., Lu, T.: Efficient and secure content-based image retrieval with deep neural networks in the mobile cloud computing. Comput. Secur. **128**(103163), 1–13 (2023). https://doi.org/10.1016/j.cose.2023.103163
9. Cheon, J.H., Kang, M., Kim, T., Jung, J., Yeo, Y.: High-throughput deep convolutional neural networks on fully homomorphic encryption using channel-by-channel packing. ePrint Archive, Paper 2023/632, pp. 1–18 (2023). https://eprint.iacr.org/archive/2023/632/20230504:000428
10. Dathathri, R., et al.: CHET: an optimizing compiler for fully-homomorphic neural-network inferencing. In: Proceedings of the 40th ACM SIGPLAN Conference on Programming Language Design and Implementation, pp. 142–156. ACM, New York (2019). https://doi.org/10.1145/3314221.3314628
11. Krizhevsky, A., Nair, V., Hinton, G.: CIFAR-10 (Canadian Institute for Advanced Research). http://www.cs.toronto.edu/~kriz/cifar.html. Accessed 06 June 2023

12. He, K., Zhang, X., Ren, S., Sun, J.: Deep residual learning for image recognition. In: Proceedings of 2016 IEEE Conference on Computer Vision and Pattern Recognition, pp. 770–778. IEEE (2016). https://doi.org/10.1109/CVPR.2016.90
13. Lee, J., et al.: Privacy-preserving machine learning with fully homomorphic encryption for deep neural network. IEEE Access **2022**(10), 30039–30054 (2022). https://doi.org/10.1109/ACCESS.2022.3159694
14. Elfwing, S., Uchibe, E., Doya, K.: Sigmoid-weighted linear units for neural network function approximation in reinforcement learning, pp. 1–18. arXiv arXiv:1702.03118v3 (2017). https://doi.org/10.48550/arXiv.1702.03118
15. Boemer, F., Lao, Y., Cammarota, R., Wierzynski, C.: NGraph-HE: a graph compiler for deep learning on homomorphically encrypted data. In: Proceedings of the 16th ACM International Conference on Computing Frontiers, pp. 3–13. ACM, New York (2018). https://doi.org/10.1145/3310273.3323047

Design and Performance Evaluation of a Two-Stage Detection of DDoS Attacks Using a Trigger with a Feature on Riemannian Manifolds

Yang Lyu, Yaokai Feng(✉), and Kouichi Sakurai

Kyushu University, 744 Motooka, Nishi-ku, Fukuoka 819-0395, Japan
fengyk@ait.kyushu-u.ac.jp, sakurai@inf.kyushu-u.ac.jp

Abstract. The DDoS attack remains one of the leading attacks today. To reduce the number of resource-consuming detection algorithm calls, the trigger-based two-stage detection approach has been proposed. In such systems, trigger mechanisms, including trigger features and threshold update algorithms, play an important role in detection performance. It is also important what features are used in the second stage of detection. In this study, 1) we introduce a Riemannian manifold metric (work) as a trigger feature for the first time since it was proven that traffic data is a Riemannian manifold; 2) we propose a new mechanism to update the trigger threshold based on historical flow data and the feedback of the second-stage detection results; 3) the feature selection algorithm ECOFS is used for the second stage detection. Experimental results using public datasets show that our proposal calls much less of the second stage detection than the latest trigger-based two-step detection systems.

1 Introduction

1.1 Cyber Attacks

Today, our heavy reliance on the internet exposes us to increasing cyber threats. Up to 122 million malwares in total were reported by Kaspersky for 2022, 6 million more than the previous year [1]. The DDoS (Distributed Denial of Service) attacks aiming to overwhelm the victims by surpassing their processing capacities are still one of the current main attacks. Microsoft's Azure DDoS Protection team noted a record DDoS attack in November 2022, peaking at 3.47T bps [2]. DDoS attacks also often cause significant damage, for example, AMCA's data breach in 2019 impacted 25 million hosts, leading to bankruptcy and legal repercussions [3].

1.2 Two-Step Detection of Cyber Attacks

Machine learning methods, such as SVM (Support Vector Machine), RF (Random Forest), and ANN (Artificial Neural Network), are commonly employed for attack detection [4–6].

L. Barolli (Ed.): AINA 2024, LNDECT 202, pp. 133–144, 2024.
https://doi.org/10.1007/978-3-031-57916-5_12

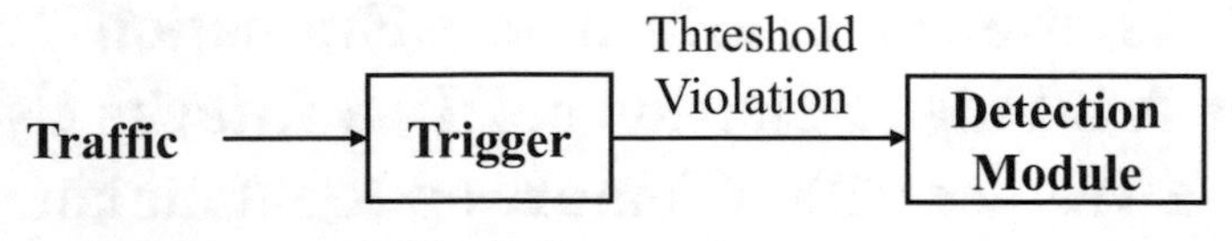

Fig. 1. System structure.

As network attacks has been greatly growing in complexity, the detection systems also become complicated and computationally expensive. Thus, how to decrease the number of the invocations of the complicated detection algorithms without affecting the detection performance becomes an important issue. Trigger-based two-stage detection method (shown in Fig. 1) has been proposed, in which the complicated detection module (i.e. the second-stage detection) is invoked only when it is considered necessary by the trigger [6–8].

1.3 Our Contributions

Our contributions in this study is as follows:

1) A metric (work) of Riemannian manifold is used for the first time as a trigger feature. The traffic data of the network has been proved to be an Riemannian manifold in recent studies [9];
2) Based on the feedback of the recent detection result of the second-stage and the recent trigger history, a novel mechanism for threshold updating in the trigger is proposed;
3) A feature selection algorithm called ECOFS is used to determine what features should be used in the second-stage detection, which successfully reduces the model training time while enhancing/keeping the detection performance;
4) Several different ML (Machine Learning) models are tested as the second-stage detection;
5) The behaviors of our proposal are examined using a public dataset and the experiment result indicates that our proposal improves the detection performance and significantly decrease the number of invocations of the resource-intensive second-stage detection model.

2 Background

2.1 Feature Selection Algorithm

Feature selection falls into three categories: filter methods, wrapper methods, and embedded methods. As one of filter methods, the algorithm ECOFS (Efficient Correlation-Based Feature Selection) has been proved efficient [10], which operate independently of model training. They offer high computational efficiency and are model-independent, making them particularly suitable for large datasets [11]. Thus, it is used in this study to select efficient features for the second-stage detection models.

ECOFS simplifies the complex pairwise computations in the CFS (Correlation-based Feature Selection) algorithm and eliminates the need for manually setting parameters. Equation (1) illustrates the correlation calculation, and Eq. (2) showcases the core formula of the ECOFS algorithm.

$$SU(X, Y) = 2\left[\frac{H(X) - H(X \vee Y)}{H(X) + H(Y)}\right], \tag{1}$$

$$J_{ECOFS} = SU_{f_i \in F}(f_i, c) - \max_{f_s \in S}(SU(f_s, f_i)|), \tag{2}$$

where f_i belong to the original feature set and f_s belong to the selected feature subset.

The formula $\max_{f_s \in S}(SU(f_s, f_i)|)$ denotes the maximum value representing the assessed redundancy between the feature *fi* and the chosen feature f_s. The underlying principle they introduced is that if a feature's contribution to the class exceeds its redundancy with the selected features, we classify the feature as "beneficial" and retain it. In this study, we utilize the feature selection algorithm to enhance the model's training efficiency and performance.

2.2 Extracting Features from Riemannian Manifolds

The concept of a Riemannian manifold has been introduced to cyber attack detection, treating internet traffic data as such [9]. In that study, the network traffic data (packet count, packet size, flow duration) are proven to follow a Riemannian manifold, with the concept of "work" defined for flow data.

Distance on the Riemannian manifold of flow data is defined as the Euclidean distance from the origin (0, 0, 0) to the detection value (packet count 'p_c', packet size 'p_s' and flow duration 't'). Force on the Riemannian manifold of flow data is defined as the factor that brings the data from the origin to the detection value. That is, force is the cause of the distance. Therefore, flow data can be likened to uniformly accelerated motion from the origin to the detection point. That is, the distance $d = \sqrt{(p_c^2 + p_s^2)} = a \times t^2/2$ and the force $F = a \times m = 2d/t^2$, where a is the acceleration, t is the time duration of the flow and m is the mass (m is set to 1). The work W that is used in our trigger as a feature can be calculated as $W = F \times d$. Thus,

$$W = F \times d = \frac{2d^2}{t^2} = \frac{2(p_c^2 + p_s^2)}{t^2}. \tag{3}$$

3 Related Work

P. Guo and N. Li [12] compute dataflow entropy to verify legitimate users, while M. Thottan et al. [13] employ wavelet analysis for effective anomaly detection and localization, using a time series formed from selected network features. K. Giotis et al. [14] use OpenFlow and SFlow simulators with entropy-based detection but face limitations in following flow distribution.

To enhance DDoS detection, You et al. [15] propose two-stage models with trigger modules to reduce resource consumption. However, fixed thresholds in You et al.'s approach limit effectiveness. Wang et al. [6] design a two-stage system using a dynamic threshold trigger module but only considers packet count, with fixed adaptation in all cases and verification solely on simulated data.

J. David and C. Thomas [16] use a dynamic threshold algorithm for the four attributes of network traffic, computing mean, variance, and coefficient β over a window size K and overlap interval. This method, while effective for flow data detection over a fixed period, is not a continuous monitoring mechanism, leading to increased monitoring delay as the time range expands. The length of the moving window K affects both monitoring delay and accuracy.

As the latest two-stage detection system, Niu et al. [17] propose a two-stage detection algorithm based on threshold auto-update, utilizing two threshold features and implementing automatic adjustment to balance accuracy while reducing the invocation frequency of the two-stage detection. However, this approach still faces issues:

1) Threshold updates lag behind traffic changes, leading to a higher false positive rate.
2) The trigger employs only two features, insufficient for comprehensive traffic behavior description.
3) Features used in the two-stage detection model lack rationality as they haven't undergone selection through a feature selection algorithm.

4 Our Proposal

Our proposal highlights the following key points:

1) Expand trigger features to four, incorporating Riemann manifold characteristics using streaming data and introducing 'work of flow data';
2) Dynamically update trigger thresholds in real-time based on historical data and the feedback of the recent detection result of the second-stage detection;
3) Two scenarios for second-stage detection: triggered by the alarm from the trigger or activated by the anti-false alarm mechanism due to prolonged trigger silence;
4) Introduce variable β in the threshold formula, adjusting it based on feedback from second-stage detection results to enhance trigger sensitivity.

4.1 Proposal Details

Our proposal is shown in Fig. 2. The main parts are explained as follows.

1) The trigger employs four flow data features: A1 (work of flow data based on a Riemannian manifold), A2 (unique source IP addresses per unit time), A3 (ratio of unique source IP addresses to unique destination IP addresses per unit time), and A4 (ratio of unique source IP addresses to unique protocol numbers per unit time). Aggregated over time units (T seconds), the system computes mean and standard deviation within a window size K from the initial time. Initial thresholds are generated based on calculations, with the sliding window's right end serving as the detection point.

2) The system conducts detection at each unit time, sequentially evaluating the four features. After completing feature determination, the system shifts the window's starting position one unit time to the right. For the new window, it computes mean and standard deviation of the features as thresholds for the next detection round.
3) If all four features indicate a potential attack in a given unit time, a DDoS attack possibility is identified. This triggers a second-stage detection for further investigation. If the second-stage detection confirms an attack, the current sensitivity settings remain suitable, and no adjustment to the coefficient β is necessary. However, if the second-stage detection deems it "normal" (not an attack), it implies that the current sensitivity of the thresholds is too high, necessitating an increase in the β value. In addition, to prevent repetitive alarms, when the trigger is activated continuously twice, the sensitivity of the trigger will be reduced.
4) To decrease false alerts, if the trigger remains consistently "normal" for 180 consecutive rounds, the second-stage detection is activated. If the second-stage detection confirms "normal," the current threshold sensitivity is adequate, and no β adjustment is needed. In case of an "attack" result, indicating low sensitivity, β is decreased. The β range is 0 to 2, and any adjusted value falling below or exceeding this range is constrained to the minimum or maximum value.
5) For the second-stage detection model, several ML modes will be investigated in the experiment. A feature selection algorithm is used to decide the proper features.

Our trigger mechanism is illustrated in Algorithm 1.

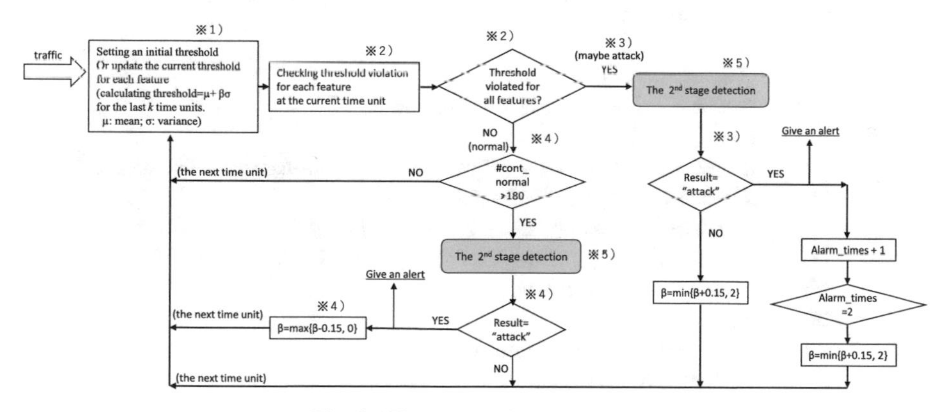

Fig. 2. Flowchart of our proposal.

4.2 Discussion

The advantages of the proposed approach are as follows:

1) The initial threshold settings are based on historical data to eliminate human-induced uncertainties.
2) The introduction of features based on the concept of Riemannian manifold in the trigger increases the number of features (four features in total), leading to a higher frequency of threshold updates (every second).

3) The system incorporates a negative feedback control mechanism, which helps maintain the sensitivity of the thresholds at an appropriate level.

Challenges in the proposed approach include the system invoking second-stage detection after 180 consecutive "normal" detections. If this count is too small, it may lead to increased calls to the two-stage detection model, causing a continuous decrease in the β value during the "attack" stage. This can make the trigger overly sensitive, resulting in frequent false positives even during the "normal" stage. Conversely, setting this count too high increases the risk of missing attack reports. Further discussion is needed to refine the negative feedback mechanism to address these issues.

Algorithm 1 Proposed trigger algorithm.

```
1: Set
        T←Sampling interval; K←Window Size
        Th←Threshold;        β←Thresholding factor
        BAR←Beta_adjustment_range;  μ  ←mean;  σ  ←Variance;
        DCT←Detection_counter_time;
2:      Initialize β = 1, K = 40, BAR = 0.15, DCT = 180
3:      Within the window size, calculate the μ and σ of four features
        A1, A2, A3 and A4
4:      Calculate the thresholds for four features using Th_i = μ_i + βσ_i
5:      Extract the feature value of the last element within
        the window.←test_i
6: for (i = 0; i < 4; i ++)
7:      if test_i > Th_i:      A_i = 1
8:      else:      A_i = 0
9: if A1 through A4 are all equal to 1 :      is_attack = 1
10:     else:      DCT + 1
11:     if is_attack == 1 or DCT == 180:  invoke Second-stage detection module
12:     if is_attack == 1 and DCT != 180
13:        if detection_result = 'attack':  Alarm_times + 1
14:              if Alarm_times == 2:  β = β + BAR; Alarm_times = 0
15:        else:   β = β + BAR
16:     else
17:        if detection_result = 'normal':  pass
18:        else:   β = β - BAR
19:     DCT == 0
20: Slide the sliding window forward by one position until the end of the dataset
```

5 Experiment

The public CICDDoS2019 dataset [18] is used, featuring both "normal" traffic and 13 distinct DDoS attack types. The dataset undergoes conditional filtering, excluding rows with all-zero or NaN values. Further filtering selects rows labeled 'BENIGN' with flow durations between 100 and 600 and 'DrDoS_DNS' with flow durations between 1 and 10. Subsequently, 1806 data points are randomly extracted, including six BENIGNs and

1800 DrDoS_DNS points. To maintain balance, BENIGN data is evenly inserted among DrDoS_DNS data. The total duration of this experiment is 3644 s. There are six time slots, and they are alternately arranged.

5.1 Feature Selection Results

The ECOFS feature algorithm was employed for the second-stage detection and the result is illustrated in the following Fig. 3. Three features were selected from the original dataset: 'Inbound', 'Min Packet Length', 'Flag Count'.

```
Time spent on feature selection:  0:01:27.913709
Selected features:
Index(['Inbound', 'Min Packet Length', 'SYN Flag Count'], dtype='object')
```

Fig. 3. The results selected by the ECOFS algorithm.

5.2 Parameter Configuration

Before delving into experimental results, let's outline the experiment parameters. The sliding window is set to 40 s; β is initially set to 1, being adjusted in the range of 0 and 2 with an adjustment step of 0.15. Based on the fact that the average attack duration is about 6 min, K is set to 180. In fact, automatically decision of the K value also seems to be possible, which is one of our future works.

5.3 Experiment Result

5.3.1 Investigation of the Effect of the Feature Selection for the Second-Stage Detection

NN (Neural Network), RF (Random Forest) [19], and SVM (Support Vector Machine) models are investigated separately, with and without feature selection. Note that, considering the character of the RF, no further feature selection algorithm for it. The results of the experiment are shown in Table 1. Note that, for NN and RF models, the training and testing datasets consists of 4,907,479 records for each label. For SVM model the dataset has 200,000 records for each label.

Table 1 indicates a significant impact of feature selection on model training efficiency. In addition, feature selection plays a significant role in preventing overfitting. Figures 4 illustrate the loss curves of the NN model without or with feature selection.

The left figure in Fig. 4 is the case without feature selection, the loss curve initially decreases but then gradually increases, indicating potential overfitting in the NN model after a certain number of iterations. In contrast, the right figure, utilizing feature selection, demonstrates a steady decrease in the loss curve without rebounds. The final loss curve value tends to approach 0.005, which suggesting that, with the help of feature selection, the NN model achieves higher accuracy on the training data.

Table 1. Model performance indicators

Module Name	Accuracy	Precision	Recall	F1-score	Training time (H:M:S)
NN (no FS)	0.863	0.785	1.0	0.8798	0:12:26
NN (FS)	0.998	0.997	0.999	0.998	0:05:21
SVM (no FS)	0.979	0.985	0.973	0.979	2:44:44
SVM_(FS)	0.987	0.996	0.978	0.987	0:12:04
RF_FS	1.0	1.0	1.0	1.0	0:31:10

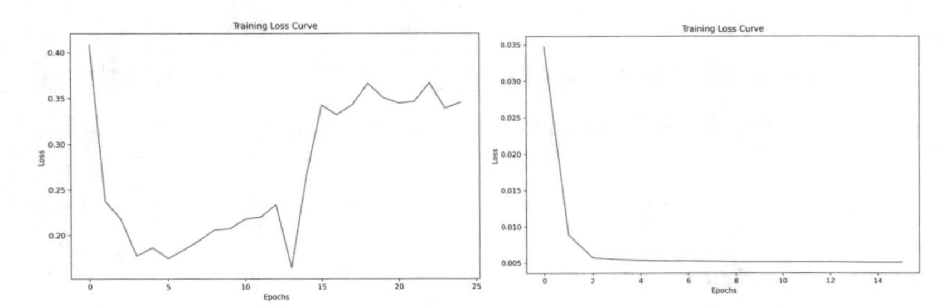

Fig. 4. The training loss curve of the NN module without feature selection and the training loss curve of the NN module with feature selection.

Table 2. Overall time consumption for each model

Module name	Overall time consumption of the system (μs)
NN_FS	805
SVM_FS	802
RF	859

We also compared the time consumption for the detection among NN, SVM and RF, with feature selection. The result is shown in Table 2.

From Tables 1 and 2, as well as Fig. 4, it can be observed that the feature selection algorithm employed in our study, by selecting the most correlated and informative features with the target classes, effectively reduced the dimensionality of the dataset. This process led to a decrease in model complexity and computational costs, mitigating the risk of overfitting and enhancing predictive performance.

While the Random Forest (RF) model outperformed the feature-selected Neural Network (NN) and Support Vector Machine (SVM) models in performance metrics such as accuracy, precision, recall, and F1 score, it incurred higher time consumption. This difference becomes more pronounced in devices with limited computational resources.

The NN and SVM models, with the support of the feature selection algorithm, experienced a substantial reduction in computational costs, coupled with an improvement in performance metrics.

After comprehensive evaluation, we ultimately selected the Neural Network (NN) model with superior performance metrics as the secondary detection model for this experiment.

5.3.2 The Investigation of Computing Resource Consumption

Table 3. Comparison of the number of second-stage detection calls and the number of threshold updates.

Approaches	The number of calls for the second-stage detection
The latest study [17]	356
This study	34

From Table 3, it can be observed that, this study reduced the number of second-stage detection calls to less than one-tenth of the latest study [17]. Note that, in this study, the number of adjustments of β (including the cases that the adjustment is zero) is the same as the number of second-stage detection calls.

In Table 4, due to the increased number of features used in the trigger (four features in our trigger and two features in the latest approach) and the more often threshold updates in our proposal, our trigger needs more average time consumption for one single trigger. However, as our study significantly reduces the invocation frequency of the second-stage detection, the average time consumption of the entire system for per detection is reduced to about one third, which is the most important for applications.

Table 4. Comparison between the latest approach and our proposal on the average time consumption for per single triggering in the trigger and the total time consumption of the entire system for per single detection.

Approaches	Average time consumption for per trigger (μs)	Average time consumption of the entire system for per detection (μs)
The latest study [17]	483	2496
This research	645	805

5.3.3 The Investigation of Detection Performance of the Entire System

The investigation results are presented in Tables 5 and 6. Note that, the effect of the "work" as a trigger feature on the detection performance is also investigated. That is, the performance comparison is conducted among the latest study, our study using or not using the trigger feature “work”.

Tables 5 and 6 demonstrate the fact that our proposal requires significantly less computational resource consumption (Table 3) without any sacrifice in detection performance. In fact, our proposal achieves almost the same or slightly better performance compared to the latest study. And it is evident that the trigger feature “work of flow data” contributes significantly to improving the trigger’s performance and finally improving the detection performance of the entire system.

Table 5. Confusion matrices of the detection result for the three approaches.

Approaches	TP	FP	FN	TN
The latest study [17]	2195	36	24	1378
Our proposal with the usage of “work”	2151	33	68	1411
Our proposal without usage of “work”	1965	2	254	1412

Table 6. Performance Metrics for the Three Approaches.

Approaches	Accuracy	Precision	Recall	Specificity
The latest study [17]	0.983	0.984	0.989	0.975
Our proposal with the usage of “work”	0.980	0.998	0.969	0.998
Our proposal without usage of “work”	0.929	0.999	0.885	0.998

6 Conclusion and Future Work

Two-stage detection approach with a triggering mechanism has been proposed to reduce the number of unnecessary invocations of the complicated detection algorithm. This paper introduced the related work and pointed out the problems in the state-of-the-art approach and presented our proposal. After explaining our proposal in detail, its behaviors were examined using a public dataset. The experiment results indicates that our proposal in this study needs far less invocations of the complicated detection models while roughly keeping (even slightly improving) the detection performance. Our main contributions include: introducing a metric of Riemannian manifold as a trigger feature, improving the threshold update mechanism in the trigger, based on the feedback of the recent detection result of the second-stage and the recent trigger history, and using a feature selection algorithm for the second-stage detection. In the future, we will try to automatically determine some parameters and to examine the behaviors of our proposal using more datasets.

Acknowledgement. The first author is supported by the WISE program (MEXT) at Kyushu University, and the second and last authors receive support from the Japan Society for the Promotion of Science (JSPS) and the Department of Science and Technology (DST), Government of India, under the Japan-India Cooperative Science Program (2022–2024).

References

1. Kaspersky: Cybercriminals attack users with 400,000 new malicious files daily - that is 5% more than in 2021 (2022). www.kaspersky.com/about/press-releases/2022_cybercriminals-attack-users-with-400000-new-malicious-files-daily---that-is-5-more-than-in-2021. Accessed 26 Jan 2023
2. The Hacker News: Microsoft mitigated record-breaking 347 billion malicious requests in 2021, January 2022. https://thehackernews.com/2022/01/microsoft-mitigated-record-breaking-347.html. Accessed 28 Jan 2023
3. Hao, Z., Feng, Y., Koide, H., Sakurai, K.: A sequential detection method for intrusion detection system based on artificial neural networks. Int. J. Network Comput. **10**, 213–226 (2020)
4. Shams, E.A., Rizaner, A.: A novel support vector machine based intrusion detection system for mobile ad hoc networks. Wireless Netw. **24**(5), 1821–1829 (2018)
5. Khraisat, A., Gondal, I., Vamplew, P., Kamruzzaman, J.: Survey of intrusion detection systems: techniques, datasets and challenges. Cybersecurity **2**(1), 20 (2019)
6. Wang, T., Feng, Y., Sakurai, K.: Improving the two-stage detection of cyberattacks in SDN environment using dynamic thresholding. In: 15th International Conference on Ubiquitous Information Management and Communication (IMCOM), pp. 1–7 (2021)
7. Guo, D., Wang, Y., Luo, X.: A SDN-based multiple mechanism DDoS attack detection trigger algorithm. In: International Conference on Urban Engineering and Management Science (ICUEMS), pp. 729–735 (2020)
8. Ashraf, J., Latif, S.: Handling intrusion and DDoS attacks in software defined networks using machine learning techniques. In: National Software Engineering Conference, pp. 55–60 (2014)
9. Liu, Z., Hu, C., Shan, C.: Riemannian manifold on stream data: Fourier transform and entropy-based DDoS attacks detection method. Comput. Secur. **109**, 102392 (2021)
10. Wang, W., Du, X., Wang, N.: Building a cloud IDS using an efficient feature selection method and SVM. IEEE Access **7**, 1345–1354 (2018)
11. Lyu, Y., Feng, Y., Sakurai, K.: A survey on feature selection techniques based on filtering methods for cyber attack detection. Information **14**(3), 191 (2023)
12. Guo, P., Li, N.: Self-adaptive threshold based on differential evolution for image segmentation. In: 2nd International Conference on Information Science and Control Engineering, pp. 466–470 (2015)
13. Thottan, M., Ji, C.: Statistical detection of enterprise network problems. J. Netw. Syst. Manage. **7**(1), 27–45 (1999)
14. Giotis, K., Argyropoulos, C., Androulidakis, G., Kalogeras, D., Maglaris, V.: Combining OpenFlow and sFlow for an effective and scalable anomaly detection and mitigation mechanism on SDN environments. Comput. Netw. **62**, 122–136 (2014)
15. You, X., Feng, Y., Sakurai, K.: Packet in message based DDoS attack detection in SDN network using OpenFlow. In: 2017 Fifth International Symposium on Computing and Networking (CANDAR), pp. 522–528 (2017)
16. David, J., Thomas, C.: Efficient DDoS flood attack detection using dynamic thresholding on flow-based network traffic. Comput. Secur. **82**, 284–295 (2019)
17. Niu, M., Feng, Y., Sakurai, K.: A two-stage detection system of DDoS attacks in SDN using a trigger with multiple features and self-adaptive thresholds. In: 2023 17th International Conference on Ubiquitous Information Management and Communication (IMCOM), pp. 1–8 (2023)

18. Doriguzzi-Corin, R., Siracusa, D.: FLAD: adaptive federated learning for DDoS attack detection (2022). arXiv preprint arXiv:2205.06661
19. Gaur, V., Kumar, R.: Analysis of machine learning classifiers for early detection of DDoS attacks on IoT devices. Arab. J. Sci. Eng. **47**(2), 1353–1374 (2022)

Discovering Personally Identifiable Information in Textual Data - A Case Study with Automated Concatenation of Embeddings

Md Hasan Shahriar[1], Abrar Hasin Kamal[1], and Anne V. D. M. Kayem[2,3](✉)

[1] Institute of Computer Science, University of Potsdam, Potsdam, Germany
{md.hasan.shahriar,abrar.hasin.kamal}@uni-potsdam.de
[2] Hasso-Plattner-Institute for Digital Engineering, University of Potsdam, Potsdam, Germany
[3] Department of Computer Science, University of Exeter, Exeter, UK
a.v.kayem@exeter.ac.uk

Abstract. Discovering personal identifying information (PII) in unstructured data is an important pre-processing step in enabling privacy preserving machine learning as well as compliance with data protection legislation such as GDPR (General Data Protection Regulation). However, PII discovery in unstructured data, and textual data in particular, is a challenging problem. One of the primary causes is the fact that representations of PII in textual data, do not follow standard grammatical representations. For example, in social media scenarios, textual data can include irregular expressions such as slang and emoticons that must be interpreted contextually to determine whether or not they qualify as PII. The problem is further compounded by the fact that PII discovery algorithms, for the most part, lean on machine learning models that are processing intensive in nature. In scenarios where real-time responses are required to determine whether or not PII exists with in a dataset, this impacts response time negatively. This paper reports on results from a series of experiments in which the Automated Concatenation of Embeddings (ACE) framework was employed to support PII discovery in textual data. Our results show that such architectures that are processing (GPU) intensive are not suitable for handling PII detection in large unstructured or semi-structured data.

Keywords: Personal Identifying Information Discovery · Textual Data · Privacy

1 Introduction

Unstructured data presents an invaluable information resource to service providers, providing insights that drive successful service delivery [35]. In fact,

L. Barolli (Ed.): AINA 2024, LNDECT 202, pp. 145–158, 2024.
https://doi.org/10.1007/978-3-031-57916-5_13

statistics indicate that about 80–90% of the textual data available on the Internet is represented in an unstructured format [22].

Until very recently, the standard approach to processing unstructured and semi-structured textual data was to employ several hours of employee time to categorise data. Most existing work in the field of textual data processing includes techniques ranging from linguistic pattern analysis [2], to ones based on deep learning models [1,24,37], and more recently large language models [5]. While methods of processing textual data now abound in the literature, we are also faced with the fact that unstructured data is growing almost exponentially in the light of the increased digital footprint. As such, and in the light of emerging privacy legislation governing digital representations of data, processing textual data to identify and remove elements that disclose personal information is an important problem. Having more automated methods for processing unstructured data in a time efficient manner, has brought to the fore several text processing techniques ranging from ones based on analysing linguistic patterns to ones based on deep learning methods [1,37], and more recently large language models [5].

In the light of emerging privacy and data protection regulations such as the GDPR: General Data Protection Regulation [7], discovering personal information in unstructured data and, textual data in particular, is an important pre-processing step in ensuring that organisations comply with data protection regulations. Processing such datasets is made challenging by the fact that representations of PII in unstructured data and, textual data in particular, do not follow standard linguistic representations. Textual data can often include include irregular expressions such as slang and emoticons that must be interpreted contextually to determine whether or not they qualify as PII, and grammatical structures are not always adhered to during formulations of the data.

PII discovery algorithms for unstructured data must however, be able to process large volumes of data (in the order of petrabytes) in a time efficient manner, without raising a high percentage of false positives. One approach to doing so, is to lean on machine learning models such as ones based on deep learning to enable PII discovery. Such deep learning methods are however time and processing intensive in nature, and so, in scenarios where real-time responses are required to determine whether or not PII exists with in a dataset, it makes sense to assess whether or not the performance and accuracy enhancements offered by a given model provide a viable cost-benefit tradeoff. In this paper, we consider a case-study of the Automated Concatenation of Embeddings (ACE), which offers potentially state-of-the-art performance but is impractical to implement in a processing resource constrained scenario.

We show through empirical results, the implications of applying processing intensive deep learning models to PII discovery in data and highlight the fact that frameworks that are processing (GPU) intensive are not suitable for scenarios that involve concatenations of transformer embeddings.

Our contributions can be summarised as follows: (1.) We show that by creating a PII labelled dataset and accounting for contextual information it is possible to achieve a high F1-score for PII discovery in unstructured textual data; and

(2.) Employing automated concatenation of embeddings on textual data, requires GPU intensive processing to obtain high performance results [32]. However, such models are impractical for standard textual data processing scenarios involving PII discovery because the volume of embeddings grows exponentially with the size of the dataset and the occurrence of PII.

The rest of the paper is structured as follows. In Sect. 2, we present related work in the field of personal identifying information (PII) discovery in textual data. We follow this in Sect. 3, with a discussion of existing deep learning approaches to PII discovery in textual data and present our approach based on the Automatic Concatenation of Embeddings (ACE) for Structured Prediction. In Sect. 4, we present and discuss results from experiments based on three datasets namely EnronPII, WikiPII, and Hybrid PII (HPII). We conclude with a summary of the contributions of the paper and a discussion of future work, in Sect. 5.

2 Related Work

As mentioned before, daily generated data on the internet via different mediums currently stands at a mammoth 120 zettabytes. According to the literature, 95% of the entire volume exists in the form of unstructured data. Textual data can also exist in this form via status updates, emails, blogs etc. on the internet. The core challenge in any form of unstructured data lies in its lack of organisation. This leads to the data being inconsistent, unpredictable and highly varying, which makes it hard to analyze. In our context, unstructured textual data can contain several pieces of sensitive information at any position of the text. With the volume of data we expect to see, and the lack of pointers to indicate the existence of PII, the task of discovering PII in textual data becomes a pressing task that needs to be addressed.

The understanding of the properties of textual data before devising a solution to capture PII, is also an important step that has to be prioritised. In this paper, we narrow down the scope of this step to understand how PII can exist in textual data. Any piece of information that can be used to identify an individual tends to be unique in nature. For instance, the `name/email/address` of an individual is highly likely to be different from another individual. Furthermore, the reference to an individual in a body of text may only occur a few times compared to the other more common words in the text. This helps us infer that PII in unstructured textual data exists as sparse representations, which aligns with past studies in this domain [13,17].

Personally Identifying Information (PII) can be described as "any information about an individual maintained by an agency, including (1) any information that can be used to distinguish or trace an individual's identity, such as name, social security number, date and place of birth, mother's maiden name, or biometric records; and (2) any other information that is linked or linkable to an individual, such as medical, educational, financial, and employment information" [19]. Explicit PIIs are those that are directly linked to an individual, such as their name, address, phone number, and email, and can be easily used to identify and track an individual [9]. Quasi-identifiers, on the other hand, are attributes that individually, may not directly identify an individual, but when combined, provide information to enable one-to-one re-identifications. For instance, the age of an individual might be perfectly safe to share in isolation, but when combined with other information such as gender, postcode, and race; might increase the potential for "corectly" linking to an individual. Similarly, depending on the granularity of sensitive attributes such as medical information or financial data, these can also can also be used to identify an individual based on contextual information.

While quasi-identifiers and sensitive attributes are important for privacy protection, detecting explicit PIIs is a crucial first step because they are the most vulnerable and easily identifiable pieces of information. By detecting and protecting explicit PIIs, we can minimize the risk of identity theft and other malicious activities. Additionally, once explicit PIIs are detected and protected, further analysis can be conducted to detect quasi-identifiers and sensitive attributes.

Methods of discovering PII in structured data, were pioneered by traditional approaches such as regular expressions and rule-based approaches that rely on predefined patterns and rules. However, regular expressions and rule-based approaches are not well suited to detecting new or unknown PIIs or variations in the format and structure of PII in very large datasets [6]. This can lead to a high number of false positives or missed detection, reducing overall accuracy. Additionally, these approaches require significant manual effort to develop and maintain rules and may struggle to handle the complexity and variability of natural language. Recent studies have explored the use of machine learning [34] and deep learning methods [8] for PII detection and have shown promise in overcoming these limitations. Deep learning models can learn to automatically identify and classify PII in unstructured text without the need for specific rules. Further research in the field of deep learning for PII detection in unstructured textual data can lead to the development of more accurate and efficient methods, thereby contributing to the improvement of data privacy and security.

Transformer embeddings, that are derived from pre-trained models [20] such as **BERT** (Bidirectional Encoder Representations from Transformers) [4], **RoBERTa** (Robustly optimised BERT approach) [16], and XLNet [36] can encode rich contextual information for each token in a sequence. This facilitates a deeper understanding of the semantics and context in the data, when compared to standard embeddings which have been shown to be very effective in a variety of natural language processing tasks. For instance, machine-driven translation [14], text summarisation [15], text classification [36], and language understanding [16] and **NER** [33].

3 Deep PII Discovery - Contextual Relationships

The understanding of the properties of textual data before devising a solution to capture PII, is also an important step that has to be prioritised. In this paper, we narrow down the scope of this step to understand how PII can exist in textual data. Any piece of information that can be used to identify an individual tends to be unique in nature. For instance, the name/email/address of an individual is highly likely to be different from another individual. Furthermore, the reference to an individual in a body of text may only occur a few times compared to the other more common words in the text. This helps us infer that PII in unstructured textual data exists as sparse representations, which aligns with past studies in this domain [13,17].

3.1 Tokenization

Tokenization is the process of breaking down a sequence of text into meaningful units (known as tokens), such as words, terms, symbols, or characters. Tokens can be as small as an individual character or as long as a whole word, depending on the chosen tokenization method. Tokenizers often rely on heuristics, considering factors such as punctuation, white spaces, and character sequences to split the text corpus into tokens [21].

For instance, in the sentence "John Doe went to Berlin to present his paper," the tokens could be the individual words: "John," "Doe," "went," "to," "Berlin," "to," "present," "his," and "paper." Alternatively, if a sub-word tokenization method is used, the word "present" might be split into subword tokens like "pre" and "sent".

3.2 Tagging Schemes and Named Entities

We employed a tagging scheme and selected named entities in the annotation process. The datasets were labeled in **BILOU** tagging scheme as it outperforms the more widely used **IOB** [23]. The **BILOU** tagging marks with B-LABEL for the first token, I-LABEL for any inner tokens (if present), L-LABEL for the last token, and U-LABEL for the unit token where 'LABEL' is the tag. Any non-entity is tagged with "O".

We decided to annotate four explicit **PII**s in the datasets to support a comparative analysis (see Sect. 4) and also to align with previous work [11,25]. PERSON, EMAIL, LOCATION, and PHONE are pervasive across various domains, making them relevant for a wide range of applications and scenarios.

- PERSON: Personal names are fundamental identifiers, and their detection is paramount for privacy preservation.
- EMAIL: Email addresses are widely used for communication and are sensitive pieces of information often targeted in privacy breaches.
- LOCATION: Identifying locations is vital, especially in contexts where the disclosure of someone's location can pose security risks or privacy concerns.
- PHONE: Phone numbers, being unique to individuals, are critical **PII** elements. Their detection is essential for safeguarding personal information.

Furthermore, in the **BILOU** scheme, tokens can take 21 states (O, B-PERSON, I-PERSON, U-PERSON, etc.) [23], allowing us to measure precision, recall, and F1-score effectively for each **PII** type. In our case, there are 14 tags in total, as all non-**PII**s are tagged with "O" and EMAIL has only single token entities. Focusing on this set of **PII** tags helps manage the complexity of the model, making it more interpretable and efficient.

3.3 Hybrid Annotation Process

Dataset Annotation is a critical phase in our **PII** detection research. We employed both automated and manual annotation processes. We used SpaCy and Stanza, two of the best state-of-the-art pre-trained **NER** models [29] in this annotation. To annotate the data we followed the following step-wise process:

1. WikiPII contains BIRTHPLACE, BIRTHDATE, PARENTS, SPOUSES, CHILDREN, and EDUCATION tags. After manual lookup, we realized there were many false negatives, such as names were not annotated properly in this dataset, similar to the baseline research. As a result, we removed all the **NER** tags.
2. All documents in both datasets are tokenized by SpaCy's **RoBERTa**-base pre-trained transformer model [12].
3. As shown in Table 1, we used **Regex** to identify and extract entities representing phone numbers and email addresses. This required formulating a series of regex rules to distinguish a phone number from an arbitrary string of numbers e.g. a bank account number or tracking ID.
4. WikiPII does not contain phone numbers and email addresses. To make both datasets consistent, we have randomly generated phone numbers and email addresses for each document: "Call" + `[NAME]` + "on" + `[PHONE]` + " . " + "Send email to " + `[EMAIL]` + " ."
5. In the next step, we employed the **Bi-LSTM** based pre-trained model Stanza to label NAME and LOCATION in the texts.
6. There were still some entities not captured by Stanza. We use SpaCy to supplement the annotation process [12]. SpaCy labels each entity as **IOB** tags. These tags label each token in the text, indicating whether it is the beginning of an entity (B), inside an entity (I), or outside any entity (O). However, it also produces other tags, like DATE, LANGUAGE, and LAW. We removed all these tags except those representing names and locations and replaced them with the 'O' (outside) tag. We have also included name prefixes (Dr., Mr., Ms.) into the PERSON tag.
7. The **IOB** tags are subsequently converted to the **BILOU** tagging scheme. This conversion optimizes the representation of entity spans in the annotated text, enhancing the effectiveness of subsequent **NER** models [23].
8. The labels were then manually checked and re-annotated in case of misclassification as manual labeling is considered the gold standard [10].

Here is an example of the labelled data in Fig. 1:

Table 1. Regex used for **PII** detection.

<table>
<tr><th>Entity</th><th>Regex Rules</th></tr>
<tr><td>PHONE</td><td>((?:\+\d{1,2}[-\.\s]??|\d{4}[-\.\s]??)?(?:\d{3}[-\.\s]??\d{3}[-\.\s]??\d{4}|\s*\d{3}[-\.\s]??\d{4}|\d{3}[-\.\s]??\d{4}))</td></tr>
<tr><td>EMAIL</td><td>[\w.+-]+@[\w.-]+\.[\w-]+</td></tr>
</table>

```
(1, 'Phillip', 'B-PERSON'),
(1, 'K', 'I-PERSON'),
(1, 'Allen', 'L-PERSON'),
(1, 'sent', 'U-O'),
(1, 'email', 'U-O'),
(1, 'from', 'U-O'),
(1, 'phillip.allen@enron.com', 'U-EMAIL'),
(1, '.', 'U-O')
```

Fig. 1. Example of the labeled email data.

Studies over recent years have shown the importance of understanding context in data. With advancements, such as Attention method in Natural Language Processing [30], deep learning models have achieved new feats in understanding the human language better. The context vector in these methods, help the model to understand how each word in a body of text are related to each other, and what they mean in the context of the text.

Context in the domain of PII detection is equally important, as it helps to define what can be termed as PII, and how it can be categorized. An example could be the word "Brooklyn", in the sentence "Brooklyn and his friends will be going to the LIU Brooklyn vs. Manhattan game tonight". Without context, it is impossible to understand that the first instance of the word refers to a person, while the second instance could be referring to an organization, or a location, which are less specific to an individual. Accounting for context in the data can give a sense of the degree of PII existing in a body of text and help in classifying PII containing text according to its degree of risk.

Several tagging schemes have therefore been developed to help in reducing the ambiguity introduced by such words. The simplest one is known as IO, where entity tokens are prepended with a 'I-' tag, and all non-entity tokens receive the 'O' token. Models trained on data labeled using the IO scheme are the most simple in terms of complexity, compared to the other schemes. Further studies have shown the BILOU tag to be the most effective, and is also the scheme used in this study [23]. We discussed the rules of this tagging scheme in details in Sect. 3.2.

3.4 Automated Concatenation of Embeddings

Pretrained contextualized embeddings are highly effective for structured prediction tasks. These embeddings provide powerful representations for words which can in turn be leveraged for PII discovery. However, while concatenating different types of embeddings can yield even better word representations, and consequently PII representations, the challenge lies in selecting the best combination of embeddings. This choice often depends on the specific task and the available set of candidate embeddings. Additionally, the increasing variety of embedding types increases the complexity of the selection process [32].

One approach to addressing the issue of optimising the combinations of embeddings selected to determine the existence of PII in a dataset, is the Automated Concatenation of Embeddings (ACE) framework [32].

ACE works by automating the process of finding optimal concatenations of embeddings for structured prediction tasks. It is inspired by recent advances in neural architecture search and employs a controller to samples different concatenations of embeddings. The controller's decisions are influenced by its current belief regarding the effectiveness of individual embedding types for a given task. The controller then updates its beliefs based on the received rewards.

In order to optimize the controller's parameters and compute rewards, the ACE framework applies strategies from reinforcement learning [28]. The reward is determined based on the accuracy of a task model, which takes the sampled concatenation as input and is trained on a task-specific dataset.

Wang et al. [32] presented empirical results from experiments conducted on six different tasks and a total of 21 datasets. The results demonstrate that ACE outperforms state-of-the-art baseline methods. Importantly, ACE achieves state-of-the-art performance when fine-tuned embeddings are used in all evaluated tasks.

We now show, through experimentation, that the ACE model is not well-suited to the task of PII discovery in large datasets, and discuss some of the reasons for this.

4 Experiments and Results

In this Section we present our results from applying the ACE model to address the problem of PII detection in large unstructured data, specifically textual data. We employed three datasets namely: an Enron email dataset, a WikiPII dataset, and a hybrid dataset composed of a combination of the Enron email and WikiPII dataset.

4.1 Datasets

Since labelled dataset for PII detection are scarce. We created a subset of labelled data drawn from the datasets (Enron email and WiikiPII) based on the unstructured text contained in them. We now provide a description of each of the datasets to provide some context for the experiments and results obtained.

The Enron email dataset [3] is a large collection of unstructured text data, consisting of thousands of emails exchanged between Enron employees. This dataset can be useful for PII detection using Named entity recognition (NER).

NER can identify and classify different types of entities, in this case PIIs. We labelled this dataset in BILOU notation as it outperforms the more widely used BIO [23]. The BILOU notation marks with B-XXXX for the first token, I-XXXX for any inner tokens (if present) and L-XXXX for the last tokens.

We also used WikiPII dataset. It contains following tags: BIRTH PLACE, BIRTH DATE, PARENTS, SPOUSES, CHILDREN, EDUCATION. However, we made few adjustments to make it similar to previous dataset, so that we can compare both. We updated BIRTH PLACE to LOCATION; PARENTS, SPOUSES, CHILDREN to PERSON NAME; removed BIRTH DATE, EDUCATION tags; included synthetic PHONE NUMBER and EMAIL (as Wikipedia does not contain Phone, Email addresses); and Converted tags to BILUO notation.

We created a third dataset named: IIPII combining the datasets described in EnronPII and WikiPII.

4.2 Results and Analysis

Baseline and Goal. In the ACE paper [32], Wang et al. used a dataset of 4 languages from the CoNLL 2002 and 2003 shared task [26,27] with standard split. They sampled 2000 training sentences on the CoNLL English Named Entity Recognition (NER) dataset and obtained 0.941 F1-score in XLM-R+Fine-tune and 0.946 F1-score when using ACE+Fine-tune as such achieving current state-of-the-art results in Named entity recognition.

Our goal in this experiment was to obtain a similar result using our selected datasets described in Sect. 4.1 focused on PII discovery in data to support tasks such as data security/protection.

Fine-Tuning. To fine-tune the embeddings, we employed a batch size of 128 tokens and AdamW [18] to optimize the model with a learning rate of 1e-4. We set the maximum training epoch to 3. We used 70% of each dataset for training and 30% for evaluation. We fine-tuned BERT, RoBERTa and XLNet with each datasets in Sect. 4.1. The results are presented in Table 2; we achieved 0.94 F1-Score with XLNet on the larest (HPII) dataset.

ACE. As a next step, we proceeded to setup the ACE library [31] provided by Wang et al. [32]. In the beginning, installing libraries from the **requirements.txt** caused `"Subprocess-exited-with-error"` during PyYAML (a YAML parser and emitter for Python) installation.

Subsequent errors generated were `"Spacy version incompatible" (v3.3.3)` with other libraries. We upgraded **Pip**, the package installer for Python and installed up-to-date versions of the libraries. However, this does not resolve all the issues. Notably, several dependency conflicts arose with: nltk (v7.1.2), regex (v2019.12.20), requests (v2.22.0), scipy (v1.4.1), sklearn

Table 2. Comparison of NER with BERT, RoBERTa, XLNet models and EnronPII, WikiPII, HPII datasets. Best result 0.94 F1-Score in XLNet+Fine-tune on HPII dataset, which is comparable to the ACE+Fine-tune in CoNLL.

Model	Dataset	Precision	Recall	F1-Score
BERT	EnronPII	0.91	0.93	0.92
RoBERTa	EnronPII	0.92	0.94	0.93
XLNet	EnronPII	0.93	0.95	0.94
BERT	WikiPII	0.85	0.90	0.87
RoBERTa	WikiPII	0.84	0.87	0.85
XLNet	WikiPII	0.86	0.89	0.88
BERT	HPII	0.92	0.94	0.93
RoBERTa	HPII	0.92	0.94	0.93
XLNet	HPII	0.94	0.94	**0.94**

(v0.0), wikipedia2vec (v1.0.5). Errors also arose on the packages: segtok, tabulate, torch, tqdm, transformers. Therefore, iwe concluded that the ACE library might be broken and unstable.

Next, we got errors on "Wikipedia2vec" which does not build. We manually downloaded it's repo from Github and built it with Pip. There were more missinng libraries: bpemb, deprecated, pytorch_transformers and flair.

Unable to use the ACE library, we attempted to use the **train.py** library directly, to determine if the issue was mainly due to the supporting libraries and incompatibilities with our experimental setup. This however, led to further issues (errors) such as: **ModuleNotFoundError: No module named 'flair.list_data'**. In the end, we concluded that the open issue on github (Issue oped by user CodeAKrome on github on June 28, 2023): "`Install fails all python versions tested 3.6, 3.7, 3.8, 3.9, 3.10, 3.11. Repo Unusable #57`" was valid also for our datasets.

Flair Embedding. Upon inspection, we found that ACE uses the Flair library. We decided to replicate recreate the concatenation of transformer embeddings and achieve better results and address the issues arising. We used Colab Pro with 15GB GPU RAM during training with minimal setting, i.e. 2 embeddings. The results are displayed in Table 3.

These experiments show that, concatenation of transformer embeddings are processing intensive and are not well suited to handling datasets that raise a large number of embeddings. Furthermore, while concatenation of embeddings might improve the prediction performance by a small margin, their dependence on reinforcement learning strategies and complex architectures due to intricate hyperparameter configurations, requires significant GPU memory usage [32].

Table 3. The concatenation of two transformer embeddings results in out of memory error immediately on training. To check if the process is correct, we tested generic embeddings, GloVe + Flair (news-forward) with the full sized dataset. However, the same error persists after two epochs. Next, to determine if the issues were due to the size of the dataset, we selected a very small dataset and observed that the whole process completes without any errors.

Embeddings	Dataset (Size)	Training Status
BERT + XLNet	HPII (Full Dataset)	CUDA out of memory immediately on training
BERT + RoBERTa	WikiPII (Full Dataset)	CUDA out of memory in the first epoch during training
GloVe + Flair (news-forward)	HPII (Full Dataset)	CUDA out of memory in third epoch during training
GloVe + Flair (news-forward)	HPII (1% Dataset)	Training completes

5 Conclusion

In this paper, we discussed the issue of discovering personal identifying information (PII) in unstructured data and more specifically textual data. We highlighted the fact that deep learning approaches offer a viable solution to PII discovery in textual data, but that the accuracy of these approaches is hindered by irregular representations. We showed that while complex framework like ACE can under high performance computing conditions, offer better performance than simpler models, they are in fact impractical to apply to large textual datasets, especially in real-time settings where processing resource constraints persist.

As future work, we are currently working on performance enhancements to state-of-the-art encoder models like BERT and XLNet, and comparing these to large language models.

Acknowledgements. The authors grateful acknowledge the support and funding of Cimpress GmbH, in particular Srinivas Kannepalli, Jannik Podlesny, as well as David Reich from University of Potsdam for their feedback. Many thanks also, to the reviewers for their comments and insights.

References

1. Ahmad, F., et al.: A deep learning architecture for psychometric natural language processing. In: ACM Trans. Inf. Syst. 38(1) ISSn: 1046-8188 (2020). https://doi.org/10.1145/3365211 URL: https://doi.org/10.1145/3365211
2. Allen, J.F.: Natural language processing. In: Encyclopedia of Computer Science. GBR: John Wiley and Sons Ltd., pp. 1218–1222. isbn: 0470864125 (2003)
3. Cohen, W.,: Enron email dataset. carnegie mellon university, 2015. url: https://www.cs.cmu.edu/~enron. (Accessed: 8 Mar 2023)
4. Devlin, J., et al.: BERT: Pre-training of Deep Bidirectional Transformers for Language Understanding. In: CoRR abs/1810.04805 (2018). arXiv: 1810.04805. URL: http://arxiv.org/abs/1810.04805

5. Dong, J.: Natural language processing pretraining language model for computer intelligent recognition technology". In: ACM Trans. Asian Low-Resour. Lang. Inf. Process. (2023). issn: 2375-4699. https://doi.org/10.1145/3605210. URL: https://doi.org/10.1145/3605210
6. Garfinkel, S.: De-identification of personal information. (2015), p. 30. https://doi.org/10.6028/NIST.IR.8053
7. DSGVO Germany. General data protection regulation (GDPR) (2018). URL: https://gdpr-info.eu
8. Gillette, J.B., et al.: Data protections for minors with named entity recognition. In: 2022 IEEE International Conference on Big Data (Big Data). (2022), pp. 3315-3323. https://doi.org/10.1109/BigData55660.2022.10021086
9. Hamza, R., Zettsu, K.: Investigation on privacy-preserving techniques for personal data. In: Proceedings of the 2021 Workshop on Intelligent Cross-Data Analysis and Retrieval. ICDAR '21. Taipei, Taiwan: Association for Computing Machinery (2021), pp. 62–66. isbn: 9781450385299. https://doi.org/10.1145/3463944.3469267
10. Hassan, S.U., Ahamed, J., Ahmad, K.: Analytics of machine learning-based algorithms for text classification. In: Sustainable Operations and Computers **3** (2022), pp. 238–248. issn: 2666-4127. https://doi.org/10.1016/j.susoc.2022.03.001. URL: https://www.sciencedirect.com/science/article/pii/S2666412722000101
11. Hathurusinghe, R.: Building a personally identifiable information recognizer in a privacy preserved manner using automated annotation and federated learning (2020). https://doi.org/10.20381/ruor-25235. URL: http://hdl.handle.net/10393/41011
12. Honnibal, M., et al.: SpaCy: industrial-strength natural language processing in python. In: (2020). https://doi.org/10.5281/zenodo.1212303. Accessed 08 Mar 2023
13. Kulkarni, P., Cauvery, N.K.: Personally identifiable information (PII) detection in the unstructured large text corpus using natural language processing and unsupervised learning technique. In: Int. J. Adv. Comput. Sci. App. **12**(9) (2021). https://doi.org/10.14569/IJACSA.2021.0120957. URL: http://dx.doi.org/10.14569/IJACSA.2021.0120957
14. Lample, G., Conneau, A.: Cross-lingual language model pretraining (2019). arXiv: 1901.07291 [cs.CL]
15. Lewis, M., et al.: BART: denoising sequence-to-sequence pre-training for natural language generation, translation, and comprehension (2019). arXiv: 1910.13461 [cs.CL]
16. Liu, Y., et al.: RoBERTa: a robustly optimized BERT pretraining approach (2019). https://doi.org/10.48550/ARXIV.1907.11692. URL: https://arxiv.org/abs/1907.11692
17. Liu. Y., et al.: Automated PII extraction from social media for raising privacy awareness: a deep transfer learning approach. In: 2021 IEEE International Conference on Intelligence and Security Informatics (ISI) (2021), pp. 1–6. https://doi.org/10.1109/ISI53945.2021.9624678
18. Loshchilov, I., Hutter, F.: Decoupled weight decay regularization (2019). arXiv: 1711.05101 [cs.LG]
19. McCallister, E., Grance, T., Scarfone, K.: Guide to protecting the confidentiality of personally identifiable information (PII), pp. 2–1 (2010). https://doi.org/10.6028/NIST.SP.800-122
20. McCann, B., et al.: Learned in translation: contextualized word vectors. In: Proceedings of the 31st International Conference on Neural Information Processing Systems. NIPS'17. Long Beach, California, USA: Curran Associates Inc., pp. 6297–6308 (2017). isbn: 9781510860964

21. Mohan, V.: Text mining: open source tokenization tools: an analysis. In: **3**, pp. 37–47 (2016)
22. MongoDB.: Unstructured Data. Dec (2023). url: https://www.mongodb.com/unstructured-data
23. Ratinov, L., Roth, D.: Design challenges and misconceptions in named entity recognition. In: Proceedings of the Thirteenth Conference on Computational Natural Language Learning. CoNLL '09. Boulder, Colorado: Association for Computational Linguistics, pp. 147–155. (2009) isbn: 9781932432299. https://doi.org/10.3115/1596374.1596399
24. Sarikaya, R., Hinton, G.E., Deoras, A.: Application of deep belief networks for natural language understanding. In: IEEE/ACM Trans. Audio, Speech and Lang. Proc. **22**(4), pp. 778–784 (2014) issn: 2329-9290. https://doi.org/10.1109/TASLP.2014.2303296. URL: https://doi.org/10.1109/TASLP.2014.2303296
25. da Silva, C.J.A.P.: Detecting and protecting personally identifiable information through machine learning techniques (2020). URL: https://hdl.handle.net/10216/129033
26. Sang, E.F.T.K.: Introduction to the CoNLL-2002 shared task: language-independent named entity recognition. In: COLING-02: The 6th Conference on Natural Language Learning 2002 (CoNLL-2002) (2002). URL: https://aclanthology.org/W02-2024
27. Sang, E.F.T.K., De Meulder, F.: Introduction to the CoNLL-2003 shared task: language-independent named entity recognition. In: Proceedings of the Seventh Conference on Natural Language Learning at HLT-NAACL 2003, pp. 142–147 (2003) URL: https://aclanthology.org/W03-0419
28. Tziortziotis, N., Dimitrakakis, C., Blekas, K.: Cover tree bayesian reinforcement learning. In: J. Mach. Learn. Res. **15**(1), pp. 2313–2335 (2014) issn: 1532-4435
29. Vajjala, S., Balasubramaniam, R.: What do we really know about state of the art NER? (2022). https://doi.org/10.48550/ARXIV.2205.00034. URL: https://arxiv.org/abs/2205.00034
30. Vaswani, A., et al.: Attention is all you need. In: Adv. Neural Inf. process. syst. **30** (2017)
31. Wang, X., Jiang, Y.: Automated concatenation of embeddings for structured prediction (2022). URL: https://github.com/Alibaba-NLP/ACE
32. Xinyu, W., et al.: Automated concatenation of embeddings for structured prediction. In: ArXiv abs/2010.05006 (2020). URL: https://api.semanticscholar.org/CorpusID:222290783
33. Xinyu, W., et al.: Automated concatenation of embeddings for structured prediction. In: Proceedings of the 59th Annual Meeting of the Association for Computational Linguistics and the 11th International Joint Conference on Natural Language Processing (Vol 1: Long Papers). Online: Association for Computational Linguistics, Sept., pp. 2643–2660 (2021) https://doi.org/10.18653/v1/2021.acl-long.206. URL: https://aclanthology.org/2021.acl-long.206
34. Wei, Y-C., Liao, T-Y., Wu, W-C.: Using machine learning to detect PII from attributes and supporting activities of information assets. In: J. Supercomputing 78 , pp. 9392–9413 (2022). https://doi.org/10.1007/s11227-021-04239-9
35. Woo, S., et al.: I've got your packages: harvesting customers' delivery order information using package tracking number enumeration attacks. In: Proceedings of The Web Conference 2020. WWW '20. Taipei, Taiwan: Association for Computing Machinery, 2020, pp. 2948-2954. isbn: 9781450370233. https://doi.org/10.1145/3366423.3380062. URL: https://doi.org/10.1145/3366423.3380062

36. Yang, Z., et al.: XLNet: generalized autoregressive pretraining for language understanding (2019). https://doi.org/10.48550/ARXIV.1906.08237. URL: https://arxiv.org/abs/1906.08237
37. Zini El, J., Awad, M.: On the explainability of natural language processing deep models. In: ACM Comput. Surv. 55.5 (Dec. 2022). issn: 0360-0300. https://doi.org/10.1145/3529755. URL: https://doi.org/10.1145/3529755

Towards User-Oriented Steganography

Urszula Ogiela[1(✉)] and Marek R. Ogiela[2]

[1] Faculty of Computer Science, AGH University of Krakow, 30 Mickiewicza Avenue, 30-059 Kraków, Poland
ogiela@agh.edu.pl

[2] Cryptography and Cognitive Informatics Laboratory, AGH University of Krakow, 30 Mickiewicza Avenue, 30-059 Kraków, Poland
mogiela@agh.edu.pl

Abstract. Steganographic protocols play an important role in computer security systems and the transmission of strategic data. The security of these methods is guaranteed by the way secret information is placed on data storage media. This work will describe new approaches in which user characteristics will also be used to place data in the container. This will allow the creation of a new class of steganographic procedures dedicated to specific people or groups of users.

Keywords: Personalized security protocols · secret hiding · user-oriented steganography

1 Introduction

One way to ensure the security of data by hiding it in digital media is to use steganography algorithms [1]. Steganography is the science of hiding information in digital media that does not arouse any suspicion and thus can be transmitted without restriction, not subject to special security controls or analysis. Digital steganography refers to the use as containers for transmitting information, digital media, which are most often images or sound files [2].

This paper will describe new developments in the field of digital staganography, which will target the creation of algorithms that use personal data or user profiles. This will allow the creation of a new class of steganographic methods, i.e. personalized steganography, which, like cognitive cryptography, will be based on characteristics or data that characterize individuals [3]. The new methods will be characterized by:

- They can use a single container for hiding information, or a collection of several independent containers (images), which together will share the hidden secret. Secret placement algorithm
- The algorithm for hiding the secret will additionally consider the user's information or preferences, contained in the so-called PSV (Personalized Security Vector) which allows only the user to reproduce the secret.

The remainder of this paper will describe methods that will allow hiding information in a single or multi-part data carrier, considering the feature vector associated with the user and the properties of the container.

L. Barolli (Ed.): AINA 2024, LNDECT 202, pp. 159–165, 2024.
https://doi.org/10.1007/978-3-031-57916-5_14

2 Creation of Personalized Security Vectors

Personalized Security Vectors will need to be defined to introduce personalized steganography algorithms. To this end, the possibility of performing artificial intelligence-based analysis of users' personal biometric and behavioral patterns is also emerging. Such feature vectors containing unique data representing a given user can be used to create personalized steganographic keys designed to hide secrets in storage media. Such keys will determine how data is dispersed in a single or multi-element container.

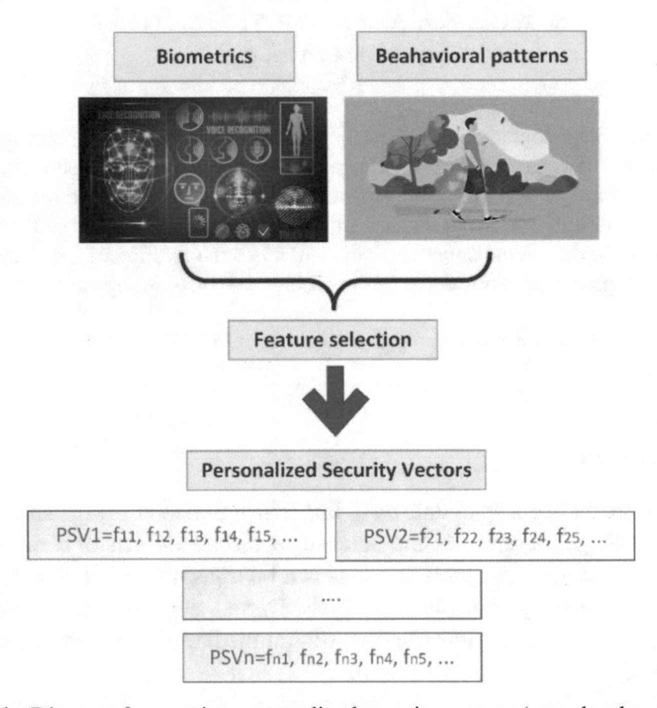

Fig. 1. Diagram for creating personalized security vectors (own development).

The cognitive or artificial intelligence algorithms used are designed to select the most characteristic and representative features that define the protocol user. It is also possible to make multiple selections to create different vectors of personal characteristics by analyzing multiple personal parameters in the form of biometric data, behavioral traits or biomedical data, but also analyzing online profiles from available user accounts [4]. Selection of the most representative features can be carried out using machine learning methods, data Fusion, or cognitive systems designed to extract user features [5]. The general scheme for creating Personalized Security Vectors is shown in Fig. 1.

Once a personalized security vector has been evaluated, it is possible to use it in steganographic procedures as a key for secret distribution [6].

3 Secret Hiding in One Container

The first example of the use of Personalized Security Vector is to use it as a key sequence to distribute secret data in a single container. Such steganographic algorithms allow secret information to be dispersed and hidden in a way that depends on the owner of the PSV [7].

The PSV vector as a sequence of values can determine irregular intervals (offsets) between successive pixels of the image carrier, in which the next secret bits of information will be placed. The placement of secrets itself in the pixels of the container determined with PSV can be carried out by various embedding methods including the simplest LSB or modifications of this technique, in which only selected color components will be considered.

As mentioned earlier for each user it is possible to generate and apply several different personalized keys, which further creates the possibility of placing many different secrets in a single data carrier. Such secrets can be placed using similar algorithms, or each in a different way, using a different key.

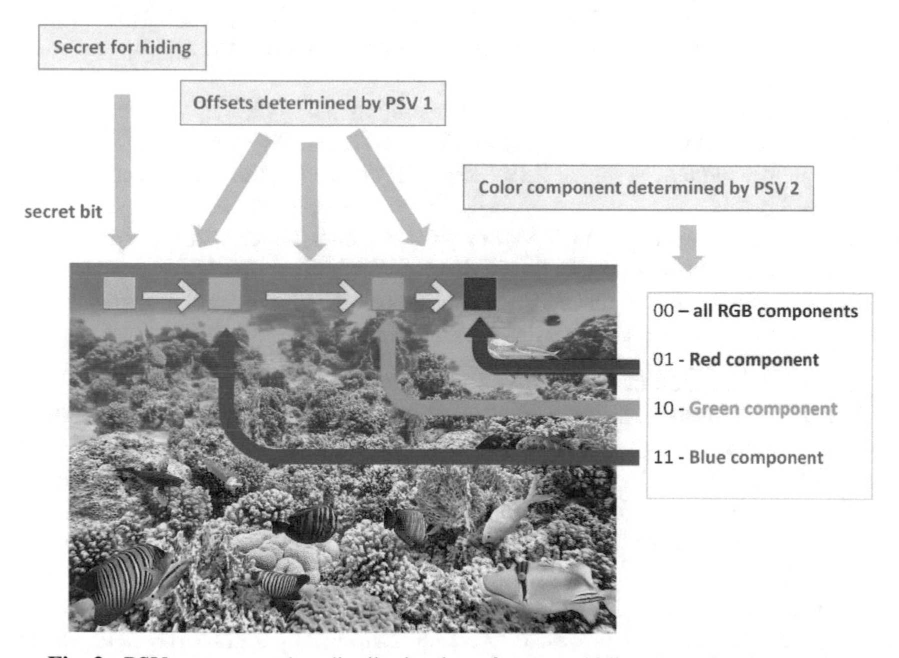

Fig. 2. PSV vectors used as distribution keys for secret hiding (own development).

It is also very interesting to use two keys of different PSVs, selected for a single user to place a single secret. In this case, the first key will determine the offsets between consecutive pixels in which the secret bits will be placed, while the second PSV vector will be the key whose bit values will be used to determine which color component of the modified pixel can be modified. This procedure is shown in Fig. 2.

Regarding the security of the presented algorithm, it is easy to see that the described method guarantees a high level of security, mainly due to the use of various personal characteristics derived from PSV. In the case of a diverse selection of data for secret hiding procedures and the possibility of random distribution of secrets in the storage media, practically the only effective cryptanalytic attack on such systems can be a brute force attack [8].

4 Secret Distribution over Several Containers

Another interesting application of PSV feature vector is the concealment of secret information with its simultaneous division and placement in several image carriers. In this procedure, one or two independent PSV personal vectors (one for determining offsets and the other for determining the color components of the modified pixels) can be used to disperse the secret data, as in the procedure described in an earlier section. On the other hand, another PSV personal vector will be needed to divide the secrets among several selected containers. In this case, the entire procedure for placing the secret data will be implemented as follows:

1. A set of containers is selected in which the secret data will be placed. This set must be ordered so that in subsequent steps the containers that have the appropriate indexes are modified.
2. The secret data is converted to a bit representation and in ole steps the individual bits of the secret will be placed using the PSV1, PSV2 and PSV3 keys.
3. Multiple-bit sequences of the PSV1 key will determine the offsets for subsequent modified pixels.
4. The two-bit sequences of the PSV2 key will determine which color component of the RGB color representation will be modified by the hidden bits.
5. Multi-bit sequences of the PSV3 key, on the other hand, will be used to index containers and determine the order in which they are modified.

The general method of operation of this procedure is presented in Fig. 3.

In terms of security of the presented algorithm, too, it can be noted that the described method guarantees a high level of security due to the use of different vectors of PSV personal characteristics. In the case of the use of multiple containers for the distribution of hidden information and the possibility of random distribution of secrets in each storage medium, it seems that the only effective cryptanalytic attack on such systems is a brute force attack.

Another security issue of procedures using PSV personalized vectors is their generation from personal features, as well as their secure storage by users. In order to guarantee such security, it seems necessary to use classical symmetric encryption algorithms, e.g. AES, which will allow complete and secure encryption of personal features and created keys by a given user and using only one private key. Such encryption techniques guarantee computational security, which effectively prevents effective cryptanalytic attacks [8].

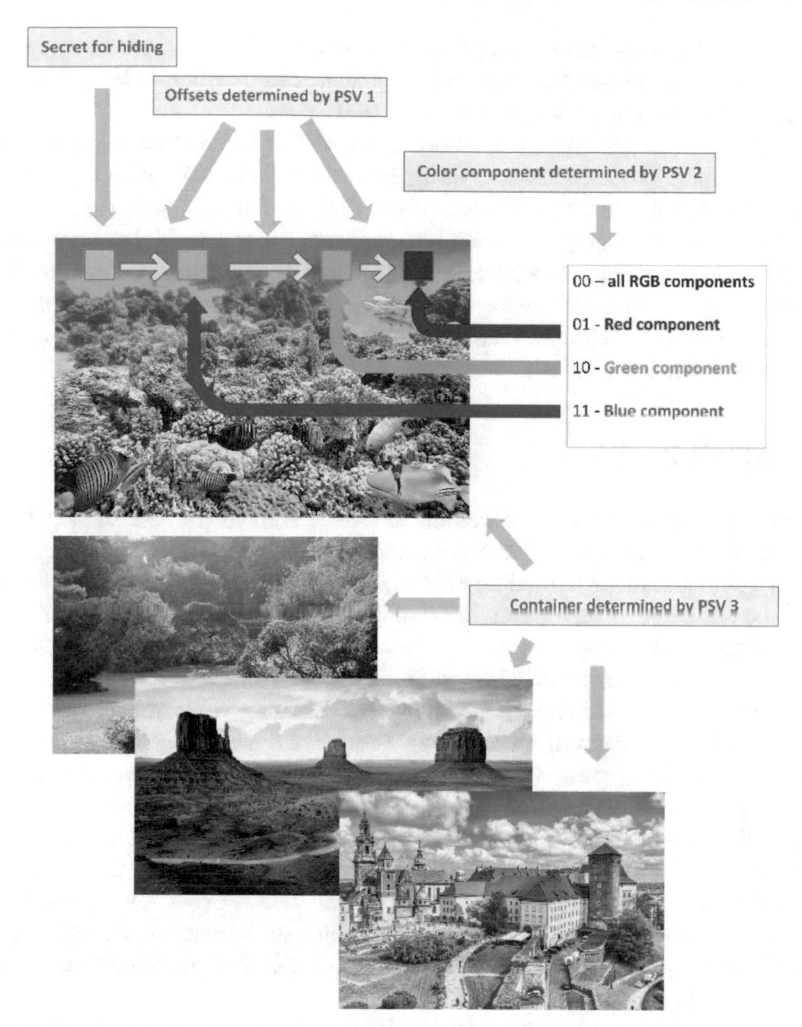

Fig. 3. Secret distribution using PSV vectors in multiply containers (own development).

5 Areas of Application

User-centric steganography methods can have a number of important applications. One of them is the creation of personalized security procedures dedicated to specific users or groups of people [9]. The most important feature of such solutions is that the protocols created are completely dependent on certain selected characteristics of the participants, and they can only work properly if the people for whom they were developed participate.

Another important area is the task of hiding and dispersing secret or strategic data in distributed or cloud-based computer systems. Often there is a need to disperse computer resources across multiple nodes so that the data stored there is completely resistant to local cyber-attacks, which even if successful will not allow the secret data to be restored.

To achieve just such resistance, in addition to hiding the data using steganography procedures, it is also necessary to divide and disperse it across several independent storage media using the multiple PSV keys described in the paper.

The described steganographic methods can also find application in the division of secret information. For such tasks, cryptographic threshold partitioning techniques are usually used, which allow the division of data into a defined number of component shares. However, cryptographic techniques are characterized by higher computational complexity, as they are based on advanced mathematical algorithms. Secret partitioning using multiple personalized PSV keys has a linear complexity, which is only due to the use of one of the PSV vectors as a hash function that disperses bits of secret information among several data carriers [10].

6 Conclusions

The paper discusses selected algorithmic issues related to the use of personal and behavioral characteristics to create new procedures for user-oriented steganography.

Such techniques make it possible not only to establish secure communication channels for the exchange of secret data, but also to create advanced solutions to ensure the security of data and computer systems. The solutions proposed in this paper are based on the use of a special personalized vector of features, which are selected from a number of available biometric characteristics of a person, as well as a set of behavioral characteristics [11–13]. Selection of the most representative features is possible through the use of artificial intelligence and machine learning algorithms, but also through the use of cognitive systems [14]. The wide range of feature choices makes it possible to create sets of personalized PSV keys for individual users, which can then be used to place different secrets in a single carrier, or a single secret dispersed in several independent containers.

The described steganographic procedures are resistant to cyberattacks, which makes them a convenient and secure solution especially due to the ever-increasing number of users of Internet systems and services, as well as the global digitization of services and the growing number of data transmissions.

The article also shows how the methods described can influence the development of new branches of cyber security and IT security, such as personalized cryptography and cognitive cryptography.

The personalized steganography techniques described in this paper can also be further improved especially in the direction of creating information hiding methods that will consider, in addition to personal characteristics, the properties of the containers or the properties of the hidden and shared data [15, 16]. In order to create such algorithms, it will be necessary to use knowledge engineering and semantic feature extraction algorithms that determine both the original content of the data media and the semantic properties of the hidden data.

Acknowledgments. The research project was supported by the program „Excellence initiative – research university" for the AGH University of Krakow.

References

1. Ogiela, M.R., Koptyra, K.: False and multi-secret steganography in digital images. Soft. Comput. **19**(11), 3331–3339 (2015). https://doi.org/10.1007/s00500-015-1728-z
2. Ogiela, L., Ogiela, M.R., Ko, H.: Intelligent data management and security in Cloud Computing. Sensors **20**(12), 3458 (2020). https://doi.org/10.3390/s20123458
3. Ogiela, M.R., Ogiela, U.: Secure information splitting using grammar schemes. In: Nguyen, N.T., Katarzyniak, R.P., Janiak, A. (eds.) New Challenges in Computational Collective Intelligence, pp. 327–336. Springer Berlin Heidelberg, Berlin, Heidelberg (2009). https://doi.org/10.1007/978-3-642-03958-4_28
4. Ogiela, U., Takizawa, M., Ogiela, L.: Classification of cognitive service management systems in cloud computing. Lect. Notes Data Eng. Commun. Technol. **12**, 309–313 (2018). https://doi.org/10.1007/978-3-319-69811-3_28
5. Ogiela, L., Ogiela, M.R.: Advances in Cognitive Information Systems, Cognitive Systems Monographs, Cosmos 17. Springer-Verlag, Berlin Heidelberg (2012). https://doi.org/10.1007/978-3-642-25246-4
6. Sardar, M.K., Adhikari, A.: A new lossless secret color image sharing scheme with small shadow size. J. Vis. Commun. Image Represent. **68**, 102768 (2020). https://doi.org/10.1016/j.jvcir.2020.102768
7. Ogiela, L.: Cognitive systems for medical pattern understanding and diagnosis. In: Lovrek, I., Howlett, R.J., Jain, L.C. (eds.) KES 2008. LNCS (LNAI), vol. 5177, pp. 394–400. Springer, Heidelberg (2008). https://doi.org/10.1007/978-3-540-85563-7_51
8. Ferguson, N., Schneier, B.: Practical Cryptography. Wiley, New York (2003)
9. Ogiela, L.: Data management in cognitive financial systems. Int. J. Inf. Manage. **33**(2), 263–270 (2013). https://doi.org/10.1016/j.ijinfomgt.2012.11.008
10. Pizzolante, R., Carpentieri, B., Castiglione, A., Castiglione, A., Palmieri, F.: Text compression and encryption through smart devices for mobile communication. In: Seventh International Conference on Innovative Mobile and Internet Services in Ubiquitous Computing, pp. 672–677 (2013). https://doi.org/10.1109/IMIS.2013.121
11. Albus, J.S., Meystel, A.M.: Engineering of Mind – An Introduction to the Science of Intelligent Systems, A Wiley-Interscience Publication John Wiley & Sons Inc, New York (2001)
12. Branquinho, J. (ed.): The Foundations of Cognitive Science. Clarendon Press, Oxford (2001)
13. Perconti, P., Plebe, A.: Deep learning and cognitive science. Cognition **203**, 104365 (2020). https://doi.org/10.1016/j.cognition.2020.104365
14. Ogiela, L.: Towards cognitive economy. Soft. Comput. **18**(9), 1675–1683 (2014)
15. Zhang, H., Liu, F., Li, B., Zhang, L., Zhu, Y., Wang, Z.: Deep discriminative image feature learning for cross-modal semantics understanding. Knowl.-Based Syst. **216**, 106812 (2021). https://doi.org/10.1016/j.knosys.2021.106812
16. Weiland, L., Hulpuş, I., Ponzetto, S.P., Effelsberg, W., Dietz, L.: Knowledge-rich image gist understanding beyond literal meaning. Data Knowl. Eng. **117**, 114–132 (2018). https://doi.org/10.1016/j.datak.2018.07.006

AI-Based Cybersecurity Systems

Marek R. Ogiela[1(✉)] and Lidia Ogiela[2]

[1] Cryptography and Cognitive Informatics Laboratory, AGH University of Krakow, 30 Mickiewicza Avenue, 30-059 Krakow, Poland
mogiela@agh.edu.pl

[2] Faculty of Computer Science, AGH University of Krakow, 30 Mickiewicza Avenue, 30-059 Krakow, Poland
logiela@agh.edu.pl

Abstract. Cybersecurity systems that guarantee the confidentiality and integrity of sensitive data use strong cryptography techniques based on public key infrastructure and data encryption systems. These techniques use protocols and user keys used to encrypt data and transmit it over VPNs or unsecured transmission networks. This work will describe new possibilities of using AI to create new solutions in the field of cybersecurity of data and computer systems. In this respect, AI techniques can be used to analyze the security of procedures used, as well as to create new protocols focused on specific systems, services or users, which will offer a higher level of data security and enable better protection against data leaks and cyberattacks.

Keywords: AI-Based security · cybersecurity · cryptographic protocols · cognitive cryptography

1 Introduction

Data security issues in modern computer systems and Internet services are of increasing importance, mainly due to the growing volume of transmitted data, as well as the need for frequent access to guarded data or Internet services [1, 2]. Cybersecurity issues are among one of the most important areas of modern computing, which has become global in scope and enables the acquisition, storage, and analysis of huge data sets generated by network users, and IoT devices [3]. Therefore, newer and newer solutions aimed at guaranteeing the confidentiality of data and Internet transmissions are being sought. These solutions must constantly improve and consider a number of factors such as the personalization of solutions, the increasing number of cyberattacks, cryptanalysis of protocols by bots and many others. One of the directions in which the development of intelligent solutions for the security of computer systems is taking place is the creation of solutions that make use of the personal characteristics of users or semantic information of classified or transmitted data. Such algorithms form a new area of modern cryptography called personalized or cognitive cryptography [4–6].

In recent years, a number of solutions have emerged in this field aimed mainly at the tasks of user or party authentication, as well as the generation of strong encryption keys

L. Barolli (Ed.): AINA 2024, LNDECT 202, pp. 166–173, 2024.
https://doi.org/10.1007/978-3-031-57916-5_15

for cryptographic procedures or protocols. Personalized steganography solutions allowing the transmission of data using special user-oriented protocols have also appeared [7].

As a natural consequence of the use of such procedures, there will be a wider use of advanced artificial intelligence (AI) algorithms also to create secure and effective cryptographic protocols.

Thus, the subject of this paper will be issues of using AI techniques and cognitive systems in cybersecurity procedures and operations. Of the many possible issues, the most noteworthy will be to find answers to such questions as:

- How can AI help create personal security keys and data exchange protocols?
- How can AI secure the correct functioning of cryptographic protocols?
- Is it possible to detect weaknesses in a system or protocol given to cyberattacks?
- Can data stolen by hackers be further secured by AI so that it cannot be read even after decryption?

The following sections of the paper will discuss selected possibilities of using AI algorithms to create advanced and secure algorithmic solutions aimed at the tasks of guaranteeing the confidentiality and integrity of stored or transmitted data, as well as enabling the implementation of secure authentication procedures for users or computer systems.

2 AI in Creation of Personalized Security Solutions

The first of the areas of use of AI for cybersecurity applications will be the creation of user-centric, intelligent and personalized cryptographic protocols. Among the many possible applications of AI, the first worth mentioning is the possibility of performing AI-enabled analysis of personal biometric and behavioral patterns, which, as unique data representing a given user, can also be used to create secure cryptographic keys and methods for hiding secrets, in which personal characteristics will be the key to dispersing the data.

AI algorithms in such applications have the task of selecting the most unique, yet representative, characteristics that define a given protocol user. To do this, classical AI techniques based on artificial neural networks or pattern classification algorithms can be used to select the most appropriate personal characteristics [8–11]. AI methods in this case can make multiple selections of personal feature vectors, from larger personal data records containing many different personal parameters in the form of biometrics, behavioral traits or biomedical data [8]. Training a classifier to select the best set of features for a given user can be done using traditional supervised machine learning methods [12].

Once a personalized set of features has been selected, it is possible to use it in cybersecurity procedures. As mentioned earlier, one of the most obvious applications is the creation of personalized encryption keys or authorization codes (Fig. 1). In order to create such keys, it is necessary to recode a selected set of personal features (e.g., using a selected hash function) into randomized bit strings of a specified length. For the personal portions of the keys determined in this way, it is then necessary to generate the random bit strings so that the created keys have the required length, i.e. for symmetric block ciphers several hundred bits, and for asymmetric cryptographic procedures several thousand bits.

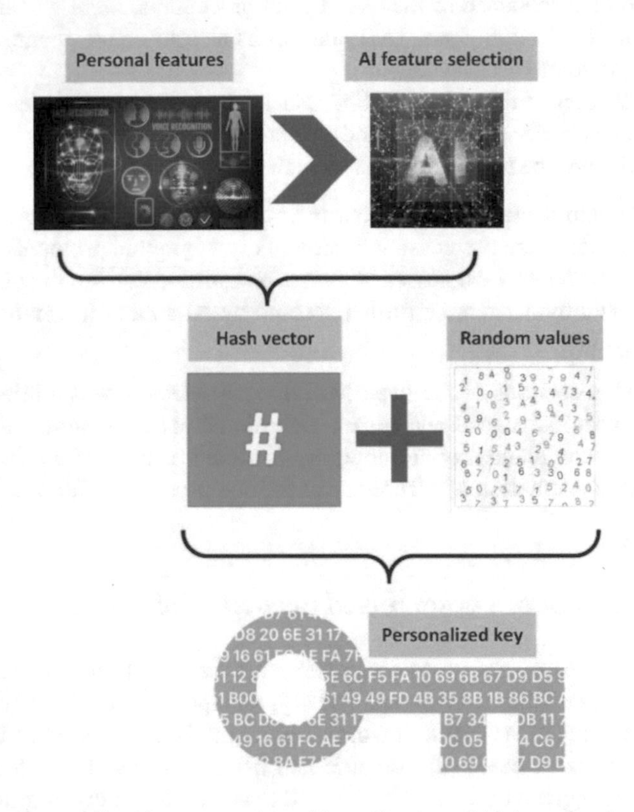

Fig. 1. Personalized key generation with application of AI (own development).

It is worth noting that the described method of personalized key generation also allows to create sets of keys that will all belong to a selected user.

In addition to personal key generation procedures, user characteristics selected by AI algorithms can also be used to create intelligent authorization and authentication procedures. As an example of such solutions, CAPTCHA authorization codes can take the form of images or questions thematically related to users' experiences or preferences [4]. Artificial intelligence algorithms can, in such procedures, generate appropriate sets of visual data related to the interests or thematic knowledge possessed by the user, but they can also generate sets of specialized questions targeted to the experiences or expectations of protocol participants.

Another important example of applications of AI-selected personal feature vectors for a particular person is the use of these features as distribution keys for secret data in visual media [13]. Such steganographic algorithms will make it possible to hide data in a way that is completely dependent on the owner of the distribution key (Fig. 2). There can be multiple personalized keys for a single person which further creates the possibility of placing many different secrets in a single medium, but each in a different way, using a different personal key.

Fig. 2. Personal features as distribution keys for secret hiding (own development).

Analyzing the security issues and the resistance of the presented algorithms to cyberattacks, it is easy to see that the described methods guarantee a very high level of security. Due to the use of different personal characteristics and supplementing them with random data in the case of encryption keys, as well as random choices of data sets for authentication procedures and random distributions of secrets in storage media, the only effective cryptanalytic attack on such systems may turn out to be a brute force attack.

3 AI in Cyberattacks Detection

Among the important issues of using AI techniques in the tasks of guaranteeing data confidentiality and strengthening cybersecurity procedures is the possibility of defending against cyberattacks. There are many approaches to detecting threats supported by artificial intelligence algorithms. Among them, the most important are monitoring of users' online behavior, analysis of network events and incidents and also, analytical studies of the history of attacks carried out.

The use of artificial intelligence to detect cyberattacks is always related to the analysis of certain patterns. AI techniques are mainly used as classifiers and advanced pattern detection methods. The possibility of using such techniques to detect cyberattacks will therefore be related to the analysis of feature vectors that determine various activities related to the security of systems, data and user actions.

Executed cyberattacks allow the creation of signatures or feature vectors describing abnormal behavior and intrusions into systems or data. The collection and use of such signatures and vectors will allow, in the future, a quick comparison of abnormal activities with the accumulated description of vectors of previous events and quick identification of unauthorized activities. Analytical tasks conducted by artificial intelligence can indicate with a high degree of accuracy the type of attack being conducted and identify its source based on signatures of previous incidents.

One of the simpler examples of the application of AI algorithms to detect security incidents can be the analysis of login data for computer systems or Internet mail, considering the frequency and addresses of logins. A cyberattack carried out at an unusual time from a distant country can be easily identified by comparing to previous login data. To do so, however, it is necessary to constantly monitor login data and analyze its frequency, session duration and the IP addresses from which it occurred.

Similar applications can apply to attacks on distributed data stored and processed in a computer cloud. Such data is usually protected using traditional cryptographic algorithms, but in order to protect against man-in-the-middle attacks, it is possible to use additional AI protections allowing to hash such data with personalized keys obtained by methods described in a previous section.

Thus, it can be seen that modern security methods can make significant use of artificial intelligence techniques, both in terms of creating new solutions to guarantee the security of systems and data, but also in terms of preventing cyberattacks.

4 AI-Based Locking Schema

This section will describe an AI-based locking schema protocol designed to provide additional security for encrypted data using a mixing method with data semantics or personalized vectors.

In order to create such a protocol that will enable additional data security, making it secure even if the encryption algorithm used is broken, it becomes necessary to use AI techniques. The functioning of such a protocol can be described as follows (Fig. 3):

- The classified or hidden data is analyzed by AI methods for its content. The result of such analysis is the determination of a semantic vector that defines this data [4, 14].
- For the analyzed data, a personal feature vector depending on the author, sender or recipient of the classified data is also determined.
- Both feature vectors i.e. semantic and personal are used to shuffle the analyzed data by means of the so-called semantic-personal key.
- The input data altered in this way can then be made secret using a selected encryption or hiding procedure.
- The data can then be deposited in a repository or transferred to another user or location.
- To restore the data, it is necessary to decrypt or read it using the original key.
- The reconstructed data is finally read using the semantic-personal key.

In terms of security, the presented protocol is completely resistant to the most common cryptanalytic attacks. In the case of a man-in-the-middle attack, even if the encryption key is captured, it will only allow decryption of the shuffled data, that is, it will

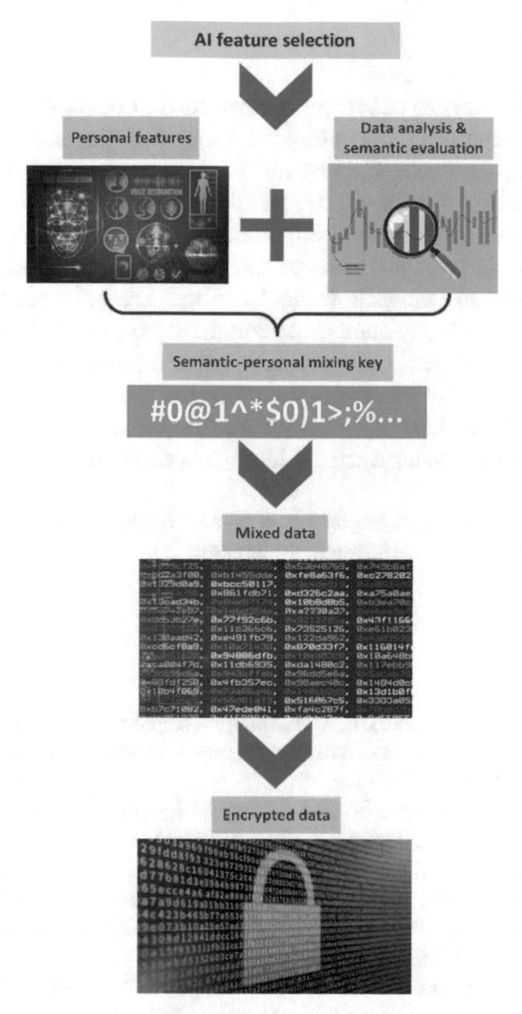

Fig. 3. AI-based locking schema protocol (own development).

be completely unreadable due to its shuffling with a semantic-personal key completely dependent on the content of the data and its owners.

In the case of a brute force attack, it will similarly be possible to decrypt the shuffled data, albeit over a significantly longer period of time, but reconstructing it without knowledge of the semantic-personal keys will be virtually impossible due to its randomized nature. Cryptanalysis of such ciphertexts would require checking each time whether the reconstructed content has any semantic meaning, and this is practically impossible to do in a brute force attack.

5 Conclusions

This paper discusses how to use advanced artificial intelligence techniques to create modern and secure solutions for ensuring the security of data and computer systems, as well as countering cyberattacks. Cybersecurity issues are extremely important due to the ever-increasing number of users of Internet systems and services, as well as the global digitization of services and the increasing number of data transmissions [15, 16].

The paper presents the main areas of the use of artificial intelligence methods in modern cryptography, and shows how these methods can influence the development of new branches of cryptography such as personalized cryptography and cognitive cryptography. It is also shown how AI algorithms make it possible to counter cyberattacks and incidents involving data leakage or unauthorized access.

The paper also shows how to create secure data security protocols using semantic-personal keys determined using artificial intelligence methods.

Acknowledgments. The research project was supported by the program „Excellence initiative – research university" for the AGH University of Krakow.

References

1. Ogiela, U., Takizawa, M., Ogiela, L.: Classification of cognitive service management systems in cloud computing. Lect. Notes Data Eng. Commun. Technol. **12**, 309–313 (2018). https://doi.org/10.1007/978-3-319-69811-3_28
2. Ogiela, L.: Data management in cognitive financial systems. Int. J. Inf. Manage. **33**(2), 263–270 (2013). https://doi.org/10.1016/j.ijinfomgt.2012.11.008
3. Ogiela, L., Ogiela, M.R., Ko, H.: Intelligent data management and security in cloud computing. Sensors **20**(12), 3458 (2020). https://doi.org/10.3390/s20123458
4. Ogiela, L., Ogiela, M.R.: Advances in Cognitive Information Systems, Cognitive Systems Monographs, Cosmos 17. Springer, Heidelberg (2012)
5. Ogiela, L.: Towards cognitive economy. Soft Comput. **18**(9), 1675–1683 (2014)
6. Ferguson, N., Schneier, B.: Practical Cryptography. Wiley (2003)
7. Ogiela, M.R., Ogiela, U. Secure Information Splitting Using Grammar Schemes, Studies in Computational Intelligence, vol. 244, pp. 327–336. Springer, Heidelberg (2009). https://doi.org/10.1007/978-3-642-03958-4_28
8. Ogiela, L.: Cognitive systems for medical pattern understanding and diagnosis. In: Lovrek, I., Howlett, R.J., Jain, L.C. (eds.) Knowledge-Based Intelligent Information and Engineering Systems. LNAI, vol. 5177, pp. 394–400, Springer, Heidelberg (2008). https://doi.org/10.1007/978-3-540-85563-7_51
9. Albus, J.S., Meystel, A.M.: Engineering of Mind – An Introduction to the Science of Intelligent Systems. A Wiley-Interscience Publication John Wiley & Sons Inc. (2001)
10. Branquinho, J. (ed.): The Foundations of Cognitive Science. Clarendon Press, Oxford (2001)
11. Perconti, P., Plebe, A.: Deep learning and cognitive science. Cognition **203**, 104365 (2020). https://doi.org/10.1016/j.cognition.2020.104365
12. Zhang, H., Liu, F., Li, B., Zhang, L., Zhu, Y., Wang, Z.: Deep discriminative image feature learning for cross-modal semantics understanding. Knowl.-Based Syst. **216**, 106812 (2021). https://doi.org/10.1016/j.knosys.2021.106812

13. Ogiela, M.R., Koptyra, K.: False and multi-secret steganography in digital images. Soft. Comput. **19**(11), 3331–3339 (2015). https://doi.org/10.1007/s00500-015-1728-z
14. Weiland, L., Hulpuş, I., Ponzetto, S.P., Effelsberg, W., Dietz, L.: Knowledge-rich image gist understanding beyond literal meaning. Data Knowl. Eng. **117**, 114–132 (2018). https://doi.org/10.1016/j.datak.2018.07.006
15. Pizzolante, R., Carpentieri, B., Castiglione, A., Castiglione, A., Palmieri, F.: Text compression and encryption through smart devices for mobile communication. In: Seventh International Conference on Innovative Mobile and Internet Services in Ubiquitous Computing, pp. 672–677 (2013). https://doi.org/10.1109/IMIS.2013.121
16. Sardar, M.K., Adhikari, A.: A new lossless secret color image sharing scheme with small shadow size. J. Vis. Commun. Image Represent. **68**, 102768 (2020). https://doi.org/10.1016/j.jvcir.2020.102768

Enhancing Mobile Crowdsensing Security: A Proof of Stake-Based Publisher Selection Algorithm to Combat Sybil Attacks in Blockchain-Assisted MCS Systems

Ankit Agrawal(✉), Ashutosh Bhatia, and Kamlesh Tiwari

Birla Institute of Technology and Science, Pilani, Rajasthan, India
{p20190021,ashutosh.bhatia,kamlesh.tiwari}@pilani.bits-pilani.ac.in

Abstract. In a blockchain-assisted Mobile CrowdSensing (MCS) System, individuals can generate as many blockchain identities as they desire, facilitating the execution of a Sybil attack. A Sybil attack can significantly impact such a system due to incorporating a reward mechanism and a majority-based data validation mechanism. An attacker can launch a Sybil attack with selfish or malicious intentions to maximize benefits from the system or to narrow down the reputation of the data requester (subscriber) and the system. Consequently, a Sybil attacker can discourage honest data collectors (publishers) and subscribers from participating, impeding the system's potential success. In this paper, we propose a Sybil attack prevention cum avoidance mechanism to narrow down the effect of it in the blockchain-based MCS systems while maintaining the system's requirements. The proposed mechanism incorporates a novel randomized publisher selection algorithm, leveraging the Proof-of-Stake (PoS) concept to render executing a Sybil attack costly and impractical. The simulation results show the effectiveness of the proposed mechanism.

1 Introduction

Mobile CrowdSensing (MCS) has become a widely embraced paradigm that leverages end-users' capabilities equipped with mobile devices containing various sensors to collect and report data [1]. The widespread availability of mobile phones, coupled with their computing and storage capacities, has rendered MCS a practical and viable solution in modern times. An MCS system typically consists of three key entities: publishers (end-users who provide the data), subscribers (entities requiring specific data), and a crowdsensing platform as shown in Fig. 1. Several blockchain-based MCS systems [2,3] have been proposed to solve the centralization issues while fulfilling the system's requirements, such as trust establishment, quality assurance, user privacy protection, fairness in distributing the rewards, and security against attacks. Blockchain's distributed nature and consensus mechanisms ensure security against specific attacks, e.g., DDoS attacks and Sybil attacks at the blockchain's network level. Additionally, since each user is identified by a pseudonymous identity (blockchain address) not directly linked to personal information, blockchain preserves identity privacy. However, Sybil

L. Barolli (Ed.): AINA 2024, LNDECT 202, pp. 174–186, 2024.
https://doi.org/10.1007/978-3-031-57916-5_16

attacks can also be launched at the application/system level built over the blockchain. It threatens the MCS system due to its characteristics, such as incorporating a reward mechanism and the attacker's malicious intentions.

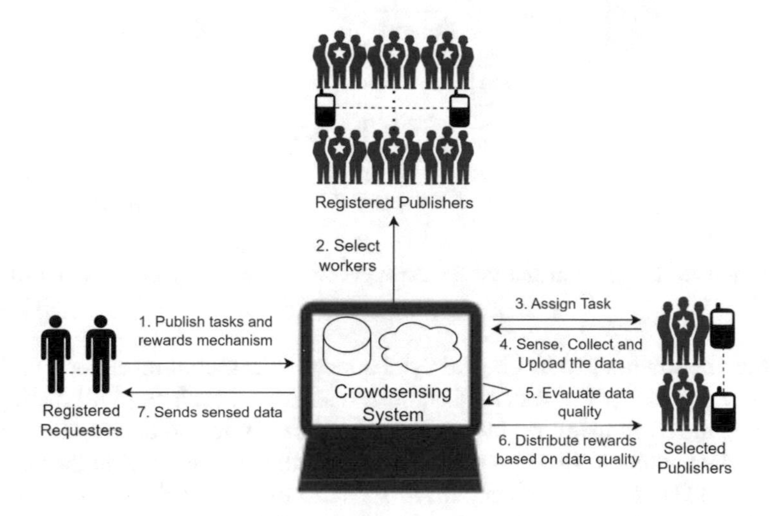

Fig. 1. Generic mobile crowdsensing system

When an attacker controls multiple Sybil identities and manipulates the MCS system by pretending to be numerous distinct participants, it is known as a Sybil attack. Launching a Sybil attack becomes easy in blockchain-based MCS systems as the attackers can create as many Sybil identities as they want. The Sybil attacker can be broadly of two types: external and internal. An external attacker forges the user's blockchain identities to participate in the system to launch a Sybil attack, thus treating an external attacker as an Internal attacker. The internal attackers are the publishers who may be interested in launching a Sybil attack either to disrupt the services generated by the subscribers based on the collected data or to get more benefits. Thus, we categorize the internal attacker as malicious and selfish.

The malicious attacker is the one who submits the wrong data multiple times using multiple Sybil identities. The impact of malicious attackers could be less if they are not in the majority; as such, Sybil identities could be detected while performing the data validation algorithm. Otherwise, it could be more dangerous, as most of the data validation algorithms discussed in the literature are based on the majority principle. Such a situation may only occur if the attacker is able to bypass the selection process. On the other hand, a selfish attacker submits the correct data multiple times to get more rewards. Creating multiple accounts may increase the chance of getting selected to participate in a particular task while decreasing the chance of honest publishers, leading to free-riding attacks [4] or gaining control over a task. Such situations may discourage honest publishers from participating in the system. Thus, the publisher selection phase of the blockchain-based MCS system plays a crucial role in handling the effect of the Sybil attack.

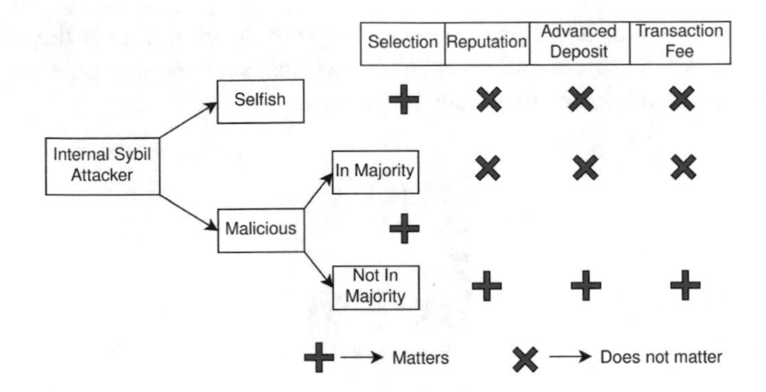

Fig. 2. Usefulness of selection and Sybil parameters on different categories of Sybil Attackers

We find the solutions to handle the Sybil attack in the literature, either on a centralized or decentralized MCS system. Centralized solutions, such as [5–8] mainly focus on detecting the Sybil identities by capturing network traffic and device fingerprinting. Moreover, the authors in [9, 10] proposed authentication mechanism in the centralized scenario and [11–13] in the decentralized scenario to prevent the MCS system from external attackers. However, such authentication mechanisms, except [13], assume that either the trusted third or certification authority is Sybil-free or the user can have only a single identity. The authors in [13] utilized a private blockchain to authenticate. The detection or prevention using such mechanisms over a public blockchain may raise privacy concerns, as it may involve personal and secret information. This may conflict with the privacy-focused principles of MCS systems. Moreover, these mechanisms do not prevent the system from Internal Sybil attackers.

Few articles consider the impact of internal Sybil attackers in Blockchain-based MCS systems. For instance, [2, 14–16] restrict publishers by depositing advanced caution money in the form of system tokens before participating in a task. The advanced deposit and the blockchain's transaction fee make the Sybil attack expensive, as for each Sybil identity, the attacker has to deposit the caution money. Moreover, this caution money is used to penalize the malicious publishers once detected in an MCS system's data validation phase. Thus, it discourages malicious publishers from launching the Sybil attack. However, such techniques fail if the attacker becomes the majority after bypassing the publisher selection process. Consequently, the attacker can bypass the data validation algorithm. Thus, the publisher selection process can be essential in controlling this situation. Moreover, these techniques can not handle the selfish behavior of Sybil's attackers. From the perspective of publisher selection, [2, 16] use reputation to select reliable publishers. In addition to the reputation, [3] selects the publishers based on the bidding they make. And, [15] randomly selects the publishers.

Figure 2 reflects the usefulness of the selection process and other parameters utilized in the literature on different categories of Sybil attackers. The considered parameters matter and can only help if the malicious attacker is not in the majority. The parameter, reputation, considers the reliability of the publishers while selecting them. The parameter advanced deposit or bidding can discourage the malicious attacker (not in the majority). However, few participants can control the task or the system if only reputation is considered or a small group of publishers accumulates most tokens. Moreover, restricting honest publishers from depositing caution money for each task can discourage them from participating in multiple tasks simultaneously. Similarly, focusing solely on randomness in the selection process can prevent participants from exerting control over the tasks or the system. However, it does not consider the reliability of the publishers and does not provide fairness in the selection process. Therefore, an effective publisher selection process can only handle the effect of selfish and malicious (in the majority) attackers.

In this paper, we propose a Sybil attack prevention cum avoidance scheme that utilizes the concept of a Proof-of-Stake (PoS) mechanism to make the Sybil attack expensive for attackers and allow honest publishers to stake amounts for a blockchain address, not for a task. The proposed scheme offers a novel randomized publisher selection algorithm based on the staked amount, CoinAge, and reputation to narrow down the effect of selfish and malicious Sybil attackers. By imposing stake requirements and other parameters, the proposed scheme increases the probability of selecting honest and committed publishers, while it may discourage malicious publishers from attempting Sybil attacks. The generated results show the effectiveness of the proposed scheme in blockchain-based MCS systems.

2 Proposed Mechanism for Sybil Prevention

In this section, we first discuss the generic workflow of a task in a blockchain-based MCS system. Then, we discuss the proposed Sybil attack prevention cum avoidance mechanism utilizing the concepts of Proof-of-Stake (PoS), which include a novel publisher selection process.

The blockchain-based MCS system consists of the following phases: registration, task creation, publisher reservations, publisher selection, data submission, data validation, and reward distribution [17]. A sensing task is defined as grouping the following parameters: sensor type, zone ID, number of publishers required, cost per sensor value, number of sensor values, task duration (in number of days), majority, etc. [2]. The working flow of a task in a blockchain-based MCS system is illustrated in Fig. 3. We make the following assumption in the proposed mechanism: 1) The World map is divided into zones, and a unique zone ID identifies each zone. 2) The number of honest publishers interested in participating in a task is more than the number of Sybil identities. 3) The publishers can get the system tokens from the time of bootstrapping the system to participate in the tasks. 4) A data validation algorithm is available to detect the wrong data submitted by malicious publishers, if not in the majority.

To lessen the effect of Sybil's attack on blockchain-based MCS systems, we utilize the concept of the Proof-of-Stake mechanism along with other parameters. PoS mechanism has shown its potential to handle the Sybil attack in the mining process of several blockchains, such as Cardano [18]. The proposed mechanism works as follows.

- All registered publishers who want to increase the chance of getting selected for all the tasks in a particular zone raise an amount of stake of their interest. The proposed

scheme allows publishers to deposit and withdraw the staked amount if they are not participating in any task to maintain the system's feasibility.
- The publishers make reservations under the interested tasks by submitting an encrypted random number. A smart contract keeps all the reservations in a reservation pool.
- The smart contract selects the publishers using the proposed selection process described in Sect. 2.4. The publisher selection will be based on the following factors: the staked amount, CoinAge, reputation, and the submitted random numbers.
- The selected publishers submit the data. Their stakes are locked if they participate in any task, and rewards will be granted after the data validation.
- Once the data validation process is completed, the publishers whose data is accepted will be rewarded based on the reward mechanism, and the reputation will increase.
- If the data is rejected, a small portion of the publishers' staked amount will be slashed as a penalty, and the reputation will decrease.

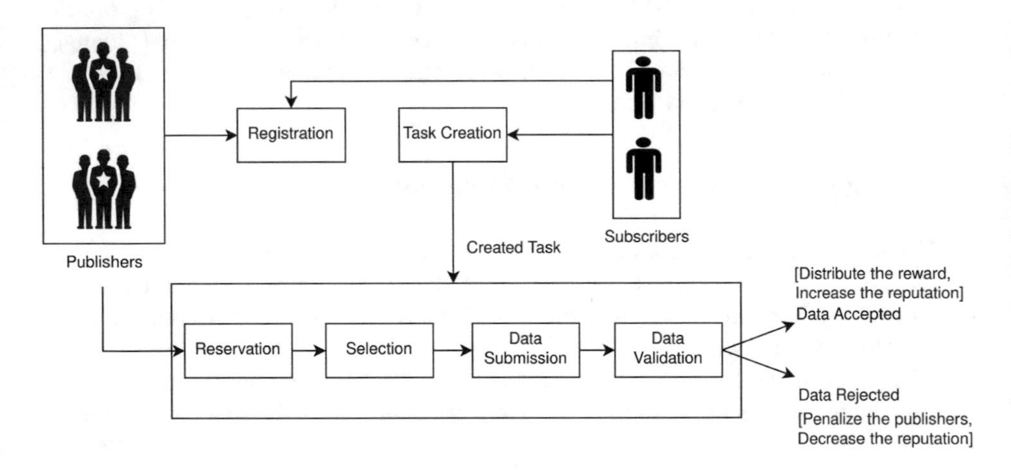

Fig. 3. Working flow of a blockchain-based MCS system

We first discuss the reasons and effectiveness of the considered parameters (staked amount, CoinAge, reputation, and randomness), then the registration process, stake deposit and withdrawal process, publisher reservation process, and finally, the publisher selection process.

Staked Amount: The staked amount is a global parameter instead of a local parameter in a particular task. The publishers can stake some tokens corresponding to a blockchain address or identity. The staked amount helps to prevent Sybil attacks by making it prohibitively expensive for an attacker to create multiple identities and participate in the tasks. However, there are several concerns while considering only the staked amount. 1) The attacker can stake large tokens for a short period of time by taking advantage of the withdrawal of the amount if not participating. 2) The reliability of the publishers is not reflected. 3) It enables the attacker to take control of the tasks if they accumulate large tokens.

CoinAge: CoinAge was a concept used in earlier versions of PoS, such as the PoS system used in PeerCoin [19]. Coin-age refers to the amount of time that the coins have been held in a wallet to determine which users were eligible to create new blocks. The idea was that users who had held more coins for longer were considered more trustworthy. We use this concept in our scenario and redefine it as the time the stake has been held in a smart contract. It is also a global parameter. Our aim in introducing this is to lessen the effect of the first concern discussed above and to provide an advantage to honest publishers if they stake for longer. However, an attacker can exploit this without actively participating in the system. Still, in this case, the attacker has to hold many tokens longer.
Reputation: Reputation is an important parameter that reflects the publishers' honesty, active participation, and historical behavior in the system. The publishers have to earn their reputation by participating honestly in the system. Considering reputation can help in dealing with the disadvantages of the coin-age concept. The third concern discussed in the staked amount parameter is not solved despite the combination of the staked amount, coin age, and reputation. Few participants can still bypass the selection process and thus can control the tasks and system.
Randomness: The publisher selection process must have some randomness to ensure security and fairness. We consider randomness for two reasons: 1) Predictability opens the door for malicious attackers to bypass the PoS mechanism by staking more stakes and thus can control the task after coming into the majority. 2) Selfish attackers can earn more rewards if they know when they will be selected. Therefore, randomness in the proposed mechanism brings unpredictability in the publisher selection process.

The working principle of blockchain does not allow the generation of random numbers through smart contracts; thus, it becomes challenging to deal with random numbers. In this situation, two options can be opted: one is to ask publishers to provide the random number, and the second is to use Oracle services to get the random numbers. We prefer asking publishers to provide a random number to remove the dependability of a third party. Since the publishers provide the random value, they should not be able to select a random value that increases the chance of getting selected in the selection process. Therefore, the random value should be unpredictable and verifiable. We utilize the concept of verifiable random function (VRF) that ensures verifiability and stops publishers from manipulating the random generation process [20].

2.1 Registration

Each publisher must register in the system through the *registration()* of a smart contract. While making registration, each publisher has to submit a public key, which will be used to verify the generation process of a random number. The smart contract will fetch the sender address (blockchain address) using a built-in variable "*msg.sender*" and store it with the public key in the user pool (*U_Pool*) if not present in the system. The smart contract initiates the following parameters: reputation (*Rep*), number of current tasks participated (*NTP*), staked amount (*SA*), staked amount with interest (*SAWI*), and time (*time*). The reputation is initialized with 0.5, indicating the neutral behavior of a publisher, and the rest of the parameters are initialized with zero. *NTR*, *SA*, *SAWI*, and *time* are related to the staked amount and CoinAge parameters. The parameter *time* will only be updated when the publisher deposits or withdraws the token amount.

2.2 Deposit and Withdraw the Stakes

Depositing and withdrawing the stakes involves two considered parameters: staked amount and CoinAge. We combine the effect of the staked amount and CoinAge by utilizing the concept of the future value of money because it involves time and money. The concept of the future value of money is a financial principle that helps calculate the value of a sum of money at some point in the future, taking into account its present value, interest rate, and the time period over which it will grow or earn interest [21]. The interest rate generally varies based on the duration of holding the money. Assuming that the interest rate (IRY) is fixed yearly. Since the task could be created frequently and the time period in the number of days could be short, there is a need to deal with the days. Equation 1 is used to calculate the interest rate on a daily basis (IRD). In our scenario, we consider the future value of the money as the present value of the staked amount while selecting the publishers in the tasks. The staked amount is the actual amount belonging to a publisher. Therefore, we maintain the staked amount and the present value of the staked amount separately using two variables, SA and $SAWI$, respectively. We represent time as the stake-holding time ($time$) from the last updated staked amount. Whenever a publisher deposits or withdraws money, SA, $time$, and $SAWI$ will be updated.

$$IRD = (1 + IRY)^{1/365} - 1 \tag{1}$$

$$SAWI = SAWI * (1 + IRD)^{TDD} \tag{2}$$

Initially, when a publisher deposits the amount, the amount is added in both SA and $SAWI$ directly and the $time$ is updated with the current time. Afterward, the deposited amount will be added to SA, $SAWI$ will be updated using the last updated value of $SAWI$ and the time difference (TDD) between the last updated time and the current time, and finally, $time$ is updated with the current time value. Then, the deposited amount will be added to $SAWI$. Once $SAWI$ is updated, it will be treated as the present value of the staked amount for the future. Equation 2 is used to update $SAWI$. While withdrawing the money, the only difference is that the amount is deducted from SA and $SAWI$ after updating $SAWI$ and $time$. Note that the publishers can deposit and withdraw the stakes when they are not participating in any task.

2.3 Publisher Reservation Process

The reservation process runs in two stages. A particular time duration is specified for each stage, and each publisher has to follow the specified time duration. In the first stage, each publisher i takes the private key (Pvt_key_i) and the hash of the blockchain address (BA_i) and the task address ($H(BA_i, task\_address)$) as input of a verifiable random function (VRF) and generates an output (VRF_Out_i) that contains a random number and proof. Each publisher i encrypts the generated output using a key K_i and submits it to the smart contract. Once the output is submitted, the smart contract increments the value of NTP by 1. Note that the private key and the blockchain address correspond to a publisher, and the task address of the hash function makes each input different for different tasks, as the requirements of each task are different. The task address is determined by creating an instance of a smart contract.

In the second stage, each publisher i submits a key to the smart contract that will be used to decrypt the submitted VRF output. The smart contract decrypts the output, verifies the proof embedded in the output using the public key submitted during registration, and extracts a random number (RN_i) from the VRF output. If verification is not done correctly, that publisher won't be considered for the publisher selection process and becomes eligible for the punishment. The number of reservations (NR) the publishers make in a particular task is also maintained in the system.

Algorithm 1: Publisher Selection Process

Data: $BA, Task\_Address, Z\_ID, SWAI, Rep, RN$
Result: $True$ or $False$

1 **foreach** *interested publisher i* **do**
2 $x_i = currentTime - time_i$;
3 $TDD_i = x_i/86400$;
4 $\_SAWI_i = SAWI * (1 + IRD)^{TDD}$;
5 **end**
6 **foreach** *interested publisher i* **do**
7 calculate the effective reputation ($ERep_i$) of publisher i;
8 calculate the effective $SAWI$ ($ESAWI_i$) of publisher i;
9 $combined_i = \alpha * ERep_i + \beta * ESAWI_i$;
10 $summation\_combined_i += combined_i$;
11 **end**
12 $min\_prob = \frac{1}{NR^2}$;
13 **foreach** *interested publisher i* **do**
14 $Sel\_prob_i = min\_prob + (combined_i / summation\_combined_i) * (1 - min\_prob * NR)$;
15 $rel_i = RN_i * Sel\_prob_i$;
16 **end**
17 Sort the publishers based on the calculated reliability (rel_i) in descending order;
18 Pick the first number of required publishers from the sorted result;
19 **foreach** *publisher not selected* **do**
20 $NTP_i = NTP_i - 1$;
21 **end**
22 return $True$;

2.4 Publisher Selection Process

As discussed, we consider several parameters, such as the present value of the staked amount ($SAWI$, reflects the combination of the staked amount, coin-age), reputation

(*Rep*), and randomness (RN_i) to select publishers. In the publisher selection process, we initially compute the present value of the staked amount using Eq. 2 as described above and store it in a local variable ($\_SAWI_i$) for each publisher i. Then, we compute the effective values of the relevant parameters $SAWI_i$ and Rep_i for each publisher i using the Eqs. 3 and 4, respectively.

$$ESAWI_i = \frac{\_SAWI_i}{\sum_{k=0}^{NR} \_SAWI_i} \tag{3}$$

$$ERep_i = \frac{Rep_i}{\sum_{k=0}^{NR} Rep_i} \tag{4}$$

Subsequently, we utilize weightage parameters (α and β) to combine the effective values of Rep_i and $SAWI_i$. The combined value is used to determine the selection probability of the publishers as shown in Algorithm 1. We consider each publisher's minimum probability of selection while calculating the selection probability. It actually has several advantages: 1) It contributes to a fairer selection process, preventing any publisher from being entirely excluded. 2) It may introduce a safeguard against biased or skewed results. 3) It can make the selection process more robust and resistant to outliers or extreme values that might distort the overall system dynamics. Once the selection probability is obtained, it is multiplied by the corresponding random number assigned to each publisher to derive the reliability of the publisher. Finally, we sort the publishers based on their reliability and select the top required number of publishers. The smart contract decrements the value of NTP by 1 for the publishers who are not selected. The proposed selection process ensures fairness, so no single publisher can gain too much control over the task, even if they hold many tokens.

The selected publishers submit the data based on the subscriber's requirements. The data validation phase validates the data, and based on the results, publishers get rewarded or penalized if involved in any malicious activities. Overall, the combination of the considered parameters in the selection process and punishment mechanism helps to prevent Sybil attacks and maintain the security and reliability of the system.

3 Results and Analysis

We implement the proposed work using Python to comprehensively analyze its impact on Sybil attacks. We examine the impact of the proposed work on Sybil attackers in a scenario where the Sybil attacker joins the system after a few tasks have been initiated. For the scenario, we conduct 100 consecutive tasks with the same Sybil identities and repeat this sequence 200 times to derive an average value for the parameters subsequently discussed. In each task, 40 publishers make reservations, and the percentage distribution of Sybil and honest publishers' identities is considered in pairs, such as (40%, 60%), (30%, 70%), (20%, 80%), and (10%, 90%), respectively. We assume that each publisher initially stakes 1000 tokens corresponding to each identity. The present value of the staked amount (*SAWI*) is computed using an interest rate set at 8% yearly. In each task, the requirement is to select five publishers. The weightage parameters, denoted as α and β, utilized in determining each publisher's selection probability, are set at 50%.

The selected publishers are assumed to submit twelve sensor values for each task. The cost per sensor value is set to five. The criterion for accepting data is established with a majority threshold set at 70%. The benefit accrued by each publisher in each task, provided their data is accepted, is calculated by multiplying the number of sensor values with the cost per sensor value. The penalization mechanism is incorporated while implementing the proposed solution to reduce the effect of the Sybil attack. The penalization mechanism is applied when the data validation algorithm rejects the data. Data rejection occurs if a publisher's data deviates from the majority. This penalization method involves subtracting twice the expected amount from the task and reducing the reputation associated with an identity.

The publishers, as Internal Sybil attackers, could behave differently for different purposes, as described earlier. We discuss the proposed mechanism's effect on the attacker's Selfish and Malicious behavior based on specific parameters: 1) the average number of Sybil selected in 100 tasks, 2) the average number of tasks in which ≥ 2 Sybil selected, 3) the average number of tasks in which Sybil are selected in the majority, 4) the average number of tasks in which honest are selected in the majority, 5) the average Sybil's benefit in 100 tasks, and 6) the average honest benefit in 100 tasks.

Scenario: When Sybil Joins After a Few Tasks. The selfish conduct of a Sybil attacker encompasses the parameters 1, 2, 5, and 6 while varying the number of tasks Sybils join after. The impact of the proposed mechanism on selfish Sybil's involvement in the mentioned parameters is depicted in Fig. 4. Figures 4a and 4b demonstrate the efficacy of the proposed mechanism on parameters 1 and 2, respectively, in constraining the Sybil attacker from influencing tasks or the overall system. Consequently, the proposed mechanism governs the benefits derived from Sybil and honest identities, shown in Figs. 4c and 4d, respectively.

The malicious behavior of a Sybil attacker encompasses the parameters 1, 3, 4, 5, and 6 while varying the number of tasks Sybils join after. The impact of the proposed mechanism on malicious Sybil's involvement in the considered parameters is depicted in Fig. 5. Figure 5a reflects the effectiveness of the proposed mechanism on parameter 1 by restricting the Sybil attacker from controlling the tasks or system. Since the malicious attacker can influence the tasks only if they are in the majority, Figs. 5b and 5c illustrate the Sybil and Honest selection in the majority, respectively. Figure 5b shows that the malicious Sybils are selected in the majority on an average of 8 tasks while considering 40% participation of Sybils and thus deriving benefits solely from such tasks. However, they incur penalties for the remaining tasks, reducing Sybil benefits, as depicted in Fig. 5d. Consequently, this increases the benefits of honest publishers, as shown in Fig. 5e. Note that as the value of the number of tasks Sybils join after increases, the influence of the selfish and malicious behavior exhibited by a Sybil attacker on the considered parameters significantly diminishes.

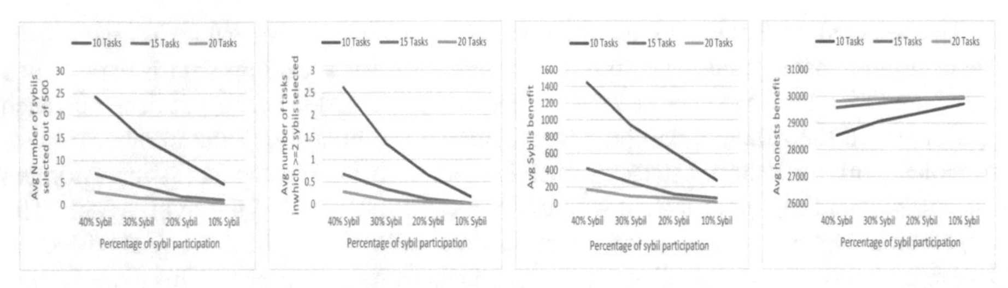

(a) Average number of Sybil selected

(b) Average number of tasks where ≥ 2 Sybil selected

(c) Average Sybil's benefit

(d) Average Honest publisher's benefit

Fig. 4. Effect of Selfish Sybil participation after few tasks on different parameters

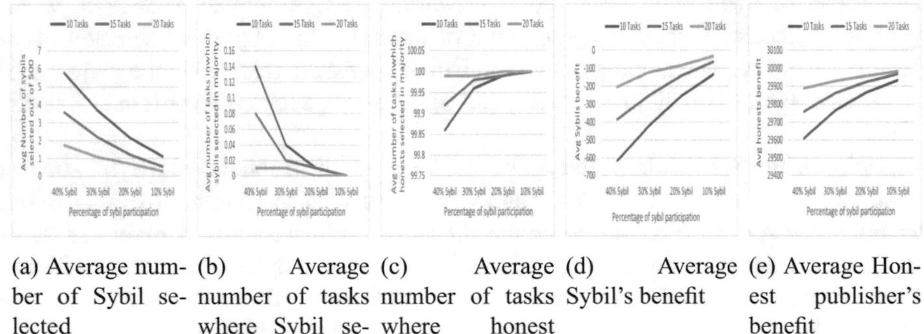

(a) Average number of Sybil selected

(b) Average number of tasks where Sybil selected in majority

(c) Average number of tasks where honest publishers selected in majority

(d) Average Sybil's benefit

(e) Average Honest publisher's benefit

Fig. 5. Effect of Malicious Sybil participation after few tasks on different parameters

Comparing the Proposed Work with Others. The severity of a Sybil attack increases when the attacker manages to manipulate tasks by avoiding the selection mechanism. The critical question revolves around how the selection mechanism can effectively identify unique publishers. The efficacy of the selection mechanism hinges on the factors incorporated into it. Certain works, such as [2,16], exclusively consider reputation as a selection factor. In a scenario where reputations are initialized with identical values, publishers selected in the first task persistently remain selected in subsequent tasks, given the selfish behavior of attackers. This leads to the dominance of a few publishers in controlling tasks, irrespective of their honesty or Sybil status. To contrast this, we evaluate our proposed selection mechanism based on the average number of unique publishers selected, as depicted in Fig. 6. The impact of the selection mechanism on Sybil attacks has been previously discussed.

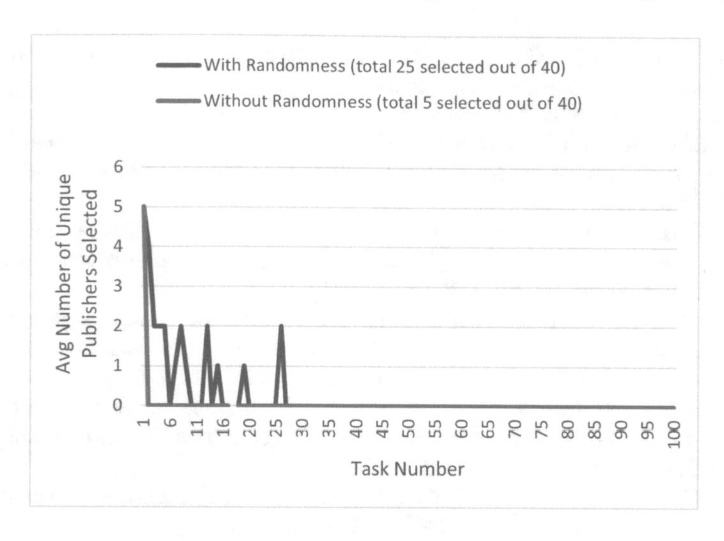

Fig. 6. Effect of randomness on controlling the tasks by few publishers

4 Conclusion

In this paper, we conduct an in-depth analysis of the impact of Sybil attacks on blockchain-assisted MCS systems. Two potential types of Sybil attackers are considered: selfish and malicious. To mitigate the influence of both selfish and malicious Sybil attackers, we introduce an innovative randomized publisher selection algorithm leveraging the PoS concept. The proposed mechanism takes into account the inherent limitations of PoS. PoS integration aims to deter Sybil attackers by increasing the cost of launching an attack and enhancing the effectiveness of publisher selection. The results demonstrate the efficacy of the proposed selection mechanism in significantly mitigating the impact of malicious and selfish Sybil attackers.

References

1. Montori, F., Bedogni, L., Bononi, L.: A collaborative internet of things architecture for smart cities and environmental monitoring. IEEE Internet Things J. **5**(2), 592–605 (2017)
2. Agrawal, A., Choudhary, S., Bhatia, A., Tiwari, K.: Pub-SubMCS: a privacy-preserving publish-subscribe and blockchain-based mobile crowdsensing framework. Futur. Gener. Comput. Syst. **146**, 234–249 (2023)
3. Xu, H., Qi, S., Qi, Y., Wei, W., Xiong, N.: Secure and lightweight blockchain-based truthful data trading for real-time vehicular crowdsensing. ACM Trans. Embed. Comput. Syst. **23**(1), 7 (2024). https://doi.org/10.1145/3582008
4. Zhang, X., Xue, G., Yu, R., Yang, D., Tang, J.: Keep your promise: mechanism design against free-riding and false-reporting in crowdsourcing. IEEE Internet Things J. **2**(6), 562–572 (2015)
5. Yun, J., Kim, M.: SybilEye: observer-assisted privacy-preserving sybil attack detection on mobile crowdsensing. Information **11**(4), 198 (2020). https://doi.org/10.3390/info11040198

6. Cui, H., Liao, J., Yu, Z., Xie, Y., Liu, X., Guo, B.: Trust assessment for mobile crowdsensing via device fingerprinting. ISA Transactions **141**, 93–102 (2023). https://doi.org/10.1016/j.isatra.2022.12.020. ISSN 0019-0578
7. Chang, S.H., Chen, Z.R.: Protecting mobile crowd sensing against sybil attacks using cloud based trust management system. Mobile Inf. Syst. **2016**, 6506341 (2016). https://doi.org/10.1155/2016/6506341
8. Lin, J., Yang, D., Wu, K., Tang, J., Xue, G.: A sybil-resistant truth discovery framework for mobile crowdsensing. In: 2019 IEEE 39th International Conference on Distributed Computing Systems (ICDCS), pp. 871–880. IEEE (2019)
9. Chen, J., Ma, H., Wei, D.S., Zhao, D.: Participant-density-aware privacy-preserving aggregate statistics for mobile crowd-sensing. In: 2015 IEEE 21st International Conference on Parallel and Distributed Systems (ICPADS), pp. 140–147. IEEE (2015)
10. Martucci, L.A., Kohlweiss, M., Andersson, C., Panchenko, A.: Self-certified sybil-free pseudonyms. In: Proceedings of the First ACM Conference on Wireless Network Security, pp. 154–159 (2008)
11. Zhu, S., Cai, Z., Hu, H., Li, Y., Li, W.: zkCrowd: a hybrid blockchain-based crowdsourcing platform. IEEE Trans. Industr. Inf. **16**(6), 4196–4205 (2019)
12. Zhong, Y., et al.: Distributed blockchain-based authentication and authorization protocol for smart grid. Wirel. Commun. Mob. Comput. **2021**, 1–15 (2021)
13. Wang, T., Shen, H., Chen, J., Chen, F., Wu, Q., Xie, D.: A hybrid blockchain-based identity authentication scheme for mobile crowd sensing. Futur. Gener. Comput. Syst. **143**, 40–50 (2023)
14. Wu, H., Düdder, B., Wang, L., Sun, S., Xue, G.: Blockchain-based reliable and privacy-aware crowdsourcing with truth and fairness assurance. IEEE Internet Things J. **9**(5), 3586–3598 (2021)
15. Yu, R., Oguti, A.M., Ochora, D.R., Li, S.: Towards a privacy-preserving smart contract-based data aggregation and quality-driven incentive mechanism for mobile crowdsensing. J. Netw. Comput. Appl. **207**, 103483 (2022)
16. Xi, J., Zou, S., Xu, G., Lu, Y.: CrowdLBM: a lightweight blockchain-based model for mobile crowdsensing in the Internet of Things. Pervas. Mob. Comput. **84**, 101623 (2022)
17. Kadadha, M., Otrok, H., Mizouni, R., Singh, S., Ouali, A.: SenseChain: a blockchain-based crowdsensing framework for multiple requesters and multiple workers. Futur. Gener. Comput. Syst. **105**, 650–664 (2020)
18. Kiayias, A., Russell, A., David, B., Oliynykov, R.: Ouroboros: a provably secure proof-of-stake blockchain protocol. In: Katz, J., Shacham, H. (eds.) Advances in Cryptology. CRYPTO 2017. LNCS, vol. 10401, pp. 357–388. Springer, Cham (2017). https://doi.org/10.1007/978-3-319-63688-7_12
19. Zhao, W., Yang, S., Luo, X., Zhou, J.: On peercoin proof of stake for blockchain consensus. In: 2021 The 3rd International Conference on Blockchain Technology, pp. 129–134 (2021)
20. Micali, S., Rabin, M., Vadhan, S.: Verifiable random functions. In: 40th Annual Symposium on Foundations of Computer Science (cat. No. 99CB37039), pp. 120–130. IEEE (1999)
21. Tandirerung, H.: Market-Determined Interest Rates and Time Value of Money (2022). SSRN: https://ssrn.com/abstract=4162248

Messages and Incentives to Promote Updating of Software on Smartphones

Ayane Sano[1,2](✉), Yukiko Sawaya[1], Takamasa Isohara[1], and Masakatsu Nishigaki[2]

[1] KDDI Research, Inc., Fujimino 356-8502, Saitama, Japan
ay-sano@kddi.com

[2] Graduate School of Science and Technology, Shizuoka University, Hamamatsu 432-8011, Shizuoka, Japan

Abstract. To improve the rate of taking security action, it is important to promote personalized approaches for each user. Related works indicate phrases and UIs of dialog messages and incentives that influence a user's action. In our previous work, we focused on smartphone users updating software, and proposed appropriate phrases of dialog messages according to the user's understanding of the updating procedure, as well as the type of software. We also analyzed appropriate incentives. However, in the terms of level of literacy, the effectiveness of the UI of dialog messages and the volume of incentives remain unclear. In this paper, we conducted a user survey to analyze appropriate UIs according to the user's understanding of the updating procedure and the appropriate volume of incentives. As a result, we confirmed different UIs are effective according to the user's understanding of the updating procedure. In addition, we found an appropriate volume of points, mobile data, and coupons in order to promote the updating of software.

1 Introduction

To prevent various cyberattacks, users need to take security action by themselves. However, it is reported that more than half of users do not continually take security action [1], because the same approaches are provided to all users even though the level of IT literacy differs according to the user. Therefore, it is important to provide personalized approaches that focus on the different level of IT literacy of each user. Related works describe the impact of phrases and user interface (UI) design of dialog messages [1–6], as well as incentives [7, 8] on a user's action. In our previous work, we focused on smartphone users updating software and proposed effective phrases of dialog messages according to the user's understanding of the update procedure, and eight types of software (OS, Communication application, Finance application, Lifestyle application, Games application, Utility application, Health application, and Entertainment application) [2]. In addition, we analyzed appropriate incentives for promoting to the updating of software. As a result, we found that the phrases about security damage incurred by not updating software is effective for the group that understands the update procedure (UUP) in the case of Communication, Games, Utility, Health and Entertainment applications. In addition, points is an effective incentive for all users in common.

L. Barolli (Ed.): AINA 2024, LNDECT 202, pp. 187–200, 2024.
https://doi.org/10.1007/978-3-031-57916-5_17

However, concerning the level of literacy of smartphone users, the effectiveness of the UI of dialog messages is not yet clear. In addition, there are no related works about the appropriate volume of incentives for smartphone users. In this paper, we propose some UI designs and verify their effectiveness by using an online survey. In the same way as in the previous study [2], we analyze the difference between UUP and the group that does not understand the update procedure (NUUP), as well as between eight types of software. We pose the following two research questions.

- RQ1: What are effective UIs of dialog messages?
- RQ2: What is the appropriate volume of incentives?

We recruited 1600 participants and classified 200 participants into eight groups who answered about each type of software. As a result of conducting the survey, we gained the following results.

- Effective UIs differ between UUP and NUUP, and between types of software. A red background and red button are the most effective UI for the NUUP.
- We found an appropriate volume for each condition. For example, appropriate points are 49.6 points to promote the updating of the Finance application for the NUUP in the case of maximum volume, 31.4 points to promote updating of the Games application for the NUUP in the case of minimum volume.

These results show that appropriate UIs of dialog messages and volume of incentives in encouraging Android smartphone users to update software. This paper is structured as follows: Sect. 2 describes related works and Sect. 3 describes proposed methods (UI design). Section 4 describes the user survey and the findings of the survey. Section 5 discusses appropriate approaches for smartphone users and the limitations of this research. Finally, Sect. 6 presents a conclusion and discusses future works.

2 Related Works

In this section, we describe approaches to updating software, our previous work, and UI design in promoting security action.

2.1 Approaches to Updating Software

There are related works [3, 9–12] about approaches to updating software. Mathur et al. [9] proposed a prototype for updating computer applications. They stated that personalizing of interfaces is required for promoting update. In our previous work [3], we analyzed effective dialog messages to promote Windows update according to the security awareness of each user. Sankarapandian et al. [10] proposed a system (TALC) that encourages users to update computer software. TALC paints graffiti on the user's desktop screen in the case of not updating software. Ndibwile et al. [11] proposed a UI design of messages that encourage smartphone users to update. They described the impact of color on a user's action. Tian et al. [12] conducted an online survey based on smartphone users' current update decisions and designed a notification scheme including user reviews. Their proposed system catches the attention of users.

2.2 Our Previous Work

In our previous work [2], we proposed effective phrases of dialog messages that promote smartphone users to update software. We chose OS and seven types of applications (Communication, Finance, Lifestyle, Games, Utility, Health, and Entertainment) because these are commonly provided in Google Play [13]. We found that appropriate phrases differ according to the user's understanding of the update procedure. For example, the most effective phrase is a phrase that shows the disadvantages of not updating software (loss-framed message: LFM) of the Health application for the UUP and a phrase that shows the advantages of updating software (gain-framed message: GFM) of the Health application for the NUUP. We understand that effective phrases also differ between types of software. Table 1 shows examples of effective phrases in our previous work [2]. The ISI (Increase Security Interest) type concerns increasing security interest. The DSC (Decrease Security Cost) type concerns decreasing security cost. The Wi-Fi type concerns not consuming mobile data by using Wi-Fi. In addition, related works [7, 8] explain that incentives affect the decision-making of users. We analyzed the effectiveness of incentives. As a result, it is clear that appropriate incentives are points, mobile data, and coupons for many users.

Table 1. Effective phrases in our previous work [2]. All phrases add the following phrases: "Updates are available" and "Please update right now".

Type	Phrase
LFM-3	If you do not update, others may access your smartphone dishonestly and leak your personal data
LFM-4	If you do not update, others may access your smartphone dishonestly and damage people around you
LFM-5	If you do not update, your smartphone bugs may not be fixed
LFM-6	If you do not update, you cannot use applications
GFM-1	If you update, you can solve security problems and improve security
GFM-3	(In the case of applications) If you update, you can fix application bugs
GFM-4	If you update, you can use your smartphone in the same way as before
ISI-2	It is important for you to update now because the attacks are sophisticated
DSC-1	(In the case of applications) If you update once every few weeks, your application will be safe
DSC-2	(In the case of OS) If you only reboot your smartphone, that is all you need to do. (In the case of applications) If you only reboot your application, that is all you need to do
Wi-Fi-1	If you use Wi-Fi, you will not consume mobile data
Wi-Fi-2	You are using Wi-Fi now. If you update now, you will not consume mobile data

2.3 UI Design in Promoting Security Action

Related works [4–6] described UI design impacts on user action. For example, appropriate UI design is different in terms of gender, age, and region [5]. Color and animation affect the user's attention [6]. There are some related works [14–20] about UI design in promoting security action. Bravo-Lillo et al. [14] focused on showing security warning messages for users and proposed some UI designs. Highlight and animation affect users' actions in response to warnings. Felt et al. [15] focused on SSL warnings and proposed new UI design. They evaluated the usability in an experiment. As a result, adherence rates improved. In addition, Felt et al. [16] proposed three indicators to gain more attention. Their design is changed by color and icon. Turland et al. [17] created a prototype system regarding Wi-Fi network selection and evaluated usability in experiments for Android users. They found that color coding influenced users' security action. Golla et al. [18] focused on two-factor authentication (2FA) and proposed UX design to improve 2FA adoption. They showed that personalization and reminders are effective approaches. Additionally, other related works [19, 20] also clarify that personalized interfaces are required to promote security actions.

3 Our Proposal Design

As described in Sect. 2.3, personalized UI design is required to promote security action. However, there are no related works about encouraging smartphone users to update software. We verify that appropriate UI design differs in terms of understanding the update procedure.

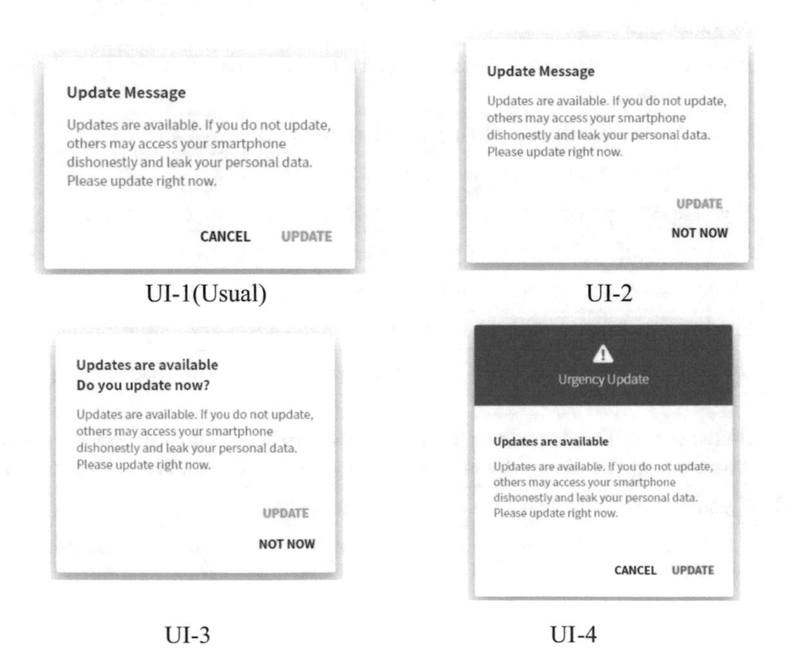

Fig. 1. Proposed UI designs.

Fig. 1. (*continued*)

To evaluate UI design, we created nine examples of UI design in reference to the related works [3, 16, 17] as well as Material Design [21]. Figure 1 shows examples of UI design in the case of LFM-3 in Table 1. These messages are shown in the Japanese language. UI-1(the usual UI), UI-2, and UI-3 are based on the Material Design [21]. In the same way in related works [16, 17], we indicated "safe" by green color and "danger" by red color.

4 Online Survey

4.1 Methodology

We conducted an online survey in Japan by using the services of a Japanese research company (Cross Marketing, Inc.) [22] from June 24th to June 28th, 2022. We recruited 1600 participants who are Android smartphones users from 20 to 69 years of age. We targeted the participants in the same conditions as our previous work [2]. Android smartphones included models such as Sony, Samsung, Sharp, and Kyocera. The participants do not set auto update and update manually or do not update at all. The participants' demographics are shown in Table 2. We classified 200 participants into eight groups who answered about OS or each application to evaluate the difference between types of software. We used the same question ("Status of understanding the update procedure") as in our previous work [2] in the screening survey and allocated each of the 100 participants to one of two groups (UUP or NUUP) by employing the user's answers to this question.

Table 2. Demographics of 1600 participants.

Age	20–29	30–39	40–49	50–59	60–69
Male	64	158	187	198	199
Female	111	164	172	170	177

The Appendix shows the questions of this survey. We showed the nine UI designs in Fig. 1 against which each of the three phrases for participants is shown. The three phrases used are the top three phrases resulting in the highest update rate in our previous work [2]; Question 1 is the highest phrase, Question 2 is the second-highest phrase and Question 3 is the third-highest phrase. If there are multiple items that match the third-highest phrase, we chose a different type from the highest phrase and the second-highest phrase. Table 3 shows the phrases that we used in this survey. In this survey, we asked about the user's action in response to 27 types of messages. In addition, the appropriate volume of incentives is unclear although previous work found the proper type of incentives to persuade users to update software. Therefore, we verified the appropriate volume of incentives. We ask about types and volume of incentives in the Question 4–Question 8. We collected the anonymized answers of the participants. In accordance with the regulations of the research company, the research company paid a fee to the participants after answering the survey. This research was exempt from ethical review in accordance with the guidelines of our office.

Table 3. Phrases that we used in this survey.

		OS	Communication	Finance	Lifestyle	Games	Utility	Health	Entertainment
UUP	Q1	LFM-6	LFM-3	LFM-6	Wi-Fi-2	LFM-3	LFM-3	LFM-3	LFM-6
	Q2	DSC-2	LFM-6	LFM-3	GFM-1	GFM-1	GFM-1	LFM-6	LFM-5
	Q3	Wi-Fi-2	DSC-2	Wi-Fi-2	LFM-5	LFM-6	LFM-5	LFM-4	LFM-3
NUUP	Q1	LFM-4	LFM-3	LFM-6	Wi-Fi-2	DSC-2	LFM-3	LFM-6	LFM-3
	Q2	LFM-6	LFM-4	GFM-1	LFM-3	LFM-3	ISI-2	GFM-1	ISI-2
	Q3	Wi-Fi-1	LFM-6	LFM-3	GFM-3	LFM-6	DSC-2	GFM-4	DSC-1

4.2 Analytical Methods

We analyzed the following two items by using HAD [23] for statistical analyses.

1. Effective UI design of dialog messages for the UUP and the NUUP
2. Appropriate volume of incentives

First, we verified Item1. In the same way as our previous work [2], we calculated the ratio of those who answered "I update right now" to each question of Question 1–Question 3. We used a t-test to compare eight UIs (UI-2–UI-9) with the usual UI (UI-1). Next, we calculated the average volume of incentives to verify Item2. Only in the case of coupons, we tallied users' answer about type of coupons.

4.3 Results

Table 4 shows the ratio of participants who answered "I update right now" to Question 1–Question 3. The upper part pertains to the UUP, and the lower part pertains to the NUUP. If the p-value is less than 0.05 and positive, the UI for comparison is effective; on the other hand, if the p-value is less than 0.05 and negative, the UI for comparison is backfire. In Table 4, the red color shows effective UIs and the blue color shows backfire UIs.

In the case of the UUP, there are effective UIs in the case of the Finance, Lifestyle, Utility, and Entertainment applications. In the case of the NUUP, there are effective UIs in the case of Finance, Lifestyle, Games, Utility and Health applications. UI-4, UI-6 and UI-8 are effective UIs for the UUP and the NUUP in common. Effective UIs are a little different according to the phrase. In the case of the Finance application of the NUUP, UI-4 is effective for LFM-6, and UI-4, UI-6, and UI-8 are effective for GFM-1.

However, the appropriate combination of phrase and UI differs between the two groups. For example, in the case of the Lifestyle application, LFM-5&UI-8 results in the highest update rate for the UUP, and LFM-3&UI-6 results in the highest update rate for the NUUP. In addition, appropriate combinations of phrases and UIs differ among software applications. For example, in the case of the UUP, LFM-5&UI-6 is an appropriate combination in the case of the Utility application, and LFM-3&UI-6 is an appropriate combination in the case of the Health application.

Table 4. Ratio of participants who answered "I update right now".

UUP		OS	Comm-unicati-on	Finance	Lifesty-le	Games	Utility	Health	Enterta-inment
Q 1	UI-1	0.130	0.180	0.202	0.170	0.180	0.180	0.180	0.204
	UI-2	0.140	0.150	0.202	0.180	0.180	0.180	0.180	0.194
	UI-3	0.120	0.170	0.232	0.210	0.180	0.190	0.170	0.143
	UI-4	0.150	0.200	0.242	0.200	0.180	0.220	0.200	0.194
	UI-5	0.120	0.150	0.202	0.160	0.160	0.180	0.170	0.194
	UI-6	0.140	0.200	0.283	0.200	0.220	0.240	0.230	0.173
	UI-7	0.120	0.140	0.202	0.160	0.170	0.180	0.140	0.204
	UI-8	0.120	0.190	0.263	0.200	0.190	0.240	0.220	0.235
	UI-9	0.090	0.160	0.212	0.170	0.160	0.170	0.170	0.163
Q 2	UI-1	0.180	0.190	0.263	0.110	0.210	0.170	0.180	0.204
	UI-2	0.180	0.200	0.273	0.140	0.170	0.160	0.190	0.214
	UI-3	0.180	0.190	0.253	0.140	0.170	0.160	0.200	0.214
	UI-4	0.180	0.200	0.323	0.160	0.230	0.220	0.210	0.255
	UI-5	0.160	0.190	0.303	0.110	0.170	0.160	0.170	0.214
	UI-6	0.170	0.190	0.313	0.170	0.240	0.230	0.180	0.245
	UI-7	0.160	0.190	0.283	0.140	0.150	0.160	0.180	0.214
	UI-8	0.150	0.200	0.323	0.180	0.190	0.220	0.190	0.265
	UI-9	0.140	0.180	0.293	0.110	0.180	0.160	0.170	0.214
Q 3	UI-1	0.190	0.170	0.242	0.190	0.280	0.180	0.170	0.245
	UI-2	0.190	0.170	0.232	0.190	0.260	0.160	0.160	0.245
	UI-3	0.190	0.170	0.222	0.200	0.260	0.180	0.160	0.265
	UI-4	0.190	0.180	0.283	0.210	0.270	0.250	0.210	0.286
	UI-5	0.180	0.150	0.253	0.200	0.230	0.160	0.140	0.255
	UI-6	0.180	0.180	0.293	0.210	0.260	0.260	0.200	0.276
	UI-7	0.180	0.150	0.232	0.200	0.240	0.180	0.150	0.255
	UI-8	0.170	0.190	0.253	0.240	0.240	0.230	0.210	0.286
	UI-9	0.170	0.150	0.253	0.200	0.240	0.180	0.150	0.265

(*continued*)

Table 4. (*continued*)

NUUP		OS	Communication	Finance	Lifestyle	Games	Utility	Health	Entertainment
Q1	UI-1	0.130	0.100	0.212	0.101	0.172	0.120	0.202	0.082
	UI-2	0.120	0.080	0.192	0.101	0.172	0.100	0.232	0.082
	UI-3	0.140	0.090	0.162	0.101	0.172	0.100	0.222	0.092
	UI-4	0.110	0.110	0.293	0.242	0.253	0.160	0.212	0.122
	UI-5	0.110	0.090	0.162	0.101	0.152	0.120	0.192	0.071
	UI-6	0.130	0.090	0.253	0.263	0.232	0.170	0.273	0.122
	UI-7	0.110	0.090	0.192	0.111	0.152	0.120	0.141	0.082
	UI-8	0.120	0.090	0.242	0.152	0.232	0.130	0.242	0.122
	UI-9	0.120	0.090	0.192	0.101	0.131	0.100	0.152	0.102
Q2	UI-1	0.150	0.100	0.162	0.141	0.192	0.130	0.172	0.061
	UI-2	0.140	0.070	0.172	0.141	0.202	0.120	0.182	0.061
	UI-3	0.150	0.080	0.182	0.162	0.192	0.110	0.192	0.051
	UI-4	0.130	0.110	0.242	0.263	0.253	0.170	0.192	0.092
	UI-5	0.120	0.070	0.162	0.152	0.182	0.130	0.182	0.071
	UI-6	0.140	0.090	0.253	0.283	0.263	0.190	0.253	0.071
	UI-7	0.120	0.090	0.182	0.152	0.172	0.130	0.162	0.071
	UI-8	0.120	0.100	0.232	0.222	0.253	0.130	0.232	0.071
	UI-9	0.140	0.090	0.192	0.172	0.172	0.100	0.172	0.061
Q3	UI-1	0.130	0.100	0.283	0.152	0.212	0.120	0.131	0.112
	UI-2	0.120	0.090	0.273	0.141	0.202	0.110	0.152	0.092
	UI-3	0.130	0.090	0.303	0.162	0.212	0.090	0.141	0.092
	UI-4	0.110	0.090	0.333	0.263	0.232	0.160	0.212	0.133
	UI-5	0.100	0.100	0.293	0.131	0.192	0.130	0.131	0.102
	UI-6	0.110	0.080	0.323	0.253	0.232	0.160	0.222	0.143
	UI-7	0.110	0.090	0.303	0.141	0.192	0.120	0.121	0.102
	UI-8	0.100	0.090	0.303	0.202	0.222	0.150	0.192	0.122
	UI-9	0.110	0.090	0.293	0.162	0.182	0.100	0.152	0.122

Table 5 shows the number of participants who answered Question 4. Table 6 shows the average volume of incentives. Table 7 shows the number of participants who answered about types of coupons. The upper part concerns the UUP, and the lower part concerns the NUUP. In Table 5, more than half of the participants in the UUP answered that points is an incentive to update. This result is the same as in our previous work [2]. In particular, more than 70% of users answered that points is an incentive to update in the case of the Communication, Finance, Lifestyle, and Health applications in the UUP. Therefore, points is an effective incentive for many users. In addition, the average volume of points is about 31.4–49.6 points in Table 6. We found that the average volume is at the same level between software and the user's understanding of the update procedure.

Table 7 shows that many users answered "restaurant", "convenience store", and "supermarket", and "drugstore". If we provide coupons, it is appropriate for users that the type of coupons is about food or daily necessities. We understand that the type of coupons differs between software. For example, in the case of the UUP, many users of the

Table 5. Number of participants who want to receive incentives by updating software (multiple choice)

Group	Incentive	OS	Communication	Finance	Lifestyle	Games	Utility	Health	Entertainment
UUP	Points	58	71	73	76	59	68	76	60
	Mobile data	25	27	23	27	32	27	23	30
	Coupons	19	28	25	36	19	33	26	26
	N/A	25	13	14	7	23	16	8	21
NUUP	Points	50	45	60	57	64	62	64	48
	Mobile data	26	20	19	26	22	27	22	27
	Coupons	18	18	23	31	19	16	18	15
	N/A	34	40	27	16	26	24	17	34

Table 6. Average volume of incentives

	OS	Communication	Finance	Lifestyle	Games	Utility	Health	Entertainment
UUP								
Points	45.1p	42.8p	35.5p	36.4p	39.0p	40.5p	40.1p	36.6p
Mobile data	1.0 GB	0.9 GB	0.6 GB	0.7 GB	0.7 GB	0.8 GB	0.8 GB	0.8 GB
Coupons	176.3 yen	251.8 yen	236.0 yen	229.2 yen	207.9 yen	166.7 yen	198.1 yen	198.1 yen
NUUP								
Points	38.9p	47.5p	49.6p	44.8p	31.4p	38.3p	35.3p	47.9p
Mobile data	0.9 GB	0.9 GB	0.9 GB	0.8 GB	1.2 GB	0.8 GB	0.7 GB	0.9 GB
Coupons	226.3 yen	180.6 yen	265.2 yen	209.7 yen	223.7 yen	137.5 yen	219.4 yen	250.0 yen

Communication application answer “convenience store” and “supermarket”, but many users of the Health application answer “convenience store” and “drug store”. Therefore, it is important to provide different coupons according to the type of software.

Table 7. Number of participants who want to receive each coupon

UUP	OS	Communication	Finance	Lifestyle	Games	Utility	Health	Entertainment
Restaurant	1	1	3	–	1	3	4	3
Convenience store	10	9	14	16	12	14	8	10
Supermarket	5	10	5	6	3	4	4	4
Drugstore	3	5	–	9	1	4	8	6
Furniture/ Interior store	–	–	–	–	–	–	–	–
Theater/ Theme park/ Tourist spot	–	–	1	2	–	1	–	1
Book	–	2	–	1	–	3	–	1
Clothing/ Miscellaneous goods/ Accessories	–	–	1	–	1	–	–	1
Home electronics retail store	–	1	–	1	–	2	2	–
Other	–	–	1	1	1	2	–	–
NUUP	OS	Communication	Finance	Lifestyle	Games	Utility	Health	Entertainment
Restaurant	2	–	3	8	4	1	1	3
Convenience store	7	9	9	15	6	7	9	8
Supermarket	2	3	7	3	3	2	3	1
Drugstore	5	6	2	5	4	2	1	2
Furniture/ Interior store	–	–	–	–	–	–	–	–
Theater/ Theme park/ Tourist spot	1	–	–	–	–	–	1	–
Book	1	–	1	–	1	2	1	–
Clothing/ Miscellaneous goods/ Accessories	–	–	1	–	1	–	2	–
Home electronics retail store	–	–	–	–	–	2	–	–
Other	–	–	–	–	–	–	–	1

5 Discussion

In this section, we describe appropriate approaches and the limitations of this paper.

5.1 Appropriate Message

We discuss appropriate UIs and messages (combination of phrase and UI) based on the result of this survey. In Table 4, effective UIs differs depending on the user's understanding of the update procedure, software, and phrase. In the case of changing only UI of message, it is important to show a different UI according to each user and each situation. In addition, there are no effective UIs in the case of the OS and Communication application for both of the two groups. Therefore, other elements are necessity to promote updating of the OS and communication application.

It is important to show a different message to each user because effective messages differ according to the user's understanding of the update procedure and also according to the software. For example, in the case of the Finance application, LFM-3&UI-4 and LFM-3&UI-8 are appropriate for the UUP, LFM-3&UI-4 are appropriate for the NUUP. We found that changing only the phrase, only the UI, both the phrase and the UI have different effects.

As described in our previous work [2], there are more effective phrases of the UUP than there are those of the NUUP. On the other hand, there are more effective UI of the NUUP than there are those of the UUP from the result of survey. Therefore, changing phrases is effective for the UUP, and changing UIs is effective for the NUUP. In particular, only in the case of changing phrases, the updating rate of the UUP is more than that of the NUUP, but in the case of changing both phrases and UI, the updating rate of the NUUP is improved to the same level as that of UUP.

These UIs, which include a red background and red button, are effective UIs to encourage smartphone users to update applications. This result indicates a similar tendency to updating the OS of computer users [3]. However, a red background and red button are not an effective UI to encourage smartphone users to update the OS. There is a difference between computer users and smartphone users in the case of updating the OS. It is a possible that the situation of use or usage environment has an effect. We will evaluate the reason for the difference between computer users and smartphone users.

5.2 Appropriate Volume of Incentives

There is no difference in the appropriate volume of points between the UUP and the NUUP, or among software. Before conducting this survey, we consider that there is a difference between the OS and other applications because updating the OS takes more time than updating other applications. However, we understand that users want to receive incentives in the case of not time but rather the volume of updating tasks. As described in Sect. 4.3, there is a difference in types of coupons between software. There is a possibility that gender and age of users affect on the result. Evaluating the effect of gender and age is future work.

5.3 Limitations

We conducted this survey for Android smartphone users of Japan. There is a possibility that different results would be obtained in the case of conducting surveys for different devices or in different countries. In this paper, we propose nine UI designs, but it is a possible that other UIs are effective. We will verify the effectiveness of other UIs in the future.

6 Conclusion and Future Works

We proposed some UIs to promote the updating of software and evaluated their effectiveness. As a result, different UIs are effective according to the user's understanding of the update procedure, software, and phrases. We found that it is important for each user to show an appropriate combination of phrase and UI. In addition, giving points is an appropriate approach for many users. In the future, we will implement demonstration system and conduct an experiment to verify the effectiveness of this proposed method.

Appendix: Questionnaire

- Q1–Q3: User action in response to each notification in Fig. 1
  1: I ignore the notification, 2: I check the notification, but I do not update, 3: I update later, 4: I update right now
- Q4: Incentives that users want to receive (Multiple choice allowed)
  1: Granting of points, 2: Granting of mobile data, 3: Granting of coupons, 4: N/A
- Q5: Volume of points that users want in order to update software
  Users answer "1" selected in Q4 ("Incentives that users want to receive")
  1: 1 point, 2: 5 points, 3: 10 points, 4: 30 points, 5: 50 points, 6: 100 points
- Q6: Volume of mobile data that users want in order to update software
  Users answer "2" selected in Q4 ("Incentives that users want to receive")
  1: 100 MB 2: 300 MB, 3: 500 MB, 4: 1 GB, 5: 1.5 GB, 6: 2 GB
- Q7: Type of coupons that users want in order to update software
  Users answer "3" selected in Q4 ("Incentives that users want to receive")
  1: Restaurant 2: Convenience store, 3: Supermarket, 4: Drugstore, 5: Furniture/Interior store, 6: Theater/Theme park/Tourist spot, 7: Book, 8: Clothing/Miscellaneous goods/Accessories, 9: Home electronics retail store, 10: Other
- Q8: Volume of coupons that users want in order to update software
  Users answer "3" selected in the Q4 ("Incentives that users want to receive")
  1: 50 yen 2: 100 yen, 3: 150 yen, 4: 300 yen, 5: 500 yen

References

1. Sano, A., Sawaya, Y., et al.: SeBeST: security behavior stage model and its application to OS update. In: Proceedings of the 35th International Conference on Advanced Information Networking and Applications, AINA, vol. 2, pp. 552–566 (2021)
2. Sano, A., Sawaya, Y., et al.: Proposal for approaches to updating software on android smartphone. In: Proceedings of the 18th International Conference on Broad-Band and Wireless Computing, Communication and Applications, BWCCA, pp. 94–108 (2023)
3. Sano, A., Sawaya, Y., et al.: Designing personalized OS update message based on security behavior stage model. In: Proceedings of the 18th Annual International Conference on Privacy, Security and Trust, PST (2021)
4. Weinschenk, S.: 100 Things Every Designer Needs to Know About People, 2nd edn. O'Reilly, Japan (2021)
5. Weinschenk, S.: 100 MORE Things Every Designer Needs to Know About People. O'Reilly, Japan (2016)

6. Johnson, J.: Designing with the mind in mind: simple guide to understanding user interface design guidelines, 2nd edn. Elsevier Inc. Impress Corporation (2014)
7. Goel, S., Williams, K., et al.: Understanding the role of incentives in security behavior. In: Proceedings of the 53rd Hawaii International Conference on System Sciences, pp. 4241–4246 (2020)
8. Goel, S., Williams, K., et al.: Can financial incentives help with the struggle for security policy compliance? In: Information and Management, vol. 58, no. 2 (2021)
9. Mathur, A., Engel, J., et al.: "They keep coming back like zombies": improving software updating interfaces. In: Proceedings of the Twelfth USENIX Conference on Usable Privacy and Security, SOUPS, pp. 43–58 (2016)
10. Sankarapandian, K., Little, T., Edwards, W.K.: TALC: using desktop graffiti to fight software vulnerability. In: Proceedings of the SIGCHI Conference on Human Factors in Computing Systems, CHI, pp. 1055–1064 (2008)
11. Ndibwile, J.D., Luhanga, E.T., et al.: A demographic perspective of smartphone security and its redesigned notifications. J. Inf. Process. **27**, 773–786 (2019)
12. Tian, Y., Liu, B., et al.: Supporting privacy-conscious app update decisions with user reviews. In: Proceedings of the 5th Annual ACM CCS Workshop on Security and Privacy in Smartphones and Mobile Devices, SPSM, pp. 51–61 (2015)
13. Google Play. https://play.google.com/store/games?hl=en-JP. Accessed 20 Nov 2023
14. Bravo-Lillo, C., Cranor, L.F., et al.: Your attention please: designing security-decision UIs to make genuine risks harder to ignore. In: Proceedings of the Ninth USENIX Conference on Usable Privacy and Security, SOUPS, pp. 1–12 (2013)
15. Felt, A.P., Ainslie, A., et al.: Improving SSL warnings: comprehension and adherence. In: Proceedings of the 33rd Annual ACM Conference on Human Factors in Computing Systems, CHI, pp. 2893–2902 (2015)
16. Felt, A.P., Reeder, R.W., et al.: Rethinking connection security indicators. In: Proceedings of the Twelfth Symposium on Usable Privacy and Security, SOUPS, pp. 1–14 (2016)
17. Turland, J., Coventry, L., et al.: Nudging towards security: developing an application for wireless network selection for android phones. In: Proceedings of the 2015 British HCI Conference, pp. 193–201 (2015)
18. Golla, M., Ho, G., et al.: Driving 2FA Adoption at Scale: Optimizing Two-Factor Authentication Notification Design Patterns, SOUPS, pp. 109–126 (2021)
19. Stokes, J., August, T., et al.: How language formality in security and privacy interfaces impacts intended compliance. In: Proceedings of the 2023 CHI Conference on Human Factors in Computing Systems, CHI, pp. 1–12 (2023)
20. Olwal, A., Lachanas, D., Zacharouli, E.: OldGen: mobile phone personalization for older adults. In: Proceedings of the SIGCHI Conference on Human Factors in Computing Systems, pp. 3393–3396 (2011)
21. Material Design: https://m2.material.io/design/guidelines-overview. Accessed 20 Nov 2023
22. Cross Marketing Inc. https://www.cross-m.co.jp/en/. Accessed 22 Nov 2023
23. Shimizu, H.: An introduction to the statistical free software HAD: Suggestions to improve teaching, learning and practice data analysis. J. Media Inf. Commun. 59–73 (2016)

A Non-interactive One-Time Password-Based Method to Enhance the Vault Security

Juarez Oliveira, Altair Santin, Eduardo Viegas([✉]), and Pedro Horchulhack

Pontifícia Universidade Católica do Paraná | Pontifical Catholic University of Parana — PUCPR, Graduate Program in Computer Science — PPGIa, Curitiba, Brazil
{juarez.oliveira,santin,eduardo.viegas,pedro.horchulhack}@ppgia.pucpr.br

Abstract. Multi-factor authentication (MFA) is recommended to access sensitive data applications. A password Vault protects secrets by storing privileged user credentials and access codes. The combination of MFA and Trusted Execution Environment (TEE) by multiple communication channels reduces the attack surface of secrets and enables secure periodic code updates from the password Vault. In this paper, we propose all these layers of protection and add a one-time password (OTP) mechanism to enhance the security of the Vault without human intervention. The expiration time of the code in the Vault remains unchanged. Finally, we show that web applications and an interactive remote shell are effectively secured by penetration testing from an adversary's point of view.

Keywords: One-Time Password · Trusted Execution Environment · Password' Vault · Mitre Att&ck

1 Introduction

Traditionally, cybersecurity focused on securing computer networks by strongly relying on cryptographic algorithms implemented in hardware and software to safeguard data storage and transportation. In response, the industry prioritized building hardware that facilitates secure software execution. Trusted Execution Environments (TEEs) enables the execution of source code that manipulates sensitive data, such as credentials, which remains encrypted within enclaves [28]. The primary purpose of these enclaves is to thwart unauthorized access to memory regions housing sensitive information.

Unfortunately, credential and session stealing are increasingly common, even in environments safeguarded by security mechanisms, including those implemented with cryptographic techniques. To address such a challenge, a widely adopted strategy involves implementing Multi-Factor Authentication (MFA) mechanisms [19]. However, MFA is susceptible to social engineering attacks, a risk that can only be mitigated through adequate user training.

L. Barolli (Ed.): AINA 2024, LNDECT 202, pp. 201–213, 2024.
https://doi.org/10.1007/978-3-031-57916-5_18

Centralized authentication, particularly in the Single Sign-On (SSO) model, is advocated as a best practice by the Center for Internet Security (CIS) Controls [4] and MITRE ATT&CK [10]. This approach is recommended due to the complexities associated with managing user credentials for each application in an enterprise environment. In the centralized Identity and Access Management (IAM) model, network administrators exert greater control over password expiration times, MFA devices, and the ability to block suspicious devices attempting unauthorized access. Notwithstanding, they also oversee users' permissions within a system. In this scenario, the IAM infrastructure can either opt for an on-premise solution (usually not recommended) or choose the preferable external alternative for the Identity Provider (IdP). This is because controlling an IAM server is challenging for an attacker due to the lateral movement difficulty, particularly when dealing with a remote and well-secured IdP domain. Through lateral movement, adversaries can pave the way for unauthorized access to a device or system via application flaws or credential theft, expanding their influence across other systems or devices and gaining control over privileged credentials. Privileged Access Management (PAM) offers effective protection, particularly through a password vault component [14].

To mitigate the risk of credential theft, imposing an expiration time for credentials (passwords/codes) is a common practice [25]. Implementing a credential expiration time can booster security but might simultaneously reduce usability and productivity. This underscores the need for adopting new best practices to ensure the safeguarding of these credentials. For example, a common industry practice involves setting a secret expiration time, often at 8 hours, implying that within this timeframe, an attacker could exploit system vulnerabilities by concealing their actions or acting as a malicious insider. Another scenario involves a hacker targeting the PAM to obtain the secret for system access, with an 8-hour window to attempt unauthorized access. This complicates the management of credential validity because if administrators reduce this time, users will need to change their secrets (passwords/codes) more frequently as soon as the validity period expires. As a result, enhancing security could potentially negatively impact productivity due to the increased frequency of token renewal.

In light of this, this paper proposes a new secure mechanism to maintain usability and productivity while also improving vault security in a non-interactive approach. Our proposed scheme combines MFA with TEE to reduce the attack surface for application secrets. Notwithstanding, we introduce an integrated approach that features the implementation of a One-Time Password (OTP) mechanism aimed at augmenting the vault security seamlessly without requiring human intervention. This innovative method enables the system administrator to maintain the expiration time of codes in the vault while not affecting the system's usability and keeping its security.

In summary, our main paper contributions are:

- A new non-interactive mechanism for vault security without compromising usability. The model combines MFA with TEE to reduce the attack surface and introduces an OTP method to seamlessly augment vault security.

– A rigorous prototype penetration testing that validates the effectiveness of our model, ensuring robust security for web applications and an interactive remote shell from an adversary's viewpoint.

2 Background

Intel Software Guard Extension (SGX) implements TEE by leveraging hardware-centric security protocols [16]. By delineating secure enclaves within the processor architecture, SGX facilitates the establishment of isolated domains where confidential computations occur with heightened security. These enclaves function as secure environments, shielding sensitive data and code from unauthorized access, even by privileged software layers. This ensures protection through cryptographic measures, employing encryption to secure the enclave's memory contents. Furthermore, robust secure initialization procedures fortify the enclave's integrity. Aligned with the overarching TEE framework, SGX stands as a concrete and effective solution for creating secure computational environments within processors.

In practical implementation, leveraging SGX involves meticulously considering enclave design, secure coding practices, and enclave attestation. Enclave design necessitates a judicious delineation of the specific computations requiring heightened security, ensuring they are encapsulated within the enclave. Secure coding practices involve adhering to SGX-specific guidelines, employing appropriate cryptographic primitives, and meticulously managing memory within the enclave. Enclave attestation, a critical component, involves verifying enclaves' integrity, confirming that they have not been compromised.

CIS Controls is a well-recognized framework that features 18 control groups designed to support the development of robust mitigation and prevention controls. As an example, the integration of Vault as a PAM system is advocated, as it paves the way for security through password and passwordless mechanisms. In particular, the framework enforces MFA for Vault access, which can be implemented using PyOTP as a standard Python library [12], while also exploring OTP variants, such as Time-Based One-Time Password (TOTP) and the HMAC-based One-Time Password (HOTP) into PAM architectures.

Apart from introducing strong cybersecurity mechanisms, frameworks also focus on identifying and addressing security vulnerabilities. As an example, Common Vulnerabilities and Exposures (CVE) is used as a comprehensive vulnerability catalog that is linked to the NIST Vulnerability Database[11]. In such a case, each identified vulnerability is mapped and measured based on their Common Vulnerability Scoring System (CVSS) score, which tracks the vulnerability impact on the system. On a similar path, the Mitre ATT&CK framework is a globally accessible knowledge base detailing adversaries' techniques and tactics, utilizing the Tactics, Techniques, and Procedures (TTP) framework [10]. In such a context, the Penetration Testing Execution Standard (PTES), which is a systematic and repeatable security assessment framework that covers various stages in the cyber kill chain, plays a crucial role [1].

3 Related Works

Wu et al. introduced SGX-UAM, a unified authentication scheme that leverages Intel SGX and OTP to address Man-in-the-Middle (MitM) and replay attacks. Despite acknowledging potential hardware costs associated with Intel SGX, the authors underscored the heightened resilience achieved in Unified Access Management. In a performance comparison, the proposed scheme demonstrated comparable authentication times to OpenID and OAuth2. However, it is noteworthy that a dedicated security evaluation was not conducted in the study [27].

The Confidential Computing Consortium assesses contemporary confidential computing technologies, with a particular focus on hardware features within TEE, such as code confidentiality. While the objective is to limit access for platform operators, it's important to note that trusted computing has limitations and may not provide a comprehensive defense against all potential attack vectors, as highlighted in the evaluation [15].

Henricks and Kettani [19] explore the merits of MFA, highlighting its strength as an additional layer beyond password authentication. Their discussion extends to potential future considerations, particularly the prospect of using human DNA characteristics for authentication. Emphasizing the importance of prudent security considerations, the authors underscore the need for careful evaluation in implementing such advanced authentication methods.

Fisher [17] underscores the significance of Vault as a central component in PAM, providing protection against a range of security threats, including credential abuse and unauthorized privilege escalation. Despite its pivotal role, Fisher highlights the importance of complementing Vault with additional secure programming techniques to attain optimal security outcomes, particularly in intricate computing environments.

4 Modeling the Vault Security Enhancement

4.1 A Non-interactive One-Time Password-Based Method

We propose a mechanism that consistently employs a Vault for storing and managing secrets, encompassing tokens, codes, and passwords. This approach minimizes the exposure surface by mitigating reliance on credentials stored in files and environment variables.

The web server, exposed to the Internet, connects to other machines via API (see Fig. 1). User authentication, in steps 1 to 3 of Fig. 1, can be an internal or external IdP, based on integration needs and adherence to internal policies.

The OTP token generation (Fig. 1, steps 7 and 11) occurs on the TEE server for MFA, preventing unauthorized PAM access and securing the OTP generator's seed. This approach, integrated into our proposal, adds complexity for potential adversaries, especially in critical environments. An alternative technique involves storing the OTP generator seed in a Vault, eliminating the need for writing local secrets to a file. In both cases, steganography is used to protect code in transit

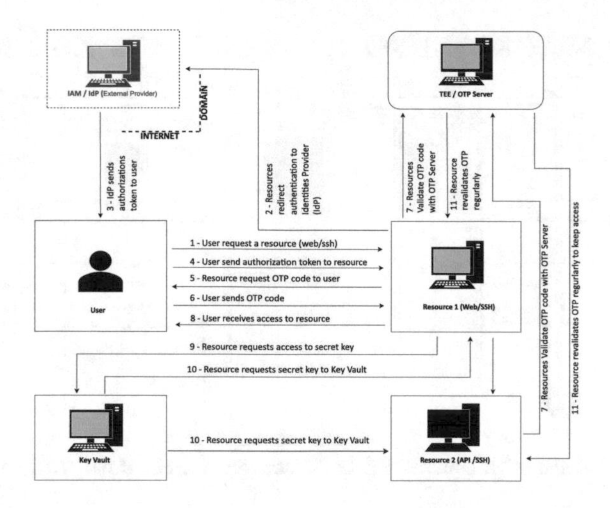

Fig. 1. Overview of our proposed enhanced credential Vault security scheme.

through APIs, highlighting the value of a TEE implementation for swift and secure control implementation.

Post-authentication, whether for a web application user, secure shell user, or other non-interactive service, the validity time of the credential or token (Fig. 1, steps 9 and 10) is crucial. It determines the vulnerability window for potential adversaries, representing the timeframe for token exploitation. Accurate calculation of this period and the inclusion of additional authentication factors beyond the password are pivotal in maintaining service security. The dual validation of credentials, performed in addition to the initial authentication process (Fig. 1, steps 1 to 3, and outlined in Fig. 2a:, detailed events from *userToken* to Data and from *SendOTPCode* to *authOTPuser*), occurs periodically on the webserver for API access (Fig. 1, steps 5 to 7 and steps 8 to 11). This dual validation enhances the overall security level and ensures that credentials or even the active session remain safeguarded against potential theft.

Continuous OTP monitoring facilitates quick responses to security issues, like terminating sessions or SSH services. Post-SSH authentication (Fig. 2b:, events from *requestUserAuth* to *SSHConnection*) in the IdP, a script for OTP verification is triggered (event *sendOTPcode*), awaiting the user's second validation (event *authOTPuser*). Access to the Vault for a single-use access credential occurs during SSH authentication

This MFA mechanism is embedded in a TEE, with a second validation during web and SSH authentication. The OTP code is transmitted non-interactively and independently after successful IdP authentication. Even if an adversary steals a credential from the Vault, they cannot use it without completing steps 1 to 4 in Fig. 1. In case of OTP-based authentication failure, the connection or session is automatically blocked, adding an extra layer of security for critical resources.

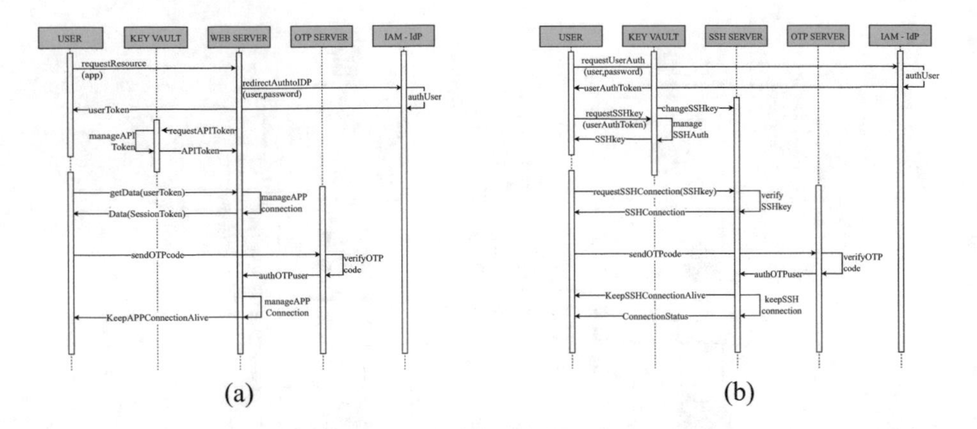

Fig. 2. a: WEB Auth with centralized OTP server. b: SSH auth with centralized OTP server.

4.2 Threat Model

The adversary's goals include reconnaissance of the environment (TA0043 in MITRE ATT&CK) [2], involving identifying services, open ports, and understanding authentication processes [22]. Once familiar with the target's characteristics and weaknesses, the attacker develops a strategy, potentially using automated tools for password cracking or social engineering tactics to acquire valid credentials from users.

The attacker's direct actions depend on the level of asset protection, including firewalls, web server restrictions, MFA, connection limits, and Web Application Firewall (WAF). To assess the adversary's capabilities, two scenarios are considered: (*i*) an Internet-based attacker with limited access to the web server and (*ii*) an internal network-based attacker with access to both SSH and web ports.

We acknowledge the potential interception of MFA authentication through keyloggers on compromised systems. External channel interception, like email or SMS, is possible without adequate service provider protection. Mitigation includes user training (TTP 1111 [21]) to remove authentication cards when not in use. Monitoring system APIs is crucial to detect malicious actions, albeit requiring additional effort [18].

In an alternative adversary action, authentication assets like password hashes, Kerberos tokens, or application OTP tokens can be stolen through improper access to RAM or configuration files. To mitigate, managing privileged accounts by enforcing the principle of least privilege (aligned with TTP 1550 [3]) is crucial. This extends to ordinary user accounts to prevent unnecessary access and privileges. Detection involves monitoring session creation and resource access logs.

In the context of Remote Services, specifically Mitre's SSH (ID: T1021.004) [13], adversaries exploit legitimate user information for unauthorized network access via the SSH protocol. APT39 is a known threat group using this technique. Mitigation includes turning off unnecessary SSH services, implementing MFA, and limiting user access. Detection involves monitoring session and user creation, and network connections. Utilizing a Vault aids in mitigation, and MFA with OTP facilitates monitoring, allowing service deactivation in case of misuse. Therefore, considering such a threat model, in this work, we want to answer the following research question: *How is it possible to provide a secure mechanism to increase the security of the Vault in a non-interactive way, aiming at non-interference in the usability of the approaches in use?*

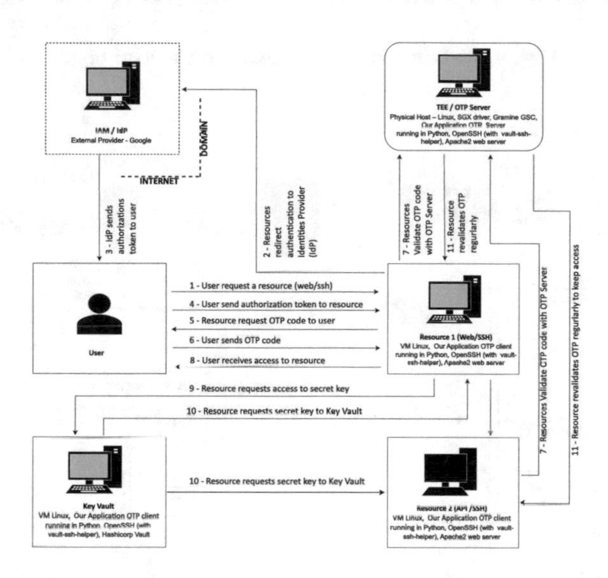

Fig. 3. Prototype Overview

5 Prototype

We implemented a prototype with a web application server on a virtual machine using Ubuntu Server 22.04 (Fig. 3). The setup includes an Apache2 web server, openSSH server, Python-based OTP client, and a Vault SSH client agent (vault-ssh-helper). The IdP is considered an external entity in this configuration.

The Vault, utilizing HashiCorp Vault solution [7], is deployed on a virtual machine with Ubuntu Server 22.04. The configuration includes a Vault SSH client (vault-ssh-helper) and a Python-based OTP client application. Integrating Vault aligns with best practices for Access Management, enhancing protection against unauthorized system access by implementing PAM, as per CIS controls.

The OTP Server application server, developed in Python with the PyOTP library [12], is on a physical machine running CentOS 8 [26]. The hardware includes a 2.70 GHz Intel(R) Core(TM) i7-7500U processor with INTEL Gramine GSC, operating within a Docker container [20]. The system utilizes Intel/SGX drivers for the TEE environment. This server also hosts a critical Python-based API and a Vault SSH client (vault-ssh-helper).

```
login as: juarez
juarez@192.168.153.154's password:
Please, input your personal OTP code:441512
OTP code sent: 441512
OTP code system: 441512
The OTP Code is correct! Welcome to this host!
juarez@webapp04:~$
```

```
login as: juarez
juarez@192.168.153.154's password:
Please, input your personal OTP code:999111
OTP code sent: 999111
OTP code system: 443822
The OTP Code is wrong. You will be desconnected. Bye bye!
```

Fig. 4. Validation screen displayed to the user

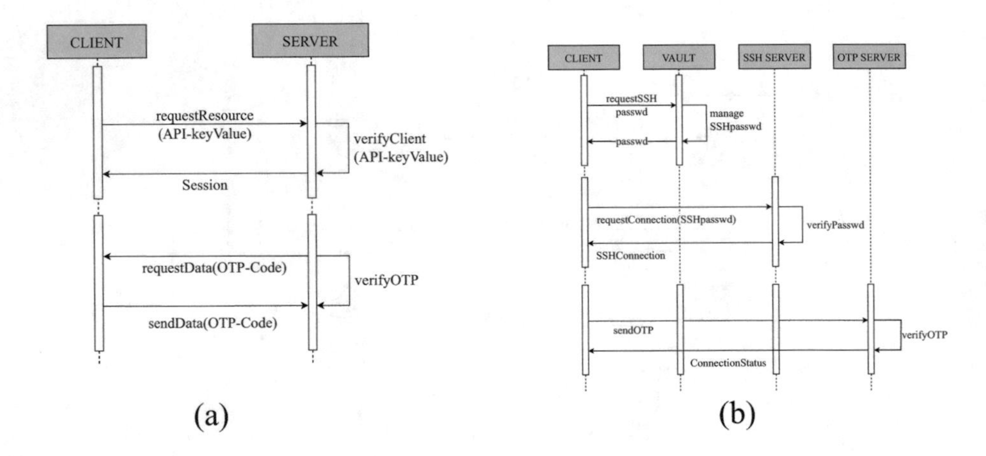

Fig. 5. a: API Authentication metaprotocol b: SSH Auth metaprotocol

Concurrently with API authentication, the user undergoes IdP authentication. A distinct OTP authentication step follows to ensure access credential integrity and validate against compromise. OTP re-validation can occur at initial login and periodically during the session.

SSH connection utilizes a one-time code/password from the Vault. After standard authentication, an internal post-script on the target server validates the OTP within the secure server (SGX-Gramine). The SSH connection finalizes only on successful secondary validation, with periodic re-validation possible. The validation screen (Fig. 4, left) displays the correct code for debugging.

The integration between the OTP application on the secure server and the Vault utilized the HVAC library [8], as shown in Fig. 5b:. This approach enables password input for the initial connection without storing it in a file or environment variable. An alternative is steganography, concealing connection information within a file for added complexity against potential attackers. If credentials are entered once, the application prompts re-entry upon each restart. A Docker container with Intel Gramine GSC [6] enhances portability and reduces dependence on the host operating system, chosen for its increased portability [9].

Installation of OTP clients and monitoring application usage was successful within a local network. For external web clients, continuous authentication without interaction is challenging, necessitating periodic entry of the OTP code. Although introducing inconvenience, this measure significantly enhances security, as depicted in Fig. 5b:.

6 Evaluation

In the case of an Internet-based attacker limited to port 443, the implemented MFA by IdP Single Sign-On substantially reduces the attack surface. Even if the attacker captures login credentials or gains access, OTP token software triggers automatic disconnection after the configured time limit, limiting the window of opportunity.

For API authentication on a separate host without external Internet access, exclusive OTP token authentication without human interaction prevents the attacker from sustaining the connection, even with authentication data (key, secret). The OTP token, transmitted with the request, can use steganography for added complexity.

An alternative in this work is sending the TOTP code out-of-band, parallel and over another channel. Every authentic code request using an API triggers the sending of a HOTP code based on the OTP code sequence. If an attacker attempts to invoke an API without the correct HOTP sequence code, the operation will not be confirmed. Our proposal includes a manual validation step for the HOTP sequence code to monitor these actions.

We have implemented manual validation of the HOTP sequence code in our proposal to track these actions. For more detail, see Fig. 6a: – verification of an OTP code from an OTP application running on the TEE, which has been successfully validated, and Fig. 6b: – verification of an OTP code from an OTP application running on the TEE, which has been successfully validated.

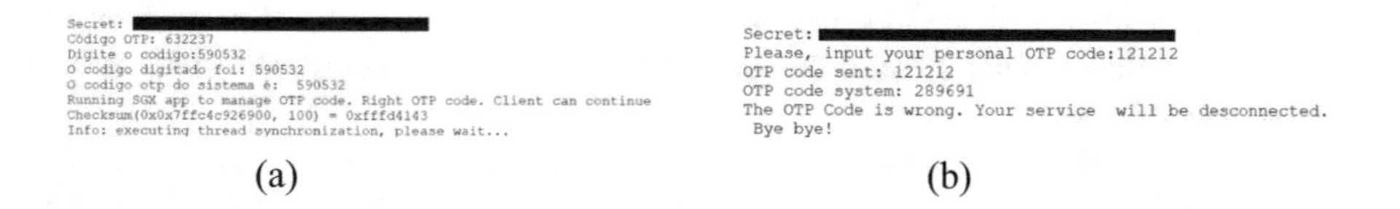

Fig. 6. a: OTP App Server – Auth OK. b: OTP App Server – Auth Not OK.

7 Security Analysis

Following cybersecurity best practices, our model emphasizes layer-based security. Procedures are structured to prevent the failure of one layer from compromising overall security. The design includes two additional layers: (*i*) robust data protection within a TEE for system applications and (*ii*) an extra authentication factor with OTP implemented out-of-band. This multi-layered approach enhances the overall security framework.

External access rules via the firewall are implemented, allowing access to TCP ports exclusively for the web service externally and for SSH, Vault, and the OTP server internally. The application client's external access is restricted to the web server, minimizing the exposure surface for enhanced security.

To evaluate the security achieved, penetration tests were conducted using established methodologies like Metasploit [23,24]. These tests aimed to identify potential attacker capabilities and assess the exposure surface.

An external NMAP scan revealed ports 443 (HTTPS) on the web server for internet connections and internal ports 4443 (OTP application), 8201 (key vault), and 22 (SSH). The accessibility of these ports indicates successful firewall-based security measures.

Additional tests simulated adversary actions using tools from PTES [1], focusing on *Customized Exploitation* and *External Footprints*. These tests provided insights into the system's resilience against exploitation and external footprinting attempts.

In assessing credential theft potential over the local network, TCPDump analysis indicated data encryption in transit, particularly over HTTPS and TLS connections to the OTP Socket, reinforcing security measures.

A memory dump test to access OTP in memory demonstrated encryption by SGX, making direct access challenging. Subsequent tests with the Volatility tool affirmed the unavailability of clear-text memory contents due to encryption.

The initial penetration test targeting SSH server configured without OTP revealed vulnerability to brute-force dictionary attacks. After access was gained, the connection was promptly revoked. Notably, there were no additional security measures on the SSH server, such as attempt limits, source IP blocks, or MFA authentication in PAM.

In a subsequent SSH server test to evaluate OTP effectiveness, a simulated insider attack with Metasploit was executed. Initial connection success was followed by a password change due to Vault's dynamic password generation. Valid SSH authentication led to OTP verification on TEE, terminating the connection. While Metasploit sessions could be created with correct credentials, shell access was not established.

A Hydra brute-force attack on the web server, assuming an Internet-based attacker on port 443, yielded no results without OTP configuration. Acknowledging potential credential acquisition through alternative means, we activated OTP as a second factor, prompting re-authentication to terminate unsuccessful connections. Balancing heightened security with usability and productivity considerations requires careful analysis due to the risk of unavailability, posing a security concern.

In assessing continuous authentication via APIs without human interaction, we tested the OTP client periodically dispatching codes. The OTP server in the TEE scrutinized the HOTP sequence's validity, successfully executing a pre-configured action for an artificially transmitted incorrect code.

Authentication check intervals vary based on the service; SSH might require a one-minute check, while databases or external API access may need longer intervals. Administrators must configure session time parameters based on specific use case requirements.

For SSH access on Linux servers, implementing tools like Fail2ban [5] or EDR protection helps thwart brute-force attempts. Additional security measures include configuring hosts.deny, hosts.allow, Apache2 web server location match rules, and applying firewall rules and host hardening practices, contributing to a layered security approach.

Vault integration has effectively reduced access to credentials traditionally stored as plain text, enhancing overall security by mitigating risks associated with long credential validity times. Secure storage in Vault aligns with best practices, reducing exposure surface and potential vulnerabilities.

Implementing multiple layers of protection challenges intruders by restricting the exposure surface and introducing temporal constraints through frequent code or password changes. This multi-layered approach reinforces overall system resilience against potential threats.

The research aimed to enhance system security without compromising productivity or user-friendliness. The proposed model successfully integrated an OTP mechanism concurrently with the conventional approach, maintaining PAM-based workflow.

While PAM credential validity remained around 8 h, the OTP mechanism introduced more frequent out-of-band validations in minutes, significantly raising the bar for attackers. Configurable time validity adapts to specific application needs, contributing to strengthened security while maintaining operational efficiency.

8 Conclusion

Implementing applications within a trusted execution environment introduces complexities for system administrators, especially in routine vulnerability management. To address this challenge, our choice of utilizing libOS Gramine helps navigate this intricate landscape.

The use of Vault for credential management aligns with established practices in major cybersecurity frameworks, underlining its significance as a robust security measure. Although adjusting session times for applications and services poses a considerable challenge, it remains imperative for overall environment protection. In our work, employing an out-of-band OTP validation strategy proved effective even with low session times, demonstrating reconfigurability. This approach fortifies security and poses challenges for potential attackers aiming to compromise the proposed security based on OTP.

The viability of the proposed mechanism aligns with project objectives, which were observed through security analyses and penetration tests. The findings indicate an enhancement in security without altering the time validity in the PAM, thereby favoring the tradeof between security, usability, and user productivity.

References

1. The penetration testing execution standard (2014). http://www.pentest-standard.org/index.php/Main_Page
2. Reconnaissance – tactic ta0043 (2020). https://attack.mitre.org/tactics/TA0043/
3. Use alternate authentication material (2022). https://attack.mitre.org/techniques/T1550/
4. CIS (2023). https://www.cisecurity.org/controls
5. Fail2ban: ban hosts that cause multiple authentication errors (2023). https://github.com/fail2ban/fail2ban
6. Gramine - a library OS for unmodified applications (2023). https://gramineproject.io/
7. HashiCorp developer (2023). https://developer.hashicorp.com/
8. HVAC – HVAC 1.2.1 documentation (2023). https://hvac.readthedocs.io/en/stable/
9. Intel software guard extensions – developer guide (2023). https://download.01.org/intel-sgx/latest/linux-latest/docs/Intel_SGX_Developer_Guide.pdf
10. MITRE ATT&CK (2023). https://attack.mitre.org/
11. National vulnerability database (2023). https://nvd.nist.gov/
12. PyOTP (2023). https://pyauth.github.io/pyotp/
13. Remote services: SSH (2023). https://attack.mitre.org/techniques/T1021/004/
14. Cheng, H., Li, W., Wang, P., Chu, C.H., Liang, K.: Incrementally updateable honey password vaults. In: 30th USENIX Security 21, pp. 857–874 (2021)
15. Consortium, C.C.: A technical analysis of confidential computing. Tech. rep. (2023). https://confidentialcomputing.io/wp-content/uploads/sites/10/2023/03/CCC-A-Technical-Analysis-of-Confidential-Computing-v1.3_unlocked.pdf
16. Fei, S., Yan, Z., Ding, W., Xie, H.: Security vulnerabilities of SGX and countermeasures: a survey. ACM Comput. Surv. **54**(6), 1–36 (2021)
17. Fisher, P.: Privileged access management (PAM) demystified (2023). https://www.oneidentity.com/what-is-privileged-access-management/
18. Geremias, J., Viegas, E.K., Santin, A.O., Britto, A., Horchulhack, P.: Towards multi-view android malware detection through image-based deep learning. In: 2022 International Wireless Communications and Mobile Computing (IWCMC). IEEE (May 2022)
19. Henricks, A., Kettani, H.: On data protection using multi-factor authentication. In: Proceedings of the 2019 International Conference on Information System and System Management. ISSM 2019, ACM (Oct 2019)
20. Horchulhack, P., Viegas, E.K., Santin, A.O., Ramos, F.V., Tedeschi, P.: Detection of quality of service degradation on multi-tenant containerized services. J. Netw. Comput. Appl. **224**, 103839 (2024)
21. Lambert, J.: Multi-factor authentication interception (2023). https://attack.mitre.org/techniques/T1111/
22. dos Santos, R.R., Viegas, E.K., Santin, A.O.: A reminiscent intrusion detection model based on deep autoencoders and transfer learning. In: 2021 IEEE Global Communications Conference (GLOBECOM). IEEE (Dec 2021)

23. dos Santos, R.R., Viegas, E.K., Santin, A.O., Tedeschi, P.: Federated learning for reliable model updates in network-based intrusion detection. Elsevier Comput. Secur. **133**, 103413 (2023)
24. dos Santos, R.R., Viegas, E.K., Santin, A.O., Cogo, V.V.: Reinforcement learning for intrusion detection: more model longness and fewer updates. IEEE Trans. Netw. Serv. Manage. **20**(2), 2040–2055 (2023)
25. Taherdoost, H.: Understanding cybersecurity frameworks and information security standards-a review and comprehensive overview. Electronics **11**(14), 2181 (2022)
26. Viegas, E., Santin, A., Bachtold, J., Segalin, D., Stihler, M., Marcon, A., Maziero, C.: Enhancing service maintainability by monitoring and auditing SLA in cloud computing. Clust. Comput. **24**(3), 1659–1674 (2020). https://doi.org/10.1007/s10586-020-03209-9
27. Wu, L., Cai, H.J., Li, H.: SGX-UAM: a secure unified access management scheme with one time passwords via intel SGX. IEEE Access **9**, 38029–38042 (2021)
28. Xia, K., Luo, Y., Xu, X., Wei, S.: SGX-FPGA: trusted execution environment for CPU-FPGA heterogeneous architecture. In: 2021 58th ACM/IEEE DAC (Dec 2021)

Causal Inference to Enhance AI Trustworthiness in Environmental Decision-Making

Suleyman Uslu[1], Davinder Kaur[1], Samuel J Rivera[2], Arjan Durresi[1](✉), and Meghna Babbar-Sebens[2]

[1] Indiana University -Purdue University Indianapolis, Indianapolis, IN, USA
{suslu,davikaur}@iu.edu, adurresi@iupui.edu
[2] Oregon State University, Corvallis, OR, USA
{sammy.rivera,meghna}@oregonstate.edu

Abstract. We present a causal model specifically designed for an environmental decision-making context, focusing on agricultural choices and the delicate balance between changes in water policy and agrarian profit. Our comprehensive causal framework incorporates various criteria, including trust sensitivity, which explains the interaction between trust and policy change, actor-specific factors such as location, AI capability, and community awareness of these factors. Additionally, we introduce "trustworthy acceptance" as a metric to measure decision-making progress. The results show that implementing community awareness interventions can increase actors' trust by up to 21%, thereby reducing the decline in trustworthy acceptance even when profits decrease. By constructing this causal model for environmental decision-making with trade-offs, we can provide a nuanced measurement of outcomes, thus illuminating the complex dynamics of such scenarios.

1 Introduction

Decision-making, particularly in natural resource management, often involves many stakeholders with diverse expertise, perspectives, and values. Reaching a consensus among these stakeholders is essential before fully embracing decisions. The absence of consensus among decision-makers has resulted in major global crises like the Global Financial Crisis in 2008 [28]. Therefore, it is imperative to have mechanisms that can identify and foster consensus among stakeholders with conflicting views, as they play a vital role in the overall decision-making process.

While various methods have been proposed to model or simulate consensus among agents with different behaviors, real-world scenarios often involve non-experts with dynamic opinions shaped by conflicting interests. For instance, landowners may seek economic profit, but environmental regulations impose constraints for sustainability [58]. Additionally, social influence theory highlights how external pressures can shape attitudes and behaviors among actors [5]. In situations requiring social consensus, social validation emerges as a key driver, where decisions align with the actions of others [4]. This psychological concept is exemplified in studies showing how social validation influences behaviors, such as volunteering among university students or energy conservation in residential communities [13,31].

L. Barolli (Ed.): AINA 2024, LNDECT 202, pp. 214–225, 2024.
https://doi.org/10.1007/978-3-031-57916-5_19

Amid trade-offs and competing interests, elucidating algorithmic solutions to the community becomes fundamental, especially as artificial intelligence (AI) tools gain prominence in water management scenarios [25,59]. Convincing the community is crucial, particularly when decisions necessitate bottom-up adoption of recommended options, and regulatory enforcement is challenging. Top-down approaches in watershed management have limitations in representing diverse interests [26]. Therefore, a multidisciplinary approach involving actors from various disciplines is essential to comprehend diverse needs, particularly in water governance systems [3].

Navigating the complexities of decision-making, particularly in natural resource management, underscores the importance of understanding the underlying reasons and root causes of decisions and outcomes [9,32]. Establishing causal models emerges as a pivotal tool, providing a structured framework to unravel the interconnected relationships among variables and identify key drivers influencing decision processes [32]. These models facilitate a more comprehensive understanding of decision-making, empowering stakeholders to effectively address conflicting interests and trade-offs in a systematic manner [9]. By discerning the impact of different factors, causal models contribute to developing informed and sustainable solutions in environmental decision-making [9,32]. As decisions often have enduring consequences, utilizing causal models becomes essential for fostering consensus, transparency, and trust among diverse stakeholders [9,32].

In the contemporary world, intricate decisions frequently involve a multitude of stakeholders [28]. In certain scenarios, experts in pertinent fields step into the role of stakeholders, participating in discussions and iterative rounds of proposals and adjustments to converge on a consensus. Dong et al. [9] proposed an approach to minimize the necessity for extensive modifications to expert solutions. Additionally, Hegselmann et al. [16] introduced a model for opinion dynamics, utilizing discrete-time intervals to represent opinions as real numbers presented by agents. However, in sectors related to natural resource management, actors may be guided by conflicting and competing interests rather than fixed deviations. This is exemplified in the behavior of landowners who are expected to prioritize revenue generation while simultaneously adhering to the environmental sustainability goals set by regulatory agencies [58].

Trust is widely acknowledged as a context-dependent concept with a pivotal role in the decision-making process [46]. However, the computational challenges associated with trust management escalate when handling numerous stakeholders, historical measurements, and trust transitivity [37]. Numerous studies [35–37,61] have delved into and investigated trust management. Ruan et al. [43] introduced a trust management framework grounded in measurement theory to assess and predict trust among entities within a network. This framework has found application across diverse domains, including trust management in social networks [41,60], cloud computing [38,39,49], the Internet of Things [40,42], identification of fake users [19], detection of damaging users [33], and crime detection [22,23].

Numerous researchers and organizations are actively involved in formulating standards and requirements on trustworthiness, explainability, acceptance, and fairness in AI decision-making systems [14,20,24]. The applicability of AI-driven methodologies in environmental models and various domains has been extensively discussed in the literature [34]. However, recent investigations have underscored that the potential risks of employing AI in environmental domains mirror those in other fields, emphasizing the need for ethical, responsible, and trustworthy AI approaches [29]. Addressing these requirements, concrete metrics have been proposed to assess such criteria, notably in vital sectors like healthcare [21] and the environment [54–56]. These studies not only introduce innovative methodologies for embedding requirements within consensus frameworks but also elucidate how such frameworks can facilitate effective collaboration between humans and machines to achieve desired decision outcomes efficiently.

Concerning the integration of trust dynamics into the management of natural resources over extended durations, we introduced a trust-centric decision support system in our prior works [50–52,57]. By leveraging a trust management framework, this system streamlined environmental decision-making processes by gauging actor trust levels, disseminating this information, and fostering negotiations among actors to attain consensus. Our findings illustrated that communities embracing trust tend to achieve consensus in fewer rounds, framing the decision-making process as a negotiation where choices evolve and ultimately stabilize. To further comprehend this dynamic, we also employed game theory and control theory in modeling the decision-making dynamics [51,53].

These investigations provided insights into overseeing decision-making processes and predicting the consensus degree based on the trust levels among actors. The outcomes underscore the advantages of incorporating actor perspectives and deploying a trust management framework, particularly in the natural resource management systems. However, understanding the underlying reasons behind decisions in environmental decision-making is crucial as it unveils the motivations, influences, and causal factors, providing valuable insights for informed policy development, sustainable resource management, and effective mitigation strategies [47].

In the realm of causal inference, two prominent paradigms, potential outcomes and structural causal models, offer distinct frameworks for modeling causality. Potential outcomes, grounded in the counterfactual framework, involve considering the outcomes that would have occurred under different treatment conditions [44]. In contrast, structural causal models provide a graphical representation of causal relationships, offering a visual tool for understanding and modeling the dependencies among variables. D-Separation, a key concept in causal graphical models, aids in identifying conditional independence relationships among variables, facilitating the discernment of causal connections [32]. This work builds a causal graphical model for environmental decision-making involving a trade-off between two outcomes.

Estimation approaches in causal inference encompass a broad spectrum of methods. Researchers employ various techniques to estimate causal effects, from traditional randomized controlled trials to observational studies. Propensity score matching, instrumental variables, and doubly robust estimators are among the many tools used to mitigate confounding and derive causal inferences from observational data [18,44]. Machine learning methods, particularly in the form of causal discovery algorithms, offer an additional avenue for estimating causal structures directly from data, enhancing the field's capacity to unravel complex causal relationships [45].

2 Environmental Decision Making and Causal Model

Water management, especially during challenges like droughts, necessitates actions such as redistributing available water, utilizing diverse sources, and sharing associated costs [30,48]. Establishing trust is vital for effective administration and sustainable partnerships in water governance, demanding swift coordination among diverse stakeholders [15]. However, challenges like knowledge disparities and limited community involvement hinder trust development [7]. Stakeholders' diverse origins, conflicting values, and competing interests further complicate matters [2,11,27].

This paper explores the nuanced water, energy, and land resources management in agrarian communities, particularly focusing on decisions on irrigation water rights. These decisions shape the natural resource management systems, impacting both the environment and economic productivity [8]. A multi-level transferable drought management framework addresses the diverse scales of decisions and their impacts in temperate agriculture [17]. It systematically explores interconnected drought impacts and responses, emphasizing social capital elements like trust.

In response to challenges, the region considers infrastructure development plans to supply Columbia River water to farmers in critical groundwater areas [6]. Our investigation focuses on a scenario where farmers with groundwater rights could potentially switch to the Columbia River as their water source. While involving trade-offs, this scenario prevents further groundwater depletion, allowing gradual replenishment. Our study primarily delves into understanding the role of trust in guiding community decisions within this specific management approach.

This study introduces a causal model for environmental decision-making focusing on the redistribution of water rights by substituting groundwater rights with river water rights. Employing an AI-based system, we generated quasi-optimal decisions for field actors tasked with choosing the decision that best aligns with their interests and the community's welfare. However, a trade-off exists among these objectives, where the most profitable solution favored by the actor might not align with the most environmentally friendly choice. In our study, trust is rooted in the environmental soundness of decisions. Consequently, decisions not aligned with environmental goals receive low trust from the community.

Despite the reliance on AI-generated decisions, two exceptions highlight the importance of human involvement in the decision-making process. Firstly, actors situated far from the river may be hesitant to replace groundwater usage with river water due to elevated costs. Secondly, if the AI-based decision generation fails to produce acceptable trade-offs and instead introduces bias, actors refrain from blindly selecting low-profit decisions. To account for these exceptions, we employed pyAgrum's `setEvidence()` method to set the evidence for each scenario. PyAgrum [1] serves as a Python interface designed to harness the capabilities of aGrUM [10,12] — a C++ library that offers cutting-edge implementations of graphical models tailored for decision-making scenarios. These models encompass a range of structures, such as Bayesian Networks, Markov Networks (Markov random fields), Influence Diagrams, Credal Networks, and Probabilistic Relational Models.

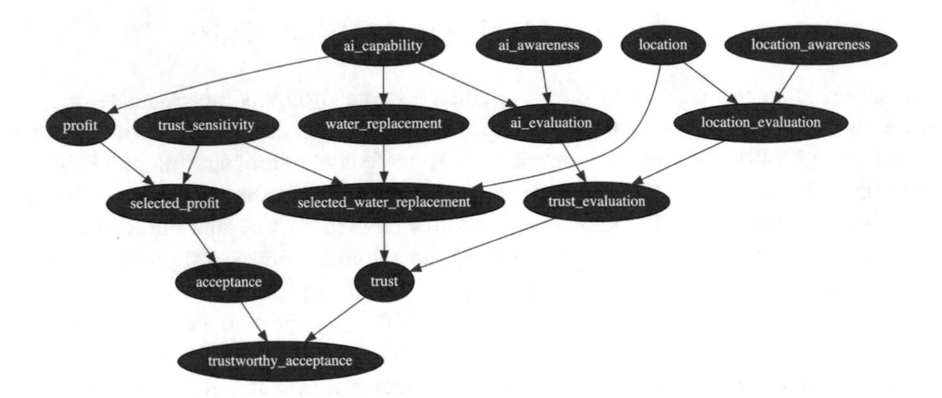

Fig. 1. The figure illustrates the causal graph for environmental decision-making, with the ultimate outcome represented as trustworthy acceptance.

Figure 1 depicts the causal graph as a Bayesian network for environmental decision-making using `showBN()` method, comprising independent and dependent nodes. Independent nodes, such as AI capability, AI awareness, location, location awareness, and trust sensitivity, operate autonomously without dependence on other nodes. For instance, an actor's location is independent of other probabilities within the causal graph. Conversely, dependent nodes, including profit, selected profit, water replacement, selected water replacement, AI evaluation, location evaluation, trust evaluation, trust, acceptance, and trustworthy acceptance, rely on the state of specific subsets of nodes, as denoted by arrows in the causal graph.

Independent nodes encompass factors like AI capability, reflecting the system's ability to generate balanced decisions without bias towards specific parameters; AI awareness, indicating community knowledge of the AI's capabilities; location, representing the actor's proximity to the river; location awareness, signaling community awareness of actor locations; and trust sensitivity, influencing an actor's inclination towards decisions garnering higher trust.

Dependent nodes cover aspects like profit, representing agricultural profit; water replacement, determining the percentage of replaced groundwater rights; AI evaluation, reflecting community perception of the AI's performance; location evaluation, indicating community considerations of actor locations; trust evaluation, summarizing community evaluations of AI and location factors; trust, determined by water replacement and trust evaluation; acceptance, influenced by profit; and trustworthy acceptance, contingent on both trust and acceptance.

Two crucial paths in the causal graph are the paths for trust and acceptance. Acceptance is tied to the profit of the selected decision, with higher profit correlating to higher acceptance, serving as an indicator of the decision-generating system's approval. Trust has two contributing paths: the water replacement percentage in the selected decision and the community's trust evaluation. Central to our research, evaluation reflects how community awareness impacts trust, ultimately influencing trustworthy acceptance [55]. Figure 2 depicts the inference using `showInference()` method showing initial probabilities for the independent nodes and the probabilities for the dependent nodes in the absence of any evidence which were determined in previous game theory scenarios [51,54].

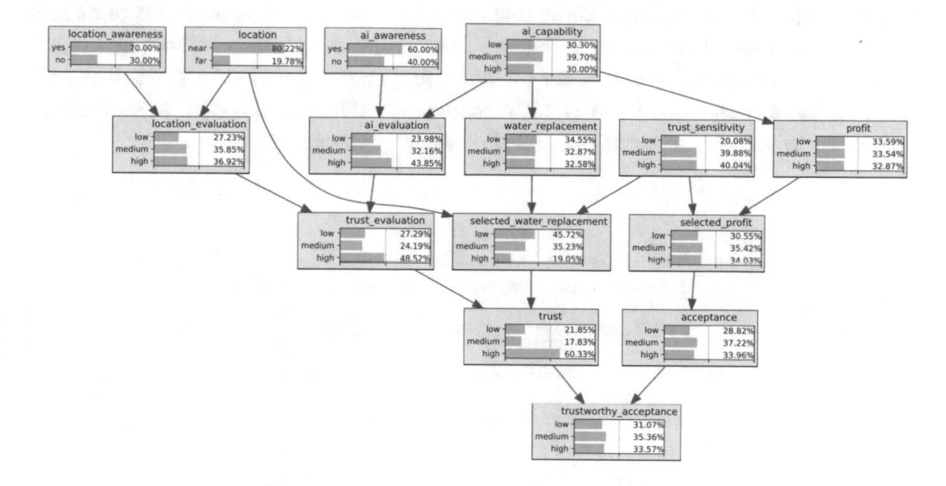

Fig. 2. The figure illustrates the initial probabilities for the independent nodes and the probabilities for the dependent nodes without evidence.

Our approach utilizes water replacement percentage as a determinant of trust, with decisions featuring higher water replacement receiving greater trust from the community. Additionally, community awareness is incorporated to democratize trust evaluation for specific factors, recognizing disparities in actor expectations based on location. For instance, an actor distant from the river is not expected to have water replacement as high as one closer to the river. Finally, trust sensitivity is introduced as an independent node within an actor's profile, influencing decisions that necessitate a trade-off between outcomes. Actors with higher trust sensitivity are more inclined toward decisions that can potentially gain higher trust.

3 Results

This section presents and deliberates on our results obtained from a decision-making scenario involving five actors tasked with making environmental decisions. The trade-off is established between profit and water replacement, with actors aiming for higher profit while their decisions are assessed based on the water replacement percentage. This percentage serves as a measure of the community's trust in each actor within this environmental context.

In our case study, two specific situations impact two actors. The first situation pertains to a location, affecting Actor 2. The goal of this environmental decision-making is to investigate the viability of replacing groundwater rights with Columbia River water rights. However, not all actors have comparable distances to the Columbia River, leading to challenges and higher costs for those located farther away, as observed in the case of Actor 2.

The second situation involves an underperforming AI-based decision generation, affecting Actor 5. The expectation is that the AI should generate solutions with sufficient resolution, enabling actors to deviate from initial decisions with minimal impact on sub-goals such as profit and water replacement percentage. However, upon examining the decisions generated for Actor 5, it becomes evident that he needs to sacrifice too much profit to achieve an acceptable level of water replacement.

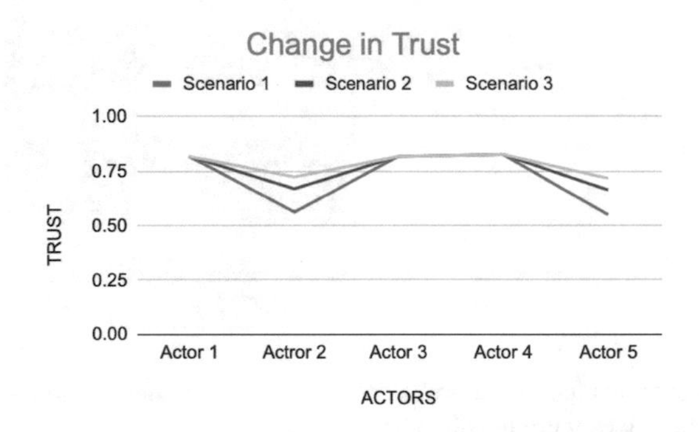

Fig. 3. In Scenario 2, the trust levels of Actor 2 and 5 rise as the community becomes aware of their situations through an intervention, leading to evaluations of their decisions based on this awareness. Subsequently, in Scenario 3, trust levels experience a further increase in a counterfactual setting where Actor 2 exhibits higher trust sensitivity, and there is an enhancement in AI capability, impacting the decisions of Actor 5.

The results of three scenarios, depicted in Fig. 3, are presented here. In the initial scenario, actors chose optimal decisions for land management, prioritizing high water replacement percentages without compromising intentional profit levels. However, special cases involving Actor 2 and 5 resulted in lower trust, indicating community disapproval due to insufficient information about their unique situations.

In the second scenario, community awareness of the specific situations impacting these two actors acted as an intervention in our causal network. Consequently, trust in these actors increased as the community evaluated their decisions in light of these situations. Expectations regarding water replacement for Actor 2 decreased, leading to higher trust evaluation. Similarly, the community comprehended the rationale behind Actor 5's decision and evaluated it accordingly.

In the final scenario, we directed the two actors to align more closely with community expectations, presenting a counterfactual scenario to explore potential policy changes. Despite Actor 2's current geographical constraints, where replacing significant groundwater rights with Columbia River water is unfeasible, the actor could exhibit higher trust sensitivity, influencing decisions that garner greater trust. In the case of Actor 5, the counterfactual scenario involved an AI with relatively higher capability than the initial system.

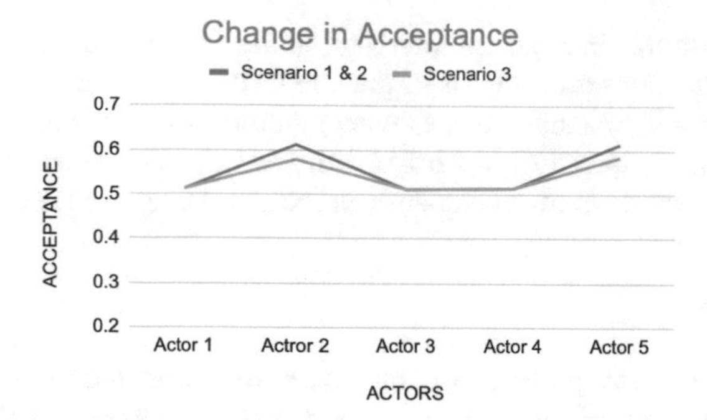

Fig. 4. Acceptance levels are higher in Scenarios 1 and 2. However, in Scenario 3, the counterfactual scenario, which includes higher trust sensitivity for Actor 2 and a more capable AI, leads to lower acceptance due to the increased emphasis on water replacement.

In addition to trust, acceptance gauges the degree of approval for the decision list presented to the actors. In the initial scenario, Actors 2 and 5 exhibit relatively higher acceptance, as they successfully chose decisions that meet their profit criteria, as illustrated in Fig. 4. This trend persists in the second scenario, where the intervention focuses solely on community awareness, resulting in trust changes. However, in the final scenario, acceptance levels for both actors increase. This is attributed to the counterfactual scenario, where Actor 2 has higher trust sensitivity, and a more capable AI system is employed for decision generation.

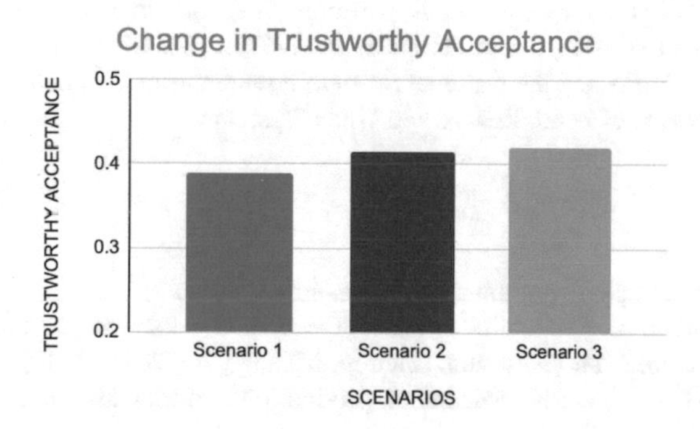

Fig. 5. Trustworthy acceptance registers a low value in Scenario 1, primarily due to the diminished trust of Actors 2 and 5. The intervention leads to an improvement in our metric in Scenario 2, aligning with the increased trust. In the counterfactual scenario, despite a decrease in trust, integrating higher trust sensitivity and a more capable AI system results in elevated acceptance. This dynamic maintains the trustworthy acceptance at a comparable level.

While acceptance is a crucial indicator of decision approval, it is equally important to consider the trustworthiness of the source providing the measurements. Our metric, trustworthy acceptance, gauges community acceptance, factoring in the trustworthiness of each actor. Illustrated in Fig. 5, an intervention that informs the community about the actors' specific situations results in an elevated level of trustworthy acceptance. Expanding on this, in a counterfactual scenario where Actor 2 exhibits higher trust sensitivity coupled with a more capable AI, the trustworthy acceptance remains comparable. This implies that the upswing in trust compensates for any decrease in acceptance.

4 Conclusions

In conclusion, our study presents a comprehensive causal model that is specifically designed for making environmental decisions, focusing on agricultural choices and the balance between water policy changes and agricultural profit. This model integrates various criteria, including sensitivity to trust, to illustrate the delicate relationship between trust and policy change. Actor-specific factors, such as location, AI capability, and community awareness, play a crucial role in shaping the decision-making landscape. Trustworthy acceptance, an important metric, serves as a measure of progress in decision-making advancements. Our findings demonstrate that strategic interventions, particularly increasing community awareness, can significantly boost actors' trust by up to 21%. Despite the challenge of diminishing profit, a decline in trustworthy acceptance can be prevented through the establishment of stronger trust. Employing and applying this causal model in the context of environmental decision-making while navigating a complex trade-off is essential for quantifying and understanding these significant outcomes.

Acknowledgements. This work was partially supported by the National Science Foundation under Grant No.1547411 and by the U.S. Department of Agriculture (USDA) National Institute of Food and Agriculture (NIFA) (Award Number 2017-67003-26057) via an interagency partnership between USDA-NIFA and the National Science Foundation (NSF) on the research program Innovations at the Nexus of Food, Energy and Water Systems.

References

1. pyAgrum (2024). https://pyagrum.readthedocs.io/en/1.11.0/
2. Benson, D., Jordan, A.: The scaling of water governance tasks: a comparative federal analysis of the European union and Australia. Environ. Manage. **46**, 7–16 (2010)
3. Brugnach, M., Özerol, G.: Knowledge co-production and transdisciplinarity: Opening pandora's box (2019)
4. Cialdini, R.B.: Influence: The New Psychology of Modern Persuasion. Morrow (1984)
5. Cialdini, R.B.: Influence: Science and Practice, vol. 4. Pearson education, Boston (2009)
6. Cook, J.R.: Columbia river supply project (2022). https://www.northeastoregonwater.org/columbia-river-supply-project
7. Davenport, M.A., Leahy, J.E., Anderson, D.H., Jakes, P.J.: Building trust in natural resource management within local communities: a case study of the midewin national tallgrass prairie. Environ. Manage. **39**, 353–368 (2007). https://doi.org/10.1007/s00267-006-0016-1
8. DeFires, R.S., et al.: Next generation earth systems science at the national science foundation (2021)
9. Dong, Y., Xu, J.: Consensus Building in Group Decision Making. Springer, Singapore (2016). https://doi.org/10.1007/978-981-287-892-2

10. Ducamp, G., Gonzales, C., Wuillemin, P.H.: aGrUM/pyAgrum: a toolbox to build models and algorithms for probabilistic graphical models in Python. In: International Conference on Probabilistic Graphical Models. PMLR (2020)
11. Edelenbos, J., Van Meerkerk, I.: Connective capacity in water governance practices: the meaning of trust and boundary spanning for integrated performance. Curr. Opin. Environ. Sustain. **12**, 25–29 (2015)
12. Gonzales, C., Torti, L., Wuillemin, P.-H.: aGrUM: a graphical universal model framework. In: Benferhat, S., Tabia, K., Ali, M. (eds.) IEA/AIE 2017. LNCS (LNAI), vol. 10351, pp. 171–177. Springer, Cham (2017). https://doi.org/10.1007/978-3-319-60045-1_20
13. Guadagno, R.E., Muscanell, N.L., Rice, L.M., Roberts, N.: Social influence online: the impact of social validation and likability on compliance. Psychol. Pop. Media Cult. **2**(1), 51 (2013)
14. Gunning, D.: Explainable artificial intelligence (XAI). Defense Adv. Res. Projects Agency (DARPA), nd Web **2**, 2 (2017)
15. Hamm, J.A., PytlikZillig, L.M., Herian, M.N., Tomkins, A.J., Dietrich, H., Michaels, S.: Trust and intention to comply with a water allocation decision: the moderating roles of knowledge and consistency. Ecol. Soc. **18**(4) (2013)
16. Hegselmann, R., Krause, U.: Opinion dynamics and bounded confidence models, analysis, and simulation. J. Artif. Soc. Soc. Simul. **5**(3) (2002)
17. Holman, I.P., Hess, T., Rey, D., Knox, J.: A multi-level framework for adaptation to drought within temperate agriculture. Front. Environ. Sci. **8**, 282 (2021)
18. Imai, K.: Causal inference with observational data. J. Am. Stat. Assoc. **99**(467), 139–190 (2014)
19. Kaur, D., Uslu, S., Durresi, A.: Trust-based security mechanism for detecting clusters of fake users in social networks. In: Barolli, L., Takizawa, M., Xhafa, F., Enokido, T. (eds.) WAINA 2019. AISC, vol. 927, pp. 641–650. Springer, Cham (2019). https://doi.org/10.1007/978-3-030-15035-8_62
20. Kaur, D., Uslu, S., Durresi, A.: Requirements for trustworthy artificial intelligence – a review. In: Barolli, L., Li, K.F., Enokido, T., Takizawa, M. (eds.) NBiS 2020. AISC, vol. 1264, pp. 105–115. Springer, Cham (2021). https://doi.org/10.1007/978-3-030-57811-4_11
21. Kaur, D., Uslu, S., Durresi, A., Badve, S., Dundar, M.: Trustworthy explainability acceptance: a new metric to measure the trustworthiness of interpretable AI medical diagnostic systems. In: Barolli, L., Yim, K., Enokido, T. (eds.) CISIS 2021. LNNS, vol. 278, pp. 35–46. Springer, Cham (2021). https://doi.org/10.1007/978-3-030-79725-6_4
22. Kaur, D., Uslu, S., Durresi, A., Mohler, G., Carter, J.G.: Trust-based human-machine collaboration mechanism for predicting crimes. In: Barolli, L., Amato, F., Moscato, F., Enokido, T., Takizawa, M. (eds.) AINA 2020. AISC, vol. 1151, pp. 603–616. Springer, Cham (2020). https://doi.org/10.1007/978-3-030-44041-1_54
23. Kaur, D., Uslu, S., Durresi, M., Durresi, A.: A geo-location and trust-based framework with community detection algorithms to filter attackers in 5G social networks. Wireless Netw., 1–9 (2022). https://doi.org/10.1007/s11276-022-03073-y
24. Kaur, D., Uslu, S., Rittichier, K.J., Durresi, A.: Trustworthy artificial intelligence: a review. ACM Comput. Surv. (CSUR) **55**(2), 1–38 (2022)
25. Khan, M.I., Sarkar, S., Maity, R.: Artificial intelligence/machine learning techniques in hydroclimatology: a demonstration of deep learning for future assessment of stream flow under climate change. In: Visualization Techniques for Climate Change with Machine Learning and Artificial Intelligence, pp. 247–273. Elsevier (2023)
26. Koontz, T.M., Newig, J.: From planning to implementation: Top-down and bottom-up approaches for collaborative watershed management. Policy Stud. J. **42**(3), 416–442 (2014)
27. Lubell, M., Lippert, L.: Integrated regional water management: a study of collaboration or water politics-as-usual in California, USA. Int. Rev. Adm. Sci. **77**(1), 76–100 (2011)

28. Maani, K.: Multi-Stakeholder Decision Making for Complex Problems: A Systems Thinking Approach with Cases. World Scientific (2016)
29. McGovern, A., Ebert-Uphoff, I., Gagne, D.J., Bostrom, A.: Why we need to focus on developing ethical, responsible, and trustworthy artificial intelligence approaches for environmental science. Environ. Data Sci. **1**, e6 (2022)
30. Montilla-López, N.M., Gómez-Limón, J.A., Gutiérrez-Martín, C.: Sharing a river: potential performance of a water bank for reallocating irrigation water. Agric. Water Manag. **200**, 47–59 (2018)
31. Nolan, J.M., Schultz, P.W., Cialdini, R.B., Goldstein, N.J., Griskevicius, V.: Normative social influence is underdetected. Pers. Soc. Psychol. Bull. **34**(7), 913–923 (2008)
32. Pearl, J.: Causality: Models, Reasoning, and Inference. Cambridge University Press (2009)
33. Rittichier, Kaley J.., Kaur, Davinder, Uslu, Suleyman, Durresi, Arjan: A trust-based tool for detecting potentially damaging users in social networks. In: Barolli, Leonard, Chen, Hsing-Chung., Enokido, Tomoya (eds.) NBiS 2021. LNNS, vol. 313, pp. 94–104. Springer, Cham (2022). https://doi.org/10.1007/978-3-030-84913-9_9
34. Rolnick, D., et al.: Tackling climate change with machine learning. ACM Comput. Surv. (CSUR) **55**(2), 1–96 (2022)
35. Ruan, Y., Alfantoukh, L., Durresi, A.: Exploring stock market using twitter trust network. In: Advanced Information Networking and Applications (AINA), 2015 IEEE 29th International Conference on, pp. 428–433. IEEE (2015)
36. Ruan, Y., Alfantoukh, L., Fang, A., Durresi, A.: Exploring trust propagation behaviors in online communities. In: Network-Based Information Systems (NBiS), 2014 17th International Conference on, pp. 361–367. IEEE (2014)
37. Ruan, Y., Durresi, A.: A survey of trust management systems for online social communities-trust modeling, trust inference and attacks. Knowl.-Based Syst. **106**, 150–163 (2016)
38. Ruan, Y., Durresi, A.: A trust management framework for cloud computing platforms. In: Advanced Information Networking and Applications (AINA), 2017 IEEE 31st International Conference on, pp. 1146–1153. IEEE (2017)
39. Ruan, Y., Durresi, A.: A trust management framework for clouds. Comput. Commun. **144**, 124–131 (2019)
40. Ruan, Y., Durresi, A., Alfantoukh, L.: Trust management framework for internet of things. In: Advanced Information Networking and Applications (AINA), 2016 IEEE 30th International Conference on, pp. 1013–1019. IEEE (2016)
41. Ruan, Y., Durresi, A., Alfantoukh, L.: Using twitter trust network for stock market analysis. Knowl.-Based Syst. **145**, 207–218 (2018)
42. Ruan, Y., Durresi, A., Uslu, S.: Trust assessment for internet of things in multi-access edge computing. In: 2018 IEEE 32nd International Conference on Advanced Information Networking and Applications (AINA), pp. 1155–1161. IEEE (2018)
43. Ruan, Y., Zhang, P., Alfantoukh, L., Durresi, A.: Measurement theory-based trust management framework for online social communities. ACM Trans. Internet Technol. (TOIT) **17**(2), 16 (2017)
44. Rubin, D.B.: Causal Inference Using Potential Outcomes: Design Decisions Modeling. Wiley (2005)
45. Spirtes, P., Glymour, C., Scheines, R.: Causation, Prediction, and Search. MIT Press Cambridge (2000)
46. Sutcliffe, A.G., Wang, D., Dunbar, R.I.: Modelling the role of trust in social relationships. ACM Trans. Internet Technol. (TOIT) **15**(4), 16 (2015)
47. Tseng, M.L.: An assessment of cause and effect decision-making model for firm environmental knowledge management capacities in uncertainty. Environ. Monit. Assess. **161**, 549–564 (2010)

48. Urquijo, Julia, De Stefano, Lucia: Perception of drought and local responses by farmers: a perspective from the Jucar river basin, Spain. Water Resour. Manage. **30**(2), 577–591 (2015). https://doi.org/10.1007/s11269-015-1178-5
49. Uslu, S., Kaur, D., Durresi, M., Durresi, A.: Trustability for resilient internet of things services on 5G multiple access edge cloud computing. Sensors **22**(24), 9905 (2022)
50. Uslu, Suleyman, Kaur, Davinder, Rivera, Samuel J.., Durresi, Arjan, Babbar-Sebens, Meghna: Decision support system using trust planning among food-energy-water actors. In: Barolli, Leonard, Takizawa, Makoto, Xhafa, Fatos, Enokido, Tomoya (eds.) AINA 2019. AISC, vol. 926, pp. 1169–1180. Springer, Cham (2020). https://doi.org/10.1007/978-3-030-15032-7_98
51. Uslu, Suleyman, Kaur, Davinder, Rivera, Samuel J.., Durresi, Arjan, Babbar-Sebens, Meghna: Trust-based game-theoretical decision making for food-energy-water management. In: Barolli, Leonard, Hellinckx, Peter, Enokido, Tomoya (eds.) BWCCA 2019. LNNS, vol. 97, pp. 125–136. Springer, Cham (2020). https://doi.org/10.1007/978-3-030-33506-9_12
52. Uslu, Suleyman, Kaur, Davinder, Rivera, Samuel J.., Durresi, Arjan, Babbar-Sebens, Meghna: Trust-based decision making for food-energy-water actors. In: Barolli, Leonard, Amato, Flora, Moscato, Francesco, Enokido, Tomoya, Takizawa, Makoto (eds.) AINA 2020. AISC, vol. 1151, pp. 591–602. Springer, Cham (2020). https://doi.org/10.1007/978-3-030-44041-1_53
53. Uslu, Suleyman, Kaur, Davinder, Rivera, Samuel J.., Durresi, Arjan, Babbar-Sebens, Meghna, Tilt, Jenna H..: Control theoretical modeling of trust-based decision making in food-energy-water management. In: Barolli, Leonard, Poniszewska-Maranda, Aneta, Enokido, Tomoya (eds.) CISIS 2020. AISC, vol. 1194, pp. 97–107. Springer, Cham (2021). https://doi.org/10.1007/978-3-030-50454-0_10
54. Uslu, S., Kaur, D., Rivera, S.J., Durresi, A., Babbar-Sebens, M., Tilt, J.H.: A trustworthy human-machine framework for collective decision making in food-energy-water management: the role of trust sensitivity. Knowl.-Based Syst. **213**, 106683 (2021)
55. Uslu, S., Kaur, D., Rivera, S.J., Durresi, A., Durresi, M., Babbar-Sebens, M.: Trustworthy acceptance: a new metric for trustworthy artificial intelligence used in decision making in food-energy-water sectors. In: AINA (1), pp. 208–219 (2021)
56. Uslu, Suleyman, Kaur, Davinder, Rivera, Samuel J.., Durresi, Arjan, Durresi, Mimoza, Babbar-Sebens, Meghna: Trustworthy fairness metric applied to AI-based decisions in food-energy-water. In: Barolli, Leonard, Hussain, Farookh, Enokido, Tomoya (eds.) AINA 2022. LNNS, vol. 450, pp. 433–445. Springer, Cham (2022). https://doi.org/10.1007/978-3-030-99587-4_37
57. Uslu, Suleyman, Ruan, Yefeng, Durresi, Arjan: Trust-based decision support system for planning among food-energy-water actors. In: Barolli, Leonard, Javaid, Nadeem, Ikeda, Makoto, Takizawa, Makoto (eds.) CISIS 2018. AISC, vol. 772, pp. 440–451. Springer, Cham (2019). https://doi.org/10.1007/978-3-319-93659-8_39
58. Walthall, C.L., Anderson, C.J., Baumgard, L.H., Takle, E., Wright-Morton, L., et al.: Climate change and agriculture in the united states: effects and adaptation (2013)
59. Zanfei, A., Menapace, A., Righetti, M.: An artificial intelligence approach for managing water demand in water supply systems. In: IOP Conference Series: Earth and Environmental Science, vol. 1136, p. 012004. IOP Publishing (2023)
60. Zhang, P., Durresi, A.: Trust management framework for social networks. In: 2012 IEEE International Conference on Communications (ICC), pp. 1042–1047. IEEE (2012)
61. Zhang, P., Durresi, A., Barolli, L.: Survey of trust management on various networks. In: Complex, Intelligent and Software Intensive Systems (CISIS), 2011 International Conference on, pp. 219–226. IEEE (2011)

Forecasting Malware Incident Rates in Higher Education Institutions

Rildo Antonio de Souza[1](✉), Vitor de Castro Silva[2], Sylvio Barbon Junior[3], and Bruno Bogaz Zarpelão[2]

[1] Security Incident Response Center (CAIS), National Research and Education Network (RNP), Campinas, São Paulo, Brazil
rildo.souza@rnp.br

[2] Computer Science Department, State University of Londrina (UEL), Londrina, Paraná, Brazil
{vitor.castro.silva,brunozarpelao}@uel.br

[3] Department of Engineering and Architecture, University of Trieste (UNITS), Trieste, Italy
sylvio.barbonjunior@units.it

Abstract. Malware is often behind cybersecurity events like data theft and service disruption. Therefore, forecasting the number of malware incidents in an organization helps security analysts, who can anticipate trends and quickly detect outbreaks. This paper delves into forecasting malware incident rates in Higher Education Institutions (HEI) using the Long Short-Term Memory (LSTM) network, recognized as the leading technique for time series forecasting based on machine learning. Education and research are among the most attacked industries, deserving special attention. In our work, weekly counts of incidents for each institution are represented as a time series. Then, a neural network is trained to take the incident rate from previous weeks as input and forecast the incident rate for the next week. The experiments used real incident data from Brazilian HEIs and investigated the performance of LSTM for multiple institution sizes, amounts of weeks used as input, and amounts of weeks for training. LSTM results were compared to the Autoregressive Integrated Moving Average's (ARIMA), a traditional statistical method. The results showed that LSTM outperformed ARIMA by a small margin. Additionally, forecasting the number of incidents in smaller institutions was harder than in larger ones.

1 Introduction

A security incident is an event that negatively impacts the confidentiality, integrity, or availability of an organization's information and systems. Among various types, this work focuses on malware incidents, which pose a continuous threat to cybersecurity. Malware infiltrates systems and devices to steal data, disrupt services, and cause financial harm. The frequency of malware incidents is upward, as reported by [12].

L. Barolli (Ed.): AINA 2024, LNDECT 202, pp. 226–237, 2024.
https://doi.org/10.1007/978-3-031-57916-5_20

A brief examination of the malware campaign landscape in the past few years provides insight into the diverse threats and targets involved in these incidents. In 2019, notable examples included Ryuk, Emotet, and Trickbot. Ryuk was particularly involved in U.S. healthcare sector attacks [14]. In 2020, Emotet and Trickbot continued to be active, while new threats like Conti and Maze appeared. Conti targeted Ireland's Health Service Executive [7], and Maze attacked various international organizations, including Cognizant, a Fortune 500 IT company [21].

2021 saw the emergence of Hive, Lockbit, and REvil as new threat actors. REvil executed a notable attack on the Colonial Pipeline in the USA [16]. Hive was used against Costa Rica's Government and Zambia's Bank [18]. Lockbit 2.0, known for its advanced exploitation techniques, posed a significant concern [19]. That year, the Apache Log4j vulnerability (CVE-2021-44228) emerged as a critical issue in November. In 2022, the malware landscape remained largely unchanged, with BlackCat added to the list [13]. The exploitation patterns were similar to the previous year, including continued attacks on Log4j and Windows OS, plus new vulnerabilities in VMware software [1].

The education and research industries account for a significant part of these incidents as indicated in CheckPoint's report [6]. Educational institutions present a concerning combination of attractive resources for attackers and difficulties in keeping a secure digital environment. Schools and universities host large amounts of personal and research data, making them highly profitable targets for cyber threats. Additionally, these organizations have become increasingly dependent on digital technologies for teaching, research and administrative activities. Cyber threats that disrupt these services may cause significant damage. Protecting these environments is particularly hard due to their diverse user base (students, teachers, researchers, and admin staff) who may operate within the institution premises or remotely. Universities, in particular, are naturally open environments, where the use of personal devices is often encouraged. On top of that, these organizations struggle to keep up-to-date security controls as they typically rely on public funds and operate within constrained budgets [11].

Organizations characterized by user diversity, extensive asset portfolios and budget constraints require tools that provide security analysts with a comprehensive perspective on the current scenario and improve their ability to utilize available resources more efficiently. Forecasting malware incident rates emerges as a valuable tool that helps to meet these needs. By knowing the expected rates of incidents, security analysts can anticipate trends, quickly detect outbreaks, and identify which assets are more or less prone to get infected. Furthermore, this management data can also be exchanged with other organizations through specialized bodies like Computer Security Incident Response Teams (CSIRTs), consequently improving overall cybersecurity conditions in the long term.

This work studies the application of the Long Short Term Memory (LSTM) network to forecast the weekly rate of incidents in Higher Education Institutions. Most current solutions for time series forecasting rely on machine learning, often outperforming statistical-based ones. Among the machine learning algorithms, LSTM stands out as particularly well-suited for time series due to its ability to learn with long-term sequences of observations. This work aims to address various questions related to the application of LSTM to forecast malware incidents: What is the LSTM's predictive capability in these scenarios? How does the institution's size affect the prediction error rate? Does increasing the number of weeks passed as input to the model reduce prediction errors? Can more accurate predictions be obtained by increasing the training period?

To carry out this analysis, real data from approximately 30,000 malware incidents in Brazilian higher education institutions over three years were collected. These data were organized into time series for each analyzed institution, and the LSTM model was trained to predict the number of occurrences in the immediately following week based on the input period. Autoregressive Integrated Moving Average (ARIMA), a statistical method traditionally used for time series forecasting, served as a baseline. Then, the accuracy results for LSTM were compared with those obtained by an ARIMA model.

The article structure includes related work (Sect. 2), material and methods with data and design details (Sect. 3), a discussion of results and answers to prior questions (Sect. 4), and concludes in Sect. 5.

2 Related Work

Over the past years, there has been an increase in the number of studies on security incident prediction. These approaches aim to provide security teams with a strategic advantage, enabling them to take proactive measures and enhance the efficiency of their operations. To achieve this goal, studies utilize varied methodologies and explore different types of neural networks and statistical models. The study in [8] proposed a framework based on bidirectional recurrent neural networks with LSTM (BRNN-LSTM) to forecast the quantity of malicious IP flows. The experiments were conducted on data collected from a honeypot. Other researchers [10] used ARIMA and exponential smoothing to predict the total number of alerts in a network. Both works use data sources that contain false positives, which can impair the estimates quality.

In [17], the use of a system employing Stacked Recurrent Neural Networks is proposed to predict information security events. According to their results, employing this type of LSTM achieved accurate predictions across various types of security events analyzed. The experiments were conducted using data collected from a commercial IPS (Intrusion Prevention System), comprising approximately 3 billion security events. By employing data from a commercial IPS, it is acknowledged that there may be a considerable presence of false positives, as evidenced in [2].

Works such as [5,20], and [4] aim to use incident records generated under analyst supervision, i.e., with a low chance of containing false positives. The study in [5] employed Bayesian state space models to predict malware incident rates. The experiments utilized records generated within the scope of the United States Department of Defense. Going in another direction, the study in [20] applied ARIMA and Bayesian networks to predict the quantity of incidents in three different ways: absolute incident rate, incident rate level (low, medium, or high), and occurrence or non-occurrence of incidents (binary series). Another study in [4] proposed combining generative models (TimeGAN) and autoencoders to handle noisy data series.

Compared to the presented works, our proposal stands out based on the following points: unlike most previous research, which uses data that lack detection and verification by security incident analysts, our study involves a large dataset comprising real malware incidents detected and analyzed within the Brazilian National Research and Education Network (RNP) by its CSIRT. This is important because using unverified incident or alert data might contaminate the analysis with false positives. Furthermore, these incidents were collected from various types and sizes of public universities, constituting a notably diverse dataset that enables a more comprehensive and thorough analysis.

3 Material and Methods

To achieve the objectives of this study, incident occurrences are organized as a time series, constructing a series for each analyzed institution. In these series, each element represents the quantity of malware incidents occurring in a specific week. The LSTM network will be employed to forecast incident occurrences for the following week based on recent week counts. In our research, we will use the supervised learning paradigm for the time series being analyzed. As per [3], the main advantages of using supervised learning in an LSTM network are related to aspects such as enabling modeling of complex sequences, providing accurate predictions, receiving clear feedback on predictions, and lastly, its ability to generalize.

The use of recurrent neural networks like LSTM has shown promise in time series applications across various domains, making this technique a valuable option for cybersecurity data analysis, as presented in [8]. This study was designed to address the following questions:

- How accurate is an LSTM-based model in forecasting the number of malware incidents that will occur in the next week, considering occurrences from previous weeks?
- How does the institution's size (and its number of weekly incidents) influence the error rate in prediction?
- Does increasing the number of weeks passed as input to the LSTM model reduce prediction errors?
- Can more accurate predictions be obtained by extending the training period used in the model?
- Is a model based on LSTM more accurate than a ARIMA-based one ?

3.1 Dataset and Institution Selection

The incident data analyzed in this study was collected over a period of 3 years in a production environment, totaling approximately 30,000 malware incidents across various types of higher education institutions in Brazil. The data is not public and was obtained through a partnership with the data custodians, the Brazilian National Research and Education Network (RNP).

Figure 1 provides an overview of how the identification, classification, and analysis of security incidents is carried out in the Brazilian National Research and Education Network. The detection of security events is performed in two distinct ways: internally, using the organization's own tools, or externally, through partners. When an event is detected, it is logged in a system for tracking and subsequently undergoes a filtering process to assess whether it is a security incident. When a security event is analyzed by the RNP security team and it is determined that it poses no risk to the clients, the event is classified as a false positive. This categorization is essential in the security monitoring process, as it allows the team to focus on real threats, optimizing efficiency in protecting the clients. For events confirmed as security incidents, the next step is to categorize them according to their nature and severity. After categorization, the incidents are prioritized so they can be handled efficiently, taking into account their urgency and impact. The prioritized incident is then assigned to the organization and/or department responsible for managing that network segment. The process culminates with the formal recording and assignment of the incident, indicating that it is ready to be addressed and resolved by the appropriate team within the organization.

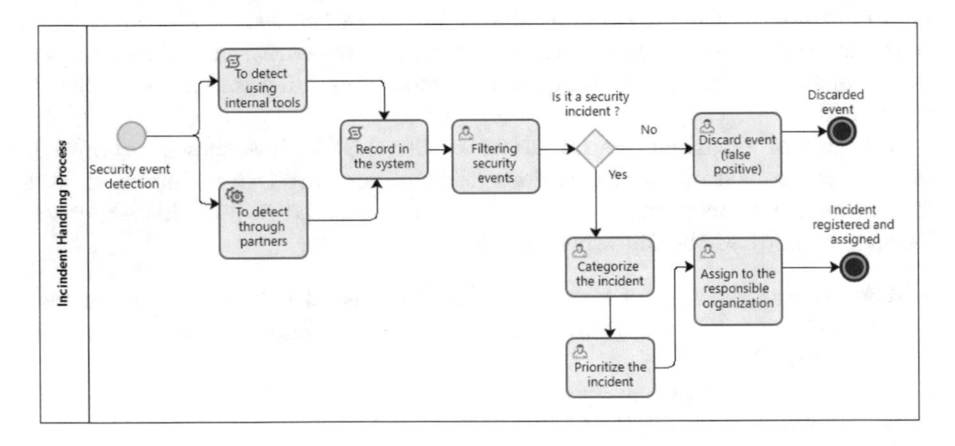

Fig. 1. Incident Handling Process in the Brazilian National Research and Education Network.

The experiments conducted do not encompass all institutions present in the dataset. Our intention in selecting a subset of institutions was to facilitate an assessment of how institution size affects incident prediction. Additionally, the dataset contains over 1000 available institutions, and collecting and carefully analyzing results for all of them would be unfeasible.

Therefore, a decision was made to choose 10 institutions divided into two equally sized groups. The first group comprises the 5 institutions with the highest number of public IP addresses and identified malware incidents per week over three years, referred to as larger-sized institutions. The second group consists of the 5 institutions with the lowest number of public IP addresses, requiring an average of at least one malware incident per week in the latter case. They are denoted as smaller-sized institutions. Institutions with fewer than one reported malware incident per week result in very sparse series and have been excluded from the scope of this work for now. Table 1 presents the details about the institutions used in our research.

Table 1. Overview of incident figures in selected institutions for a 3-year period.

Institution	Average Number of Incidents Per Week	Standard Deviation
larger_01	16.69	9.36
larger_02	70.85	90.24
larger_03	40.67	22.18
larger_04	19.23	18.56
larger_05	16.19	12.15
smaller_01	1.03	1.71
smaller_02	1.01	2.46
smaller_03	1.05	2.08
smaller_04	1.04	1.98
smaller_05	1.01	1.75

3.2 Experimental Design

The experiment was designed to evaluate the prediction performance of LSTM and ARIMA methods. Alongside LSTM, the core of this study, ARIMA was chosen to serve as a baseline for comparison since it is arguably the most adopted option when it comes to statistical methods for time series analysis [9]. In our work, we employed prequential evaluation to observe how performance evolves with increasing training sample size and number of weeks used as input. This approach was designed to emulate a scenario where the model undergoes training over varying durations, specifically 6 months, one year, and two years. The objective was to ascertain whether different training periods would enhance the model's performance. This method allowed us to systematically compare the model's effectiveness across these distinct time frames. Table 2 shows how the data was used for testing and training.

Table 2. Training and test sets.

Training Period	Test Period
April 2021 to October 2021	November to 2021 to November 2022
October 2020 to October 2021	November to 2021 to November 2022
October 2019 to October 2021	November to 2021 to November 2022

LSTM models were designed with an architecture that employs the Adaptive Moment Estimation (Adam) optimization function. Additionally, we chose the Mean Squared Error (MSE) as the loss function. Regarding the neural network model used, the activation method employed was ReLU (Rectified Linear Unit) with a hidden layer. Additionally, we varied the epochs between 100 and 500 with an increment of 100, and the values of the units ranged from 10 to 60 with an increment of 10. The specific hyper-parameters for LSTM are presented in Table 3. These variations in hyper-parameter values and network topologies aimed to find combinations that exhibit the best results in terms of prediction accuracy.

To apply the ARIMA model, the auto-arima functionality from the Pmdarima library [15] was used, as it automatically adjusts its parameters. This approach was adopted because parameter preparation ends up being a time-consuming task. The auto-arima technique performs various procedures automatically, simplifying and speeding up the process by finding the best parameters for each input data.

Table 3. Specific hyper-parameters of LSTM.

Parameter	Experimental Choice
Units	10, 20, 30, 40, 50, 60
Units Increments	10
Epochs	100, 200, 300, 400, 500
Epochs Increments	100
Dropout	0

The metrics used to assess the models' predictive capability were the Mean Absolute Error (MAE) and the MAE/mean ratio. MAE was chosen due to its intuitive nature and ease of interpretation, being calculated according to equation (1), where n represents the total number of observations in the time series, y_i denotes the actual values in the time series, $\hat{y}_i$ indicates the model predictions for each point in the time series, and i is an index that individually goes through each point in the sample, starting from 1 to n.

$$MAE = \frac{1}{n}\sum_{i=1}^{n}|y_i - \hat{y}_i| \quad (1)$$

The MAE/mean ratio was applied to allow a comparison between the accuracy achieved for different institutions. It provides a relative measure of the average error compared to the average magnitude of observed values. To obtain the ratio, MAE is divided by the mean calculated according to the formula $(\sum_{i=1}^{n} y_i)/n$, where we have a time series with n elements and y_i represents the i-th real element of the time series.

4 Results

Table 4 presents the scenarios where LSTM achieved the best MAE results in each of the analyzed institutions. The top five rows present the larger institutions, while the bottom half of the table refers to smaller institutions. The results for the MAE/mean ratio indicate that, for larger institutions, the forecasts achieved reasonable results, ranging from 0.17 to 0.29. These values mean that the average error of the predictions for the weekly incident rate in large institutions represents 17% to 29% of the average number of incidents. However, for smaller institutions, the average errors are much higher, with the MAE/mean ratio ranging from 0.39 to 1.60. The best MAE/mean value analyzed across all institutions is highlighted in Table 4 and is in bold.

Table 4. Best results for the MAE metric considering each institution.

Institution	Hidden Layer Units	Epochs	Training Period	Input Weeks	MAE	MAE/mean
larger_01	50	100	2 years	7	5.37	0.25
larger_02	20	100	2 years	5	35.02	0.26
larger_03	10	300	2 years	6	12.55	0.24
larger_04	10	100	2 years	4	8.11	0.29
larger_05	30	400	2 years	5	4.17	**0.17**
smaller_01	40	100	2 years	4	0.82	0.82
smaller_02	10	400	1 year	7	1.92	0.65
smaller_03	40	200	2 years	4	1.06	0.39
smaller_04	50	400	2 years	7	0.08	1.60
smaller_05	10	100	6 months	5	0.25	0.89

These MAE and MAE/mean results address the first two questions presented in Sect. 3, since they reveal the prediction accuracy of the LSTM models and underscore the impact of the institution's size on their predictive capability. Smaller institutions obtained significantly inferior results compared to those obtained for larger institutions. This is due to the fact that smaller institutions have much noisier series of weekly incident rates, as they have few reported incidents that spread more irregularly over time.

The results presented in Table 4 also allow us to establish initial answers to the other questions posed in Sect. 3, regarding the impact of varying the number of weeks used as input and the period of time for training. With the exception of the 6-week input, all other input configurations shared an equal number of occurrences in the best results ranking. The 4-week, 5-week, and 7-week inputs each contributed to three instances of the best results. Figure 2 provides more detailed information on the impact of the number of weeks used as input on the prediction accuracy. It displays the best MAE/mean achieved for each input configuration in each institution. The figure indicates that there is no apparent correlation between the MAE/mean ratio and the number of weeks used as input.

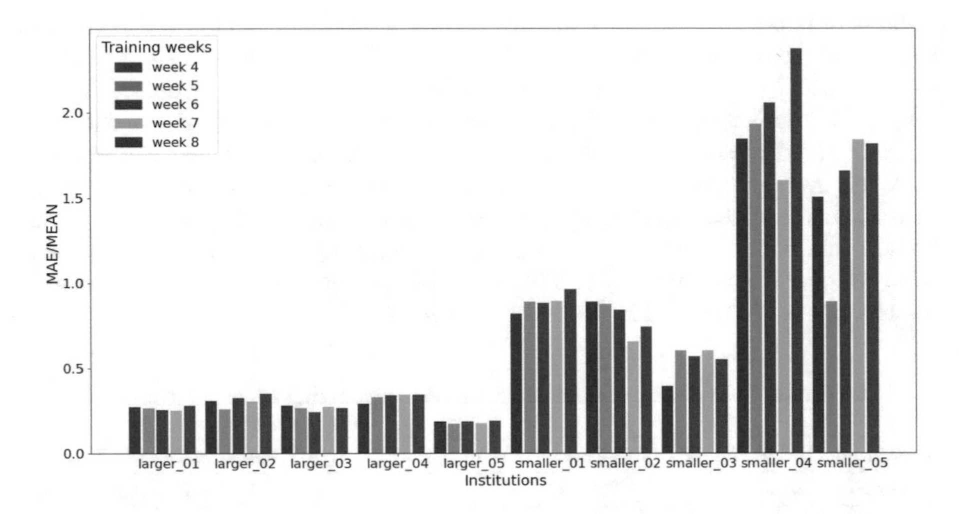

Fig. 2. Best results obtained for MAE/Mean considering different numbers of weeks as input.

Also according to the results in Table 4, the 2 year configuration for the training set size yielded the best results for 8 out of 10 institutions. Although this initially seems a clear trend, these results deserve further examination. Figure 3 shows the best MAE/mean ratio reached for each training set size configuration in each institution. It becomes apparent that larger institutions exhibit relatively minor variation in MAE/mean ratio when the training set size is changed. The figure shows that the 2-year configuration outperformed the other ones in larger institutions by a small margin. In smaller institutions, changing the training set size had a higher impact than in larger ones, but it remains uncertain whether increasing the training set size can improve the MAE/mean ratio results.

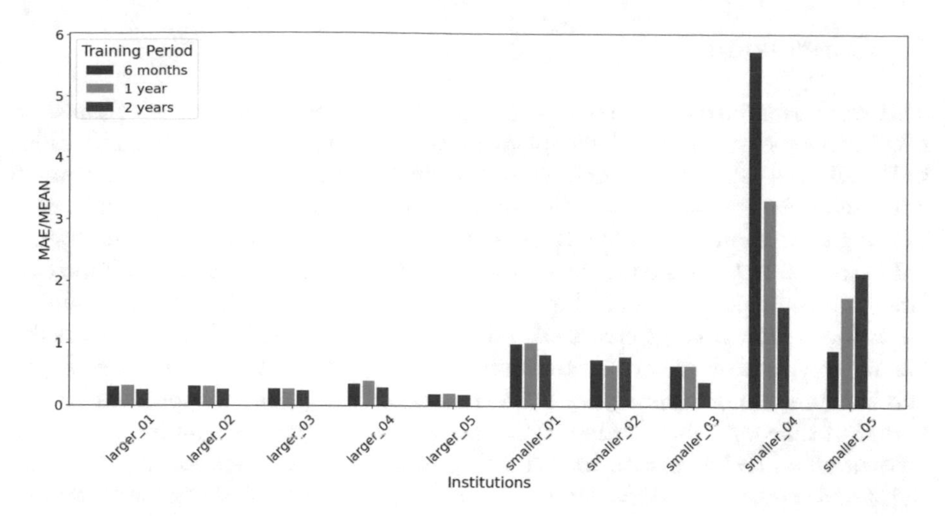

Fig. 3. Best results obtained for MAE/Mean considering different training set sizes.

The subsequent results are analyzed to evaluate and compare the accuracy of forecasts generated by LSTM and ARIMA models, aiming to identify which of the two techniques exhibited superior predictive capability. Table 5 presents the best MAE results achieved by LSTM and ARIMA models for each institution. MAE results yielded for LSTM models were around 10% lower than those for ARIMA models in all institutions except for smaller_02, where the MAE reduction reached 34%. These findings demonstrate that LSTM were consistently more accurate than ARIMA in forecasting the malware incident rate. Nevertheless, it is important to note that given the LSTM's status as a cutting-edge machine learning technique the difference may be considered relatively modest.

Table 5. Best MAE results for LSTM and ARIMA models

Institution	Best LSTM - MAE	Best ARIMA - MAE
larger_01	5.37	6.11
larger_02	35.02	38.8
larger_03	12.55	14.12
larger_04	8.11	9.14
larger_05	4.17	4.57
smaller_01	0.82	0.92
smaller_02	1.92	2.94
smaller_03	1.06	1.17
smaller_04	0.08	0.09
smaller_05	0.25	0.28

5 Conclusion

This work examined the use of a recurrent neural network architecture based on LSTM to forecast malware incident rates in public higher education institutions in Brazil. The experiments used a dataset comprising real incident data from 10 Brazilian universities. The selection of institutions was based on their scale, considering their number of public IP addresses and volume of incidents. The results indicate that larger institutions, characterized by a higher incident volume and less variable behavior, tend to have lower forecasting errors. In contrast, smaller institutions show average errors exceeding 100% of the incident rate. The results did not reveal a clear correlation between the number of weeks used as input and the forecasting accuracy for both smaller and larger institutions. As for the period of time used for training, different results emerged for smaller and larger institutions. Smaller institutions did not present a clear relationship between the training period and the models' accuracy. In contrast, larger institutions achieved slightly better results when the training series encompassed 2 years of data. Finally, the error rates for LSTM models were compared to those obtained with ARIMA models. LSTM outperformed ARIMA models by a modest margin, achieving lower errors by 10%.

As future work, it is proposed to investigate the behavior of the LSTM network when adding more hidden layers. Additionally, exploring other recurrent neural network architectures, such as GRU networks and Transformer memory networks, is suggested. Analyzing other variables related to malware incidents, such as malware type and attack origins, is also proposed to enrich the model and obtain more comprehensive insights.

References

1. 2021 Top Routinely Exploited Vulnerabilities. https://www.cisa.gov/news-events/cybersecurity-advisories/aa22-117a. Accessed 05 Jan 2024
2. Alahmadi, B.A., Axon, L., Martinovic, I.: 99% false positives: a qualitative study of SOC analysts' perspectives on security alarms. In: 31st USENIX Security Symposium (USENIX Security 22), pp. 2783–2800, USENIX Association, Boston, MA (2022). https://www.usenix.org/conference/usenixsecurity22/presentation/alahmadi
3. Alzubaidi, L., et al.: Review of deep learning: concepts, CNN architectures, challenges, applications, future directions. J. Big Data **8**, 1–74 (2021)
4. Amin Mahmood, S.H., Abbasi, A.: Using deep generative models to boost forecasting: a phishing prediction case study. In: 2020 International Conference on Data Mining Workshops (ICDMW), pp, 496–505 (2020). https://doi.org/10.1109/ICDMW51313.2020.00073
5. Bakdash, J.K., et al.: Malware in the future? Forecasting of analyst detection of cyber events. J. Cybersecurity **4**(1), tyy007 (2018). https://doi.org/10.1093/cybsec/tyy007
6. Check Point Research Reports a 38% Increase in 2022 Global Cyberattacks. https://blog.checkpoint.com/2023/01/05/38-increase-in-2022-global-cyberattacks/. Accessed 05 Jan 2024

7. Conti Ransomware: Inside One of the World's Most Aggressive Ransomware Groups. https://flashpoint.io/blog/history-of-conti-ransomware/. Accessed 05 Jan 2024
8. Fang, X., Xu, M., Xu, S., Zhao, P.: A deep learning framework for predicting cyber attacks rates. EURASIP J. Inf. Secur. **2019**, 1–11 (2019)
9. Hameed, S., et al.: Deep learning based multimodal urban air quality prediction and traffic analytics. Sci. Rep. **13**(1), 22181 (2023). https://doi.org/10.1038/s41598-023-49296-7
10. Husák, M., Komárková, J., Bou-Harb, E., Čeleda, P.: Survey of attack projection, prediction, and forecasting in cyber security. IEEE Commun. Surv. Tutorials **21**(1), 640–660 (2019). https://doi.org/10.1109/COMST.2018.2871866
11. Lallie, H.S., Thompson, A., Titis, E., Stephens, P.: Understanding cyber threats against the universities, colleges, and schools (2023). arXiv preprint arXiv:2307.07755
12. Malware Statistics & Trends Report. https://www.av-test.org/en/statistics/malware/. Accessed 05 Jan 2024
13. New BlackCat ransomware. https://www.kaspersky.com/blog/black-cat-ransomware/44120/. Accessed 05 Jan 2024
14. Novinson, M.: The 10 Biggest Ransomware Attacks of 2019. https://www.crn.com/slide-shows/security/the-10-biggest-ransomware-attacks-of-2019/. Accessed 05 Jan 2024
15. pmdarima: ARIMA estimators for Python. https://alkaline-ml.com/pmdarima/. Accessed 05 Jan 2024
16. Ransomware Spotlight: REvil. https://www.trendmicro.com/vinfo/us/security/news/ransomware-spotlight/ransomware-spotlight-revil. Accessed 05 Jan 2024
17. Shen, Y., Mariconti, E., Vervier, P.A., Stringhini, G.: Tiresias: predicting security events through deep learning. In: Proceedings of the 2018 ACM SIGSAC Conference on Computer and Communications Security, CCS 2018, pp. 592–605. Association for Computing Machinery, New York, NY, USA (2018). https://doi.org/10.1145/3243734.3243811
18. The Good, The Bad and the Ugly in Cybersecurity - Week 21. https://www.sentinelone.com/blog/the-good-the-bad-and-the-ugly-in-cybersecurity-week-21-3/. Accessed 05 Jan 2024
19. Understanding Ransomware Threat Actors: LockBit — CISA — cisa.gov. https://www.cisa.gov/news-events/cybersecurity-advisories/aa23-165a. Accessed 05 Jan 2024
20. Werner, G., Okutan, A., Yang, S., McConky, K.: Forecasting cyberattacks as time series with different aggregation granularity. In: 2018 IEEE International Symposium on Technologies for Homeland Security (HST), pp. 1–7 (2018). https://doi.org/10.1109/THS.2018.8574185
21. What is maze ransomware? Definition and explanation — kaspersky.com. https://www.kaspersky.com/resource-center/definitions/what-is-maze-ransomware/. Accessed 05 Jan 2024

Detecting Malicious Android Game Applications on Third-Party Stores Using Machine Learning

Thanaporn Sanamontre, Vasaka Visoottiviseth(✉), and Chaiyong Ragkhitwetsagul

Faculty of Information and Communication Technology, Mahidol University, Nakhon Pathom, Thailand

thanaporn.sa4@student.mahidol.ac.th, {vasaka.vis, chaiyong.rag}@mahidol.edu

Abstract. Due to Android's flexibility in installing applications, it is one of the most popular mobile operating systems. Some Android users install applications from third-party stores even though they have the official application store, Google Play. These third-party stores usually have the mod version and the self-proclaimed original applications, which can be repackaged applications. Applications on these third-party stores can introduce security risks because of the non-transparent alteration and uploading processes. In this research, we inspect 492 Android applications from ten third-party stores for repackaged applications using information of APK files and a token-based code clone detection technique. We also classify repackaged applications as benign or malicious and categorize malicious applications into twelve malware categories. For the malware classification, we use machine learning techniques, including Random Forest, Decision Tree, and XGBoost, with the CCCS-CIC-AndMal-2020 Android malware dataset. Finally, we compare the results with VirusTotal, a well-known malware scanning website.

Keywords: Android Security · Android Malware Detection · Repackaged Applications · Machine Learning

1 Introduction

Android OS has the most market share in the mobile OS from the first and second quarter of 2023 [1]. Based on Statista [2], The application category Android users downloaded the most is the game category. Usually, users download game applications from Google Play, the official application store for Android devices. To protect users from malicious applications, Google Play has a feature called Google Play Protect [3], which automatically detects and protects Android devices from potentially harmful applications from both Google Play itself and from other sources. However, some people download Android applications from untrusted sources because the applications are not available on Google Play or are regionally restricted, or they need to use the other version of applications. Unfortunately, downloading games from third-party stores poses security risks due to their non-transparent alteration and uploading processes and lack of security features. Malicious actors can repackage applications, turn them into malware, and trick

L. Barolli (Ed.): AINA 2024, LNDECT 202, pp. 238–251, 2024.
https://doi.org/10.1007/978-3-031-57916-5_21

users to install these altered malicious applications. Installing them can lead to malicious activities, such as collecting sensitive data without user consent, stealing users' credentials, injecting hidden malware, adware, or backdoor, and compromising and harming the devices. Therefore, detecting these malicious applications from third-party stores is essential.

In this paper, we inspect ten selected third-party stores and their game applications including the self-proclaimed original and the mod versions of games. Regarding the self-proclaimed original game applications, we investigate whether they are repackaged applications by comparing hash values, a token-based code clone detection, and other attributes of the APK files. Next, we classify and categorize both repackaged and mod versions of game applications into twelve malware categories using machine learning techniques, including Random Forest, Decision Tree, and XGBoost, with the CCCS-CIC-AndMal-2020 Android malware dataset [4, 5].

The contributions of this paper are as follows.

- We inspect a total of 557 selected game applications: 65 original games from Google Play and 492 games from selected third-party stores. Among 492 applications, there are 196 self-proclaimed original applications, which we investigate whether they are repackaged.
- We use machine learning techniques with a recent Android malware dataset to classify altered game applications, including the repackaged version and mod versions of games, whether they are benign or malicious applications. Malicious applications are classified into twelve categories based on malware characteristics.
- We compare the results with VirusTotal [6], a famous malware scanning website.

This research paper's outlines are as follows. Section 2 describes the background and related works. The proposed work details are in Sect. 3. Section 4 shows the experiments and results. Lastly, we conclude and discuss our findings in Sect. 5.

2 Background and Related Work

This section provides the definitions, the background, and the related work.

2.1 Types of Altered Android Game Applications

On third-party Android application stores, there are games stating they are the same as the original ones on Google Play and altered games. Some modifications are with malicious intentions, while some of them are not. The alteration processes of applications affect their appearance, source code, functionalities, and behaviors. There are two categories of altered applications based on owners' authorization. Authorized altered applications have the owners' approval, such as games with updates and UI improvement. However, most application alterations do not have the proper authorization from the owners and violate the copyrights and distribution rights. Games are modified for some reasons, for example, accessing the premium features without paying, using the advertisement-free applications, and cheating the games by gaining the advantage against them. Examples of altered game applications are cracked games, repackaged applications, and mod versions of applications. Below are the definitions of the repackaged applications and the mod versions of applications.

- **Repackaged Applications**

Third-party developers decompile an original APK file, do unauthorized alteration, recompile, repackage it into a Repackaged Application, and distribute it on third-party stores. These repackaged applications have significant security risks because the alteration usually contains malicious intents. Malicious actors use these applications' appearance and functionalities as baits to trick users to download and install them. On third-party stores, applications claiming they are the same as the original ones on Google Play are called "Self-Proclaimed Original Application." Therefore, users will not know if they download "Repackaged Applications."

- **Mod Version of Game Applications**

Mod versions of game applications are created by enthusiasts and third-party developers who modify the original applications and then distribute them to third-party stores. They may modify the games to enhance gameplay, add new features, or cheat games. Even though mod versions of games are usually not modified with malicious intent, using applications from untrusted sources always carries security risks. Third-party stores always tell users that the games are the "Mod Versions" in their descriptions. Thus, users always know whether they install the mod versions.

2.2 CCFinderSW

Semura et al. [7, 8] proposed CCFinderSW, a token-based code clone detection tool. CCFinderSW uses a lexical analysis mechanism. Users can select programming languages such as Python, Java, and Ruby by selecting options for comments and reserved words of each language. CCFinderSW can detect Type-1 and Type-2 clone pairs. According to Roy and Cordy [9], Type-1 pairs are pairs with identical code fragments, while Type-2 pairs are similar to Type-1, but the variable names, values, and identifiers are changed.

2.3 Related Work

To prevent downloading malicious applications, some researchers proposed approaches for detecting Android malware. Chen et al. [10] proposed an Android malware detection with text-based classification and static analysis technique. After extracting static features, they converted them into word vectors for text classification using the BiLSTM processing model. They also used the DPCNN and Fasttext algorithms for performance comparison. However, the BiLiSTM has the highest accuracy score at 97.47% and the F1-score at 97.43%. Jaiswal et al. [11] provided a study of differences between benign, malicious, and clone applications based on system calls and behaviors. They also proposed a gaming malware detection system using dynamic analysis techniques. They found that some system calls, such as *stat64* and *llseek*, are more frequently called in malicious games. Moreover, some system calls that are not in the original games, such as *stat64*, *pread63*, and *brk*, but appear in the clone ones. The clone applications also ask for high-level permissions that the original ones do not require.

Several papers use the CCCS-CIC-AndMal-2020 dataset for Android malware detection. DIDroid [12] provided a detection system using 2D images with a deep learning algorithm based on CNN. They extracted static features, created vectors, selected the features, converted them into 2D gray images, and used the convolutional layers for training and testing models. DIDroid has an accuracy of 93.36%. The other provided work regarding this dataset is the EntropLyzer. The EntropLyzer [13] is an entropy-based detection system using dynamic features. Its entropy algorithm is Shannon entropy. They used Naive Bayes, Support Vector Machine, Random Forest, and Decision Tree algorithms for training the models. They analyzed and compared the entropy values of their features before and after rebooting and visualized their behavioral changes. They found that Decision Tree with the F1-score of 98.3% has the best performance for classifying malware categories.

3 Proposed Work

This section provides an explanation of the workflow, including the repackaged application detection and the malware detection and categorization.

3.1 List of Target Applications

We select top 200 free Android game applications for this study. We collected the Android games from the following ten third-party stores based on their popularity:

- APKAward
- APKCombo
- APKMody
- APKPure
- APKVision
- HappyMod
- LiteAPKs
- LuckyModAPK
- ModYolo
- PlayMods

The following are the criteria for selecting the target applications.

- Original games must have at least five third-party stores with the same versions.
- If the third-party stores have both self-proclaimed original and mod versions of applications, we will select the mod versions as our target.
- If there are many mod versions of an application, we will select the mod version with the latest updated date as our target.
- We will filter out the selected applications with the updated date and versions conflicted with the original ones on Google Play.

After the collection and filtering, we retrieve 557 games for this study. We classify target game applications into three categories: original, self-proclaimed original, and mod versions. Table 1 shows the amount of target applications in each category.

Table 1. Numbers of target applications in each category.

Type of Applications	Amount	Description
Original Applications	65	Filtered games from the top 200 free game applications on Google Play
Self-Proclaimed Original Applications	196	Applications on third-party stores which claimed that they are the same as the original ones on Google Play
Mod Version of Applications	296	Modified games on third-party stores for adding features and cheating games
Total	557	

3.2 Analysis Workflow

We propose an approach for inspecting the self-proclaimed original games whether they are repackaged. Then, we classify and categorize mod versions and the repackaged games whether they are in one of twelve malware categories or benign. Therefore, our proposed approach consists of two main processes: (1) the repackaged application detection and (2) the malicious application detection and categorization. Figure 1 shows the workflow of the proposed approach in detail. The analysis steps of the proposed approach are as follows.

1. Download and decompile the target applications.
2. Select a self-proclaimed original game and pair it with the original one for repackaged application detection processes.
3. Conduct the repackaged application detection on the pair.
4. Sort the result from the experiment. We will store the repackaged games for the malicious applications detection and categorization section. However, if the game is not a repackaged, record it as a "Benign Original Application."
5. If any self-proclaimed original games are remained, select a new one and repeat the process. If not, go to the next step.
6. Conduct malicious application detection and categorization on the repackaged and mod versions of games.
7. Sort the result from the experiment. If the application is benign, record it as a "Benign Application" with its type. However, if the application is malicious, classify it as a "Malicious Application" with its malware category.

As shown in Fig. 2 (a), there are three possible inputs and four main possible outputs. Three inputs include the original applications from Google Play, the self-proclaimed original, and the mod versions of games. On the other hand, four main possible outputs are as follows:

- Benign Original Applications (The same as the original games)
- Benign Repackaged Applications (The repackaged games but not malicious)
- Benign Mod Version Applications (The mod versions of games but not malicious)

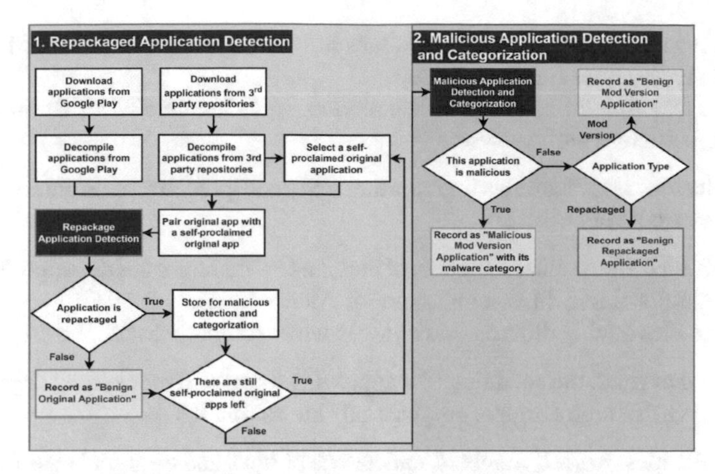

Fig. 1. Analysis workflow of the proposed approach

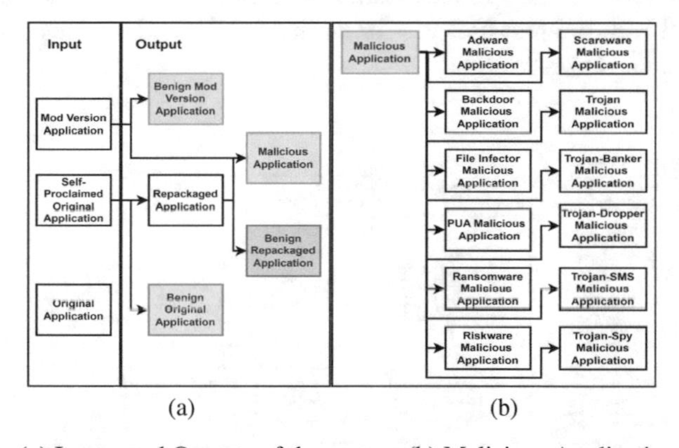

Fig. 2. (a) Inputs and Outputs of the system (b) Malicious Application Output

- Malicious Applications (The malicious games with possible twelve categories)

In case of malicious applications, they are categorized into one of twelve malware categories as depicted in Fig. 2(b).

3.3 Repackaged Application Detection

This section explains the method used to detect repackaged applications from the self-proclaimed original games in detail. Figure 3 illustrates the workflow of the repackaged application detection process. The following steps are how the process works.

1. Compute and compare the APK files' hash values.
2. Get and compare the number of files in both APK files.
3. Get and compare the names of all files in both APK files. We store all files with different names and additional files, except files for Android Application Binary Interface (ABI).

4. Compute and compare all file hash values in both APK files. We store all files with different hash values for the next step.
5. Conduct token-based code clone detection using CCFinderSW [7, 8] on the JAVA files with different hash values.

The following are conditions for reporting that a self-proclaimed original application is a "Repackaged Application."

1. The APK files' hash values, numbers of files, and filename are not the same. Moreover, not all additional and different files are for Android ABI.
2. There are files having different hash values when comparing with the original one.

On the other hand, the conditions for approving that a self-proclaimed original game application is a "Benign Original Application" are as follows.

1. Their APK files' hashes, numbers, and names of files, and each file's hash values are the same compared to the original APK file.
2. Even though the APK files' hashes, numbers of files, and filenames are not the same, all additional and different files are for Android ABI, and all files' hash values are the same as their original ones.

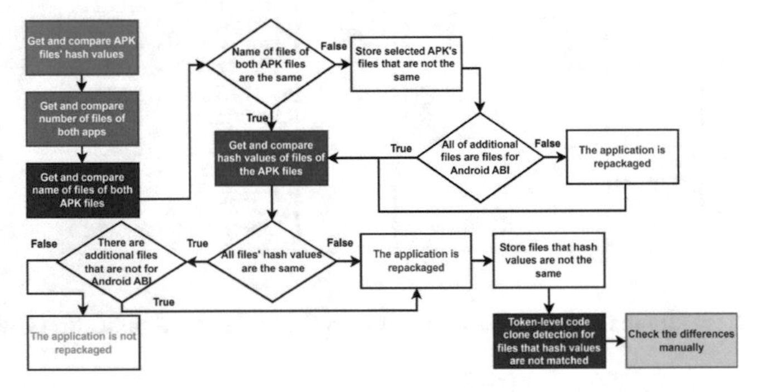

Fig. 3. Workflow of the Repackaged Application Detection Process

3.4 Malicious Application Detection and Categorization

This section presents the method used for detecting the malicious repackaged and mod versions of games and categorizing them into twelve categories. We use codes and unique lists provided by AndroidAppLyzer [14] for feature extraction and the CCCS-CIC-AndMal-2020 Android malware dataset [4, 5] for training the machine learning models. Characteristics of this dataset are as shown in Table 2. The machine learning algorithms include Random Forest, Decision Tree, and XGBoost.

Table 2. Number of malware samples in the CCCS-CIC-AndMal-2020 dataset.

Malware Categories	Numbers of Families	Numbers of Samples
Adware	48	47,210
Backdoor	11	1,538
File Infector	5	669
No Category	–	2,296
PUA	8	2,051
Ransomware	8	6,202
Riskware	21	97,349
Scareware	3	1,556
Trojan	45	13,559
Trojan-Banker	11	887
Trojan-Dropper	9	2,302
Trojan-SMS	11	3,125

Note that this dataset also contains 200k benign Android applications. The following steps are the workflow of the malicious application detection and categorization process.

1. Extract and create vectors of the APK file's 9,503 static features. There are 9,491 features from the provided unique lists, including permissions, actions, and categories. The rest are the additional information on the APK file as listed below.

- Numbers of Icons
- Numbers of Audio
- Numbers of Videos
- Size of the App
- Numbers of Activities
- Numbers of Meta-data
- Numbers of Services
- Numbers of Permissions
- Numbers of Categories
- Numbers of Actions
- Numbers of Providers

- Numbers of Receivers

2. Train the machine learning models with the CCCS-CIC-AndMal-2020 Android malware dataset. However as from Table 2, our number of samplings in each category of dataset is unbalanced. Therefore, for training we data-sampled each label in the dataset into 3,200 and 15,000 samples for training the model.
3. Use the models for malware detection and categorization on the target applications.

4 Experimental Results

4.1 Hardware and Software Specifications

We used Oracle VM VirtualBox version 7.0.4 to virtualize Kali Linux version 2023.2 VM with 2-core CPU and 8-GB RAM for conducting the repackaged application detection. Kali Linux has GNU bash 5.2.15, Python 3.10.9, JADX 1.4.7, and CCFinderSW 1.0 installed. For malicious application detection and categorization, we used Katana GF66 11 UG, CPU 8 cores, and RAM 16 GB with Windows 10.

4.2 Repackaged Application Detection

We experimented on the pair of a self-proclaimed original game and its original one on Kali Linux. We used JADX tool to decompile the APK files, SHA256 for the hash algorithm, and CCFinderSW to inspect Java files with different hash values from their original ones. The information of APK files used in the experiment are (1) The hash value of APK file, (2) numbers of files in each APK file, (3) filenames, and (4) hash values of each file. The expected outputs are the "Benign Original Application" and the "Repackaged Application." However, there is the "Exceptional Application," which are applications that has the same numbers of files, the same filenames, the same hash values of each file, but the hash values of APK files are not the same.

We conducted experiments on 196 self-proclaimed original games from third-party stores with 65 original games from Google Play. Table 3 shows results based on each third-party repository. We found 93 repackaged applications out of 196 self-proclaimed original games, which is 47.45%. We also found 19 exceptional applications. Only 84 applications are benign original applications. The store with the highest percentage of repackage applications is ModYolo which has 100%, while the store with the lowest percentage of application is LiteAPKs which has 33.33%.

Figure 4 depicts the results of the repackaged application detection in the pie chart. In conclusion, there are 93 repackaged applications out of 492 targets from third-party stores, which is 18.90%. The benign original and the exceptional applications are 17.07% and 3.86% out of the targets, respectively.

Table 3. Repackaged applications on each third-party repository.

Third-Party Stores	Self-Proclaimed Original Applications	Repackaged Applications	Benign Original Applications	Exceptional Applications	Repackaged Application Percentage
APKAward	9	7	1	1	77.78%
APKCombo	56	24	28	4	42.86%
APKMody	–	–	–	–	0.00%
APKPure	65	22	35	8	33.85%
APKVision	11	6	2	3	54.55%
HappyMod	14	6	6	2	42.86%
LiteAPKs	9	3	5	1	33.33%
LuckyModAPK	–	–	–	–	0.00%
ModYolo	6	6	0	0	100.00%
PlayMods	26	19	7	0	73.08%
TOTAL	**196**	**93**	**84**	**19**	**47.45%**

* APKMody and LuckyModAPK do not have any self-proclaimed original games.

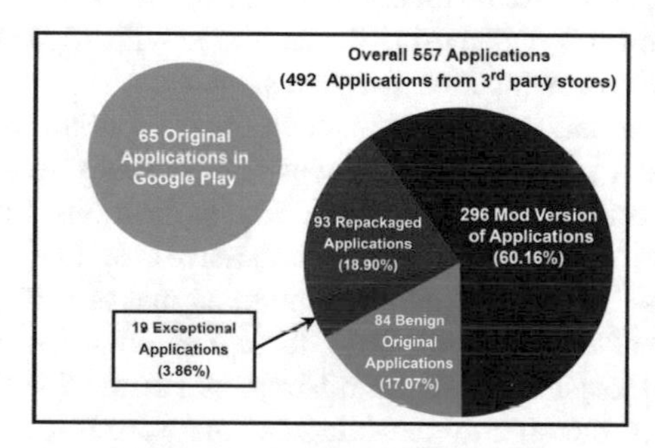

Fig. 4. Target applications after Repackaged Application Detection

4.3 Malicious Application Detection and Categorization

We experimented on 389 game applications, 93 of which are repackaged games and 296 of which are mod versions of games. Regarding extracting static features, we used the AndroidAppLyzer's code [14]. We then run feature selections for each machine learning algorithm including Random Forest (RF), Decision Tree (DT), and XGBoost (XGB) by using the CCCS-CIC-AndMal-2020 Android malware dataset to classify applications as malware or benign. As the results, the maximum number of features after the feature selections are 743 features. For training the model, to balance the dataset from each category, we data-sampled each label in the dataset into with 3,200 and 15,000 samples, respectively. Table 4 shows each algorithm and its performance. Criteria for performance evaluation of each algorithm are the accuracy and the F1-score. The algorithm with the highest accuracy and F1-score for both 3,200 and 15,000 samples is the Random Forest algorithm. It has accuracy and F1-score of 91.80% and 91.85% for 3,200 samples. Regarding 15,000 samples, it has accuracy and F1-score of 94.78% and 94.83%. Using 15,000 samples resulted in higher accuracy and F1-score, but all three algorithms have similar performances.

Table 4. Algorithms and their performances.

ML Algorithms	3,200 Samples		15,000 Samples	
	Accuracy	F1-Score	Accuracy	F1-Score
Random Forest (RF)	91.80%	91.85%	94.78%	94.83%
Decision Tree (DT)	90.29%	90.31%	93.98%	94.03%
XGBoost (XGB)	90.99%	91.04%	93.12%	93.17%

We then tested 389 game applications with the trained models. Table 5 shows the malware classification results of each machine learning algorithm. There are 13 outputs: one benign and twelve malware categories. The decision tree algorithm detected the most numbers of malware, with 84 malwares for 3,200 samples and 39 malwares for 15,000 samples. Using the decision tree with 3,200 samples, the most numbers of malware is the SMS category. However, when using decision tree with 15,000 samples, Adware is the category with the most number of malware samples.

We then verified our malware detection results by comparing with the APK scanning results with VirusTotal [6], a popular online malware-scanning website containing results from 62–65 security vendors. However, from 389 applications, there are 37 APK files we could not upload to VirusTotal due to their excessive size. Therefore, we could verify only 352 applications. We counted an application as malware if one or more security vendor on VirusTotal detected the malware. Table 6 summarizes our machine learning results and VirusTotal results. Results from VirusTotal stated that 57.33% of the targets are malware. However, the highest percentage of malware from our machine learning results is only 21.59%, when using the decision tree algorithm with 3,200 samples.

Table 7 compares the percentage of the detection when compares the results from our machine learning (ML) to the VirusTotal. According to Table 7, for 3,200 samples DT algorithm could detect the highest percentage of malware applications with 22.87%, while 15,000 samples, DT algorithm could detect only 9.42%.

Table 5. Malware classification results of each algorithm.

Label	3,200 Samples			15,000 Samples		
	RF	DT	XGB	RF	DT	XGB
Adware	0	14	0	0	30	0
Backdoor	0	0	0	0	0	0
Banker	2	6	2	1	2	2
Benign	383	305	381	387	350	384
Dropper	0	3	0	0	0	0
File Infector	0	0	0	0	0	0
PUA	0	8	0	0	0	3
Ransomware	0	3	0	0	0	0
Riskware	4	0	0	1	5	0
SMS	0	49	0	0	0	0
Scareware	0	1	2	0	1	0
Spy	0	0	0	0	0	0
Trojan	0	0	4	0	1	0
TOTAL (Malware)	**6**	**84**	**8**	**2**	**39**	**5**

Table 6. Results of the experiment and VirusTotal.

Label	VirusTotal	3,200 Samples			15,000 Samples		
		RF	DT	XGB	RF	DT	XGB
Benign	129	383	305	381	387	350	384
Malware	223	6	84	8	2	39	5
Malware Percentage	57.33%	1.54%	21.59%	2.06%	0.51%	10.03%	1.29%

Table 7. Comparison and percentage of the results and VirusTotal.

Label	3,200 Samples			15,000 Samples		
	RF	DT	XGB	RF	DT	XGB
ML Result = Benign VirusTotal = Benign	124/129 (96.12%)	101/129 (78.29%)	123/129 (95.35%)	128/129 (99.22%)	122/129 (94.57%)	128/129 (99.22%)
ML Result = Benign VirusTotal = Malware	222/223 (99.55%)	172/223 (77.13%)	221/223 (99.10%)	222/223 (99.55%)	202/223 (90.58%)	222/223 (99.55%)
ML Result = Malware VirusTotal = Benign	5/129 (3.88%)	28/129 (21.71%)	6/129 (4.65%)	1/129 (0.78%)	7/129 (5.43%)	1/129 (0.78%)
ML Result = Malware VirusTotal = Malware	1/223 (0.45%)	**51/223** **(22.87%)**	2/223 (0.90%)	1/223 (0.45%)	**21/223** **(9.42%)**	1/223 (0.45%)

5 Conclusion and Discussions

Installing applications from third-party stores comes with security risks. Users do not know whether games on third-party Android application stores are benign, repackaged, or malicious. Therefore, we proposed a method to detect the repackaged applications from the self-proclaimed original games. We found that 93 applications are repackaged applications, which is 47.45% of the self-proclaimed original games. Then, we detected and categorized repackaged applications and mod versions of games whether they were malicious by using three machine learning techniques including Random Forest, Decision Tree, and XGBoost with the CCCS-CIC-AndMal-2020 Android malware dataset. We also used VirusTotal for the result verification. The Random Forest algorithm has the highest accuracy and F1-score at 94.78% and 94.83%, using 15,000 samples, respectively. The decision tree algorithm detected the most numbers of malware which is 84 from 389 application or 21.59%. However, VirusTotal detected 223 malwares from 389 targets, which is 57.33%. The decision tree with 3,200 samples has the highest percentage of malware detection which is 22.87% when comparing to the VirusTotal detection results. However as mentioned earlier, VirusTotal uses the scanning results from around 62–65 security vendors, but among our total of 223 applications identified as malware by VirusTotal, 138 applications or around 61.88% of them are detected as malware by only 1–2 virus security vendors. Therefore, some results from VirusTotal may be false positives and need further investigation. For the future work, to increase the accuracy and performance of this research, we should also use the dynamic analysis technique.

Acknowledgments. This research project was partially supported by Faculty of Information and Communication Technology, Mahidol University.

References

1. Statcounter. Mobile Operating System Market Share Worldwide. https://gs.statcounter.com/os-market-share/mobile/worldwide/#monthly-202301-202306
2. Statcounter. Number of mobile app downloads in the Google Play Store in 1st quarter 2021 and 1st quarter 2022, by category. https://www.statista.com/statistics/1331430/google-play-app-downloads-by-category
3. Google Play. Use Google Play Protect to help keep your apps safe and your data private. https://support.google.com/android/answer/2812853?hl=en
4. University of New Brunswick. CCCS-CIC-AndMal-2020. https://www.unb.ca/cic/datasets/andmal2020.html
5. D'hooge, L.: CCCS-CIC-AndMal-2020. https://www.kaggle.com/datasets/dhoogla/cccscicandmal2020
6. VirusTotal. https://www.virustotal.com/gui/home
7. Semura, Y., Yoshida, N., Choi, E., Inoue, K.: CCFinderSW: clone detection tool with flexible multilingual tokenization. In: 2017 24th Asia-Pacific Software Engineering Conference (APSEC), pp. 654–659 (2017)
8. Semura, Y.: CCFinderSW. https://github.com/YuichiSemura/CCFinderSW
9. Roy, C.K., Cordy, J.R.: A survey on software clone detection research. In: School of Computing TR 2007–541 (2007)
10. Chen, M., Zhou, Q., Wang, K., Zeng, Z.: An android malware detection method using deep learning based on multi-features. In: 2022 IEEE International Conference on Artificial Intelligence and Computer Applications (ICAICA), pp. 187–190 (2022)
11. Jaiswal, M., Malik, Y., Jaafar, F.: Android gaming malware detection using system call analysis. In: 2018 6th International Symposium on Digital Forensic and Security (ISDFS), pp. 1–5 (2018)
12. Rahali, A., Lashkari, A.H., Kaur, G., Taheri, F., Gagnon, F., Massicotte, F.: DIDroid: android malware classification and characterization using deep image learning. In: 2020 10th International Conference on Communication and Network Security (ICCNS 2020), pp. 70–82 (2020)
13. Keyes, D.S., Li, B., Kaur, G., Lashkari, A.H., Gagnon, F., Massicotte, F.: EntropLyzer: android malware classification and characterization using entropy analysis of dynamic characteristics. In: 2021 Reconciling Data Analytics, Automation, Privacy, and Security: A Big Data Challenge (RDAAPS), pp. 1–12 (2021)
14. Lashkari, A.H.: AndroidAppLyzer. https://github.com/ahlashkari/AndroidAppLyzer

Neural Network Innovations in Image-Based Malware Classification: A Comparative Study

Hamzah Al-Qadasi[1], Djafer Yahia M. Benchadi[2(✉)], Salim Chehida[1], Kazuhiro Fukui[2], and Saddek Bensalem[1]

[1] Univ. Grenoble Alpes, CNRS, Grenoble INP, Verimag, 38000 Grenoble, France
{hamzah.al-qadasi,salim.chehida,saddek.bensalem}@univ-grenoble-alpes.fr
[2] University of Tsukuba, Tsukuba, Ibaraki 305-8577, Japan
djafer@cvlab.cs.tsukuba.ac.jp, kfukui@cs.tsukuba.ac.jp

Abstract. As the digital landscape continuously evolves, the field of cybersecurity faces the growing challenge of effectively detecting and classifying malware. Malware visualization is an emerging technique that involves converting malware's binary structure into images. This technique uses deep learning as an appealing approach and has been providing promising results. However, the intricate effects of recent deep neural network models on classification performance need to be explored. This paper presents a comprehensive study of malware classification using recently developed models such as ConvNeXt V1 and V2. We initiate our investigation by examining the effectiveness of various neural network architectures in malware detection. We compare traditional models such as VGG16 and ResNet18 with recent ones like ConvNeXt V1 and V2. Our results demonstrate that ConvNeXt V2 significantly outperforms all other examined neural networks in both effectiveness and efficiency. For instance, the ConvNeXt V2 model achieves remarkable F1 scores, excelling on the Malimg and MaleVis datasets with scores of 99.47% and 98.91%, respectively. Our study also includes a comparative analysis between the ConvNeXt V2 model and the recent research works in this area. This analysis reveals that ConvNeXt model substantially outperforms all malware classification methods, achieving an F1 score improvement of 0.1% to 6%. In brief, this research contributes valuable insights to the ongoing efforts aimed to advance malware classification techniques.

1 Introduction

Malware is harmful software designed for unwanted actions that frequently alters its form to evade detection. Therefore, accurate identification and classification are crucial for maintaining computer security. Existing literature outlines two

H. Al-Qadasi and M. Benchadi Djafer Yahia—These authors contributed equally to this work.

L. Barolli (Ed.): AINA 2024, LNDECT 202, pp. 252–265, 2024.
https://doi.org/10.1007/978-3-031-57916-5_22

primary approaches for malware detection and classification. The first, a static approach, focuses on analyzing machine or assembly codes to identify patterns that indicate malicious behavior. The second, a dynamic approach, involves executing malware within a safe virtual environment to detect malicious behavior from the resulting execution trace. Both approaches share a common challenge: the intricate and time-consuming process of extracting features from malware.

Recent research has increasingly focused on leveraging vision-based Deep Neural Networks (DNNs) for malware classification. This approach begins with various malware datasets, created by transforming malware binaries into images. Key contributions include Tekerek [1] with DenseNet-121 achieving a 97.76% F1 score on BIG2015 dataset. Also, Wong [2] used ShuffleNet and DenseNet-201 with SVMs for a 95.01% F1 score on MaleVis dataset. Lastly, Atitallah [3] combined ResNet18, MobileNetV2, and DenseNet161 for a top F1 score of 98.70% on MaleVis dataset. Collectively, these studies demonstrate the growing trend and effectiveness of deep learning in image-based malware classification.

While DNNs have significantly advanced malware classification, minimizing errors, i.e., false negatives and false positives, remains crucial. False negatives can allow malware to breach systems, leading to data loss and financial damage. In contrast, false positives can disrupt operations and damage the credibility of security systems. As a result, achieving a high F1 score in malware detection is essential, even with small improvements. Therefore, it is imperative to continually explore and test innovations in DNN models within this domain of malware classification.

Our study aims to explore the recent innovations of DNNs. Thus, our work is structured around two main experiments. In the first experiment, we evaluate a range of twelve neural network architectures, including traditional and recent advanced models. This evaluation aims to study the impact of neural network design on malware classification. More specifically, we evaluate the ability of these models to classify malware accurately in challenging datasets, e.g., Malimg and MaleVis. The second experiment carefully investigates the capabilities of the most effective DNN in the first experiment, i.e., ConvNeXt V2, against a broader spectrum of six datasets. This comprehensive approach allows us to assess the model's robustness and benchmark it against existing state-of-the-art (SOTA) techniques in malware classification.

In summary, this paper presents the following key contributions:

- We extensively analyze various DNN architectures, including traditional and most recent models, to study their efficacy in malware classification.
- We conduct a comparative study to evaluate the ConvNext V2 model against the recent research in malware classification across various challenging datasets.
- We critically discuss the applicability and limitations of traditional data augmentation techniques in malware byteplot analysis.

The rest of this paper is structured as follows: we give an overview of the image-based malware approach and the ConvNeXt model in Sect. 2. We introduce our approach in Sect. 3. The experimental setup and evaluation are given in Sect. 4 and Sect. 5. Lastly, Sect. 6 discusses related work and Sect. 7 presents the conclusion and future work.

2 Background

Section 2.1 provides an analysis of the Malware PE Source Structure. Then, we explore the transformation of malware from source code to image format in Sect. 2.2. Finally, Sect. 2.3 presents the ConvNeXt V2 model, highlighting its design improvements for efficient image processing.

2.1 Malware Source Structure

Malware often manipulates the Portable Executable (PE) file format, which is fundamental to executable files (EXE) and Dynamic Link Libraries (DLLs) in Microsoft Windows systems. This manipulation aims to disrupt or gain unauthorized access to computer systems. Thus, understanding the PE file format is crucial for effective malware analysis and detection.

Figure 1 illustrates an overview of the PE file format. This file is structured into two main parts: the *Header* and the *Sections*. The Header includes the DOS Header, PE Header, Optional Header, and a Sections table. These components are essential for the dynamic linker, which maps the file into memory. The Sections part of the PE file contains Code, Import, and Data, further divided into subsections like .text, .data, .idata, etc. [4]. This format encapsulates PE32 for 32-bit and PE32+ for 64-bit executables.

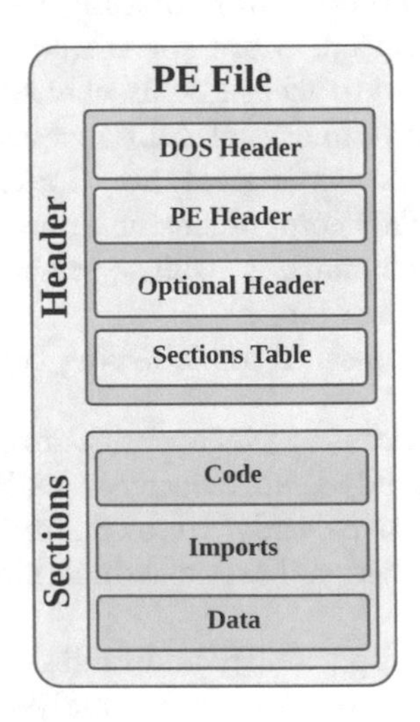

Fig. 1. Portable Executable file structure.

2.2 Malware from Source Code to Image

A raw byte binary stream of a PE file is the starting point for converting malware source code into an image. The stream, composed of binary sequences, is divided into 8-bit substrings. Each substring, an 8-bit unsigned integer, represents values between 0 and 255. These integers are then organized into a two-dimensional (2D) array, forming a 2D grayscale image. Moreover, the width of the image is computed according to the file size as introduced by Nataraj et al. [5]. The height varies based on the amount of data in the malware file. In other words, a larger malware file results in a 2D image with more rows. This method ensures that the resulting image effectively represents the structural patterns of the binary data. Lastly, the 2D array can be stored as a grayscale image using one of the visualization Python libraries, e.g., OpenCV. All the steps involved in the conversion process are depicted in Fig. 2. Alternatively, color images are increasingly used for more detailed representations of malware families. As a result, some studies have explored using RGB encoding to generate images from malware dump files [6].

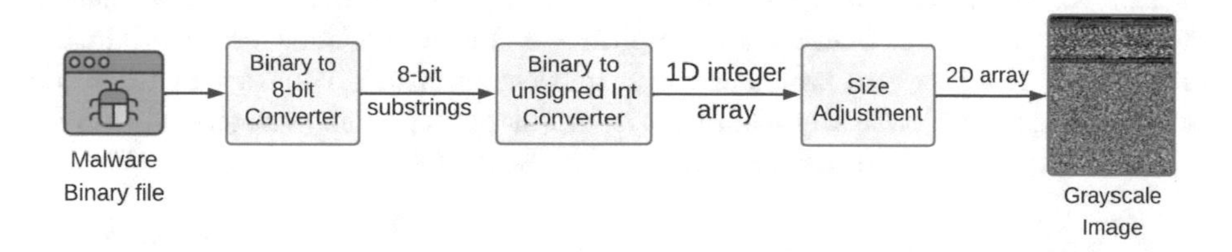

Fig. 2. Conversion of source PE code to grayscale image.

2.3 ConvNeXt

The ConvNeXt V2 model [7] is an upgrade from its previous version, ConvNeXt V1 [8]. The most significant change is how it learns from images, using the Fully Convolutional Masked Autoencoder (FCMAE) architecture. As shown in Fig. 3, the architecture includes a ConvNeXt encoder based on sparse convolution and a simplified ConvNeXt block decoder. This autoencoder design is asymmetrical. The encoder works solely with visible pixels, while the decoder uses these encoded pixels and mask tokens to rebuild the image.

Another key improvement in ConvNeXt V2 is the addition of a new layer known as Global Response Normalization (GRN). This layer helps the model to distinguish different features in an image better. It solves a problem where the model might overlook various aspects of an image, especially when learning from partially hidden or obscured images. This improvement enhances ConvNeXt V2's ability to recognize and understand different elements in pictures. Consequently, it becomes a powerful tool for tasks involving image recognition and processing.

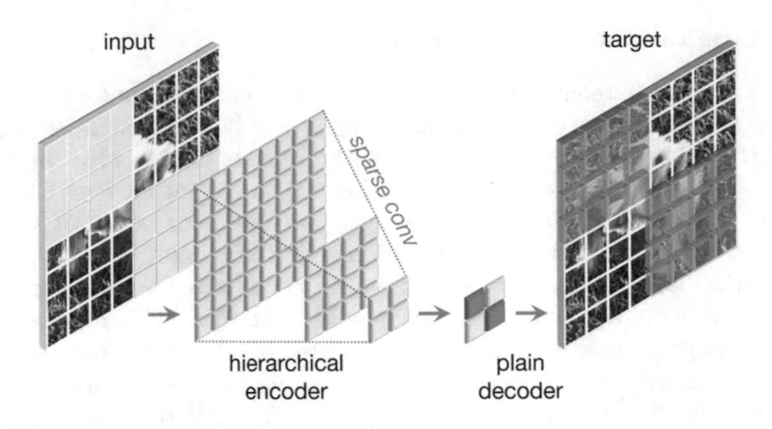

Fig. 3. FCMAE framework [7].

3 Approach

Section 3.1 describes the approach that carefully selects malware classification models and datasets. Section 3.2, explains a 3-fold stratified cross-validation method for training and validating our models to ensure robust and unbiased results. Finally, we critically examine the applicability of the existing data augmentation techniques in malware analysis in Sect. 3.3.

3.1 Model and Dataset Selection

In our research, we focus and experiment on two main objectives. First, we explore the evolution of neural network architectures and their impact on malware classification. Second, we deeply investigate the capabilities of the most effective neural network in the first experiment and compare it with the existing research works across various challenging datasets. This comprehensive evaluation aims to produce accurate and reliable conclusions in the field.

In the first experiment, we focus on a range of pre-trained neural networks. The selection of neural networks includes a spectrum from established models like VGG16 [9] and ResNet18 [10] to the latest advanced ones such as ConvNeXt V1 [8] and ConvNeXt V2 [7]. Additionally, the inclusion of models such as VGG16 and ResNet18 establishes a baseline for tracking the evolution of deep learning architectures over time. We also gain insights into how the progression and recent improvements in deep learning impact malware detection. We test 12 neural networks against two datasets: *Malimg* [5] and *MaleVis* [11]. These datasets represent a significant challenge in achieving a high F1 score in earlier studies.

In the second experiment, we select the best-performing neural network in the first experiment, i.e. the ConvNeXt V2 nano model. Then, we rigorously test this model with the other existing malware classifiers on six diverse malware datasets. This range of datasets offers a comprehensive evaluation of the model's effectiveness. It is worth mentioning that datasets such as *Malimg*, *Blended* [12], and *Microsoft* [13] feature highly imbalanced classes. Furthermore, datasets such as *Blended*, *Malimg*, and *MaleVis* present more significant challenges where the number of malware families exceeds 20 classes. Finally, *Virus-MNIST* [14] is recognized as one of the most challenging and largest datasets, with current studies yet to exceed an F1 score of 89%. A diverse dataset selection leads to an in-depth and accurate assessment of the model's robustness.

3.2 Cross Validation

In this research, we adopt a 3-fold stratified cross-validation method in all experiments for training and validating our neural networks, a crucial practice in deep learning. Cross-validation entails dividing the dataset into several segments, or *folds*. In our approach, one fold is used as the validation set, and the remaining folds serve as the training set. This method is essential for evaluating the neural networks' generalizability. Traditional random cross-validation can lead to skewed results in imbalanced datasets, as the distribution of classes may not be accurately represented. We address this challenge by applying stratified k-fold validation, which guarantees that each fold reflects the overall dataset's class distribution. This strategy is particularly effective in mitigating biases and enhancing the validity of the evaluation process. After training the network on each fold, we calculate performance metrics, e.g., F1 score, for each fold. Then, we average all F1 scores across all folds to obtain the overall average F1 score, namely, *macro F1 score*.

3.3 Data Augmentation Issue

In many existing studies [1,15,16], the practice of applying data augmentation to malware byteplot images has been common among researchers, yet without sufficient validation. Consequently, the authenticity of these images in reflecting the original data is questionable. Such an approach can lead to significant challenges in accurately classifying malware:

1. **Binary Sequence Integrity:** In conventional images, transformations maintain the identity of objects. Contrarily, malware byteplots require the preservation of byte order to maintain their operational code and data structure. Spatial transformations, such as rotations or flips, corrupt the binary sequence. This results in skewing the malware's true nature and impeding accurate classification.
2. **Representational Feature Accuracy:** In the byteplots, each grayscale or RGB intensity level corresponds to a specific byte value, which is essential for representing the original malware code. Transforming these levels with color variations or added noise can falsely insert non-existent features into the analysis. Such distortions could compromise the learning process, leading to the misclassification of software as malicious or vice versa.

In summary, data augmentation in malware byteplot analysis must be carefully tailored. More specifically, augmentation techniques must avoid adding randomness that changes the fundamental binary characteristics. This will preserve the data integrity and the reliability of the classification outcomes.

4 Experimental Setup

In Sect. 4.1, we describe the datasets and SOTA baselines used in our experiments. Section 4.2 details the evaluation metric for assessing our study. Finally, Sect. 4.3 outlines the research questions.

4.1 Dataset Description

Table 1 describes the datasets used in our study. This includes the number of malware families and total samples per dataset. Moreover, *Approach 2* and *Approach 3*, represent the recent and SOTA research methodologies we select for each dataset. These approaches will be analyzed in Sect. 5, providing a comparative framework for our study.

Table 1. Malware datasets and corresponding research methods.

Dataset	# Classes	# Samples	Approach 2	Approach 3
Virus-MNIST [14]	10	51,880	ECOC-SVM [2]	FACILE [14]
Malimg [5]	25	9,339	EfficientNetB4 [17]	Kernel Subspace [18]
Microsoft BIG15 [13]	9	10,868	ResNet-50 [19]	DenseNet-121 [1]
MaleVis [11]	26	13,760	ECOC-SVM [2]	Fused 3 CNNs [3]
Dumpware [6]	11	4,294	DenseNet-121 [1]	Kernel Subspace [18]
Blended [12]	25	13,747	ResNet50 [12]	NA

4.2 Evaluation Metrics

Our study emphasizes the F1 score as the primary metric for evaluating neural networks. In malware detection, both precision and recall hold equal importance. Thus, the F1 score offers a more balanced assessment of a model's performance. It is critical in situations where false positives and false negatives carry significant consequences. Precision, recall, and the F1 score are defined as follows:

$$\text{Precision} = \frac{TP}{TP + FP}, \quad \text{Recall} = \frac{TP}{TP + FN}, \quad F1 = 2 \times \frac{\text{Recall} \times \text{Precision}}{\text{Recall} + \text{Precision}}$$

Here, TP are True Positives, FP are False Positives, and FN are False Negatives.

Accuracy alone is misleading in malware classification, as it cannot distinguish between the types of classification errors. For instance, a model with high accuracy could still misclassify critical malware instances, leading to severe real-world implications. Conversely, the F1 score enhances the robustness of our evaluation by considering both precision and recall. This balance is crucial, as missing a malicious file (low recall) or wrongly identifying a benign file as malicious (low precision) can have significant consequences.

4.3 Research Questions

- ❶ **Neural Networks Effectiveness:** How effective are the recent neural networks in malware classification compared to the existing neural networks?
- ❷ **Comparative Approach Analysis:** How effective is the best neural network in **RQ.1** compared to the existing research work in malware classification?
- ❸ **Neural Networks Efficiency:** How efficient is the top-performing network in **RQ.1** in compared to the recent research works in terms of model size and training convergence?

5 Experimental Evaluation

This section addresses the research questions outlined in Sect. 4. We conducted our experiments on a laptop with RTX A4000 Mobile GPU, 2.80 GHz, and 8 GB RAM. We conduct our experiments on top of the fastai library [20] v2.7.13. The source code for our experiments is available at this repository[1]. This ensures transparency and reproducibility of our results.

5.1 RQ1: Neural Networks Effectiveness

Figure 4 demonstrates that ConvNeXt V2 model emerges as the top-performing neural network. It achieves an impressive average F1 score of 99.47% on the Malimg dataset and 98.91% on the MaleVis dataset. Conversely, on the Malimg dataset, Inception V3 lags, with an average F1 score of 93.01%, significantly smaller than the leading models. In the MaleVis dataset, ResNet18 and MobileNet V2 also fall short, with average F1 scores of 96.48% and 96.29%, respectively.

In comparison, VGG16 also shows promising results with F1 scores of 99.39% on Malimg and 98.77% on MaleVis. Although it is outperformed by the ConvNeXt models, VGG16 surpasses other architectures. This can be attributed to the dataset's nature and the complexity of the models. The relatively small datasets used in this experiment are unlikely to cause overfitting when a model with a smaller number of parameters, e.g., VGG16, is used. In the case of larger and more complex models, there is a risk of overfitting to such datasets which undermines the performance of these models on unseen data.

The ultimate goal is to obtain an optimal model that achieves a 100% F1 score, as any misclassification may have costly and catastrophic consequences

[1] https://github.com/halqadasi/Malware-Classification.

in safety and security-critical applications. In these applications, the difference between 98% and 99% F1 scores means the difference between a secure and a compromised system. As a result, many studies effortlessly strive to achieve minor improvements to approach the ideal scenario where every instance of malware is accurately identified.

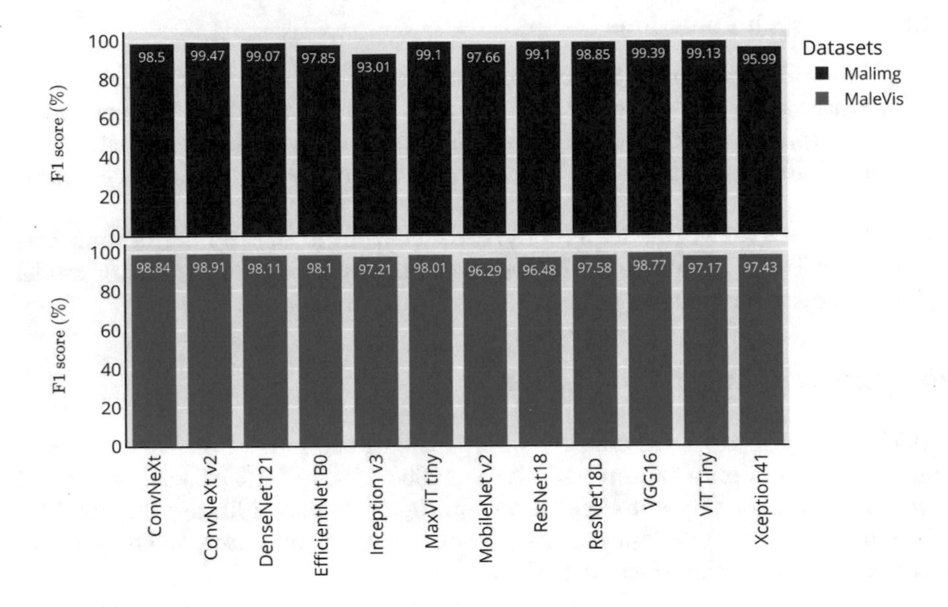

Fig. 4. F1 scores of various neural networks with Malimg and MaleVis Datasets.

> **RQ 1 Answer :**
>
> The recently developed ConvNeXt V2 model outperforms the existing neural networks in malware classification. ConvNeXt V2 model achieves near-optimal F1 scores on two challenging datasets: Malimg and MaleVis.

5.2 RQ2: Comparative Approach Analysis

We select the most recent SOTA works in malware classification for comparison with the ConvNeXt V2 model. We select the two most recent significant research works for each dataset, labeled as Approach 2 and Approach 3 as detailed previously in Sect. 4.1.

The ConvNeXt V2, as the best-performing model identified in RQ.1, demonstrates significant effectiveness in malware classification across several datasets. For example, in the Virus-MNIST dataset, there is a significant margin, up to 6%, in the performance between ConvNeXt V2 and other approaches. In the Blended dataset, it achieves an F1 score of 99.30%, exceeding ResNet50's [12]

(95.00%). In the Dumpware dataset, the ConvNeXt V2 has an F1 score of 99.5%, outperforming DenseNet-121 [1] and Kernel Subspace [18], which score 95.82% and 96.20% respectively, showing its robustness in diverse data scenarios (Fig. 5).

In the MaleVis dataset, the model achieves competitive performance with a higher F1 score of 98.91% compared to ECOC-SVM's [2] 95.37% and Fused 3-CNN's [3] 98.7%. In the Malimg dataset, ConvNeXt has an F1 score is 99.5%, which is much better than EfficientNetB4 [17] and Kernel Subspace [18]. These results across various datasets highlight the consistent performance of ConvNeXt V2. Also, this highlights the importance of ongoing advancements in neural network design to tackle challenges in the cybersecurity domain effectively.

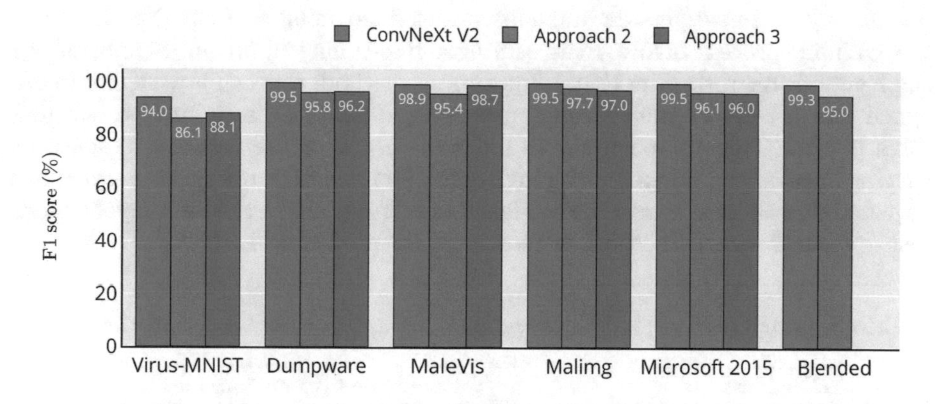

Fig. 5. Comparison of ConvNeXt V2 with the other recent malware classifiers.

> **RQ 2 Answer :**
>
> The ConvNeXt v2 model significantly outperforms the recent research approaches in malware classification with higher F1 scores across six malware datasets.

5.3 RQ3: Neural Networks Efficiency

In the field of malware classification, it is vital to prioritize the accuracy and security of the model over the model efficiency due to the security risks associated with misclassifications. However, the efficiency of models like the ConvNeXt V2 nano remains an important factor to consider. The ConvNeXt V2 nano model stands out with its relatively lower parameter count of 15.6 million. Moreover, the ConvNeXt V2 nano model demonstrates greater efficiency than DenseNet 121 and ResNet50, which have parameter counts of 25.98 million and 25.56 million, respectively. The compact size of ConvNeXt V2 nano suggests a more efficient use of computational resources. Additionally, the model's training process is significantly faster which requires only 12 epochs to converge. This is substantially lower than the 100 epochs needed for DenseNet 121 [1] and 30 epochs for EfficientNetB4 [17], as shown in Table 2. This indicates a faster learning capability and a shorter overall training time.

On the Virus-MNIST dataset, ConvNeXt V2 nano achieves a 94% F1 score which outperforms FACILE's score of 88.1%. This is noteworthy considering that ConvNeXt V2 has a larger model size, with 15.6 million parameters versus FACILE's 0.125 million. Despite its size, ConvNeXt V2 nano demonstrates efficient learning by reaching high accuracy in just 12 epochs which matches FACILE's training duration. This balance of high accuracy and efficient training highlights the effectiveness of ConvNeXt V2 nano in malware classification.

The rapid convergence of the ConvNeXt V2 nano is largely attributed to the pretraining on ImageNet-12k and subsequent fine-tuning on ImageNet-1k. Also, the training process involves the strategic fine-tuning approach supported by the Fastai framework. The Fastai fine-tuning method, especially with one freeze epoch and a total of twelve training epochs, leverages discriminative learning rates [21,22]. This approach allows different layers of the network to learn at varying speeds, i.e., different learning rates. The earlier, more generic layers are adjusted slightly to preserve the pre-learned features, while the later layers, more task-specific layers are subject to more significant changes.

Table 2. Comparison of model size and training duration across diverse classifiers.

Classifiers	#Parameters (in million)	Epochs
DenseNet 121 & CycleGAN [1]	25.98	100
EfficientNetB4 [17]	19.34	30
ResNet50 [12]	25.56	18
ECOC-SVM [2]	20.01	NA
Fused 3 CNNs [3]	43.19	35
FACILE [14]	**0.125**	100
ConvNeXt V2 nano	15.6	**12**

RQ 3 Answer :

The ConvNeXt V2 nano model demonstrates notable efficiency compared to recent research work in terms of model size, and training convergence.

6 Related Work

As previously discussed in Sect. 5.2 and Sect. 5.3, we have explored various neural network methodologies in malware classification. These methodologies have been compared in terms of performance.

Tekerek et al. [1] implemented DenseNet 121 alongside CycleGAN to classify malware images in datasets such as Dumpware and Microsoft BIG15. Wong et al. [2] combined ShuffleNet and DenseNet-201 with ECOC-SVM to address challenges in class distinctions across multiple datasets. Danish Vasan et al. [17] utilized a fine-tuned EfficientNetB4 to enhance the performance. Djafer Benchadi et al. [18] explored Kernel Subspace Method (KSM) based on image-set dimensional reduction for malware classification. Finally, Atitallah et al. [3] fused three CNN models (ResNet18, MobileNetV2, and DenseNet161) with a random forest voting strategy. This approach focuses on generalization in image-based malware classification.

Compared with these techniques, our study investigates the advanced recognition capabilities of the ConvNeXt V2 model [7]. The enhanced structure of the ConvNeXt V2 model establishes a leading position in malware classification across diverse benchmarks.

7 Conclusion and Future Work

In this study, we achieved significant progress in malware classification by rigorously evaluating the ConvNeXt V2 nano model. Our analysis revealed that this model outperforms both traditional and newer neural networks in effectiveness. Furthermore, the model is more operationally efficient than other recent research work. The performance of the ConvNext V2 nano model is attributed to the smaller model size and the quick convergence time. This effective balance between model efficiency and effectiveness in critical security scenarios is a significant contribution of our research. Lastly, our findings suggest that traditional data augmentation techniques are ineffective in malware classification, as they can compromise the integrity of binary sequences.

Future research should focus on integrating the ConvNeXt V2 nano model with real-time malware detection systems to evaluate its practical effectiveness. Moreover, it is crucial to develop advanced data augmentation techniques that respect the unique nature of malware data. Such techniques must ensure the integrity of binary sequences and representational features. Lastly, given the rapid advancements in neural network architectures, there is a pressing need for continual evaluation and comparison with emerging models. This ongoing analysis is essential to maintain SOTA performance in malware classification.

Acknowledgements. This work was supported in part by the Japanese Ministry of Education, Culture, Sports, Science and Technology (MEXT) scholarship. Moreover, this project has also received funding from the European Unions Horizon 2020 research and innovation programme under grant agreement No 956123.

References

1. Tekerek, A., Yapici, M.M.: A novel malware classification and augmentation model based on convolutional neural network. Comput. Secur. **112**, 102515 (2022)
2. Wong, W., Juwono, F.H., Apriono, C.: Vision-based malware detection: a transfer learning approach using optimal ECOC-SVM configuration. IEEE Access **9**, 159262–159270 (2021)
3. Atitallah, S.B., Driss, M., Almomani, I.: A novel detection and multi-classification approach for IoT-malware using random forest voting of fine-tuning convolutional neural networks. Sensors **22**(11), 4302 (2022)
4. Pietrek, M.: An in-depth look into the win32 portable executable file format, part 2. MSDN Mag. (2002)
5. Nataraj, L., Karthikeyan, S., Jacob, G., Manjunath, B.S.: Malware images: visualization and automatic classification. In: The 8th International Symposium on Visualization for Cyber Security, pp. 1–7 (2011)
6. Bozkir, A.S., Tahillioglu, E., Aydos, M., Kara, I.: Catch them alive: a malware detection approach through memory forensics, manifold learning and computer vision. Comput. Secur. **103**, 102166 (2021)
7. Woo, S., et al.: ConvNeXt V2: co-designing and scaling convnets with masked autoencoders. In: Proceedings of the IEEE/CVF Conference on Computer Vision and Pattern Recognition, pp. 16133–16142 (2023)
8. Liu, Z., Mao, H., Wu, C.-Y., Feichtenhofer, C., Darrell, T., Xie, S.: A convnet for the 2020s. In: Proceedings of the IEEE/CVF Conference on Computer Vision and Pattern Recognition, pp. 11976–11986 (2022)
9. Simonyan, K., Zisserman, A.: Very deep convolutional networks for large-scale image recognition. arXiv preprint arXiv:1409.1556 (2014)
10. He, K., Zhang, X., Ren, S., Sun, J.: Deep residual learning for image recognition. In: Proceedings of the IEEE Conference on Computer Vision and Pattern Recognition, pp. 770–778 (2016)
11. Bozkir, A.S., Tahillioglu, E., Aydos, M., Kara, I.: Catch them alive: a malware detection approach through memory forensics, manifold learning and computer vision. Comput. Secur. **103**, 102166 (2021)
12. Shaik, A., Pendharkar, G., Kumar, S., Balaji, S., et al.: Comparative analysis of imbalanced malware byteplot image classification using transfer learning. arXiv preprint arXiv:2310.02742 (2023)
13. Ronen, R., Radu, M., Feuerstein, C., Yom-Tov, E., Ahmadi, M.: Microsoft malware classification challenge (2018). arXiv:1802.10135
14. Zou, B., Cao, C., Wang, L., Fu, S., Qiao, T., Sun, J.: FACILE: a capsule network with fewer capsules and richer hierarchical information for malware image classification. Comput. Secur. **137**, 103606 (2024)
15. Catak, F.O., Ahmed, J., Sahinbas, K., Khand, Z.H.: Data augmentation based malware detection using convolutional neural networks. PeerJ Comput. Sci. **7**, e346 (2021)
16. Marastoni, N., Giacobazzi, R., Dalla Preda, M.: Data augmentation and transfer learning to classify malware images in a deep learning context. J. Comput. Virol. Hack. Tech. **17**, 279–297 (2021)
17. Mitsuhashi, R., Shinagawa, T.: Exploring optimal deep learning models for image-based malware variant classification. In: 2022 IEEE 46th Annual Computers, Software, and Applications Conference (COMPSAC), pp. 779–788. IEEE (2022)

18. Benchadi, D.Y.M., Batalo, B., Fukui, K.: Efficient malware analysis using subspace-based methods on representative image patterns. IEEE Access **11**, 102492–102507 (2023)
19. Ma, Y., Liu, S., Jiang, J., Chen, G., Li, K.: A comprehensive study on learning-based PE malware family classification methods. In: Proceedings of the 29th ACM Joint Meeting on European Software Engineering Conference and Symposium on the Foundations of Software Engineering, pp. 1314–1325 (2021)
20. Howard, J., et al.: Fast.ai (2018). https://github.com/fastai/fastai
21. Smith, L.N., Topin, N.: Super-convergence: very fast training of neural networks using large learning rates. In: Artificial Intelligence and Machine Learning for Multi-Domain Operations Applications, vol. 11006, pp. 369–386. SPIE (2019)
22. Howard, J., Ruder, S.: Universal language model fine-tuning for text classification. arXiv preprint arXiv:1801.06146 (2018)

Enhancing Robustness of LLM-Synthetic Text Detectors for Academic Writing: A Comprehensive Analysis

Zhicheng Dou[1,2](✉), Yuchen Guo[1,2](✉), Ching-Chun Chang[1], Huy H. Nguyen[1], and Isao Echizen[1,2]

[1] National Institute of Informatics, Tokyo, Japan
{dou,guoyuchen,ccchang,nhhuy,iechizen}@nii.ac.jp
[2] The University of Tokyo, Tokyo, Japan

Abstract. The emergence of large language models (LLMs), such as Generative Pre-trained Transformer 4 (GPT-4) used by ChatGPT, has profoundly impacted the academic and broader community. While these models offer numerous advantages in terms of revolutionizing work and study methods, they have also garnered significant attention due to their potential negative consequences. One example is generating academic reports or papers with little to no human contribution. Consequently, researchers have focused on developing detectors to address the misuse of LLMs. However, most existing methods prioritize achieving higher accuracy on restricted datasets, neglecting the crucial aspect of generalizability. This limitation hinders their practical application in real-life scenarios where reliability is paramount. In this paper, we present a comprehensive analysis of the impact of prompts on the text generated by LLMs and highlight the potential lack of robustness in one of the current state-of-the-art GPT detectors. To mitigate these issues concerning the misuse of LLMs in academic writing, we propose a reference-based Siamese detector named **Synthetic-Siamese** which takes a pair of texts, one as the inquiry and the other as the reference. Our method effectively addresses the lack of robustness of previous detectors (OpenAI detector and DetectGPT) and significantly improves the baseline performances in realistic academic writing scenarios by approximately 67% to 95%.

1 Introduction

Large-scale language models (LLMs), such as OpenAI's GPT-4 [8], and Google's Pathways Language Model 2 [1], have become an integral part of our lives and jobs and are often utilized unknowingly. However, while LLMs greatly facilitate daily activities, they also pose significant security risks if maliciously exploited for attacks or deception. Consequently, with the growing popularity of LLMs, the importance of AI security has come to the forefront of people's attention [3,4,11].

Z. Dou and Y. Guo—These authors contributed equally

L. Barolli (Ed.): AINA 2024, LNDECT 202, pp. 266–277, 2024.
https://doi.org/10.1007/978-3-031-57916-5_23

Among the various security concerns, academic cheating stands out as a particularly grave issue. ChatGPT, in particular, has gained widespread popularity among college students worldwide. Consequently, universities urgently need robust detectors, which has driven continuous advancements in the field of detection technology.

Researchers in detector development have explored strategies to optimize the training set for improved model performance. Notably, Liyanage et al. [6] pioneered an AI-generated academic dataset using GPT-2, although it is considered less effective than the more advanced ChatGPT model currently available. Yuan et al. [12] proposed BERTscore, an evaluation method for filtering high-quality generated text that closely resembles human writing. Such text can be incorporated into the training set, thereby enhancing the performance of the detectors.

Researchers have also focused on optimizing the detector itself. Jawahar et al. [5] addressed the challenge of hybrid text, introducing a method to detect the boundary between machine-generated and human-written content, rather than solely distinguishing between the two. Zhao et al. [13] conducted a comprehensive survey of various LLMs, analyzing their performance across multiple dimensions, including pretraining, adaptation tuning, utilization, and capacity evaluation. They also identified potential future development directions for LLMs. Additionally, Mitchell et al. [7] proposed a model utilizing a curvature-based criterion to determine whether a given passage was generated by an LLM.

Studies examining the robustness of detectors include Rodriguez et al. [10], which investigated the impact of dataset domain on detector performance, highlighting a significant decrease in performance when the training and test datasets differ in domain. Their findings emphasized how the diversity of training sets directly affects the detector's performance. In addition, Pu et al. [9] analyzed the issue of insufficient robustness in existing detection systems by exploring changes in decoding or text sampling strategies. While previous research focused on robustness in terms of dataset domains and generation models' parameters, **this study highlights that prompt adjustments alone can significantly affect the robustness of the detector, particularly in the context of academic cheating.**

The contributions of our paper are as follows:

- **Highlighting the insufficient robustness of existing detectors through the example of academic writing cheating:** We demonstrate that solely adjusting the prompts is inadequate for ensuring the robustness of existing detectors (OpenAI detector and DetectGPT), particularly in the context of academic writing cheating.
- **Exploring the prompt-induced lack of robustness and evaluating model applicability:** We put forward a hypothesis to explain the reasons behind the lack of robustness caused by input prompts and provide a theoretical basis for our following experiments.
- **Introducing Synthetic-Siamese, an approach for detecting cheating in academic writing:** We analyze cheating in academic writing and propose a detection approach based on a Siamese network. Synthetic-Siamese exhibits

greater prompt generalization capabilities compared to existing detectors, effectively addressing the issue of insufficient robustness.

2 Asserting the Limitation of Existing Detectors

Since the release of GPT-3, OpenAI has allowed users to provide input prompts to shape the output text, enabling a wide range of functionalities. In contrast, the previous model, GPT-2, lacks prompt functionality and is irrelevant to the robustness of prompt-related issues. ChatGPT, a question-answering platform, does not offer adjustable parameters. Therefore, the output hardly changes when using the same prompt, making it unsuitable for generating large-scale datasets with diverse outputs. GPT-4, the latest LLM at the time of writing, has shown to be highly effective in terms of text quality and multi-modality. However, since GPT-4 was still in development when conducting this experiment, we did not choose GPT-4 to collect the training set, though we will show in Table 4 how our model performs on ChatGPT and GPT-4 generated text. Hence, this paper uses GPT-3 for dataset generation, which serves as the benchmark for our measurements. Throughout this paper, the term "GPT model" specifically refers to GPT-3.

For the human-written part of the dataset, we collected 500 samples of paper abstracts about Artificial Intelligence written by actual authors from the arXiv dataset [2], which is available on Kaggle[1] and covers various fields. To create the AI-generated part of the dataset, we divided it into two subsets as depicted in Fig. 1. The "Simple prompt" subset consists of 500 GPT abstracts generated by GPT-3 using the prompt "Write an abstract for a professional paper." The "Specific prompt" subset includes 500 GPT abstracts generated by GPT-3 using prompts beginning with"Write an abstract for a paper about" followed by the corresponding titles from the actual abstracts.

Among the state-of-the-art detectors available, such as ChatGPT detector and GPTZero, many lack associated published articles or datasets for reproduction. Moreover, a significant number of these detectors do not provide APIs, making it impossible to conduct batch-testing experiments. Consequently, we have chosen the RoBERTa base OpenAI Detector (OpenAI detector for short) on Hugging Face[2], a single-input binary classifier, as our target detector due to its availability and usability. At the same time, we also use the most widely used DetectGPT [7] as another target detector. The comparison of the performance between our model and DetectGPT, OpenAI detector can be seen in Table 2.

The OpenAI detector demonstrates an exceptionally high accuracy of 98% in detecting abstracts generated by **simple prompts** and 98% in identifying human-written abstracts. However, for abstracts generated by **specific prompts**, the accuracy rate drops to 70. This substantial reduction in performance by simply adding a human-written sentence to the prompt clearly indicates the limited robustness of existing detectors. Notably, specific prompts

[1] https://www.kaggle.com/datasets/Cornell-University/arxiv.

[2] https://huggingface.co/roberta-base-openai-detector.

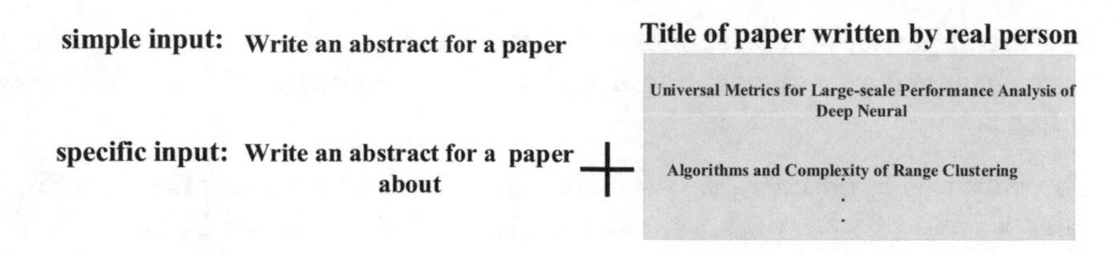

Fig. 1. Examples of a simple prompt and a specific prompt.

are commonly used in academic cheating scenarios, where students tailor their assignments or reports to meet specific requirements provided by their professors, utilizing prompts similar to the specific prompts used in this study.

3 Hypothesis for Prompt-Induced Lack of Robustness

In our hypothesis, we first divide the prompt into two parts: the **Template** describing the task summary and the **Content** containing specific human information, which is denoted by **X**. The Content here means that in addition to the text describing the task user wants the LLM to do, the prompt also contains text with practical meaning, such as the title of an actual paper, written by real people.

- For the **Template** part, prompts for the same task can be divided into different Template variants depending on how task is expressed. Template variants include *"Directly use requirement"* which is the specific prompt we designed before. *"Another expression"* is where the student expresses the meaning of the requirement using different wording. The *"Double GPT"* variant involves using the generation model (GPT) twice, where the student modifies Content X using GPT before generating the article. Lastly, the *"Many → one"* variant simulates a common plagiarism method where the student collects five articles written by people about idea X and combines them into a new article. These Template variants allow us to evaluate the detector's performance in detecting different manipulative strategies employed by students. Examples of each Template variant are shown in Fig. 2
- For the **Content** part, prompts can be divided into different levels depending on the length and complexity of the Content. The levels range from including only the field name to including the title, summary of the abstract, and the entire abstract, denoted as 0, 1, 2, and 3, respectively.

Our **hypothesis** can be explained as follows:

- Within the prompt, only component X, which is Content, influences the characteristics of the generated articles and contributes to the limited robustness observed in existing detectors.
- When component X remains at a certain level, the generated articles exhibit similar characteristics regardless of the other parts of the prompt as shown by the red horizontal line in Fig. 2.

- As the complexity and level of detail in the X component increase, it becomes more challenging to detect the generated articles as shown by the blue arrow in Fig. 2.

To verify this hypothesis, we designed the following experiments: First Fig. 2 contains a total of four variants and four levels, for a total of 16 squares. We use the prompt corresponding to each square to generate 100 machine-generated texts for a total of 16 sets of test data. Then, we use each set of data to test the accuracy of the OpenAI detector and fill in the corresponding position in Table 1.

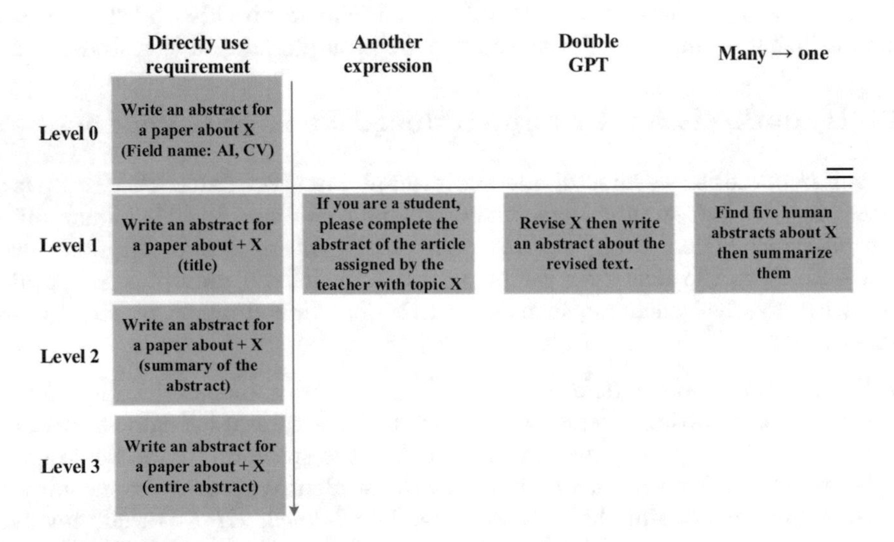

Fig. 2. The prompts can be categorized into different levels based on the length and complexity of the Content part and different Template variants based on how generation task is expressed in the Template part

Table 1. Accuracy of the original OpenAI detector (before fine-tuning) in different levels. X denotes the Content incorporated into the prompts.

X level	Directly use requirement	Another expression	Double GPT	Many → **one**
level 0 (X = Field name)	100%	100%	99%	86%
level 1 (X = Title)	70%	74%	53%	72%
level 2 (X = Summary of abstract)	34%	24%	20%	29%
level 3 (X = Entire abstract)	11%	17%	7%	11%

The results in Table 1 are consistent with our hypothesis. In summary, the X component of the prompt plays a crucial role in the characteristics of the generated articles and poses challenges for detection, particularly as it becomes more intricate and detailed. At the same time, the results in Table 1 also show how the specific prompt affects the accuracy of the OpenAI detector. The accuracy remained acceptable at level 1 but deteriorated significantly at **level 2** and above.

4 Synthetic-Siamese

Synthetic-Siamese consists of two key components. First, we analyze potential academic cheating scenarios and develop a **cheating model** specifically tailored to address these instances of cheating. Then we propose a **detection system** designed to identify instances of academic cheating based on our developed model.

4.1 Student Cheating Model

As depicted in Fig. 3, the model comprises two parties: the student side and the teacher side. The teacher assigns specific topic including Template and Content (defined in Sect. 3) for an academic task, and then a potentially deceitful student utilizes these topic as prompt (as described by (2) in Fig. 3) to generate an article using a generation model such as GPT-3. Meanwhile, the teacher also proactively employs the generation model to generate an article. Then, the teacher uses a model to compare the similarities between the student's article and their own generated article to determine whether the student has cheated.

This cheating model closely resembles real-life situations where students' assignments or examinations are typically centered around specific area and come with detailed requirements from teachers. To meet these requirements, students generally use the teacher's topic as input for generating their articles. However, it is also possible that students do not directly use the input prompt same as the teacher's. For example, students may use their own topic different from teacher's as the Content of prompts (which means that the teacher may not know the level of the Content of the prompt that students may use.) or modify the teacher's Template to use a different way of describing the generation tasks to build students' own prompts (as described by (1) in Fig. 3). To verify that our model can deal with various real-life scenarios, we designed a **Level generalization test** and a **scenario generalization test** in Sect. 5

4.2 Detection System

As depicted in Fig. 3, the network structure involves the input of two articles, $\mathbf{x}$ and $\mathbf{y}$. Article $\mathbf{y}$ represents the teacher's AI-generated article, while $\mathbf{x}$ can either be a human-written article or an AI-generated one submitted by the student.

Our detector employs a pre-trained BERT network as a feature extractor, denoted as $f(.)$, which is initialized with pre-trained weights. We fine-tune it using a supervised training approach. When labeling training data, if both $\mathbf{x}$ and $\mathbf{y}$ represent AI-generated articles, the label l is assigned as 0. Conversely, if $\mathbf{x}$ corresponds to a human-written article and $\mathbf{y}$ represents an AI-generated article, the label l is set as 1.

We use cosine distance $d(.,.)$ to measure the similarity between two feature vectors $\mathbf{f_x} = f(\mathbf{x})$ and $\mathbf{f_y} = f(\mathbf{y})$, as described in Eq. 1.

$$d(\mathbf{f_x}, \mathbf{f_y}) = 1 - \frac{\mathbf{f_x} \cdot \mathbf{f_y}}{\|\mathbf{f_x}\|_2 \|\mathbf{f_y}\|_2} \tag{1}$$

The loss function utilized during training is described by Eq. 2.

$$\mathcal{L} = l d(\mathbf{f_x}, \mathbf{f_y})^2 + (1 - l)(2 - d(\mathbf{f_x}, \mathbf{f_y}))^2 \tag{2}$$

During the inference phase, our model calculates the cosine distance between the two input texts. A smaller distance indicates a higher similarity between $\mathbf{x}$ and $\mathbf{y}$. As $\mathbf{y}$ represents AI-generated text, a smaller distance indicates that $\mathbf{x}$ is more likely to be generated by AI. Conversely, $\mathbf{x}$ is more likely to be written by a real person.

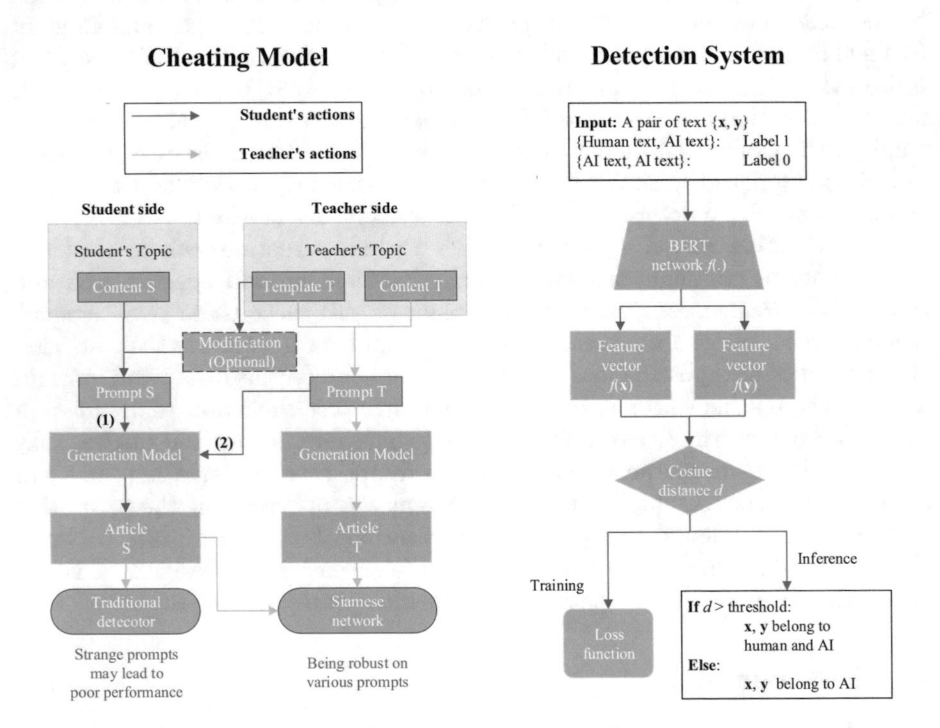

Fig. 3. The figure on the left is the Cheating Model, and the figure on the right is the Detection System. In the Cheating Model, both the student and teacher sides need to use Prompt containing Content and Template (defined in Sect. 3) as input to the generation model and generate corresponding articles. We use the two suffixes S and T to distinguish the prompts students and teachers use and the articles they get.

For the practicality of this detection system, it need two inputs: the article to be detected from students and the generated GPT article from teachers themselves. However, in most cases, Teacher's Topic are easily accessible to teachers themselves. At the same time, teachers can easily use generation models (such as ChatGPT) with low learning costs. Therefore, this detection system is practical in academic cheating.

5 Experiment Results and Discussions

We conducted the following experiments with an expanded dataset as outlined below to show that Synthetic-Siamese can achieve higher accuracy than existing detectors in various generalization scenarios.

5.1 Experimental Design

The model settings for Synthetic-Siamese are as follows:

- The training set for Synthetic-Siamese consisted of 2,000 human-written abstracts from arXiv about Artificial Intelligence and 4,000 GPT-3 generated abstracts using **level 1** specific prompts. This dataset was also employed for fine-tuning the OpenAI detector. **Content** and **Template** mentioned in the following are described in Sect. 3.
- For **teacher's articles**. As described in Sect. 4, Synthetic-Siamese requires two inputs, which are represented by (x, y) in Fig 3: one is the article submitted by the student (the text to be detected), and the other is the article generated by the teacher. Therefore, when testing Synthetic-Siamese, in addition to the text to be detected, we also need to provide Synthetic-Siamese with machine-generated text as the teacher's article. As a result, we generated 100 GPT-3 abstracts using the prompts of only *"Directly use requirement"* Template variant in level 1 as the teacher's article. We use level 1 because, in real-life scenarios, no matter what the teacher's topic is and whether it is specific or detailed, it is enough to serve as the Content of level 1 prompts. Therefore, in the following tests, we use this data set as the teacher's article.

For student articles (the text to be detected). We designed the following three generalization test sets:

- For the **level generalization test set**, we selected 100 human-written abstracts and generated 100 GPT abstracts using prompts in level 1, 2 and 3 each with the *"Directly use requirement"* Template variant that mimics different manipulative behaviors students may employ. This test set helps us verify that Synthetic-Siamese can maintain high robustness in all levels while the teacher only use the text generated by prompts in level 1 as the teacher's article and does not need to know the prompt level of the student's article.

- For the **scenario generalization test set**, we selected 100 human-written abstracts and generated 100 GPT abstracts using prompts with four different Template variants each in level 1 as the student's article which need to be detected to simulate how a student might evade detection by describing the task in a different way. In addition to the prompt of teacher article and the student article using different Template variants, We also aim to make the prompt of the two use different Content. Therefore, to simulate a student evading detection by using different human information from the teacher's topic. This test set helps us verify that Synthetic-Siamese can maintain high robustness when student attempts to evade detection.
- For the **model generalization test set**, we chose 50 human-written abstracts and generated 50 abstracts for each generation model comprising OpenAI's GPT-3, Perplexity's customized GPT-3.5[3], the Falcon-7B[4], ChatGPT, and GPT-4. All abstracts were generated using level 1 and level 2 prompts. This test set enables us to assess the Synthetic-Siamese's ability to generalize across different generation models, providing insights into their performance and adaptability in diverse AI-generated text scenarios.

5.2 Level Generalizability

In level generalization test set, we simulated that teachers always use prompts in level 1, while students use level 1, 2 and 3. As presented in Table 2, the OpenAI detector exhibited a significant drop in performance of **level 3** prompts, even after fine-tuning, with a maximum accuracy of only 66.5%. Although DetectGPT is a widely used detector now, its maximum accuracy in level 3 is only 50.5%. This is because DetectGPT scores the text to be detected and determines whether the text was written by a person or GPT based on whether the score is greater than 50% (the boundary between humans and machines). As a result, in certain situations (such as the text generated by the specific prompt mentioned in this paper), DetectGPT tends to produce ambiguous results, i.e., the score is in the middle.

In contrast, Synthetic-Siamese demonstrated much greater generalizability, achieving the accuracy of 95% in all three levels. This level generalizability demonstrates that in academic cheating scenarios, Synthetic-Siamese can effectively detect the students' usage of GPT, regardless of the level of Content that students use.

At the same time, to prove that our proposed Synthetic-Siamese has better level generalizability than traditional single-input network detectors. Using the same training set, we trained a single-input traditional classification model (with the same Bert as Synthetic-Siamese following a three-layer classifier). As presented in Table 2, the performance of the single-input model is worse than that of Synthetic-Siamese. As a result, the Synthetic-Siamese model structure is the root cause of better level generalizability.

[3] https://www.perplexity.ai/.
[4] https://falconllm.tii.ae/.

Table 2. Accuracy of the detectors on the Level generalization test set with level 1, 2 and 3 prompts.

Level of the prompt	OpenAI detector (original)	OpenAI detector (fine-tuned)	DetectGPT (original)	Synthetic-Siamese	Single-input Model
Level 1	85.0%	98.5%	100.0%	95.0%	99.0%
Level 2	66.5%	89.0%	50.5%	95.0%	85.5%
Level 3	54.5%	66.5%	50.5%	95.0%	58.5%

5.3 Scenario Generalizability

In the scenario generalization test set, we simulated possible methods that students may use to avoid detection when using GPT. We separately tested the accuracy of Synthetic-Siamese when the student used the same Content as the teacher's prompt but a different Template than the teacher's, and when the student used both different Content and Template than the teacher. As presented in Table 3, regardless of whether the Content or the Template of prompts used by students or teachers are consistent, Synthetic-Siamese maintains high accuracy.

Template and Content in the Prompt will affect the generated text under human observation. But for models, they don't affect the distinction between human-written and machine-generated. This scenario generalizability demonstrates that Synthetic-Siamese can still perform well even if the student uses different prompts than the teacher to evade detection.

Table 3. Accuracy of Synthetic-Siamese on the scenario-generalization test set. The Template variant of prompts used by teacher is *"Directly use requirement"* while the student use four different Template variants. The upper line represents the result when Content of teacher's prompts and student's prompts are same, while the lower line represents the result when they are different.

Content of Teacher's and Student's Prompt	Template variant of Student Prompt			
	Directly use requirement	Another expression	Double GPT	Many → **one**
Same	95.0%	95.0%	95.0%	95.0%
Different	95.0%	95.0%	95.0%	95.0%

5.4 Model Generalizability

Although GPT has become mainstream, students may utilize several other LLM-based text-generation models to avoid detection. To assess Synthetic-Siamese's effectiveness, we conducted tests using the model generalization test set. It is important to note that teacher's articles were only generated by GPT-3.

The results of the tests are presented in Table 4. The DetectGPT and the original OpenAI detector struggled to perform effectively in most cases, while the fine-tuned OpenAI detector achieved the better accuracy except when dealing with text generated by Falcon-7B using **level-2** prompts. Synthetic-Siamese performed highly on the GPT-3 and customized GPT-3.5 but showed limited generalizability when faced with Falcon's generated text. We hypothesized that when training our Siamese-based detector with the proposed cheating model, the detector learned to identify authorship information. It distinguished GPT as one author and humans as another. When a new "author" (Falcon-7B) emerged, the detector struggled to assign its text to either human or GPT.

For ChatGPT and GPT-4, the performance of Synthetic-Siamese is better than the other three detectors. Traditional detectors such as OpenAI detector are more suitable for **closed set detection**, so if we use level 1 prompts to fine-tune the OpenAI detector and then use level 1 to test, the performance of the OpenAI detector can significantly improve. However, for level 2, the performance of the fine-tuned OpenAI detector drops significantly. In contrast, Synthetic-Siamese has excellent **cross-level open set detection** capabilities. Regardless of the level, Synthetic-Siamese maintains high accuracy. This model generalizability demonstrates that Synthetic-Siamese can effectively detect the students' usage of generation models, no matter which model students use.

Table 4. Accuracy of the detectors on the text generated by different LLMs. OpenAI detector, a binary classifier, only needs one input. Besides the query text, our detector requires the corresponding generated text (from the teacher) as an anchor. Within each cell, the upper number represents the result on **level-1** prompts, while the lower number represents the result on **level-2** prompts.

Detector	Source of input text				
	GPT-3	**Falcon-7B**	**Perplexity**	**ChatGPT**	**GPT-4**
Synthetic-Siamese	92.0% 92.0%	88.0% 48.0%	100.0% 100.0%	100.0% 79.0%	100.0% 100.0%
OpenAI detector (original)	85.0% 66.5%	71.5% 55.5%	80.0% 74.0%	60.5% 53.5%	63.0% 54.0%
OpenAI detector (fine-tuned)	98.5% 89.0%	90.5% 54.5%	98.5% 98.5%	98.5% 56.0%	97.0% 83.5%
DetectGPT	55.5% 52.5%	50.0% 50.0%	64.0% 51.0%	51.5% 50.0%	52.0% 52.5%

6 Conclusion

This study addresses the issue of academic cheating facilitated by LLMs, which are widely utilized in contemporary contexts. By examining the RoBERTa Base OpenAI Detector as a case study, we identified potential limitations in the robustness of existing detection methods. Additionally, we conducted an in-depth analysis and presented a hypothesis highlighting the role of Content (X factor) in prompts contributing to the detector's lack of robustness. We then formulated a cheating scenario in academic writing and proposed Synthetic-Siamese, a detection approach which determine whether a student cheated based on the similarity of the teacher's and student's texts. Our experimental results conclusively demonstrated that Synthetic-Siamese exhibits much greater prompt generalization capabilities compared to that of the OpenAI detector.

Acknowledgments. This work was partially supported by JSPS KAKENHI Grant JP21H04907, and by JST CREST Grants JPMJCR18A6 and JPMJCR20D3, Japan.

References

1. Anil, R., et al.: PaLM 2 technical report. arXiv preprint arXiv:2305.10403 (2023)
2. Clement, C.B., Bierbaum, M., O'Keeffe, K.P., Alemi A.A.: On the use of arxiv as a dataset. arXiv preprint arXiv:1905.00075 (2019)
3. Crothers, E., Japkowicz, N., Viktor, H.: Machine generated text: a comprehensive survey of threat models and detection methods (2023)
4. Greshake, K., Abdelnabi, S., Mishra, S., Endres, C., Holz, T., Fritz, M.: Not what you've signed up for: compromising real-world llm-integrated applications with indirect prompt injection (2023)
5. Jawahar, G., Abdul-Mageed, M., Lakshmanan L.V.S.: Automatic detection of machine generated text: a critical survey (2020)
6. Liyanage, V., Buscaldi, D., Nazarenko, A.: A benchmark corpus for the detection of automatically generated text in academic publications (2022)
7. Mitchell, E., Lee, Y., Khazatsky, A., Manning, C.D., Finn, C.: Detectgpt: zero-shot machine-generated text detection using probability curvature (2023)
8. OpenAI. GPT-4 technical report. arXiv preprint arXiv:2303.08774 (2023)
9. Jiameng, P., et al.: Deepfake text detection: limitations and opportunities (2022)
10. Rodriguez, J., Hay, T., Gros, D., Shamsi, Z., Srinivasan, R.: Cross-domain detection of GPT-2-generated technical text. In: Proceedings of the 2022 Conference of the North American Chapter of the Association for Computational Linguistics: human language technologies, pp. 1213–1233. Seattle, United States, July (2022). Association for Computational Linguistics
11. Stiff, H., Johansson, F.: Detecting computer-generated disinformation. Int. J. Data Sci. Analytics **13**, 05 (2022)
12. Yuan, W., Neubig, G., Liu, P.: BARTScore: evaluating generated text as text generation (2021)
13. Zhao, W.X., et al.: A survey of large language models (2023)

Cognitive Blind Blockchain CAPTCHA Architecture

Nghia Dinh[1], Huy Tran Tien[1], Viet-Tuan Le[2], Huu-Thanh Duong[2], Lidia Ogiela[3], and Vinh Truong Hoang[2](✉)

[1] VSB Technical University of Ostrava, 17. Listopadu 15/2172, 708-33 Ostrava-Poruba, Czech Republic

[2] Faculty of Information Technology, Ho Chi Minh City Open University, 97 Vo Van Tan Street, Ho Chi Minh 722000, Vietnam
Vinh.th@ou.edu.vn

[3] AGH University of Krakow, 30 Mickiewicza Ave, 30-059 Kraków, Poland

Abstract. CAPTCHAs are a safe bet because they are used by so many other websites. They minimize risk from websites simply by integrating them. However, repeating that CAPTCHA creates a bad user experience and spends millions of hours of human brain cycles to resolve recurrent CAPTCHAs is meaningless. Furthermore, because CAPTCHA providers target users with advertising, some consumers have expressed worries regarding privacy promises. We also ran into problems in nations where CAPTCHA services were occasionally unavailable. As a result of productivity, blocking, and privacy concerns, numerous academics have pondered building new CAPTCHA architecture over the years. Rather than seeking to replace CAPTCHA with a single alternative unilaterally, we developed an adaptive blind CAPTCHA architecture that improves both individual and organizational privacy and efficiency while still permitting interoperability with existing CAPTCHAs. Based on blockchain and blind token technology, this solution protects users' privacy and sensitive data while simultaneously delivering high usability, quick deployment, and protecting online services from automated harmful bots.

Keywords: CAPTCHA · cognitive · blockchain · blind token · security

1 Introduction

CAPTCHA [1] (Completely Automated Public Turing Test to Tell Computers and Humans Apart) or HIP (Human Interactive Proof) is an automatic security mechanism that creates and scores AI (Artificial Intelligence) tests that humans can solve but are beyond the capabilities of current computer programs in determining whether the user is a human or a computer program. Denial of Service (DoS) assaults by malevolent automated programs have become a big problem with the growth of Web services. CAPTCHA has evolved as an important way for distinguishing between humans and hazardous automated computers. However, as AI has evolved, the web has progressed from simple CAPTCHAs based on text recognition to detecting things from images. This causes several serious issues for human Internet users:

L. Barolli (Ed.): AINA 2024, LNDECT 202, pp. 278–288, 2024.
https://doi.org/10.1007/978-3-031-57916-5_24

- Productivity: Time is lost, as is concentration on the task at hand, and this is frequently exchanged for frustration.
- Accessibility: Users are assumed to have the necessary physical and cognitive abilities to complete the tests, which may not be true. A visual impairment, for example, may make performing a CAPTCHA-solving task impossible.
- Privacy: User behavioral data, cookies, and sensor data are sent to remote servers.
- Device compatibility: Sensor-based CAPTCHA schemes are difficult to implement on the majority of users' devices because they require sensors found only on smartwatches, tablets, or smartphones.

CAPTCHA, which is still employed by many websites, is an efficient mechanism for differentiating real human users from bots by creating obstacles in front of their visitors. They transfer abuse responses to a third party and remove the risk from websites with simple integration. Anyone who has run a high-performing online business, on the other hand, would tell you that it's not something you should undertake until you have no other choice. That is why businesses spend so much time refining the performance and look of their websites and applications in order to keep customers delighted. Many CAPTCHA replacements have been explored, but none have achieved the same level of popular adoption as CAPTCHAs. The fundamental reason is that only one solution was tried to replace CAPTCHA. It's challenging to cover all attack scenarios with a single solution, especially when some of them are false positives or false negatives. Rather than attempting to replace CAPTCHA with a single alternative, we proposed a platform called the blind CAPTCHA platform, which would still reuse CAPTCHA to varying degrees while leveraging trust information from users and resolving all productivity, blocking, and privacy concerns. The platform will now decide whether to offer a visual puzzle or another means of showing humanity to users based on the client's evidence information during challenges. As a result, we may fine-tune the challenge's difficulty and avoid revealing visual riddles to human requests while providing more difficult problems to non-human users. We validate the user's PoP (Proof of Presence) in the blind CAPTCHA by using True Identity, a highly trusted program that can retrieve credential device information and interact with the user. This information is likewise encrypted and included in blind messages to avoid content leaking. Following the completion of our challenges, we assess the signals gathered. If we are satisfied that the user is likely human based on the combination of those signals, we take no further steps and redirect the user to the destined page with no interaction necessary. However, if the signal is weak, we may show the user with a visual puzzle CAPTCHA to verify their humanity.

The research is structured as follows: Section 2 describes similar works, while Sect. 3 describes the planned blind architecture. Section 4 then goes into greater detail about security analysis. Finally, we address the evaluation and conclusion of the strategy.

2 Related Works

Moni Naor [2] was the first to offer theoretical ways for telling computers apart from humans. The most extensively used technology has been text-based CAPTCHAs [2]. Popular text-based methods were broken by attacks employing image processing, pattern recognition, and Machine Learning (ML) techniques. As a result, image classification and recognition CAPTCHAs are considered more secure than text recognition

CAPTCHAs. Following that, various picture-based CAPTCHA systems [2] developed, which used drag and drop, image selection, or slide to distinguish people from machines. In contrast, advanced CV (Computer Vision) and ML technologies contributed to the defeat of the most important image-based CAPTCHA methods. To deal with visually handicapped users, researchers proposed audio-based CAPTCHAs in addition to text-based and image-based CAPTCHAs. However, the usefulness of these approaches is restricted by language obstacles and poor usability. Furthermore, supervised learning and automated Speech Recognition (ASR) show how these systems can be used. Researchers began creating cognitive behavioral-based CAPTCHA schemes in the 2010s to generate problems depending on behavioral traits. These techniques have been demonstrated to be sensitive to bot assaults that replicate the user's behavioral pattern [2]. Recently, research directions [2] have used sensor data to generate problems that are tough for automated bots to repeat. Google has released reCAPTCHA v3, which includes a business component called reCAPTCHA Enterprise, which is built on the existing reCAPTCHA API and uses advanced risk analysis techniques to discriminate between humans and bots. While this version is intended to defend huge corporations from bot attacks, it still has many faults. Good users are frequently flagged as suspicious by reCAPTCHA and forced to go through onerous authentication procedures. This is partly because it is heavily reliant on the use of Google cookies. One of the benefits of reCAPTCHA was that it was completely free. However, reCAPTCHA Enterprise is not free, and organizations may struggle to justify the ROI of deploying this solution. For enterprises with high traffic levels, such as e-commerce sites, gambling platforms, and digital banking apps, this can be highly costly. Furthermore, reCAPTCHA Enterprise captures a wide range of user data points in order to make risk choices. This is a problem because consumer data privacy rules are becoming more stringent all over the world. Cloudflare recently decided to suspend the use of reCAPTCHA due to major privacy concerns regarding this CAPTCHA [3].

We proposed the blind CAPTCHA platform in this study, which would reuse existing CAPTCHAs to variable degrees while leveraging user-proofed trust information. The platform will now decide whether to offer a visual puzzle or another means of showing humanity to users based on the client's evidence information during challenges. During a challenge, a typical client behavior would be to return the client credential information provided by True Identity, a client-side high-trust software that can obtain the client device's credentials, which would then be re-verified for proof of work. Furthermore, the platform enables us to design rules that function as security firewalls, assess requests using ML (Machine Learning) or Heuristic algorithms, and monitor critical requests via authorization verification requests.

3 Proposed Blind Architecture

3.1 Overview

The Security Protector, as illustrated in Fig. 1, is a crucial component of this architecture since it checks and verifies each access request. The Security Verifier verifies the token collected by the Security Protector in response to a request. The agent of the website or app, such as user agent plugins or SDK libraries (Javascript for frontend, SDK library

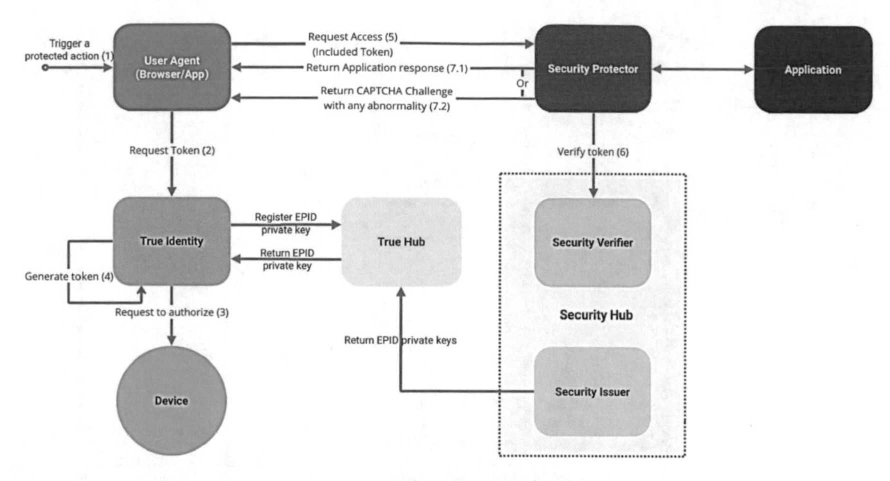

Fig. 1. CAPTCHA architecture

for mobile app), acquires the token from True Identity for each request generation. True Identity, a highly privileged software, may get credential information from a trusted device or computer for the proof of space (PoW). Security Hub will offer a private key for each device or computer that has True Identity installed. Security Hub manages public keys and private keys with Intel EPID [4], allowing a public key to control a set of private keys. The advantage in this scenario is that we can control a collection of device private keys, allowing us to easily discontinue protection actions on a customer's multi-device. True Identity collects the request challenge, issued by Security Protector, seeks authorization from the device to validate the proof of presentation (PoP), then encrypts this information with its EPID private key to obtain a signed token.

When an end-user visits a web page or uses a mobile application, the following events occur in the order shown in Fig. 2:

- The client loads the application's user interface. Furthermore, the application returns the Security Protector's challenge value, as well as the Client ID, which is the EPID public key generated by Security Hub.
- When an end-user initiates a CAPTCHA-protected action, the JavaScript API, or the mobile SDK requests a credential token from True Identity.
- True Identity returns to the client a credential token generated by using the EPID private key.
- The credential token is sent to the Security Protector for evaluation along with the Client ID and challenge.
- The Security Protector gets the Security Hub to verify the authenticity of the credential token.
- After assessing, the Security Hub returns to the Security Protector a result based on the risk evaluated for this request.
- Depending on the verdict, the Security Protector can navigate the request to the application or show a visual CAPTCHA with a varying degree to the end user.

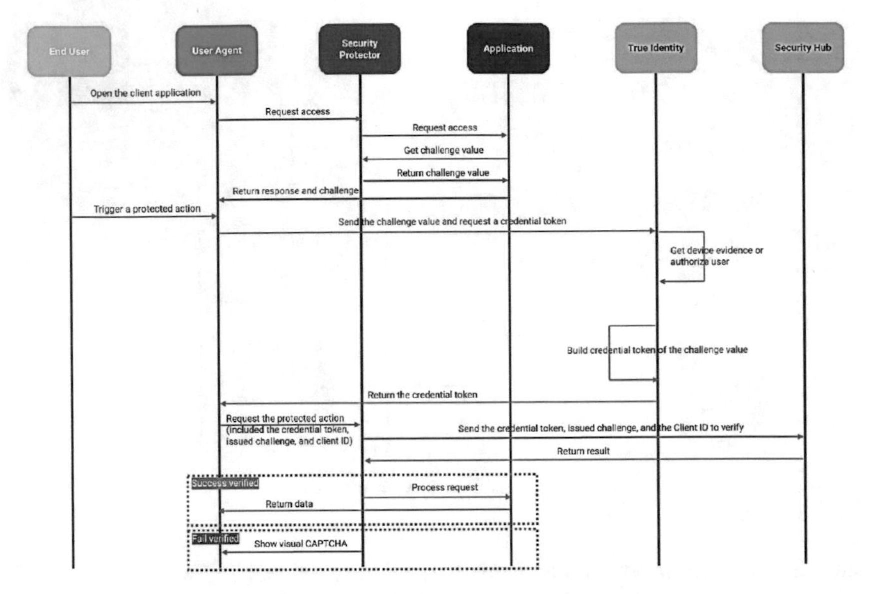

Fig. 2. CAPTCHA sequence flow

3.2 Principal Components

3.2.1 Security Protector

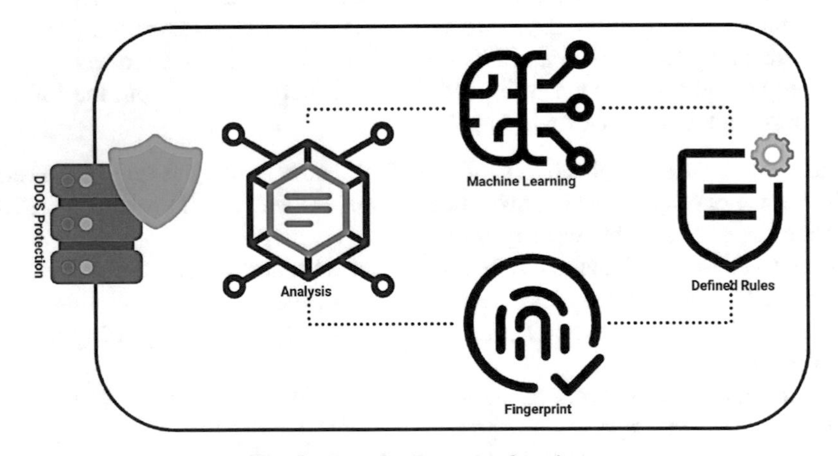

Fig. 3. Security Protector functions

Security Protector safeguards the application by discriminating between bots and people and validating signed token transfers from user clients. Aside from DDOS prevention, Security Protector uses the following strategies, as indicated in Fig. 3, to distinguish bots from humans or bad requests from good requests:

- Heuristics: To identify automated traffic, the heuristics engine [5] evaluates all requests, which are then compared to a developing database of harmful fingerprints.
- Machine learning: The great bulk of all detections, both human and bot, are handled by the Machine Learning (ML) engine [6]. This approach detects both automatic and human traffic by boosting the amount of traffic that passes through Security Protector. The machine learning engine is constantly being taught in order to increase its accuracy and react to new threats.
- Anomaly detection: The Anomaly Detection (AD) engine [5] is a detection engine that employs unsupervised learning to create a baseline of traffic and then uses that baseline to intelligently recognize outlier requests.
- JavaScript detections: The JavaScript Detections (JSD) engine detects headless browsers and other harmful fingerprints. This engine does a lightweight, undetectable JavaScript injection on the client side of any request. Wc do not collect any personally identifiable information during the procedure. Requests are either refused, challenged, or forwarded to other engines by the JSD engine.
- Defined rules: The rule engine enables customers to construct strong rules that reject, challenge, or log questionable requests based on their requirements, while legal requests are delivered to their destination.

3.2.2 True Components

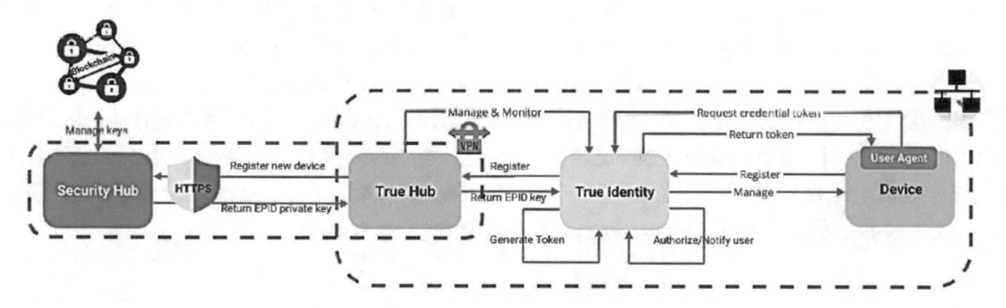

Fig. 4. True components

True components, as shown in Fig. 4, manage devices, connect with users for permission, and produce signed tokens using EPID [4]. True Identities, each with its own private key returned by True Hub, can be assigned to an organization. True Hub is a self-contained device that only talks with True Identity in a LAN or private network and interacts with Security Hub over SSH to receive the EPID private key for the newly registered True Identity. True Identity handles one or more registered devices and has access to their credentials in the event of token creation requests. True Identity can be installed directly on devices or connected to True Identity agents on LAN or private network devices. When a user performs a protected activity, the True Identity is contacted using the client's user agent's injected Javascript or SDK. The True Identity address is provided by the user agent's extension, such as Chrome or Firefox extensions, via Content Script. The extension's settings can be set to default or explicitly customized to avoid conflicts. True Identities are exclusively used for internal connections on LAN or private networks to protect data and avoid data leakage. The client then contacts True Identity and requests a signed credential token using the supplied challenge. True Identity secures user permission by providing an event popup that asks whether they accept to proceed with this activity before producing a signed credential token with its EPID private key.

3.2.3 Security Hub

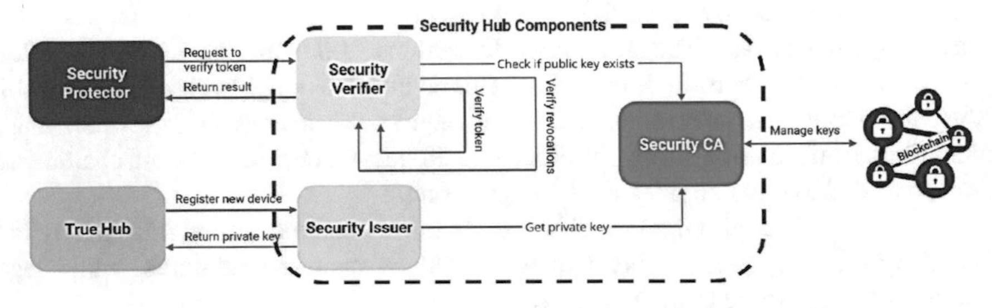

Fig. 5. Security Hub components

The Security Hub, as seen in Fig. 5, is in charge of token issuance and verification, with three unique components: the Security Issuer, the Security Verifier, and the Security CA (Certificate Authority). Key generation and administration are the responsibility of the security CA. The keys are maintained in a decentralized structure in BlockChain to promote data privacy and security, immutability, individual data control, visibility, and traceability. True Identity generates a signed credential token using the Hub's EPID private key. The Security Verifier examines the Security CA for the presence of the Client ID, which is the EPID public key, before verifying the token.

3.3 The Methods

3.3.1 Intel EPID

The CAPTCHA platform's Security Hub employs EPID [4], a revolutionary solution based on finite field arithmetic and elliptic curve encryption (ECC) that addresses all aspects of the active anonymity challenge. EPID offers a user or device with "Direct Anonymous Attestation" by introducing the concept of Group-level membership, which allows a device to be authenticated for a specified level of access while remaining anonymous. As shown in Fig. 6, EPID enables the linking of a group of membership private keys to a single public group key. This public group public key can be used to validate the signed token generated by any group member private key.

As shown in Fig. 7, the process of registering a new device's True Identity begins with the device requesting True Hub to join the ecosystem administered by the Security Hub. True Hub is a stand-alone program that communicates with the Security Hub via the Client ID, which is the EPID public key generated and registered in the CAPTCHA platform, and the secret key provided. The Security Issuer validates the Client ID and

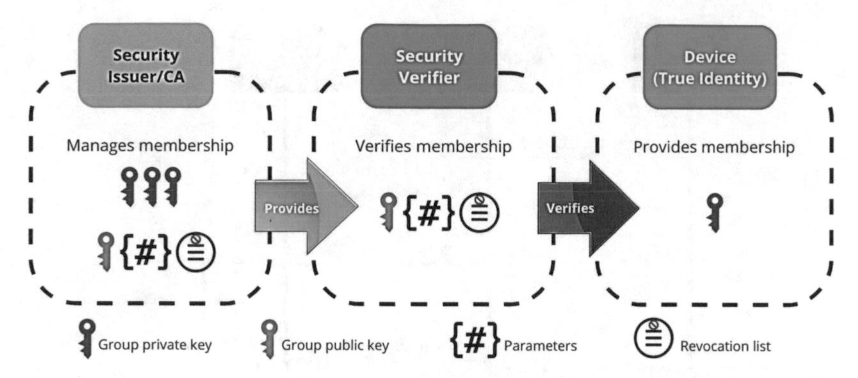

Fig. 6. Component roles

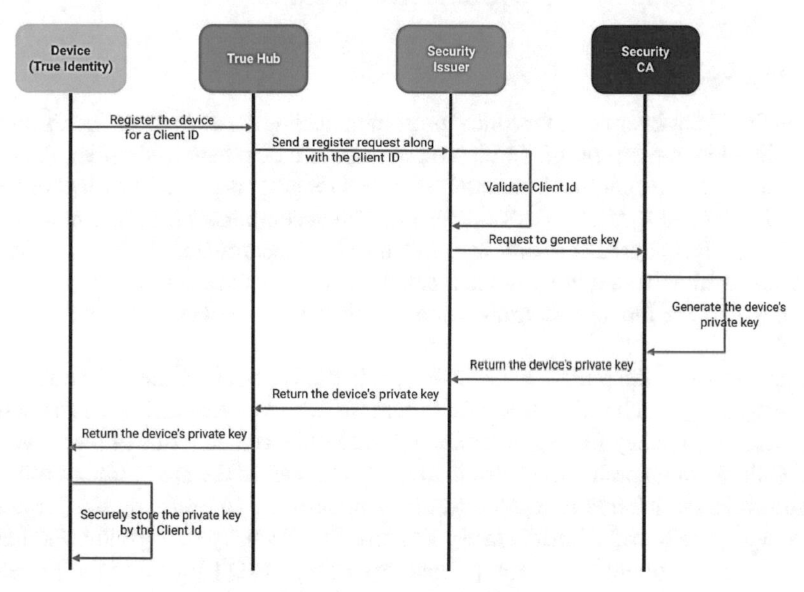

Fig. 7. Registration flow

secret key. If everything is in order, the Security Issuer will request that the Security CA issue a new EPID private key for the device.

EPID also allows us to revoke an individual device by identifying a compromised signed token or key, as illustrated in Fig. 8. If the private key of a device is compromised or stolen, this feature revokes the device and prohibits future forgery. This built-in revocation allows devices or entire groups of devices to be flagged for revocation and denied service promptly. It allows an issuer to ban a device from a group without knowing which device was blacklisted.

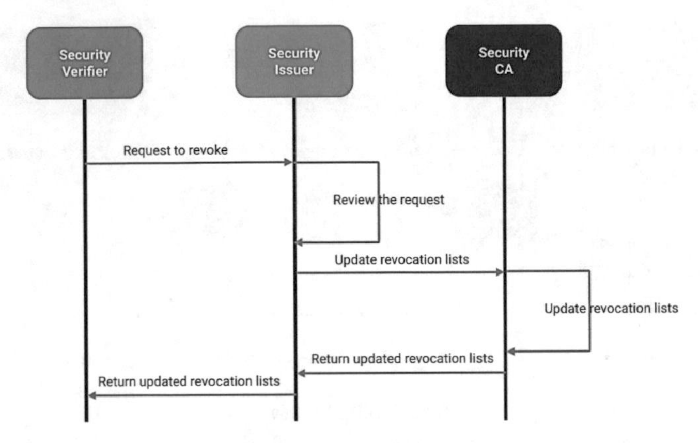

Fig. 8. Revocation flow

3.3.2 Blockchain

Blockchain [7] has been one of the most prominent technologies in recent years due to its decentralization, tamper-proof, and irreversibility. A blockchain is simply a distributed shared transaction ledger that is maintained by all member nodes of the blockchain network and is limited by the consensus process. The decentralized transaction processing of blockchain has piqued curiosity in the idea of a "decentralized digital identity" system. A user achieves anonymity in the blockchain system since the system produces the public key pair for blockchain transactions, with the private key available only to the user.

In this platform, the Merkle Patricia tree [8] is used to hold private and public keys in the Security CA, which is developed on Ethereum. The Merkle Patricia tree is based on the Patricia tree and key value generation in the Merkle tree. In terms of identifying key values with the same prefix, the Patricia tree, also known as the prefix tree or dictionary tree, surpasses the Merkle tree. As a result of integrating and refining the Merkle and Patricia trees, the Merkle Patricia tree was built. The Security CA produces a platform public key from a client's platform private key using EPID [4] during the private key issuance process, then encodes the public key to byte nibbles before putting the nibbles of the public key as key and the private key as value into the Merkel Patricia tree.

4 Security Analysis

The blind CAPTCHA platform, which employs cutting-edge security and blockchain technology, can answer all productivity, blocking, and privacy concerns. The platform has now decided to avoid providing visual riddles to human requests based on the client's PoP (Proof of Presence) during challenges, while punishing those that exhibit non-human behaviors with harder assignments. When the user's device's True Identity can certify their validity by obtaining the EPID private key from the Security Hub, the user can quickly navigate to their destination pages without encountering any CAPTCHA. It increases productivity by eliminating the need for CAPTCHAs, which create a terrible user experience and waste time. Furthermore, by providing components that are simple to set up and use in private networks, this platform can assist in resolving issues in countries where CAPTCHA services are sporadically limited.

In terms of privacy, this platform only allows users to demonstrate that they are qualified for the service. By supplying the Client ID, which is the EPID public key, the user can authenticate that they are a member of a group that has been granted access to the specified service. The Security Hub attempts to authenticate the user by verifying the Client ID's existence on the blockchain. The service provider cannot track the user because the user only needs to show that they are one of the qualified users without revealing their identity. Furthermore, by invoking an EPID platform public key, EPID device private key, or even a signed token, we can revoke a specific member of a group or the entire group to terminate services for only the selected user or group without harming non-revoked individuals and groups. Each True Identity user has a unique EPID private key, which is used to generate a signed token. The user agent demands that True Identity create a signed token when the user performs a protected operation. True Identity validates the user's PoP (Proof of Presence) by displaying an event popup that asks if they wish to perform this activity before generating a signed credential token with its EPID private key. Depending on the platform's settings, this approval procedure may be skipped or informed to the user. Following token generation, the request is forwarded to the Security Protector, together with the produced token and the challenge issued during loading, for verification of the signed token using the EPID platform public key via the Security Hub. Because each user's signature token is signed by their private key in a blind signature technique, an attacker cannot broadcast deceptive information. The Security CA also adds an organization's EPID platform public key to the blockchain. All contributed information cannot be edited and can be confirmed due to the blockchain's decentralization, tamper-proof, and irreversibility.

The EPID protocol can remove members or groups from the system. The EPID authority is notified by a verifier with reasons when a group or member needs to be removed. The authority checks the validity of the request and carries out the necessary removal actions. The EPID authority may also remove platforms or groups on its own. There are three kinds of removal requests: based on private keys, signatures, or groups. The EPID authority maintains a single list of removed groups, and for each active group, it maintains a list of removed private keys and signatures. The removal lists are centrally managed and shared with all verifiers in the system.

5 Conclusion

We proposed the blind CAPTCHA, a unique CAPTCHA platform based on blockchain and Intel EPID security, in this article. All productivity, blocking, and privacy concerns can be addressed by the blind CAPTCHA platform, which uses cutting-edge security and blockchain technology. The platform may now decide whether to present visual riddles in response to human queries based on the client's PoP (Proof of Presence) during challenges, while punishing those who engage in non-human conduct. It boosts productivity by removing the need for CAPTCHAs, which provide poor user experience. By providing components that are simple to set up and engage with in private networks, this platform can assist governments in overcoming blocking concerns. Furthermore, this platform secures a user's privacy by leveraging the notion of Group-level membership. The service provider is unable to trace them down because the user merely needs to establish that they are among the qualifying users without exposing their name. The platform keys are also added to the blockchain by this platform. All new information cannot be modified and can be authenticated due to the blockchain's decentralization, tamper-proofing, and irreversibility. Using the Merkle Patricia tree in Ethereum allows the digital certificate authority to scale the entire anonymous authentication method.

Moreover, the proposed CAPTCHA platform exhibits significant potential for growth, serving as an inspiration for future research endeavors. It facilitates end-user authentication through multiple robust factors, including "something you know," "something you have," and "something you are." Furthermore, it empowers users by granting them control over their authentication process and aids in managing the sharing of their device information with Relying Parties. By employing adaptable authenticators supporting various authentication methods, the platform harnesses emerging device capabilities and addresses evolving risks through a cohesive authentication strategy. Additionally, it utilizes existing open standards and innovations, such as FIDO2 [9], to develop protocols and APIs conducive to scalability, high security, and privacy protection.

References

1. von Ahn, L., Blum, M., Langford, J.: CAPTCHA: using hard AI problems for security (2003)
2. Dinh, N., Ogiela, L.: Human-artificial intelligence approaches for secure analysis in CAPTCHA codes. EURASIP J. Info. Security **2022**, 8 (2022). https://doi.org/10.1186/s13635-022-00134-9
3. End Cloudflare CAPTCHA. https://blog.cloudflare.com/end-cloudflare-captcha/
4. EPID. https://www.intel.com/content/www/us/en/developer/articles/technical/intel-enhanced-privacy-id-epid-security-technology.html
5. Jyothsna, V., Rama Prasad, V.V.: FCAAIS: Anomaly-based network intrusion detection through feature correlation analysis and association impact scale. ICT Express **2**(3), 103–116 (2016)
6. Araujo, A.M., de Neira, A.B., Nogueira: Autonomous machine learning for early bot detection in the Internet of Things. Digital Commun. Netw. **9**(6), 2023
7. M. Swan. Blockchain: Blueprint for a new economy. O'Reilly Media Inc., 2015
8. Patricia Merkle tree. https://ethereum.org/vi/developers/docs/data-structures-and-encoding/patricia-merkle-trie/
9. FIDO UAF overview. https://fidoalliance.org/specs/fido-uaf-v1.2-ps-20201020/fido-uaf-overview-v1.2-ps-20201020.html

DVID: Adding Delegated Authentication to SPIFFE Trusted Domains

Andrew Jessup[1], Henrique Z. Cochak[2], Guilherme P. Koslovski[2], Maurício A. Pillon[2], Charles C. Miers[2], Pedro H. B. Correia[3], Marco A. Marques[3], and Marcos A. Simplicio Jr.[3](✉)

[1] Hewlett Packard Enterprise, Texas, USA
jessup@hpe.com
[2] Santa Catarina State University, Joinville, Brazil
henrique.zc@edu.udesc.br,
{guilherme.koslovski,mauricio.pillon,charles.miers}@udesc.br
[3] Universidade de São Paulo, São Paulo, Brazil
{pedro.correia,marcomarques,msimplicio}@usp.br

Abstract. One of the challenges of cloud computing is ensuring secure access to data and resources. Identity Management Systems (IMS), which enable organizations to handle user identities, authentication, and authorization, are commonly employed for tackling this issue. Whilst OAuth 2.0, SAML, and OpenID Connect are typically used in web applications, the Secure Production Identity Framework for Everyone (SPIFFE) is today among one of the many open source IMS for cloud environments. The reason is that SPIFFE provides a secure and standardized attestation framework for authenticating cloud workloads from the moment they are instantiated. Our work extends SPIFFE's capabilities, allowing the identification not only of the workload making a request, but also of the user behind that request. For this purpose, we design a new credential called Delegated Assertion SVID (DVID), describe a proof of concept implementation, and benchmark some baseline scenarios.

Keywords: SPIFFE · SPIRE · Workload security · DVID

1 Introduction

In complex cloud environments, security mechanisms that enable the correct authentication and authorization of cloud workloads are paramount to promote adequate isolation of resources from different users [9,12]. This is usually accomplished with the aid of Identity Management Systems (IMS). Essentially, "identity management" refers to a set of policies, tools, and mechanisms employed to manage the life cycle of digital identities associated with system participants – both individuals or software/hardware components [19]. The Secure Production Identity Framework for Everyone (SPIFFE) [3] is today among the main cloud-oriented IMS solutions. In a nutshell, SPIFFE defines an open framework and set of standards for identifying workloads and securing their communications,

L. Barolli (Ed.): AINA 2024, LNDECT 202, pp. 289–300, 2024.
https://doi.org/10.1007/978-3-031-57916-5_25

thus facilitating the establishment of a zero-trust architecture among workloads in a complex and heterogeneous environment [17]. A core component of SPIFFE is the SVID (SPIFFE Verifiable Identity Document), a short lived identity document that cryptographically binds a workload to a unique identifier inside a trust domain.

Although SPIFFE offers security and flexibility for clouds, the existing SVIDs focus solely on identifying the immediate sender or receiver of a message. Many applications, however, require a richer authentication context. For example, suppose that a workload receiving a message wants to authenticate not just the calling workload, but also a principal (e.g., a human end-user) behind the request. This extra, delegated authentication context allows for more fine-grained authorization policies. The lack of standard solutions for doing so, though, is prone to introduce implementation and maintenance overheads, besides leading to insecure practices. For instance, a common method for propagating end user authentication context is to retransmit a bearer token obtained from an external entity (e.g., an OAuth JWT [8]). One risk of this approach is that it grants intermediate workloads, often maintained by different teams and organizations, access to this same bearer token. Hence, should one of these services be compromised or accidentally leak the token, user impersonation attacks might occur beyond the scope allowed for the workload that originally received the token. Also, from an interoperability perspective, using generic bearer tokens in a SPIFFE environment pushes to intermediate workloads the task of validating documents issued by external authorities, increasing verification complexity and latency.

Aiming to address the issue of delegated identity assertions in the SPIFFE framework in a systematic manner, this work proposes a new identity document for binding a user's identity token to the workload authorized to make requests on that user's behalf. This delegated verifiable identity document (DVID) consists in a JSON Web Token (JWT)-based bearer token [11] supporting non-SPIFFE principals in the context of SPIFFE-authenticated messages. With a Delegated Assertion SVID (DVID), a workload evaluating the authentication context of a message can determine if the request was initiated on behalf of a non-SPIFFE principal and, if so, the identity of that principal. This precludes the need of retransmiting unscoped (or broadly scoped) bearer tokens obtained from external Identity Providers. Consequently, it decreases the risk of accidental or malicious escalation of privilege via a token replay attack, besides facilitating validation of such tokens via a proxy (hereby called Asserting Workload) and, subsequently, by internal workloads. Besides specifying the DVID, we develop a proof-of-concept implementation to validate its capabilities in the SPIFFE ecosystem, and provide our benchmark results.

The rest of this paper is organized as follows. Section 2 gives an overview of SPIFFE and its limitations targeted by our proposal. Section 3 presents the DVID specification and its capabilities. Section 4 describes our proof-of-concept implementation and experimental evaluation. Section 5 discusses related works. Section 6 presents our considerations.

2 Background: Identity Management with SPIFFE

SPIFFE is a set of open-source specifications for bootstrapping and issuing short-lived, cryptographically verifiable identity documents (SVIDs) to workloads operating across heterogeneous environments and organizational boundaries [3]. Identities in SPIFFE take the form of a Uniform Resource Identifier (URI) containing two parts: a trust domain name, and a unique workload identifier within that domain. Each trust domain, be it an individual, organization, or environment, is responsible for managing its own workloads. In particular, this means using an independent SPIFFE infrastructure as root of trust, provisioning workloads with SVIDS that can be verified against the root public keys of that trust domain [3].

An SVID can be encoded as a X.509 certificate [2] or as a JWT [11], and is considered valid if it is digitally signed by an authority within the corresponding trust domain. Whichever the case, the SVID issuance procedure typically involves an attestation process, so only authorized workloads are able to obtain valid SVIDs via the specified API calls. Workloads can then use these identity documents when authenticating to each other, e.g., by establishing a mutually authenticated Transport Layer Security (mTLS) connection with X.509 SVIDs, or exchanging signed JWT SVIDs within secure communication channels [3]. Even though using on X.509 SVIDs is preferred when an end-to-end security tunnel can be established, JWT-based SVIDs are somewhat unavoidable when identities need to be carried over multiple hops [3]. This later case includes scenarios where a load balancer stands between two workloads, so they cannot connect directly to each other. Similarly, JWTs are useful when a certain authentication context (e.g., a user identity) needs to be conveyed to multiple workloads involved in the processing of a given request. This is commonly the case, for example, of services that require authorization from users to access their personally identifying information.

This need for handling user identities and assertions among services is tackled by many IMS in the Internet, including OpenID [16], OAuth 2.0 [8], and Security Assertion Markup Language (SAML) [1]. The current SPIFFE specification, however, is such that SVIDs carry only the identities of workloads sending and receiving requests, lacking procedures to deal with end-users behind those requests. It is useful, thus, to conceive a method for combining user- and workload-related contexts into a SPIFFE credential, which can then be conveyed to all workload instances handling the corresponding request.

3 Extending SPIFFE Identities: DVID

The goal of the delegated verifiable identity document (DVID) is to enable support for non-SPIFFE principals in the context of SPIFFE-authenticated messages. Essentially, the proposed framework allows a SPIFFE workload W obtain a DVID in exchange for an IdP-issued credential (in our prototype, an OAuth access token) received from an end-user U. The resulting JWT-based document, issued by an Asserting Workload after validating user token, binds together W's and U's identities, asserting that W is entitled to act on behalf of U for a given period. Inside a SPIFFE trust domain, the Asserting Workload may be considered a Trusted Third Party (TTP), so it is assumed to have received and validated the user token. Aiming to support multiple domains, though, an Asserting Workload may also prove the validity of the corresponding user token without revealing its full contents, using a zero-knowledge proof (ZKP) [5], as further described in Sect. 4.1.

A DVID's header is structured identically to an x5c header [10], and aims to establish a SPIFFE context and unambiguously identify the Asserting Workload as the root of trust for the DVID: (i) the SPIFFE-ID in the certificate encoded as the first element in the array (i.e., the leaf certificate) must match the value of the issuer identity (*iss* field explained below); and (ii) the certificate encoded as the first element in the array must correspond to the private key used to generate the DVID signature. The payload includes a set of mandatory claims associating an end user with a given workload:

- Issuer (*iss*): the issuer claim must be set to the SVID of the Asserting Workload that generated the DVID.
- Subject (*sub*): represents the identity about the assertion that is being created, taking the form of a valid SVID.
- Expiration Time (*exp*): the expiration time of the generated document.
- Delegated Principal Authority (*dpa*): identifies the authority that issued the original access token used to authorize the DVID minting. Its public key must be used to validate the originally issued user token.
- Delegated Principal (*dpr*): the non-SPIFFE principal (e.g., an end-user) on behalf of which the workload is authorized to act.

The DVID contains information about the user (e.g., username, email) and about the workload to receive the delegated permissions (e.g., Subject Workload). The validation process of this token involves verifying the credential issued by the external l Identity Provider (IdP), by checking the X.509 certificate and its certification chain, which make up the group of certificates obtained by the trusted domain. This document allows to: (i) identify the user behind the request; and (ii) the service instance that was used as the entry point for that request, receiving the original access credential, seeks to increase the expressiveness of the credentials used within the SPIFFE / SPIRE. This additional type of delegated authentication context allows the workload receiving a message to apply a finer-grained authorization policy than would otherwise be possible.

The Proof of Concept (PoC) was developed to assess the viability of performing authentication using the DVID, and its main flow is presented in Fig. 1. This scenario exemplifies a request to get data stored inside a database managed by the application within a SPIFFE environment from a user external to the cluster.

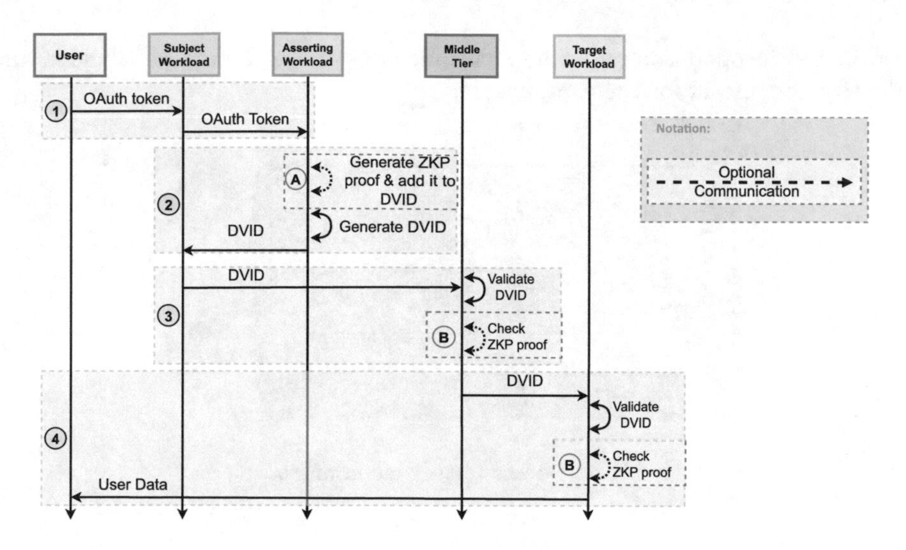

Fig. 1. DVID main flow.

Except for processes **A** and **B** (discussed in Sect. 4.1), this flow is as follows:

1. We use the front-end as Subject workload in this scenario. After receiving a proper OAuth token, it calls the Asserting Workload to mint a DVID.
2. The Asserting Workload decodes and validates the OAuth token, and then mints a DVID from it. The SPIFFE-ID from the requesting and Asserting workloads become, respectively, the *sub* and *iss* claims.
   - The optional procedure [**A**] generates a RSA-ZKP, allowing the Asserting Workload to prove that a given DVID corresponds to a valid OAuth without revealing the OAuth token itself.
3. In its way to a Target workload, the request and DVID pass through one or more intermediary workloads (Middle-Tier), which may include load balancers or other services
4. To validate an DVID, workloads check its signature and expiration date. The *sub*, *dpr* and *dpa* claims are also inspected for authorization purposes and, if valid, the Target Workload sends back the data requested by the user.
   - The optional procedure [**B**] allows validation of the RSA-ZKP proof.

4 Experimental Evaluation

The experiment was conducted through the proof-of-concept operation on a server machine with the following specifications: GNU/Linux Ubuntu Server operating system (version 20.04.6), 192GB of RAM, 4TB of HDD storage, and CPU model Intel(R) Xeon(R) CPU E5-2620 2.5Ghz 24 cores. The proof of concept components are individually contained in their own Docker container, and each container is limited to 1 CPU core and 128MB RAM each (Fig. 2). During the experiments, the SPIFFE Runtime Environment (SPIRE) environment is considered in operation as an application, aiming of generate certificates and identity documents for the components.

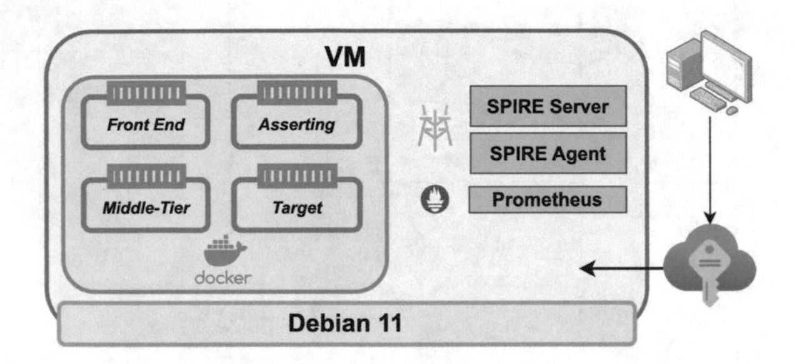

Fig. 2. Docker testing environment.

Prometheus is employed as a monitoring tool to capture performance metrics related to the operation of different components running in Docker containers to eventually generate a baseline. Establishing it involves a systematic process of gathering and analyzing data to define the standard behavior of the application and the cost of functions to generate and validate a DVID, shown in Fig. 1.

The purpose of our PoC is to show the feasibility of the DVID document, under a use case of the SPIFFE pattern based on the SPIRE implementation. The information related to the consumption of computational resources of the main components of the application was collected, to determine the minimum resources demanded when executing the proposed SPIFFE extended pattern. We define two use cases:

1. Base Case: explores the PoC without the offline credential validation through the RSA-ZKP. Simply put, this case does not consider the necessity to generate and validate the proof, and on Fig. 1, it does not make use of the optional procedures.
2. RSA-ZKP: support offline credential verification by locally storing the resulting proof, thus requiring the use of the RSA-ZKP generation and validation. Thus, both optional procedures (Fig. 1) are now utilized and added to the cost of each function (generate and validate).

The main focus of these use cased is confirm the viability to delegate a principal authentication, an OAuth bearer token, within a SPIFFE environment, and endorse the cost to generate and validate the DVID security document through a basic proof of concept. Furthermore, our results can be employed as a comparative basis, or even develop future extended/adapted applications or comparisons between new security documents. Thus we provide information about: (i) memory consumption; (ii) processor consumption; (iii) execution time of generate and validate functions; and (iv) payload of the DVID document with the RSA-ZKP proof is presented.

To collect the data, data scrapping from Prometheus on (i), (ii), and (iv) occurs at intervals of 50 milliseconds. This interval strikes a balance between collecting frequent updates and minimizing unnecessary load on the monitored system, by allowing Prometheus to capture metrics frequently enough to provide a near-real-time view of the system's behavior without overwhelming the system. Moreover, a 50-millisecond interval can be suitable for detecting short-lived spikes or fluctuations in metrics. This granularity is often beneficial in identifying and addressing transient issues that might occur within the monitored systems. Lastly (iii) is captured from the Golang time package inserted directly in each function.

4.1 ZKP: Ensuring Correct Data Access by Workload

The fundamental idea of zero knowledge is the consideration of a scenario where a prover is proving a theorem for a probabilistic polynomial-time verifier [5]. In other words, a proof system is considered zero knowledge if it allows a party to prove that it knows something without revealing what it is to another party. Zero-knowledge systems consist of efficient interactive and non-interactive proofs that have the remarkable property of producing nothing but the validity of the claim [4].

In this work, zero-knowledge proof (ZKP) is proposed to allows the Asserting Workload to prove to any requesting workload that a given DVID corresponds to a valid OAuth, without revealing any information of the OAuth token itself. Asking for this proof should be optional in most scenarios, and, it is up to each workload to decide if they want to request it (within a trust domain this usually would not be necessary). Our PoC uses an RSA-ZKP based on [18], so two extra claims are present in the DVID:

- ZKP (*zkp*): stores the result of the Asserting Workload function that generates the ZKP. This function uses an OAuth token as the parameter.
- OAuth Message (*oam*): stores solely the payload of the OAuth token, eventually used during the ZKP validation. If this value is null, the *zkp* claim receives an empty string.

Besides checking the previous claims and public signature, it is required to validate the RSA-ZKP proof, formerly created during the minting process. It is expected the DVID minting to increase proportionally to the use of computational resources if the production of RSA-ZKP proof is required. Similarly, the validation process is also expected to be longer since the veracity of the proof has be checked.

4.2 Results and Analysis

Fig. 3 presents the data about CPU and memory consumption. Despite the anticipation of heightened CPU usage due to the utilization of RSA-ZKP, the observed outcomes contradicted these expectations. As depicted in Fig. 3(a), the RSA-ZKP procedure does not lead to a notable impact on CPU resources usage. This unexpected similarity in performance between scenarios using and not using RSA-ZKP raises questions about its effective utilization in this context. Likewise, the examination of memory consumption depicted in Fig. 3(b) echoes a similar narrative. Also, regarding memory demands due to RSA-ZKP integration, the results exhibited negligible variance compared to scenarios where this procedure is absent. This discrepancy between anticipated and observed memory utilization prompts a closer examination of the protocol's implementation, the hardware aspects. A probable reason for this may be the option to store locally on a HDD while the Golang garbage collector manages memory by cleaning up unused resources after each function call.

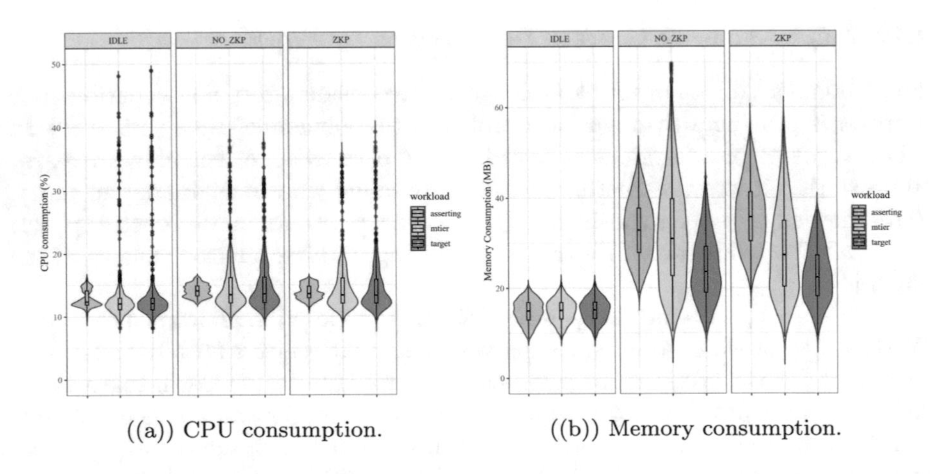

((a)) CPU consumption. ((b)) Memory consumption.

Fig. 3. Resources consumption.

By not including the optional procedure on the Asserting Workload (label **A** and **B** on Fig. 1), it takes 221.80±37.22ms to create the document, while the extra procedure leads to a cost of 461.07±77.78ms. In summary, just the RSA-ZKP computational adds around 240ms to generate the DVID. The execution time of 270.13±76.13ms for validating a document containing the RSA-ZKP proof also highlights a substantial increase compared to the validation process without the proof, which takes 11.61±0.84ms. This significant difference in execution times emphasizes the considerable computational overhead introduced by the validation of RSA-ZKP proofs within this cloud environment.

This stark contrast in validation times suggests that the inclusion of RSA-ZKP adds a substantial computational burden, taking approximately 259ms solely for proof validation. In a cloud environment, in which efficiency and scalability are paramount, this increased execution time for validation could significantly impact system responsiveness and overall throughput. The fact that merely validating a proof incurs such a time cost underscores the importance of optimizing the RSA-ZKP validation process to make it more efficient within this specific cloud infrastructure. Minimizing computational overhead while ensuring the integrity and security of the verification process becomes crucial for maintaining optimal performance in such environments.

The payload cost of the proof is also non-trivial. Typically, a ZKP involves an interactive exchange between a challenger and a prover to establish the validity of a claim. However, in several ZKPs implementations, including the one discussed herein, efforts are made to render these protocols non-interactive. This adaptation often involves replacing the interactive challenger with a randomly generated hash function. Fundamentally, the ZKP operates by having the challenger generate a challenge"c," which the prover uses as an input to produce a proof "v" associated with a specific bit. For a prover lacking knowledge of the original value, the likelihood of correctly guessing this bit stands at 50% (i.e., guessing either 0 or 1).

In our specific ZKP design, this process iterates 80 times, equivalent to ensuring a 128-bit security level. Consequently, the resulting *zkp* structure comprises an assortment of key-value pairs encompassing 80 challenges c and their respective proofs v. This construction is engineered to uphold stringent security standards while preserving the essential zero-knowledge property. It ensures that the prover demonstrates knowledge of a secret without disclosing any information pertaining to the secret itself. Such an approach is vital for maintaining confidentiality while validating assertions within cryptographic frameworks. So each$c$ and v costs 512 bytes. By iterating 80 times, the payload size ends with an extra 81920 bytes traversing the network.

In cloud environments, where data transfer and resource utilization are critical, this amplified payload could pose significant challenges. The increased network traffic due to larger payloads can lead to higher latency, slower response times, and increased data transfer costs, especially when dealing with a considerable number of transactions or communications simultaneously. Furthermore, the computational overhead involved in processing these extensive payloads can strain the cloud infrastructure. Verifying and processing such large volumes of data can consume considerable computational resources, potentially impacting the overall system performance and scalability. To answer this, other security algorithms also may be considered instead of the RSA-ZKP to provide better hardware consumption.

5 Related Works

The literature includes a few proposals where a user's token context can be combined with identities from internal workloads via continually exchanged bearer tokens. Examples include WS-Trust [15] and OAuth 2.0 with refresh token rotation [8, Sec. 10.4]. The main disadvantages of this approach, though, are the overhead placed at the token-issuing authority, and the latency imposed by repeating the issuance procedure.

Those concerns are avoided in existing commercial applications, in particular in Microsoft Identity platform Entra with "On-Behalf-Of flow" [14] and in Google Identity and Access Management (IAM) [7]. Table 1 presents a comparison based on key points identified in its documentation.

Table 1. Comparison of user token context integration in related works.

	User to workload delegation	Open-source	Offline credential/ token validation
Microsoft Entra	Yes	No	No
Google IAM	Yes	Yes	No
SPIFFE DVID	Yes	Yes	Yes

Microsoft Entra is a proprietary multi-cloud IAM service. It enables the development of applications where users and customers authenticate using their Microsoft identities issued by a trusted identity provider. It contains both authentication and authorization methods using standard approaches like OAuth 2.0 and OpenID Connect. Microsoft Entra's "On-Behalf-Of flow" (OBO) [14] shares similarities with our work, since it also tackles a scenario in which a web API, using an identity other than its own, makes a call to another web API. This process, known as delegation in OAuth, aims to transmit a user's identity and permissions along the request chain. For example, for a middle-tier service to send authenticated requests to a downstream service, it must obtain an OAuth 2.0 access token from the Microsoft identity platform. This bearer token is then consumed for authentication and authorization purposes [13]. While this flow is similar to our proposal, the token validation procedure is always performed online, via the Microsoft identity platform.

A similar approach from Google is the service account impersonation. When an authenticated principal, such as a user or another service account, authenticates as a service account to gain the service account permissions, it is called impersonating the service account. Impersonating a service account lets an authenticated principal access whatever the service account can access. Only authenticated principals with the appropriate permissions can impersonate service accounts [7]. Through the authentication via the service account itself or on behalf of Google Workspace or Cloud Identity users through a domain-wide delegation, applications gain access to resources by associating the service account with the running resource, and granting it IAM roles. One way to get the authentication is with the service account key, eventually consumed to obtain an OAuth 2.0 access token. This access token is one of the many ways Google coded to create what it defined as a short-lived credential.

This short-live credential conceptually is very similar to our own, although it accommodates a bigger context and scope. The most important idea here is that it can achieve the same end-result, to have an user context with an OAuth 2.0 access token bound with authenticated workloads, eventually being possible to know what principal, or user, initially made the initial request. However, this credential also does not provide any offline credential validation. On both Microsoft and Google cases, the concept to hold an offline validation poses security risks like compromised access awareness delays and limited revocation control. It also introduces operational challenges, such as managing credential renewal and potential exposure of stored credentials if not done correctly.

Finally, Google IAM platform is open-source with a BSD-3-Clause license, providing access to the IAM API and the complete library to be analyzed [6], granting developers and security enthusiasts to gain a deeper understanding of the whole technical process.

6 Considerations and Future Work

This work sought a foundation by exploring the basic concepts introduced by the SPIFFE standard and relating them to other identity solutions that are frequently used and highly requested in cloud environments. The proposed solution expands the concept of identity, federated identity, and transitive identity in SPIFFE, by providing ways to examine the authentication context of a message during workload evaluation, making it feasible to identify whether the request was initiated by a non-SPIFFE principal and, in that case, ascertain the identity of that principal.

Other algorithms like Elliptic Curve Digital Signature Algorithm (ECDSA) may offer a compelling alternative to RSA due to several key advantages. It achieves the same level of security as RSA but with significantly shorter key lengths, resulting in faster cryptographic operations and smaller digital signatures. ECDSA's superior security-performance trade-off makes it ideal for resource-constrained devices and applications where computational efficiency is paramount. As future work, other security algorithms also may be considered instead of the RSA-ZKP to provide better hardware consumption.

Acknowledgements. This work was supported by Hewlett Packard Enterprise (HPE), and in part by the Brazilian CNPq (grant PQ 304643/2020-3 and 311245/2021-8), FAPESP (grant 2020/09850-0), and CAPES (Finance Code 001). Special thanks for the discussions and contributions to the work: Adriane Cardozo (HPE), Andrew Harding (VMware), Caio Milfont (HPE), Eugene Weiss (Sentima), Evan Gilman (SPIRL), João Ambrosi (HPE), and Yogi Porla (Stealth Startup). This work was funding by FAPESC, UDESC, USP and developed at LabP2D/LARC.

References

1. Campbell, B., Mortimore, C., Jones, M.: RFC 7522: Security assertion markup language (SAML) 2.0 profile for OAuth 2.0 client authentication and authorization grants. Technical report, Internet Engineering Task Force (IETF) (2015)
2. Cooper, D., Santesson, S., Farrell, S., Boeyen, S., Housley, R., Polk, W.: RFC 5280: Internet X.509 public key infrastructure certificate and certificate revocation list (CRL) profile. Technical report, IETF (2008)
3. Feldman, D., et al.: Solving the bottom turtle: a SPIFFE way to establish trust in your infrastructure via universal identity (2020). ISBN 978-0-578-77737-5
4. Goldreich, O., Krawczyk, H.: On the composition of zero-knowledge proof systems. SIAM J. Comput. **25**(1), 169–192 (1996)

5. Goldwasser, S., Micali, S., Rackoff, C.: The knowledge complexity of interactive proof systems. SIAM J. Comput. **18**(1), 186–208 (1989)
6. Google: Google golang IAM package (2024). https://pkg.go.dev/google.golang.org/api-/iam/v1
7. Google: Service accounts overview. https://cloud.google.com/iam/docs/service-account-overview (2024)
8. Hardt, D.: RFC 6749: The OAuth 2.0 authorization framework. Technical report, Internet Engineering Task Force (IETF) (2012)
9. Hu, V., Iorga, M., Bao, W., Li, A., Li, Q., Gouglidis, A.: NIST SP 800–210: General access control guidance for cloud systems. Technical report, NIST (2020)
10. Jones, M., Bradley, J., Sakimura, N.: RFC 7515: JSON Web Signature (JWS). Technical report, Internet Engineering Task Force (IETF) (2015)
11. Jones, M., Bradley, J., Sakimura, N.: RFC 7519: JSON web token (JWT). Technical report, IETF (2015)
12. Li, H., Dai, Y., Tian, L., Yang, H.: Identity-based authentication for cloud computing. In: Jaatun, M.G., Zhao, G., Rong, C. (eds.) CloudCom 2009. LNCS, vol. 5931, pp. 157–166. Springer, Heidelberg (2009). https://doi.org/10.1007/978-3-642-10665-1_14
13. Microsoft: Access tokens (2024). https://learn.microsoft.com/en-us/entra/identity-platform/access-tokens
14. Microsoft: Microsoft identity platform and OAuth 2.0 On-Behalf-Of flow. https://learn.microsoft.com/en-us/entra/identity-platform/v2-oauth2-on-behalf-of-flow (2024)
15. OASIS: WS-Trust 1.4. OASIS standard incorporating approved errata 01. Technical report, Org. for the Advancement of Structured Information Standards (2022)
16. OpenID: Openid specifications (2022). https://openid.net/developers/specs
17. Rose, S., Borchert, O., Mitchell, S., Connelly, S.: Zero trust architecture. Technical report, National Institute of Standards and Technology (NIST) (2020)
18. Shafi, G.: Lecture notes on cryptography (2008). http://www-cse.ucsd.edu/users/mihir
19. Tracy, K.: Identity management systems. IEEE Potentials **27**(6), 34–37 (2008)

Transparent Encryption for IoT Using Offline Key Exchange over Public Blockchains

Mamun Abu-Tair[1(✉)], Unsub Zia[1], Jamshed Memon[1], Bryan Scotney[1], Jorge Martinez Carracedo[1], and Ali Sajjad[2]

[1] School of Computing, Ulster University, Belfast, UK
m.abu-tair@ulster.ac.uk
[2] British Telecom, London, UK

Abstract. Internet of Things (IoTs) framework involves of a wide range of computing devices that rely on cloud storage for various applications. For instance, monitoring, analytics, surveillance and storing data for later processing within other applications. Due to compliance with security standards and trust issues with third-party cloud storage servers, the IoT data has to be encrypted before moving it to cloud server for storage. However, a major concern with uploading encrypted IoT data to cloud is the management of encryption keys and managing access policies to data. There are several techniques that can be used for storing cryptographic keys used for encryption/decryption of data. For instance, the keys can be stored with encrypted data on the cloud, a third-party key storage vault can be used for storing keys or the keys can stay with client so that they could download and decrypt the data by themselves. In case of encryption keys leakage, the data stored on the cloud storage could be compromised. To resolve the challenge of key management and secure access to data in third-party cloud storage, an end-to-end transparent encryption model has been proposed that securely publishes the cryptographic keys in a blockchain ledger. The data is encrypted at edge gateway before it is transmitted to cloud for storage. The user does not require cryptographic keys to access data; a seamless process involves the client proving their identity to a crypto proxy agent built upon zero trust security principles, ensuring continuous verification.

1 Introduction

The IoT technology has transformed several aspects of life by incorporating innovative and intelligent networking systems. The concept of smart devices has led to a revolution shift from traditional world to smart world comprising of smart home, smart cities, and industry 4.0 amongst others. Due to undeniable benefits of IoT assisted living, a rapid increase in adoption of IoT technology can be seen [1]. The recently published annual report of Cisco (2018-2023) provides insights to the increasing growth rate of the internet connecting users, devices, and machines [2]. With reference to the Cisco's report, the worldwide ratio of machine to machine (M2M) connections and devices will rise, approximately up to 14.7 billion by 2023, which is much greater in comparison to internet users. The machine to machine (M2M) communication with integrated IoT have gained popularity by contributing towards various applications for example, health

L. Barolli (Ed.): AINA 2024, LNDECT 202, pp. 301–313, 2024.
https://doi.org/10.1007/978-3-031-57916-5_26

care monitoring, camera scrutiny, smart asset tracking services, smart meters, and traffic control. According to the research conducted by McKinsey [3], it is expected that by 2025, the number of IoT connected devices will reach up to 1 trillion. Since IoT and its applications have become prevalent systems representing as an indispensable part of our daily lives, this raises serious concerns about privacy and security vulnerabilities. As tremendous amount of information is being shared every nano second over the heterogeneous network, indirectly creates opportunities for the cyberattacks to exploit security vulnerabilities that cannot be ignored. The IoT devices are resource constrained in terms of low powered with limited computing and storage capabilities [4]. These limited resources expose IoT framework to several security risks allowing malicious individuals to retrieve critical information from the device and collapse the network. As the developments in IoT technologies are continuously accelerating and escalating, one should be aware of the severity caused by different types of cyber-attacks.

In another report conducted by renowned company Hewlett-Packard (HP) stated that to cope with the growing demand of IoT products, security aspects are often compromised during the designing of these products. Approximately, 70% of the IoT based devices are potentially vulnerable and on average, there are 25 vulnerabilities per IoT product due to the weak credentials, un-encrypted networks, and lack of granular user access permissions [5]. This immense distribution of connected devices over the network and the increased deployment of IoT devices has raised their potential towards exploitable vulnerabilities. Consequently, in proportion to the increasing number of IoT devices and services, significant research efforts are required to minimize the rapidly emerging cyber-attacks. The number of IoT devices will continue to grow at an exponential rate, adoption of new and secure mechanisms to mitigate the threats large- scale attacks must be considered, and small-scale attacks are often ignored. No doubt, the former attacks can cause serious damage, but the latter attacks should not be ignored as they are more treacherous and may go undetected for a long period of time causing devastating effects on the reliability of the network. Therefore, adequate, and reliable security measures should be developed that can cope with the competences of constrained technology, ensuring data integrity and confidentiality [6]. Another revolution in the field of internet is the advent of Web 3.0, where blockchain would be used to host internet [7]. We utilize this opportunity to propose a novel security model for securing IoT data using blockchain to store cryptographic keys. The proposed transparent encryption architecture provides remedy for specific problem area, where the IoT data has to be stored in an untrusted cloud storage. In such scenarios there are three main challenges, i) safe storage of data, ii) key management and iii) access policy for the data. The proposed model provides solution to above mentioned challenges by encrypting data at rest (transparent encryption) and storing the cryptographic keys to the blockchain. By doing this, the encrypted data can be stored in any untrusted cloud storage, moreover, the client can access the data without requiring the keys. This architecture has been filed as patent by British Telecom [8]. The remainder of this paper is organized as follows. Section 2 builds motivation of the paper based on the problem statement. Section 3 covers relevant prior art and inventions. The proposed idea and system architecture has been introduced in Sect. 3. Section 4 summarizes the contributions made by this research. The proof-of-concept implementation design and experiments is explained in Sect. 5. The summary of this work and future directions have been identified in Sect. 6.

1.1 Motivation

Identity management and secure key distribution have always been challenging goals to attain when designing cybersecurity protocols. However, with the speedy adoption of IoT and cloud technology, these goals are much more harder to achieve.

IoT framework is a complex network, comprising of different types of sensors and actuators, producing large amounts of data that is analysed using edge gateway and cloud servers [9]. Too many moving points in a system make it vulnerable to attacks. For instance, it is difficult to manage and store the encryption/decryption keys, in cases where the data is has to be accessed by different parties. Moreover, another problem that arises is the management of user identities. Since, each user might need access to the data for a specific time window, makes it difficult to revoke the old keys and encrypt the data using new keys.

The walled garden approach to secure networks does not suffice in the case of IoT framework, rather modern day approaches based on zero trust principles are needed [10]. In situations, where the data is stored on third party cloud storage and there are multiples users accessing the data make it challenging to ensure the integrity of data while maintaining the legitimacy of cryptographic keys used to encrypt data. There is a need of smarter solution, that could allow any number of users to access the data without the need to generate new keys every time and without the need to trust third party cloud or the users. The presently available solutions are either forced to trust the third party cloud server, or have separate key for every user, or in worst cases have to re-encrypt data with a new key every time. The existing solutions that do not require re-encrypting data mostly rely on public key infrastructure, which opens doors to quantum attacks. Thus, neither of the aforementioned approaches provide secure services, efficiency and convenience as in a single package.

1.2 Contributions

In this paper, a novel architecture has been proposed for IoT frameworks to resolve identity and key management problems, enabling secure and seamless access to data. The main contributions made by the proposed model are listed as follows:

- **Transparent Encryption** Transparent data encryption is usually referred for securing data at rest, for instance Microsoft uses it to secure SQL Server, Azure SQL Database, and Azure Synapse Analytics data files. In this paper, the proposed architecture ensures that the data stored in third-party cloud (at rest) is encrypted and the encrypted keys are kept safe.
- **Secure Key Management** Secure management of cryptographic keys is a crucial task, as compromised keys can put the entire system, including the data, at risk. This paper proposes the use of blockchains as a key escrow system to store encrypted keys. Since blockchains serve as a distributed ledger that is accessible from anywhere, key distribution is no longer a problem. The legitimate party can access the keys and thus decrypt the data stored on third-party cloud storage.

- **Seamless Access to data** The uniqueness of this study is the seamless access to data throughout the IoT data lifecycle. The IoT edge server is responsible for encrypting the data and pushing the encryption keys on to the blockchain based distributed ledger system. While on the decryption end, the crypto proxy agent ensures that the data requested by client is provided in unencrypted form, conditional to successful identity check.
- **Zero Trust Policy** The beauty of proposed architecture is that it is built upon zero trust security principles, adhering to explicit verification, assuming breach and least privileged access. Each client is verified explicitly by the crypto proxy agent and is given a limited access to data based on his eligibility conditions.

2 Related Work

Several data encryption processes exist where a sender encrypts data and stores it on a cloud service and the relevant decryption key is sent to the intended recipient via a side-channel, e.g., by email, SMS or even postal service. The receiver then downloads the encrypted data from the cloud service and decrypts it using the shared key.

In some variants of the above-mentioned process, the decryption key is sent to a trusted third-party key storage vault. The client must authenticate itself with this key management service to retrieve the decryption key and then download and decrypt the data from the cloud service. In [11], the authors proposed an offline data access scheme, where the client downloads data from cloud using privilege keys that are generated using hash for the particular user shared by the owner of the data. However, the limitation with this approach is the exposure to encryption/decryption keys. According to the zero trust security principles, there is no trusted entity, client and cloud both must be treated with zero trust and encryption/decryption keys must be kept secret. In another study, a scheme Cloud stash has been proposed to resolve the key management problem in cloud-based scenarios [12]. They propose to split every file into secret shares and distribute them across different cloud storage, however the file can only be reconstructed with the threshold set number of shares retrieved back. Cloud computing is mainly preferred in situations with bulk data storage; however, the proposed idea seems impractical for large amounts of data. Considering the advent of Web 3.0 [13], a hot area of research is focusing using blockchains to get rid of conventional old school key sharing schemes for securing IoT data [14]. A very basic example could be to encrypt user file and distribute it across decentralized peer blockchain network [15]. However, more advanced solutions have been already proposed to secure IoT data and access management. A study similar to our work also proposes decentralized access management of data stored in cloud but using blockchain based on temporal dimension [16]. However, the way they have used blockchain to verify user is different from our proposed solution. They have used blockchain to perform transaction using attribute-based encryption, whereas we use public blockchains just to store encrypted credentials. Another relevant prior art is a patent for titled as Transparent Proxy of Encrypted Sessions [17]. A client and a server are configured to trust a certificate of an intermediate proxy device. The proxy device intercepts the session requests between the client and the server and establishes a client-proxy session and a proxy-server session which appears transparent to the client and server. The main differences of this patent with our invention are that in our invention the client does not perform online or interactive communication with the server and the Crypto Proxy Agent does not intercept the network traffic between the client and the server. Our proposed model is novel in terms of access management for data stored in third party storage, as the client or third-party cloud never gain access to the encryption keys, yet distributed identity service (DIDS) allows session based access to the clients based on successful verification. Proxy re-encryption is another solution widely used to establish access policies for clients to access data stored in third party cloud storage. In a recently published research, a new kind of conditional proxy re-encryption has been proposed for secure cloud storage [18]. The main idea is to have a semi-trusted crypto proxy agent similar to one in our approach, however once the proxy has necessary re-encryption keys from the delegator, it can transform all ciphertexts for a delegator to ciphertexts for a delegatee. Transforming the ciphertexts and generating individual keys for every delegatee is a resource greedy task. It is equivalent to encrypting the data with new encryption keys for every client. However, in our case the encryption key for the data remains same, the crypto proxy agent just decrypts it for the clients that are able to confirm their identity via DIDS. Wang et al., recently proposed a blockchain-based access control system for secure cloud storage [19]. The proposed approach rely on smart contracts using Ethereum blockchain, that interfaces for data retrieval and storage. The data owner is in charge of developing and executing smart contracts, uploading encrypted files, specifying access control rules, appointing attribute sets to data users, and appending valid access duration. The data user accessing a cloud server-stored encrypted file, needs key from the smart contract which is needed to decrypt the encrypted file. There are few proposals for secure service guarantee from cloud storage is based on Ciphertext-Policy Attribute-Based Encryption (CP-ABE). [20] However, due to CP-inherent ABE's "all-or-nothing" decryption characteristic, implementation of the protocol may result in an unavoidable security breach known as the misuse of access credentials (i.e., decryption privileges). In a recent study, the authors proposed CP-ABE-based storage model for safe data access and storage for Internet of Things applications. [21] The proposed model adds an attribute authority management (AAM) module to cloud, that serves as an agent and offers a user-friendly access control while drastically reducing the storage overhead of public keys. This approach is completely different from our proposed architecture and despite the claimed benefits, CP-ABE approaches are yet susceptible to attacks such as chosen plain text [22] and dictionary attacks [23]. In many earlier techniques, it was impossible for the cloud provider to confirm whether a downloader could decrypt the data. Due to this, everyone with access to the cloud storage should have these data available.

Another way to provide secure access control and avoid re-encryption of data, is to deploy public key infrastructure based solutions [24]. However, despite the ease of key distribution and access control public key based solutions provide, they also bring along several security weaknesses. In general, searching is much more difficult if users routinely encrypt the data. This conflict is settled using public-key encryption with keyword search (PEKS). On the other hand, because keywords have low entropy, it is susceptible to keyword guessing attacks (*KGA*) [25].

3 Proposed System

An integrated key escrow and data decryption scheme in which Data Encryption Keys (DEK) encrypted by per-client Key Encryption Keys (KEK) are published on a public blockchain. The blockchain stores the encrypted DEKs as assets whose ownership can be transferred to other members of the blockchain. A Crypto Proxy Agent (CPA) deployed close to a Client's network, upon authenticating various claims by the Client, re-constructs the Client-specific KEK to decrypt the encrypted DEK. Once in possession of the DEK, the CPA downloads its corresponding encrypted data from an untrusted cloud storage service, decrypts and transmits it to the Client. The proposed approach has following unique selling points:

1. As only one DEK is created and used, the data does not have to be decrypted and re-encrypted again each time it is needed to be shared with a new Client. Only a new client specific KEK is generated.
2. The Clients never get to know the DEK, so the same DEK can be used repeatedly to encrypt different data sets. Using a unique DEK for different data sets can create a large number of encryption keys, which can be difficult to manage in a Smart Home IoT environment.
3. As the IoT EG is usually not a high-powered computer/device, a "create a DEK once" and "encrypt data once" model is more suitable for it.
4. Unlike standard key escrow solutions, the Key Encryption Key or the Data Recovery Field is never attached to the encrypted data. So even if the encrypted data is subject to a ransomware attack, the keys are not lost.
5. The data owner and data consumer (Client) do not have to communicate with each other interactively (offline)
6. Blockchains allow a client to transfer the ownership of a transaction (in this case the encrypted KEKs) to other clients. Therefore, should the client want to share its KEK with another entity, it does not have to retrieve the KEK from the blockchain and send it securely to other entity using a separate protocol.

3.1 Architecture Design

The identities of the Clients are managed through an external Distributed Identity Service (DIDS), which is not the scope of proposed idea, not part of proposed solution and is shown in red in the main diagram. However, the DIDS process relevant to this invention should work as below:

- A member Client (that has subscribed to the DIDS) can request the DIDS to generate temporary private/public key pairs y,g^y for it.
- The DIDS signs and publishes the latest version of its clients' temporary public keys $(g^{y_1}, g^{y_2}, \ldots, g^{y_n}$ for n Clients) on the public blockchain.

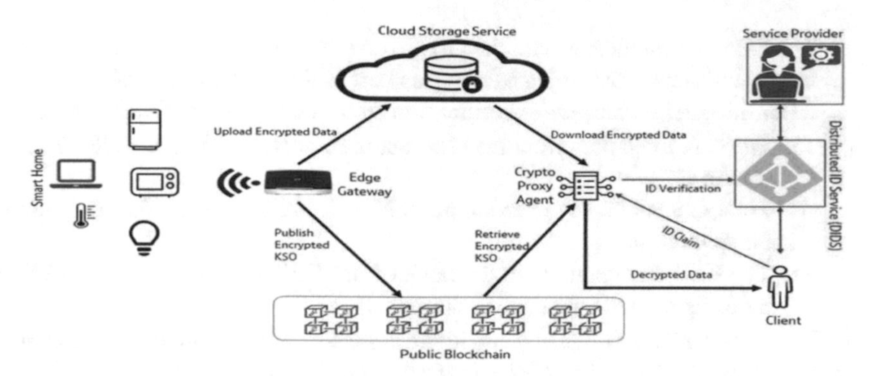

Fig. 1. The Transparent Encryption system for Edge devices in IoT environments using offline key mechanism of public blockchains.

The detailed flow description of the main process is as follows:

1. The Smart Home IoT sensors are connected to the Edge Gateway (EG); all sensors' data passes through the EG.
2. The EG does the following:
   a. The EG encrypts the sensor data of interest with a Data Encryption Key (DEK), i.e., $AES(DEK,M) = C$.
   b. EG sends the encrypted data C to the third-party cloud storage service and receives a URI (Uniform Resource Identifier) of the uploaded encrypted data as a result of this operation.
   c. A Client-specific Key Encryption Key (KEK) is generated by the EG in the following way:
      i. The EG creates a temporary private/public key pair x,g^x and publishes the temporary public key g^x on the public blockchain.
      ii. The EG queries the blockchain to get the latest public keys of all member Clients $(g^{y_1}, g^{y_2}, ..., g^{y_n})$.
      iii. The EG then generates the Client KEKs as: $KEK_1 = Hash(EG_{ID}||Client_{id}||g^{y_1x})$, where EG_{ID} and $Client_{id}$ are the identities of the EG and a particular Client, and g^{y_1x} is computed by the EG using its private key x and a Client's public key g^{y_1}.
   d. The EG wraps the DEK using the Client-specific KEK to generate a Key Security Object (KSO), i.e.,$AES(KEK, DEK) = KSO$
   e. The EG signs and publishes the $Hash(URI)$,KSO key-value pair on the public blockchain.
3. The Client accesses the decrypted data through a Crypto Proxy Agent (CPA) according to following steps:
   a. The Client verifies their identity to the CPA by sharing their Verifiable Claim with the CPA (the claim contains information like the Client's identifier/certificate, ID of the EG that is the source of the encrypted data and URI of the encrypted data stored on the cloud service that the Client wants to access).

b. The CPA communicates with the DIDS to verify the Client's claim and receives the Client's temporary private key y_1 as part of the successful verification reply.
c. After the Client's successful verification from the DIDS, the CPA retrieves
   i. The encrypted data from the cloud storage service, using the URI obtained in step a.
   ii. The EG's public key g^x from the public blockchain, using the ID of the EG obtained in step a.
   iii. The KSO associated with the hash of the URI from the public blockchain, by using the URI obtained in step a.
d. Using the Client's public key and identifier information from the claim, the CPA constructs the KEK, i.e., $KEK_1 = Hash(EG_{ID}||Client_{id}||g^{xy_1})$
e. The CPA decrypts the KSO using the KEK constructed in the previous step, and obtains the DEK as a result, i.e., $DEK = AES^{-1}(KEK_1, KSO)$
f. The CPA finally decrypts the encrypted data needed by the Client using the DEK and sends it to them, i.e., $M = AES^{-1}(DEK, C)$

4 Proof of Concept Implementation

The proposed idea has been implemented using MATRIX edge, which is an Edge gateway built on top of an opensource platform known as EdgeX foundry [26]. MATRIX has been designed by British Telecom Innovation Ireland Centre (BTIIC) team and is usual choice for testing or experimentation. Figure 2 presents the sequence flow of the process of encryption and storage of the encrypted data. As shown in Fig. 1, the Edge gateway is responsible for encrypting all the data generated by the IoT devices, which in our case is Matric Edge. The Matrix Edge generates a unique key for the encryption. This key is generated for each device associated with a predefined encryption session/window.The Matrix Edge then forwards the encrypted data to a third-party storage service, and the key to the DLT storage. In the next step, the MATRIX Edge renews/generates a new key for each encryption session/window and use it for the encryption.

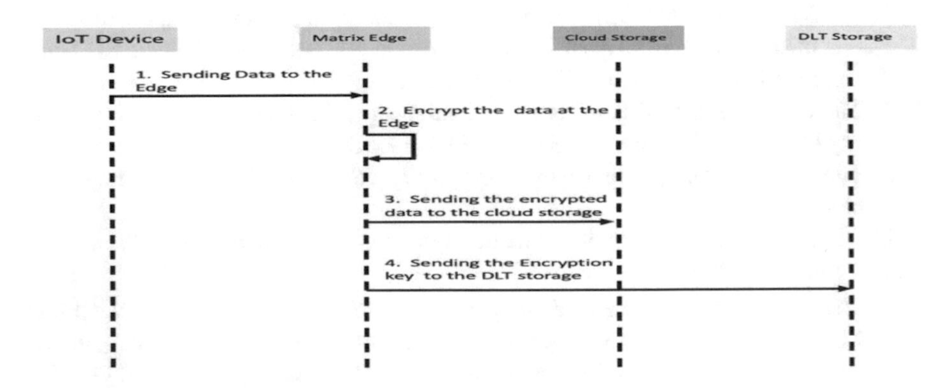

Fig. 2. Sequence diagram of the encryption process.

The process of retrieving the encrypted data is presented in Fig. 3. The user requests specific data from the Crypto Proxy by providing the ID of the device and the session/window of the encrypted data. The Crypto Proxy will retrieve the specific data and key to decrypt the data and display it to the user. Figure 4 shows the proposed system in operation in which the user can choose one of the sensors attached to the MATRIX edge for the encryption, and the Distributed Ledger Technology (DLT) address (to store the encrypted key). By default, we choose the Amazon AWS storage services to store all encrypted data generated by the sensors. Figure 5 shows the interface for retrieving data, in which the Crypto Proxy will consider the user input to retrieve the data from the different components of the system and display the requested data to the user.

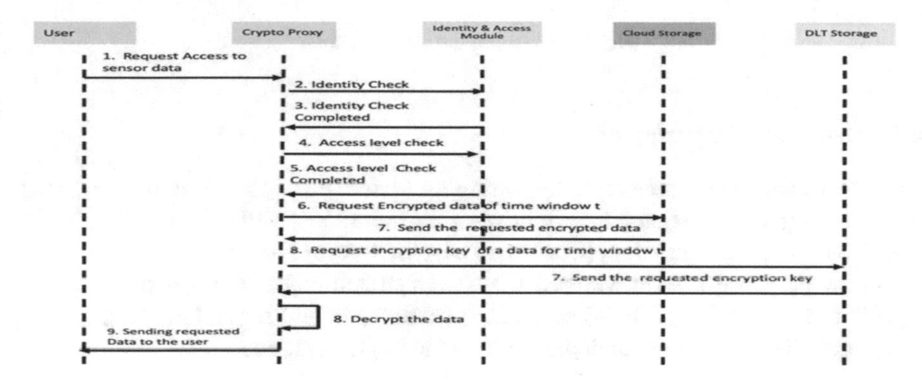

Fig. 3. Sequence diagram of retrieving the data.

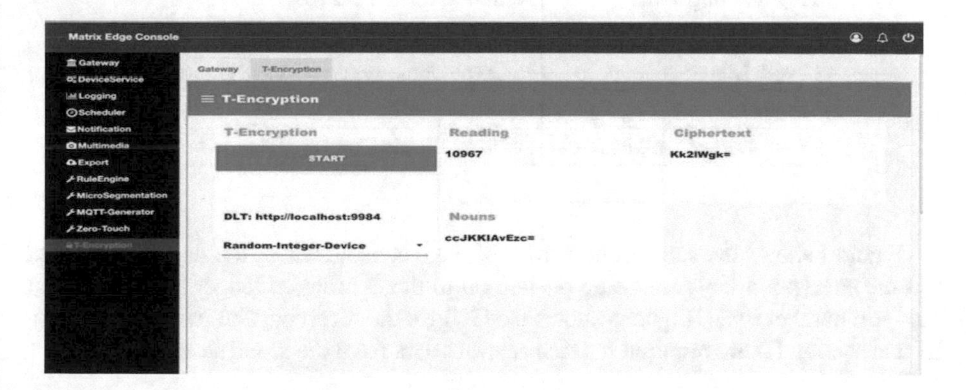

Fig. 4. MATRIX Edge Transparent Encryption Module.

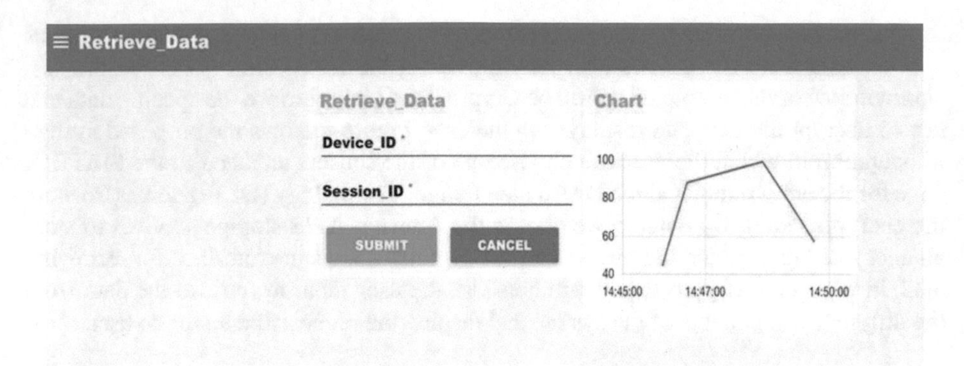

Fig. 5. Retrieving the Data Module.

4.1 Design and Experiments

The experiments were performed by setting up Matrix Edge on a desktop server with the following specifications: Intel(R) Core (TM) i5-8250U CPU @1.80 GHz, 16 GB RAM, 64-bit operating system (Ubuntu 20.04.4 LTS Focal Fossa).

In the proposed system we used ChaCha20 stream cipher [27] as the encryption algorithm and BigChainDB [28] as the DLT solution. Table 1 lists all the programming languages, libraries and technologies used in the implementation.

Table 1. Libraries and tools used in the implementation.

Programming Languages	Python 2.7, Python 3.5
Programming Tools	Node-red version 2.2 [29]
DLT	BigChainDB
Crypto Libraries	PyCryptodome version 3.14 [30]
Libraries	bigchaindb-driver
Cloud storage toolkits /APIs	Amazon Elastic Compute Cloud / boto3

Figure 4 shows the screenshot of Matrix Edge console where the data is encrypted and the encryption keys are being pushed on to the distributed ledger hosted as local host port number 9984. Figure 5 shows the GUI for the receiving end, where the device ID and session ID are required to fetch relevant data from the cloud storage.

5 Summary and Future Work

This paper presents a solution to key management problem for Edge based IoT networks that use third party cloud storage. Conventionally, the encryption keys are stored with the data or handed over to the client for accessing data. However, the in this paper a fool proof architecture based on zero trust security principles that provides for transparent encryption of IoT data, without requiring encryption keys to be shared with third party cloud or the client for accessing the data.

In the proposed model, an integrated key escrow and data decryption scheme has been presented where, Data Encryption Keys (DEK) encrypted by per-client Key Encryption Keys (KEK) are published on a public blockchain. The blockchain stores the encrypted DEKs as assets whose ownership can be transferred to other members of the blockchain. A Crypto Proxy Agent (CPA) deployed close to a Client's network, upon authenticating various claims by the Client, re-constructs the Client-specific KEK to decrypt the encrypted DEK. Once in possession of the DEK, the CPA downloads its corresponding encrypted data from an untrusted cloud storage service, decrypts and transmits it to the Client.

The proof of concept implementation of the idea has been done using MATRIX Edge platform, which is an Edge gateway designed by British Telecom Innovation Ireland Centre. This framework has been filed as a patent by British Telecom. In future the proposed idea would be extended to inclusion of smart contracts in blockchains to automate the process of authentication and asset transfer.

Acknowledgements. This research work was conducted under the BT Ireland Innovation Centre (BTIIC) project and was funded by Invest Northern Ireland and BT.

References

1. Sathesh, A., Smys, D.S.: A survey on internet of things (IoT) based smart systems. J. IoT Soc. Mob. Anal. Cloud **2**(4), 181–189 (2020). https://doi.org/10.36548/jismac.2020.4.001
2. Cisco, Unlocking the potential of the internet of things (2022). https://www.cisco.com/c/en/us/solutions/collateral/executive-perspectives/annual-internet-report/white-paper-c11-741490.pdf
3. Manyika, J., et al.: Cisco annual internet report (2018–2023), March 2015. https://www.mckinsey.com/capabilities/mckinsey-digital/our-insights/the-internet-of-things-the-value-of-digitizing-the-physical-world
4. Noura, H., Couturier, R., Pham, C., Chehab, A.: Lightweight stream cipher scheme for resource-constrained IoT devices. In: 2019 International Conference on Wireless and Mobile Computing, Networking and Communications (WiMob), pp. 1–8 (2019). https://doi.org/10.1109/WiMOB.2019.8923144.
5. Garber, L.: Melissa virus creates a new type of threat. Computer **32**(06), 16–19 (1999). https://doi.org/10.1109/MC.1999.769438
6. Riahi Sfar, A., Natalizio, E., Challal, Y., Chtourou, Z.: A roadmap for security challenges in the internet of things. Digital Commun. Netw. **4**(2), 118–137 (2018). https://www.sciencedirect.com/science/article/pii/S2352864817300214, https://doi.org/10.1016/j.dcan.2017.04.003
7. Lassila, O., Hendler, J.: Embracing "web 3.0". IEEE Internet Comput. **11**(3), 90–93 (2007). https://doi.org/10.1109/MIC.2007.52
8. Sajjad, A., Abu-Tair, M., Zia, U.: Transparent IoT edge encryption using offline key exchange over public blockchains (2022). https://shorturl.at/hrKN2
9. Tayeb, S., Latifi, S., Kim, Y.: A survey on IoT communication and computation frameworks: an industrial perspective. In: 2017 IEEE 7th Annual Computing and Communication

Workshop and Conference (CCWC), pp. 1–6 (2017). https://doi.org/10.1109/CCWC.2017.7868354
10. Rose, S., Borchert, O., Mitchell, S., Connelly, S.: Zero trust architecture (2020-08-10 04:08:00 2020). https://tsapps.nist.gov/publication/get_pdf.cfm?pub_id=930420, https://doi.org/10.6028/NIST.SP.800-207
11. Voundi Koe, A.S., Lin, Y.: Offline privacy preserving proxy re-encryption in mobile cloud computing. Pervas. Mob. Comput. **59**, 101081 (2019). https://www.sciencedirect.com/science/article/pii/S1574119219301488, https://doi.org/10.1016/j.pmcj.2019.101081
12. Alsolami, F., Boult, T.E.: Cloudstash: using secret-sharing scheme to secure data, not keys, in multi-clouds. In: 2014 11th International Conference on Information Technology: New Generations, pp. 315–320 (2014). https://doi.org/10.1109/ITNG.2014.119
13. Hendler, J.: Web 3.0 emerging. Computer **42**(01), 111–113 (2009). https://doi.org/10.1109/MC.2009.30
14. Singh, M., Singh, A., Kim, S.: Blockchain: a game changer for securing IoT data, in: 2018 IEEE 4th World Forum on Internet of Things (WF-IoT), pp. 51–55 (2018). https://doi.org/10.1109/WF-IoT.2018.8355182
15. Shah, M., Shaikh, M.Z., Mishra, V., Tuscano, G.: Decentralized cloud storage using blockchain. In: 2020 4th International Conference on Trends in Electronics and Informatics (ICOEI), pp. 384–389 (48184) (2020)
16. Jemel, M., Serhrouchni, A.: Decentralized access control mechanism with temporal dimension based on blockchain. In: 2017 IEEE 14th International Conference on e-Business Engineering (ICEBE), pp. 177–182 (2017). https://doi.org/10.1109/ICEBE.2017.35
17. Wang, J., Sundaresan, A., Kaza, V.B., Calia, D.: Transparent proxy of encrypted sessions, March 2012. https://image-ppubs.uspto.gov/dirsearch-public/print/downloadPdf/8214635
18. Zeng, P., Choo, K.-K.R.: A new kind of conditional proxy re-encryption for secure cloud storage. IEEE Access **6**, 70017–70024 (2018). https://doi.org/10.1109/ACCESS.2018.2879479
19. Wang, S., Wang, X., Zhang, Y.: A secure cloud storage framework with access control based on blockchain. IEEE Access **7**, 112713–112725 (2019). https://doi.org/10.1109/ACCESS.2019.2929205
20. Ning, J., Cao, Z., Dong, X., Liang, K., Wei, L., Choo, K.-K.R.: Cryptcloud^{+}+: Secure and expressive data access control for cloud storage. IEEE Trans. Serv. Comput. **14**(1), 111–124 (2021). https://doi.org/10.1109/TSC.2018.2791538
21. Xiong, S., Ni, Q., Wang, L., Wang, Q.: Sem-ACSIT: secure and efficient multiauthority access control for IoT cloud storage. IEEE Internet Things J. **7**(4), 2914–2927 (2020). https://doi.org/10.1109/JIOT.2020.2963899
22. Cui, H., Deng, R.H., Lai, J., Yi, X., Nepal, S.: An efficient and expressive ciphertext-policy attribute-based encryption scheme with partially hidden access structures, revisited. Comput. Netw. **133**, 157–165 (2018). https://www.sciencedirect.com/science/article/pii/S138912861830046X, https://doi.org/10.1016/j.comnet.2018.01.034
23. Li, L., Gu, T., Chang, L., Xu, Z., Liu, Y., Qian, J.: A ciphertext-policy attribute-based encryption based on an ordered binary decision diagram. IEEE Access **5**, 1137–1145 (2017). https://doi.org/10.1109/ACCESS.2017.2651904
24. Sukhodolskiy, I., Zapechnikov, S.: A blockchain-based access control system for cloud storage. In: IEEE Conference of Russian Young Researchers in Electrical and Electronic Engineering (EIConRus) 2018, pp. 1575–1578 (2018). https://doi.org/10.1109/EIConRus.2018.8317400
25. Nayak, S.K., Tripathy, S.: Seps: efficient public-key based secure search over outsourced data. J. Inf. Secur. App. **61** 102932 (2021). https://doi.org/10.1016/j.jisa.2021.102932, https://www.sciencedirect.com/science/article/pii/S2214212621001514
26. EdgeXFoundry: Edgex: Open source edge platform, January 2019. https://www.edgexfoundry.org

27. Nir, Y., Langley, A.: ChaCha20 and Poly1305 for IETF Protocols, RFC 7539, May 2015. https://doi.org/10.17487/RFC7539, https://www.rfc-editor.org/info/rfc7539
28. B. GmbH. Bigchaindb 2.0 the blockchain database (2018). https://www.bigchaindb.com/whitepaper/bigchaindb-whitepaper.pdf
29. O.F. Contributors: Node-red (2016). https://nodered.org
30. PyPi: Pycryptodome (2022). https://pypi.org/project/pycryptodome/

An End-to-End Approach for the Detection of Phishing Attacks

Badis Hammi[1(✉)], Tristan Bilot[2,3,4], Danyil Bazain[5], Nicolas Binand[5], Maxime Jaen[5], Chems Mitta[5], and Nour EL Madhoun[3,4,6]

[1] SAMOVAR, Télécom SudParis, Institut Polytechnique de Paris, Paris, France
badis.hammi@telecom-sudparis.eu
[2] Iriguard, 5 Rue Bellini, 92800 Puteaux, France
[3] Université Paris-Saclay, CNRS, Laboratoire Interdisciplinaire des Sciences du Numérique, 91190 Gif-sur-Yvette, France
tristan.bilot@universite-paris-saclay.fr
[4] LISITE Laboratory, ISEP, 10 Rue de Vanves, 92130 Issy-les-Moulineaux, France
{tristan.bilot,nour.el-madhoun}@isep.fr
[5] EPITA Engineering School, Paris, France
{danyil.bazain,nicolas.binand,maxime.jaen,chems.mitta}@epita.fr
[6] Sorbonne Université, CNRS, LIP6, 4 place Jussieu, 75005 Paris, France

Abstract. The main approaches/implementations used to counteract phishing attacks involve the use of crowd-sourced blacklists. However, blacklists come with several drawbacks. In this paper, we present a comprehensive approach for the detection of phishing attacks. Our approach uses our own detection engine which relies on Graph Neural Networks to leverage the hyperlink structure of the websites to analyze. Additionally, we offer a turnkey implementation to the end-users in the form of a Mozilla Firefox plugin.

1 Introduction

Phishing is one of the most common forms of cyber crime on the web, especially in the last years as the Fig. 1 shows. According to Verizon's 2023 Data Breach Investigations Report [1] 36% of all data breaches involved phishing. Also, according to Forbes[1] over 500 million phishing attacks have been reported in 2022. This number has been more than doubled compared to 2021, which is not surprising, considering that it's one of the easiest scams to execute. According to the Cybersecurity and Infrastructure Security Agency (CISA)[2], phishing is a form of social engineering in which a cyber attacker poses as a trustworthy colleague, acquaintance, or organization to lure a victim into providing sensitive information or network access. The lures can come in the form of a crafted email, text message, or even a phone call. However, the email factor remains the

[1] www.forbes.com/advisor/business/phishing-statistics/.
[2] cisa.gov/sites/default/files/2023-02/phishing-infographic-508c.pdf.

L. Barolli (Ed.): AINA 2024, LNDECT 202, pp. 314–325, 2024.
https://doi.org/10.1007/978-3-031-57916-5_27

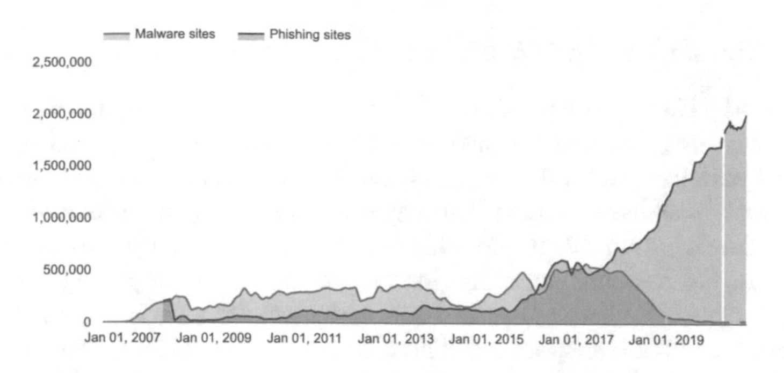

Fig. 1. Number of phishing websites observed between 2007 and 2021 according to Google Safe Browsing https://transparencyreport.google.com/safe-browsing/.

most used one. It is estimated that 3.4 billion malicious emails are sent everyday[3]. According to [2] the direct financial loss from successful phishing attacks increased by 76% in 2022. Hence, the detection of phishing attacks remains among the most important/sensitive tasks to ensure the security of web users.

Unfortunately, despite the existing academic works that aim for the detection of phishing attacks, most of the deployed/implemented solutions rely on collaborative blacklists [3,4]. We present in this paper the continuation of our previous work. In [5] we discussed the use of Graph Neural Networks for the detection of phishing attacks. More precisely, we showed that GNNs applied to the website hyperlink structure are more effective compared to traditional machine learning methods applied to features. In this work, we demonstrate that through using the semi-supervised structure of the graph built using a website's hyperlink structure, a classifier can be trained on supervised data and provide predictions on unsupervised ones.

We highlight the contributions of this paper as follows: (1) We propose an end-to-end approach for the detection of phishing attacks. Unlike the existing works, our detection engine leverages the hyperlink structure thanks to Graph Deep Learning, along with many other hand-crafted features learned with traditional Machine Learning. (2) We provide a ready to use solution, available for the end-users through a Mozilla Firefox plugin. (3) We provide the source codes of our implementation (the plugin for the client and the detection engine for the server).

To the best of our knowledge, our work is the first to introduce a comprehensive end-to-end phishing detection solution built upon graph neural networks.

2 Related Works

Despite the proposal of numerous phishing detection techniques in academia, most of currently operational solutions rely on crowd-sourced blacklists [3]. In this section we discuss the main related works in academia and commercial solutions.

[3] www.itgovernance.co.uk/blog/51-must-know-phishing-statistics-for-2023.

2.1 Academic Related Works on Phishing Detection

Traditional Techniques. The predominant method employed to identify phishing websites involves the use of blacklists. Nevertheless, this approach is associated with several limitations, namely: (1) it needs the creation and maintenance of such blacklists, making it susceptible to zero-day attacks and dependent on human intervention. (2) it demands either storage capacity (resulting in space consumption) or frequent querying (leading to time and computing resource consumption) of a blacklist. (3) crowd-sourced blacklists, such as PhishTank are centralized and lack transparency. The resource consumption problem was tackled by the Google Safe Browsing API[4]. This API is prominently employed in Chromium and serves as a fallback in Firefox. It enables clients to manage a compact local database comprising only truncated hashes of malicious Uniform Resource Locators (URLs). However, this solution remains vulnerable to zero day attacks (new phishing domain names).

Because of the limits of blacklist based approaches, different other techniques have been proposed, based on human-defined heuristics, and designed after identifying inherent characteristics of known phishing websites. Indeed, phishing websites often use patterns in the URL to make them look like legitimate domains, while being subtly different. This can be done by confusing users with slightly different names (e.g. targeting "foobar.com" using the domain name "foo-bar.com"), by using subdomains of trusted entities (e.g. "foobar.example.com") or by including keywords related to the trusted entity in the path section of the URL (e.g. "example.com/foobar" [6]). Other lexical features derived from the URL can be useful. *Sonowal et al.* [7] suggest that having symbols such as "-" and "@", or having more than three dots in the domain name is suspicious, and considers long URLs suspicious as well because they make it harder for users to read the significant part of the URL.

Machine Learning Techniques. Most state of the art approaches for phishing classification are URL-based. That is, they focus on the extraction of useful features directly from the raw URL. Some works [8] employ conventional machine learning techniques, incorporating manually designed features for prediction. In contrast, others [3] opt for deep learning methods, allowing the model to autonomously learn features. The use of deep learning offers the advantage of avoiding human-assisted feature engineering, eliminating the need for domain expert intervention. Consequently, many recent studies [9] leverage deep learning for URL classification, considering it a crucial step in the broader task of phishing classification. This importance arises from the multitude of lexical features that can be extracted from a raw URL string. *Saxe et al.* [10] introduced eXpose, a solution based on a Convolutional Neural Network (CNN). *Le et al.* [11] proposed URLNet, a framework that integrates a character-level CNN with a word-level CNN.

[4] https://developers.google.com/safe-browsing/v4.

To the best of our knowledge, the sole application of Graph Neural Networks to phishing detection is based on the HTML structure of the website [12,13]. In this approach, a graph is built from the HTML DOM and a GNN is fed with this graph. However, this method only relies on the HTML content, which could be easily stolen from benign websites in order to build prefect website copies. This method could thus be easily bypassed by cloning the HTML structure of legitimate websites.

In contrast to prior works, our approach capitalizes on the internal links structure of the website, in conjunction with the conventional features that have demonstrated success in previous approaches. By analyzing multiple phishing websites, we observed that most of them employ similar "href" patterns in <a>, <form> and <iframe> tags. These links are usually self-loops anchors (URLs starting by #) or outgoing links to external domains (usually pointing to a legitimate website like a bank or a social media). Such patterns prove valuable for phishing classification, since a neural network can be trained to discern distinct structures among websites. Malicious websites could hardly bypass this detection system because most of the outgoing links present on these websites redirect to external websites from other domain names in order to fool victims by persuading them that the website is legitimate.

2.2 Commercial Phishing Detection Solutions

Most of the commercial solutions that aim to protect users from phishing attacks are available as web browser plugins. In this section, we present the most used ones. Table 1 presents a comparison of these solutions with our work.

Google Safe Browsing is mainly available on Google Chrome web browser and relies on an updated crowd-sourced blacklist of domain names. However, to protect users' privacy Chrome sends only a fraction of the URL to be checked to Google's server, not the full URL [4]. McAfee WebAdvisor is another browser extension designed to help users browse safely by alerting them to potentially malicious or dangerous websites. Similar to the Chrome plugin, McAfee WebAdvisor relies on crowd-sourced blacklists. However, it differs from Chrome in its ability to evaluate search engine results and provide indications of website security directly in these results [14]. Norton Safe Web is also a web browser plugin that relies on a crowd-sourced blacklist for phishing detection. Furthermore, it integrates Norton's threat intelligence network, which enables it to identify other online threats, such as malware and trackers [15]. The Avast Online Security plugin, like the previously described plugins use crowd-sourced blacklists to detect phishing websites. However, it relies on a cloud-based architecture, which is continually updated, guaranteeing real-time detection of emerging threats [16]. There exist some other browser extensions like Bitdefender TrafficLight or Kaspersky Protection. However, they all rely on crowd-sourced blacklists like the previously described solutions. Indeed, to the best of our knowledge, most of the commercial solutions rely on collaborative blacklists making these solutions dependent on these lists which are often flawed [5], consequently impacting the users' experience. In our solution we rely on our own detection engine PhishGNN.

Table 1. Comparison of commercial solutions for phishing detection (✓: Yes, ✗: No)

Approach	Crowd-sourced blacklist-based	Real-time detection capability	Vulnerable to zero day attacks	Transparency regarding blacklisting criteria	Dependency on continuous human intervention	Sensitivity to human bias
Google Safe Browsing	✓	✗	✓	✗	✓	✓
Avast Online Security	✓	✓	✓	✗	-	-
McAfee WebAdvisor	✓	✗	✓	✗	✓	✓
Norton Safe Web	✓	✗	✓	✗	✓	✓
Bitdefender TrafficLight	✓	✗	✓	✗	✓	✓
Kaspersky Protection	✓	✗	✓	✗	✓	✓
Our solution	✗	✓	✗	✓	✗	✗

3 Proposed Approach and Implementation

The architecture we propose is similar to the Online Certificate Status Protocol (OCSP). As the Fig. 2 shows, the architecture includes two additional entities compared to the conventional client-server architecture; a web plugin and a detection server (PhishGNN responder). No modifications to the web client or to the web server are required. More precisely, the web plugin acts as a proxy and intercepts the HTTP request that the web browser creates. The proxy extracts the domain name from the request and sends it to the detection responder. Next, the detection engine (within the detection responder) analyses the domain name and sends a boolean response to the proxy. If the detection responder's response indicates that the domain name is a phishing domain name, then, the plugin blocks the request. However, if the detection responder's response indicates that the domain name is safe, the proxy forwards the original HTTP request to the web server. In the following we describe the different parts of the architecture and the implementation choices we made[5].

3.1 Detection Engine

Our detection engine relies on PhishGNN, an approach that we proposed for the classification of websites as phishing or benign [5]. Our detection engine leverages a Graph Neural Network (GNN) model to capture complex patterns hidden in the underlying hyperlink structure of web pages[6]. More precisely, we consider the task of phishing websites classification as a node classification task. In this context, the node to classify is a specific URL, and the other nodes represent each potential link originating from that URL. From these links, it is possible to build a graph where nodes represent URLs, and edges are the links between URLs, extracted either from <a>, <form> or <iframe> tags. Hence, the graph is structured as a rooted graph, with the root node identified as the website

[5] A video that shows the implementation of our approach (with a plugin developed on Mozilla firefox) is available on: https://youtu.be/SNik7Du3Mk8.

[6] The source code of our detection engine is available on: https://github.com/TristanBilot/phishGNN.

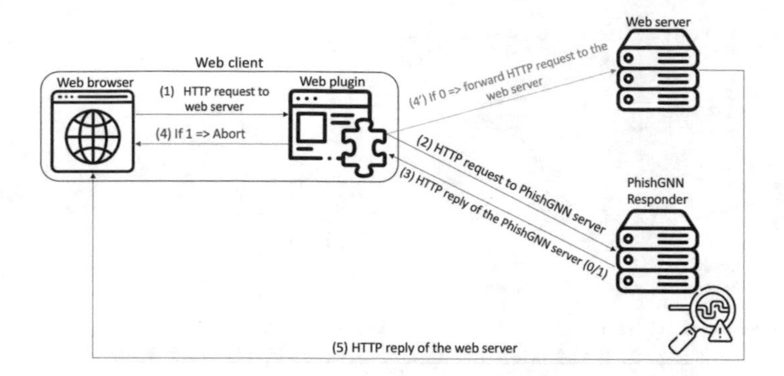

Fig. 2. System architecture of the end-to-end approach for the detection of phishing attacks

to be classified, commonly referred to as the root URL. For each root URL, a feature vector is derived, along with a vector encompassing all URLs going from the root URL (referred to as children URLs). Features are similarly extracted for these children URLs. Subsequently, the features' vectors contribute to the construction of the features matrix X. The children URLs are used to build the actual graph-structure matrix A.

In our approach, we propose training the model in a semi-supervised mode. The known labels pertain to the actual root URLs, while the unknown labels encompass every child URL-meaning it is unknown whether these URLs are phishing or not. Our approach heavily relies on the premise that having labels for every node around the root node significantly facilitates the classification of that root node. Since labels are unavailable for every child URL, we employ a random forest classifier to infer these labels. This classifier is trained on supervised examples in the dataset and is subsequently used for inference on all other examples. Following this, a GNN with message passing gathers information from classified nodes to construct embeddings. Pooling methods such as add, max, or mean are applied to these embeddings to reduce the graph dimension to a single node embedding. Finally, a linear layer is employed as the last layer for graph classification. The Fig. 3 describes how the detection engine follows two steps:

Pre-classification. initially, the graph comprises n nodes, where each node $x_i(1 \leq i \leq n)$ is a vector of d features extracted from the corresponding i^{th} URL. x_1 is the root URL node and every node $x_i(1 < i \leq n)$ represent a link coming from x_1. At this first step, a binary classifier is used to predict in a semi-supervised mode whether a node is phishing or benign, for each feature node $x_i(1 \leq i \leq n)$. The classifier is a function $g : \mathbb{R}^d \rightarrow \mathbb{B}$, that maps node features of size d to a prediction in the Boolean domain $\mathbb{B}$. After this step, the feature matrix X is transformed to a vector $\hat{\mathrm{X}}$ containing respectively zeroes and ones for legitimate and phishing predictions.

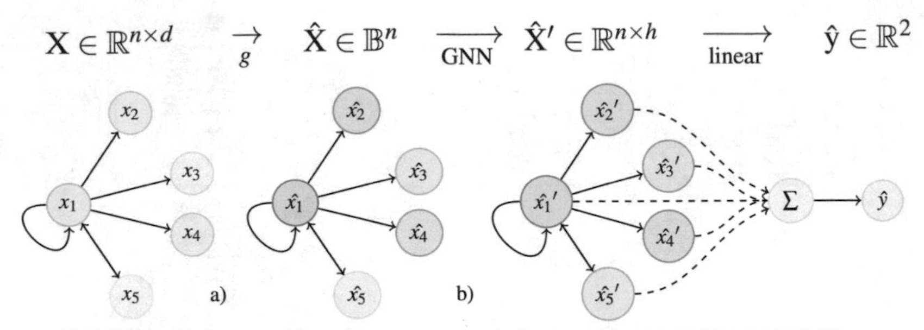

Fig. 3. PhishGNN architecture

Message-Passing. The predictions are subsequently fed into a conventional message-passing GNN with h hidden layers. This process allows for the propagation of information throughout the graph and the learning of node embeddings. The outcome is represented as a matrix $\hat{\mathrm{X}}'$ where each node is an embedding vector of size h. A pooling method is used to reduce the dimension of graph embedding to a single node of shape $1 \times h$. Subsequently, a dot product is conducted between this node and a linear layer with a shape of $2 \times h$, yielding to a vector $\hat{\mathrm{y}}$ that encapsulates the probability of belonging to each class: phishing or benign.

Once implemented, our detection engine must crawl web pages recursively to extract features for the referenced webpages. Despite the existence of multiple web crawlers, we chose to implement our own crawler in order to meet the requirements of PhishGNN. The source code of the crawler is made available in [5]. Figure 4 exhibits some of the extracted features for every URL. We classify the features that the crawler extracts as (1) lexical features, (2) content features, and (3) domain features. Table 2 describes some of these features. Once features have been extracted by the crawler, they are exported to a `CSV` file which can then be read and pre-processed in Python. The detection server caches every domain name that it analyzes. In order to avoid the issue of aging domain names to use them for phishing attacks, a configurable Time To Live (TTL) value is assigned to the cached domain names.

In summary, from the detection responder's perspective, upon receiving an incoming request, it initially extracts the domain name from the request. A hashtable lookup is then performed to determine whether this domain is already present in the cache memory. If the domain is found in the cache, the server sends the prediction (benign or malicious) back to the client. However, if the domain is not found in the cache, the following steps are executed: (1) the URL is appended to the crawler's queue. (2) The crawler retrieves the hyperlink graph structure of the webpage, along with lexical and content features, and stores them in `CSV` format. (3) A trained phishGNN model is invoked to make an inference on the

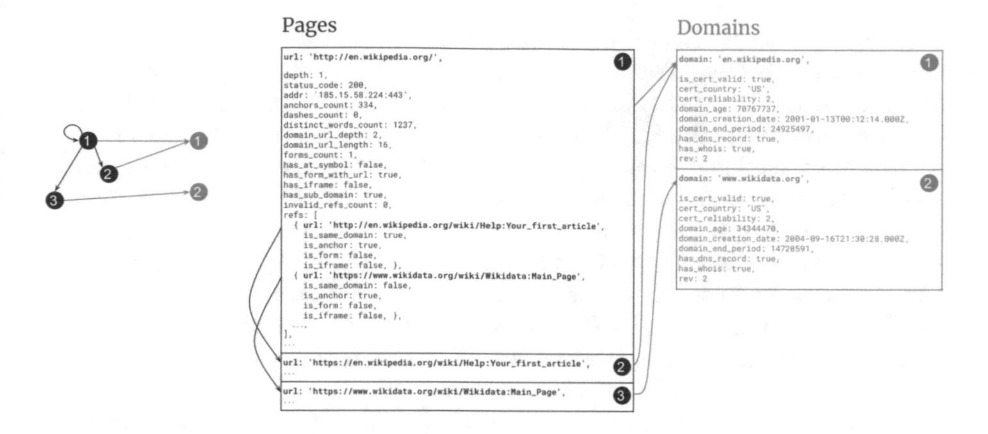

Fig. 4. Example of some extracted feature for every URL

Table 2. Classification of the features

Lexical features	Content features	Domain features
is_https (is the URL scheme "https"), is_ip_address (is the domain an IP address in any form), domain_length (length of the domain name, including subdomains and Top Level Domain (TLD)), domain_depth (number of dots in the domain name), has_subdomain (domain_depth ≥ 2), dashes_count (number of dash characters in the domain name), has_at_symbol (contains "@"), is_same_domain (false if the URL domain is not the same as the root URL)	is_valid_html (false if the response body contains HTML parsing errors), has_iframe (true if an <iframe> tag is in the page document), has_form_with_url (true if a <form> element exists with a valid, static src attribute). References are added for <a> elements with valid (i.e. statically known and leading to a valid HTTP or HTTPS URL after resolution) href attributes, <form> elements with valid action attributes, and <iframe> elements with valid src attributes	is_cert_valid (false if expired or rejected by rustls), cert_country, cert_reliability (computed using the duration of the certificate and whether its issuer is trusted), has_whois (false if WHOIS could not be resolved for the domain), domain_age (in seconds, between the last update date and the domain registry expiry date), domain_end_period (in, seconds between the date of the extraction and the domain registry expiry date)

new domain, using the graph and features collected by the crawler as input. (4) The prediction is then sent back to the client and stored in the cache memory.

3.2 Web Plugin/Proxy

We implemented the proxy as a web plugin. Web plugins offer numerous benefits. They provide customization and enhanced functionality, allowing users to tailor their browsing experience. Additionally, they support accessibility, synchronize across platforms, and when used responsibly, significantly enhance the overall browsing experience. Hence, our approach is completely transparent from the final user's perspective. We used Javascript for the development of the web plugin. Currently, we have only created a Mozilla Firefox compatible version[7]. Once the plugin intercepts the HTTP request of the client, it acts as an HTTP client towards the detection responder (which runs an HTTP server). Hence, the plugin sends an HTTP request containing the URL and waits for a boolean response. If the awaited response stands for a positive detection of a phishing website, then the plugin sends a warning to the user and blocks the HTTP request. However, if the awaited response stands for a negative detection, the plugin forwards the HTTP request made by the web client as it is to the corresponding web server. Unlike the existing commercial solutions that rely on crowd-sourced blacklists, it fully relies on PhishGNN detection engine.

Furthermore, we implemented a caching system within the extension using Firefox's *"localstorage"*[8]. e.g., if a domain is identified as non-phishing, it is cached, and subsequently, the extension refrains from intervention when the user accesses this domain in the future. This caching mechanism significantly reduces the number of requests sent.

3.3 The Communication Protocol

During the design of our end-to-end architecture, we faced the decision between two communication protocols, the HyperText Transfer Protocol (HTTP) and Message Queuing Telemetry Transport (MQTT), each of which bears its own merits. Indeed, MQTT is a reliable and fast protocol, consuming minimal bandwidth, a crucial characteristic for our extension, given its frequent requests to the remote server. With a lightweight header of only 2 bytes, our messages remain concise. The payload during client-to-server communication consists of an URL, while server-to-client responses typically convey a binary value of 0 or 1.

However, MQTT being a publish/subscribe protocol, it does not support a request/response system as such. Hence, it needs to implement an additional MQTT broker to communicate with the subscribers.

Consequently, the use of MQTT solution adds complexity to the end-to-end architecture which leads to additional delays for the end user. For these reasons, we chose to use HTTP. Hence, the detection responder (implemented in C language) implements an HTTP server that (1) waits for HTTP requests from the clients. Then, (2) triggers the detection engine, and (3) responds to the clients via an HTTP reply.

[7] The source code of the plugin is available on: https://github.com/STERN3L/Semester_III_Project-Web_Plugin.

[8] https://developer.mozilla.org/fr/docs/Mozilla/Add-ons/WebExtensions/API/storage/local.

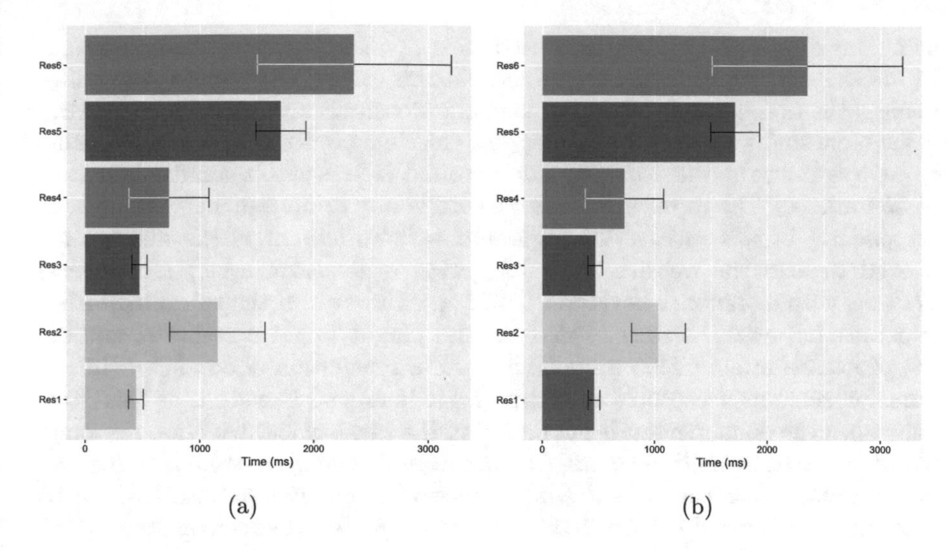

Fig. 5. Webpage loading times. Res3: time needed to load a benign website with our solution (DN in the cache); Res4: time needed to detect and block a phishing website with our solution (DN in the cache); Res5: time needed to load a benign website with our solution (DN not in the cache); Res6: time needed to detect and block a phishing website with our solution (DN not in the cache). (a) Res1: time needed to load a benign website without protection; Res2: time needed to load a phishing website without protection. (b) Res1: time needed to load a benign website using the Avast plugin; Res2: time needed to detect and block a phishing website using the Avast plugin.

4 Performance Evaluation and Discussion

In our previous work [5] we showed how our detection engine PhishGNN outperforms the existing works. Hence, in this evaluation we focus on the performance of the end-to-end approach and the impact on the users' experience.

For the evaluation of our approach, we implemented the architecture using Firefox in headless mode as client. We implemented the detection responder on an Intel Xeon Silver 4108 CPU @ 1.80 GHz with 45GB RAM and running CentOS Linux 7 64 bits. The Fig. 5.a illustrates the webpage loading times acquired during the evaluation of our detection architecture. Each value on the plot corresponds to the average value across 10 tests.

Without our architecture, the time needed to load a benign website is in average 440 ms, while the time needed to load a phishing website[9] is around 1156 ms with a standard deviation of 420 ms. We attribute this distinction to the following reasons: (1) phishing websites lack optimization, (2) certain ones execute malicious JavaScript code, and (3) they are hosted on servers/platforms with inferior performance compared to authentic commercial websites. When

[9] The phishing websites used during our experimentations were obtained from the Phishtank list.

using our approach we face three use-cases: *(1) The requested domain name is in the cache of the detection responder:* in this case, if the requested website is benign, the average time needed to load the website is around 469 ms. However, if the requested website is a phishing one, the time needed for the plugin (proxy) to get a response and stop the process is around 728 ms with a standard deviation of 350 ms. *(2) The requested domain name is not in the cache of the detection responder:* in this case, if the requested website is benign, the average time needed to load the website (after triggering a detection cycle) is in average 1712 ms with a standard deviation of 217 ms. However, if the requested website is a phishing one, the time needed for the plugin to get a response and stop the process is around 2350 ms with a standard deviation of 842 ms. While the duration remains acceptable to end-users, it is crucial to note that this occurs only when the domain name is not stored in the cache of the detection responder, which is shared by multiple users. *(3) The requested domain name is in the cache of the client:* (thanks to *localstorage* function in our implementation). In this case, the time needed for the whole process is just few microseconds.

To better understand the performances of our approach regarding the execution time, we compare it to one of the most used solutions, the Avast Online Security browser extension. The Fig. 5.b shows the webpage loading times when Avast Online Security plugin is implemented on Firefox. We can observe that our approach performs better than the Avast solution when the domain name is in the cache of the detection responder (728 ms against 1032 ms to detect and block a phishing website). However, it requires more time when the domain name is not in the cache. We recall that the Avast approach relies on crowd-sourced blacklists that are often flawed as it was discussed earlier in this paper and does not have its own detection engine.

5 Conclusion and Future Works

In this paper, we have tackled the problem of phishing attacks and proposed an end-to-end detection approach that relies on Graph Neural Networks (GNNs). As far as we know, PhishGNN represents the first application of a GNN to the hyperlink structure of websites for the task of phishing detection. Furthermore, we provided a turnkey solution for the end-users in the form of a web browser extension. The evaluation of our approach, shows its efficiency and performance towards the end-users.

The Time To Live (TTL) setting plays a significant role in the context of our proposal and influence the balance between performance and data freshness, where caching is crucial to overall performance. Therefore, for our short-term future works, we plan to conduct a study/measurement campaign to determine the optimal values for this parameter. Afterwards, we aim to incorporate privacy support for end-users. Indeed, in addition to employing hashing techniques to conceal client-requested domain names as used by Google Safe Browsing, we intend to implement bloom filters which will enable clients to download compressed lists of domain names. Furthermore, we plan to develop versions of our client-side web browser plugin for the different existing web browsers.

References

1. 2023 Data Breach Investigations Report (DBIR). Technical report, Verizon (2023)
2. 2023 State of the Phish: An in-depth exploation of user awareness, vulnerability and resilience. Technical report, Proofpoint (2023)
3. Sahoo, D., Liu, C., Hoi, S.C.H.: Malicious URL detection using machine learning: a survey. arXiv preprint arXiv:1701.07179 (2017)
4. Bell, S., Komisarczuk, P.: An analysis of phishing blacklists: Google safe browsing, openphish, and Phishtank. In: Proceedings of the Australasian Computer Science Week Multiconference, pp. 1–11 (2020)
5. Bilot, T., Geis, G., Hammi, B.: PhishGNN: a phishing website detection framework using graph neural networks. In: Proceedings of the 19th International Conference on Security and Cryptography, vol. 1 (2022)
6. Ledesma, N.C.R., Teraguchi, Y., Mitchell, J.C.: Client-side defense against web-based identity theft. Computer Science Department, Stanford University (2004)
7. Sonowal, G., Kuppusamy, K.S.: PhiDMA-a phishing detection model with multi-filter approach. J. King Saud Univ. Comput. Inf. Sci. **32**(1), 99–112 (2020)
8. Adeyemo, V.E., Balogun, A.O., Mojeed, H.A., Akande, N.O., Adewole, K.S.: Ensemble-based logistic model trees for website phishing detection. In: Anbar, M., Abdullah, N., Manickam, S. (eds.) ACeS 2020. CCIS, vol. 1347, pp. 627–641. Springer, Singapore (2021). https://doi.org/10.1007/978-981-33-6835-4_41
9. Benavides, E., Fuertes, W., Sanchez, S., Sanchez, M.: Classification of phishing attack solutions by employing deep learning techniques: a systematic literature review. In: Developments and Advances in Defense and Security, pp. 51–64 (2020)
10. Saxe, J., Berlin, K.: eXpose: a character-level convolutional neural network with embeddings for detecting malicious URLs, file paths and registry keys. arXiv preprint arXiv:1702.08568 (2017)
11. Le, H., Pham, Q., Sahoo, D., Hoi, S.C.H.: URLNet: learning a URL representation with deep learning for malicious URL detection. arXiv preprint arXiv:1802.03162 (2018)
12. Ouyang, L., Zhang, Y.: Phishing web page detection with html-level graph neural network. In: 2021 IEEE 20th International Conference on Trust, Security and Privacy in Computing and Communications (TrustCom), pp. 952–958. IEEE (2021)
13. Bilot, T., El Madhoun, N., Al Agha, K., Zouaoui, A.: Graph neural networks for intrusion detection: a survey. IEEE Access (2023)
14. Yu, S., An, C., Yu, T., Zhao, Z., Li, T., Wang, J.: Phishing detection based on multi-feature neural network. In: 2022 IEEE International Performance, Computing, and Communications Conference (IPCCC), pp. 73–79. IEEE (2022)
15. Yao, H., Shin, D.: Towards preventing QR code based attacks on android phone using security warnings. In: Proceedings of the 8th ACM SIGSAC Symposium on Information, Computer and Communications Security, pp. 341–346 (2013)
16. Sjösten, A., Van Acker, S., Sabelfeld, A.: Discovering browser extensions via web accessible resources. In: Proceedings of the Seventh ACM on Conference on Data and Application Security and Privacy, pp. 329–336 (2017)

Efficient Inner-Product Argument from Compressed Σ-Protocols and Applications

Emanuele Scala[1(✉)] and Leonardo Mostarda[2]

[1] Computer Science, University of Camerino, Camerino, Italy
emanuele.scala@unicam.it
[2] Mathematics and Computer Science, University of Perugia, Perugia, Italy
leonardo.mostarda@unipg.it

Abstract. The Inner-Product Argument (IPA) is a subroutine of well-known zero-knowledge proof systems, such as Bulletproofs and Halo. These proof systems are then applied in large cryptographc protocols for anonymous and private transactions in the public blockchain. Despite its trustless nature and logarithmic communication efficiency, IPA suffers from low computational efficiency. While not specifically aimed at optimizing the IPA, Attema et al. propose the compressed Σ-protocol theory. Their intuition is simple: the prover provides an argument for a single committed vector to the verifier, whose commitment satisfies an arbitrary linear relation. We follow this intuition, but instead we provide an argument for two vectors committed under a single compact commitment, satisfying a linear form that is the inner-product relation. Hence, we propose the compressed Σ-protocol version of the original IPA, namely the compressed Σ-Inner-Product Argument (Σ-IPA). To this end, we prove security and provide a Σ-IPA that is complete and has soundness in standard DLOG setting. Finally, we conduct an efficiency analysis showing that our IPA reduces the computational complexity of prover and verifier algorithms by a factor of 2 compared to the original IPA.

Keywords: Inner-Product Argument · Σ-protocols · Zero-knowledge · Bulletproofs · Blockchain

1 Introduction

The Inner Product Argument (IPA) is an interactive proof between prover $\mathcal{P}$ and verifier $\mathcal{V}$, which engage an argument of knowledge of two vectors of scalars satisfying an *inner product relation*. Bootle et al. in [4] introduce the Inner Product Argument secure under discrete logarithm (DLOG) assumptions. Later, Bünz et al. in [7] propose Bulletproofs (BPs), that are arguments of knowledge for range proofs and general arithmetic circuits and use the IPA as a subroutine. The authors also optimize the IPA using a *folding strategy* with recursive composition, so that the resulting proof has logarithmic size as the size of the statement behind the inner product relation increases. Another interesting property of the BP argument of knowledge is that it does not come with trusted setup, which makes it attractive for trustless cryptographic protocols. From this

L. Barolli (Ed.): AINA 2024, LNDECT 202, pp. 326–337, 2024.
https://doi.org/10.1007/978-3-031-57916-5_28

result, BP has become widely adopted, especially in public blockchain cryptographic protocols, such as anonymous and private transactions. In these contexts, it is worth mentioning Quisquis [12] and Zether [6], which propose private transactions using BP to prove that amounts and balances are non-negative. Similarly, Lelantus [14] and Monero [1] hide the values of input coins, and prove that the outputs in a spend transaction are in the range of admissible values. However, the trade-off of gaining a trustless protocol results in worse scalability and higher fees. This is because trustless comes with non-constant size of the proofs and linear verification time. An amortization strategy is proposed in ZeroMT [10] and in MTproof [17]. Here, the transaction fees can be amortized by performing multiple transfers in a single transaction equipped with an aggregate *Zero-Knowledge proof* (ZK-proof), obtained by combining the aggregation technique of BP and Σ-protocol theory. Hence, the cost that a private transaction would have for a single transfer is now spread across multiple transfers. However, it turns out that the IPA subroutine still weighs significantly on the overall ZK-proof verification time. With the aim of optimizing the IPA's verification time, some proposals arises such as the one of Bowe et al. in [5]. Here, the authors introduce a new inner-product relation which leads to an argument for polynomial commitment evaluation. Along the same lines, Bünz et al. in [8] propose an amortized succinctness introducing accumulators for polynomial commitment schemes. Applying this technique to the IPA, the verifier asymptotically results in a logarithmic cost barring a single linear time check. In [18], the new inner-product relation is directly applied to the BP, landing into constant-size proofs. In another line of research, there are proposals for the IPA based on pairing-friendly groups, landing into the *inner-pairing product* [9,11,15]. However, they may result in expensive pairing operations once applied. While not specifically aimed at optimizing the IPA, closely related is the compressed Σ-protocol theory of Attema et al. [2,3]. In their notable works, the authors reconcile the BP compression mechanism with the Σ-protocol theory. This allows the design of RPs or arithmetic circuits within an established theory and in DLOG assumptions, with the same communication complexity of BPs. Their intuition is simple: a prover $\mathcal{P}$ provides a proof of knowledge of a committed vector $\mathbf{x}$ to a verifier $\mathcal{V}$, whose commitment satisfies an arbitrary and public linear relation $L(\mathbf{x})$. We follow this intuition, but in our case we want that $\mathcal{P}$ and $\mathcal{V}$ engage a proof of knowledge of two secret vectors committed under a single compact Pedersen commitment, satisfying a public linear form corresponding to the inner-product relation. Therefore, our specific instantiation of compressed Σ-protocol lies on the problem that the IPA tries to solve. Finally, with respect to the original IPA, we observe that our interactive proof algorithms require fewer expensive computations at the same communication complexity. This leads to a more computationally efficient IPA for prover and verifier algorithms.

Our Contribution. In this paper, we propose a new argument to prove that two secret vectors, which are committed under a single compact Pedersen commitment, satisfy the inner-product relation. To this end, we develop the compressed Σ-Inner-Product Argument (Σ-IPA) following the compressed Σ-protocols theory and the BP folding strategy. We prove security and provide a Σ-IPA that is complete and has soundness in standard DLOG setting. Finally, we conduct an efficiency analysis showing that our IPA reduces the computational complexity of prover and verifier algorithms by a factor of 2 compared to the BP's IPA. The paper is organized as follows: Sect. 2 presents related work; Sect. 3 gives the cryptographic background; Sect. 4 provides an IPA with inefficient communication and related security proofs; Sect. 5 provides an IPA with logarithmic communication, improved computational cost and related security proofs; Sect. 6 analyzes the computational costs of the optimized IPA; Sect. 7 are the conclusions and future work.

2 Related Work

Bootle et al. [4] propose an inner product argument where the soundness relies on discrete logarithm assumption in prime order groups. The authors also present the notion of *witness extended emulation* from which the security property of soundness for the IPA is derived. Bünz et al. [7] optimizes the communication complexity of the IPA by a factor of 3, introducing a *folding strategy* in the recursive composition. The authors also propose Bulletproofs (BPs), zero-knowledge arguments for range proofs and general arithmetic circuits based on DLOG assumptions. Despite the proofs have logarithmic size, verification time is linear with respect to the length of the witnesses. Bowe et al. [5] propose an amortization strategy for the IPA verification time. This strategy follows a different inner-product relation, which turns out to be satisfied by a polynomial evaluation argument. Bünz et al. [8] establish a generalized result from the previous one. The authors demonstrate that any polynomial commitment scheme based on DLOG assumption has an accumulation scheme. With an accumulation scheme, the IPA verifier results in a logarithmic cost barring a final linear time opening check. Here, security is based on random oracle model. In our previous work [18], we apply the new relation of [5] to the IPA of BP and see that it could improve the communication complexity to a constant size. The security inherits that of [5], however the analysis requires further details. Another side of the research is devoted to the IPA based on pairing-friendly groups and universal setup. In that direction, Daza et al. [11] achieve logarithmic verification complexity in the circuit size based on the work of [4]. Bünz et al. [9] achieve a logarithmic-time verifier for a generalized IPA in pairing settings. Lee [15] proposes an argument of knowledge from inner-pairing products with a transparent setup, where the verifier has an asymptotic logarithmic time plus the cost for a number of pairings. However, pairing operations may result expensive once applied. Attema et al. [3] propose compressed Σ-protocols, reconciling the BPs compression mechanism with the theory of Σ-protocols and achieving the same communication complexity. The authors provide a general relation for the proof of knowledge of a vector commitment with arbitrary linear form openings. In our work, we essentially develop a compressed Σ-inner-product argument for the original inner-product relation of BPs, thus considering the security properties from the theory of compressed Σ-protocols.

3 Preliminaries

Notation. We denote with $\lambda \in \mathbb{N}$ the security parameter, PPT means probabilistic polynomial-time, and with $s \xleftarrow{\$} \mathcal{S}$ we indicate a random variable s uniformly sampled from the set $\mathcal{S}$. We consider cyclic groups of large prime order p denoted with $\mathbb{G}$, and

$|\mathbb{G}|$ is the order of the group. In every occurrence, g or h are generators of a cyclic group $\mathbb{G}$. We use the *group-generation* function $\mathcal{G}$ on input the security parameter 1^λ (written in unary) to generate the tuple $(\mathbb{G}, p, g) \leftarrow \mathcal{G}(1^\lambda)$. We use the *multiplicative notation* for group operations and scalar multiplications. Rings of integers modulo prime p are denoted with $\mathbb{Z}_p$, and the invertible elements of $\mathbb{Z}_p$ are in $\mathbb{Z}_p^*$. We denote vectors in bold, e.g., $\mathbf{a} = (a_1, ..., a_n) \in \mathbb{Z}_p^n$ is a vector of scalars and $\mathbf{g} = (g_1, ..., g_n) \in \mathbb{G}^n$ a vector of generators. We denote the *inner-product* between vectors of dimension n with $\langle \mathbf{a}, \mathbf{b} \rangle = \sum_{i=1}^{n} a_i \cdot b_i \in \mathbb{Z}_p$. The *hadamard-product* between vectors of size n with $\mathbf{a} \circ \mathbf{b} = (a_1 \cdot b_1, ..., a_n \cdot b_n) \in \mathbb{Z}_p^n$. Let n be the size of a vector $\mathbf{s}$, then $\mathbf{s}_{lo} = (s_1, ..., s_k)$ of size k and $\mathbf{s}_{hi} = (s_{k+1}, ..., s_n)$ of size $n-k$ are vector slice operations.

Assumptions. We consider groups in which the *discrete logarithm problem* (DLOG problem) is computationally intractable. The following definition is for the *discrete logarithm assumption.*

Definition 1 (DLOG assumption). We say that the discrete-logarithm problem is hard relative to $\mathbb{G}$ if for all PPT algorithm $\mathcal{A}$ there exists a negligible function **negl** such that

$$Pr\left[\begin{matrix}(\mathbb{G}, p, g) \leftarrow \mathcal{G}(1^\lambda),\ y \xleftarrow{\$} \mathbb{G}; \\ x \in \mathbb{Z}_p \leftarrow \mathcal{A}(\mathbb{G}, p, g, y)\end{matrix} \quad : g^x = y \right] \leq \mathbf{negl}(\lambda)$$

An alternative definition is the *non-trivial discrete-logarithm relation* from [7].

Definition 2 (Non-trivial DLOG relation). For all PPT algorithm $\mathcal{A}$ and for all $k \geq 2$ there exists a negligible function **negl** such that

$$Pr\left[\begin{matrix}\mathbb{G} \leftarrow \mathcal{G}(1^\lambda),\ h_1, ..., h_k \xleftarrow{\$} \mathbb{G}; \\ x_1, ..., x_k \in \mathbb{Z}_p \leftarrow \mathcal{A}(\mathbb{G}, h_1, ..., h_k)\end{matrix} \quad : \quad \begin{matrix}\exists x_1, ..., x_k \neq 0 \\ \wedge \prod_{i=1}^{k} h_i^{x_i} = 1\end{matrix}\right] \leq \mathbf{negl}(\lambda)$$

We say that there is a *non-trivial DLOG relation* between uniformly random group elements $h_1, ..., h_k$ when $\prod_{i=1}^{k} h_i^{x_i} = 1$ and each $x_1, ..., x_k \in \mathbb{Z}_p$ is non-zero. Thus, the DLOG relation assumption states that it is hard to find a non-trivial relation between randomly chosen group elements.

Commitments. We use the form of *Pedersen commitments* which can be defined over prime order cyclic groups $\mathbb{G}$. In particular, let g and h be two distinct generators and β a randomly chosen blinding factor, we compute a Pedersen commitment T to the value $t \in \mathbb{Z}_p$ as $T = g^t h^\beta$. We can also commit to multiple values at once using the *Pedersen vector commitment* variant. Here, values and generators are gathered in vectors of size n and the commitment is computed as $T = \mathbf{g}^{\mathbf{t}} h^\beta = \prod_{i=1}^{n} g_i^{t_i} \cdot h^\beta$. Pedersen commitments are computationally binding under DLOG assumption and perfectly hiding. Moreover, Pedersen commitments are *additive homomorphic*.

Zero-knowledge Relations. In the following, we give a definition for *zero-knowledge relation* and the relative notation.

Definition 3 (Zero-knowledge relation). A relation is a binary relation $\mathcal{R} \subseteq \mathcal{X} \times \mathcal{W}$, where $\mathcal{X}$, $\mathcal{W}$, and $\mathcal{R}$ are finite sets. Elements $\mathbf{x} = \{x_1, ..., x_n\}$ of $\mathcal{X}$ are called instances and $\mathbf{w} = \{w_1, ..., w_n\}$ of $\mathcal{W}$ witnesses. A relation $\mathcal{R}$ formally specifies some statements as a function $f(\mathbf{x}, \mathbf{w})$ which are satisfied if and only if $(\mathbf{x}, \mathbf{w}) \in \mathcal{R}$. We use the notation

$$\mathcal{R} : \{(\mathbf{x}; \mathbf{w}) : f(\mathbf{x}, \mathbf{w})\}$$

to specify a relation for an interactive proof between prover $\mathcal{P}$ and verifier $\mathcal{V}$, where elements in $\mathbf{x}$ are public and are known to both $\mathcal{P}$ and $\mathcal{V}$, while those in $\mathbf{w}$ are only known to $\mathcal{P}$. We say that the relation is zero-knowledge if $\mathcal{P}$ convinces $\mathcal{V}$ that the statements are true, without revealing information about $\mathbf{w}$.

Interactive Proofs and Σ-Protocols. Let $\mathcal{R}$ be a relation and $\mathcal{L}$ the corresponding NP language such that $\mathcal{L} = \{x \mid \exists w : \ (x, w) \in \mathcal{R}\}$. An interactive proof between prover $\mathcal{P}$ and verifier $\mathcal{V}$ is a conversation where $\mathcal{P}$ tries to convince $\mathcal{V}$ that an instance x belongs to the language $\mathcal{L}$ according to the specified relation $\mathcal{R}$. Such conversation is called *transcript* of the interactive proof, and the verifier can `accept` or `reject` the transcript. When the verifier $\mathcal{V}$ outputs `accept` we call the conversation an *accepting transcript* for x. An interactive proof also requires the parties to execute some algorithms, we call these algorithms *interactive protocol algorithms*.

Σ-Protocols are a class of interactive proofs well established also in the context of zero-knowledge proofs. We now give a general definition for Σ-protocol:

Definition 4 (Σ-protocol). Let $\mathcal{R} \subseteq \mathcal{X} \times \mathcal{W}$ be a binary relation. A Σ-protocol for $\mathcal{R}$ is an interactive proof $\Pi = (\mathcal{P}, \mathcal{V})$ where:

- $\mathcal{P}$ is an interactive protocol algorithm which takes as input an instance-witness pair $(x, w) \in \mathcal{R}$.
- $\mathcal{V}$ is an interactive protocol algorithm which takes as input an instance $x \in \mathcal{X}$ and outputs `accept` or `reject`.
- The interactive proof between $\mathcal{P}$ and $\mathcal{V}$ is structured so that it always works as follows:
  - $\mathcal{P}$ starts the protocol by computing a message a, called announcement, and sends a to $\mathcal{V}$;
  - Upon receiving $\mathcal{P}$'s announcement a, $\mathcal{V}$ chooses a challenge c at random from a finite challenge space $\mathcal{C}$, and sends c to $\mathcal{P}$.
  - Upon receiving $\mathcal{V}$'s challenge c, $\mathcal{P}$ computes a response z, and sends z to $\mathcal{V}$.
  - Upon receiving $\mathcal{P}$'s response z, $\mathcal{V}$ outputs `accept` or `reject`. The $\mathcal{V}$'s output must be computed strictly as a function of the instance x and the conversation (a, c, z). In particular, all $\mathcal{V}$ computations are completely deterministic except the random choice of the challenge.

We require that for all $(x, w) \in \mathcal{R}$, when $P(x, w)$ and $V(x)$ interact and follow the prescribed protocol, $V(x)$ always outputs `accept`.

Definition (4) highlights that Σ-protocols are 3-round protocols. When we execute multiple protocol instances, this leads to a multi-round Σ-protocol. Interactions between the $\mathcal{V}$ and an honest $\mathcal{P}$ produce accepting transcripts; this suggests how to verify the correctness of a conversation between $\mathcal{P}$ and $\mathcal{V}$. We can generalize the above concept with the definition of *perfect completeness* for any Σ-protocol.

Definition 5 (Perfect completeness). Let $(\mathcal{P}, \mathcal{V})$ be a Σ-protocol for relation $\mathcal{R}$. If the prover $\mathcal{P}$ follows the protocol then the verifier $\mathcal{V}$ will accept with probability 1.

Of course, non-accepting transcripts can occur if $\mathcal{V}$ interacts with a "dishonest" $\mathcal{P}^*$ who does not follow the protocol. To prevent this, we require the security properties of *special-soundness*. Furthermore, Σ-protocols are often required to have a large challenge space.

Definition 6 (k-Special-Soundness). Let $\Pi = (\mathcal{P}, \mathcal{V})$ be a Σ-protocol for relation $\mathcal{R} \subseteq \mathcal{X} \times \mathcal{W}$. We say that Π is k-special-sound if there exists a polynomial-time deterministic algorithm $\mathcal{E}$, called witness extractor, which is given as input an instance $x \in \mathcal{X}$ and k accepting transcripts (a, c_1, z_1), ...,(a, c_k, z_k) with a the common first $\mathcal{P}$'s message, $c_1, ..., c_k$ pairwise distinct $\mathcal{V}$'s challenges, $z_1, ..., z_k$ the final $\mathcal{P}$'s messages, and always outputs a witness $w \in \mathcal{W}$ satisfying $(x, w) \in \mathcal{R}$, i.e., w is a witness for x. When $k = 2$, it is simply said that Π is special-sound.

Suppose $\Pi = (\mathcal{P}, \mathcal{V})$ is a *special-sound* Σ-protocol that has a large challenge space, we say that Π acts as a *proof of knowledge*. There are alternative notions of special-soundness known in the literature: *knowledge-soundness* and *witness extended emulation* [3,16]. In the *knowledge-soundness* the difference is that the extractor only has *oracle access* to $\mathcal{P}^*$. In the *witness extended emulation*, the extractor with oracle access to $\mathcal{P}^*$ is also required to output a transcript that is indistinguishable from a conversation between $\mathcal{P}^*$ and an honest $\mathcal{V}$. In [4], it is shown that multi-round spacial-soundness implies *witness-extended emulation*. Essentially, a multi-round Σ-protocol is a $(2\mu+1)$-round interactive protocol where the verifier sends μ challenges. The special-soundness definition for multi-round Σ-protocol is a generalization of definition (6) and is given below.

Definition 7 ($(k_1, ..., k_\mu)$-Special-Soundness). Let $\Pi = (\mathcal{P}, \mathcal{V})$ be a $(2\mu+1)$-round Σ-protocol with μ verifier's challenges and for relation $\mathcal{R} \subseteq \mathcal{X} \times \mathcal{W}$. We say that Π is $(k_1, ..., k_\mu)$-special-sound if there exists a polynomial-time deterministic algorithm $\mathcal{E}$, called witness extractor, which is given as input an instance $x \in \mathcal{X}$ and a $(k_1, ..., k_\mu)$-tree of accepting transcripts and always outputs a witness $w \in \mathcal{W}$ satisfying $(x, w) \in \mathcal{R}$, i.e., w is a witness for x.

For the definition of $(k_1, ..., k_\mu)$-tree of transcripts we refer the reader to [3,4]. In this paper, we focus on $(k_1, ..., k_\mu)$-special-soundness protocols with some μ (challenges) and some set of k_i's (transcripts). Then, from [4] it follows that our protocols are *proof of knowledge*.

The following definition is useful when composing interactive proofs; the definition is revised from [2].

Definition 8 (Composable interactive proofs). Let Π_1 for relation $\mathcal{R}_1$ and Π_2 for relation $\mathcal{R}_2$ be two interactive proofs with $2\mu_1 + 1$ and $2\mu_2 + 1$ rounds respectively. Then, Π_1 and Π_2 are composable if, for an efficient computation ψ, the transcript $(\alpha_1, c_1, \alpha_2, ..., c_{\mu_1}, \alpha_{\mu_1+1})$ of Π_1 for statement x_1 is accepting if and only if the prover's final message α_{μ_1+1} is a witness for statement $x_2 = \psi(\alpha_1, c_1, \alpha_2, ..., c_{\mu_1})$ and $(x_2; \alpha_{\mu_1+1}) \in \mathcal{R}_2$. If the verifier of Π_2 accepts the proof for $\mathcal{R}_2$, then the composition $\Pi = \Pi_2 \diamond \Pi_1$ is accepted.

Moreover, if Π_1 is $\mathbf{k}_1$-special-sound and Π_2 is $\mathbf{k}_2$-special-sound, where $\mathbf{k}_1 = (k_1, ..., k_{\mu_1})$ and $\mathbf{k}_2 = (k_1, ..., k_{\mu_2})$, then $\Pi = \Pi_2 \diamond \Pi_1$ is $(\mathbf{k}_1, \mathbf{k}_2)$-special-sound [2].

Σ-protocols are commonly expected to be *public-coin* and *special Honest-Verifier Zero-Knowledge* (sHVZK). In public-coin Σ-protocols, all the messages sent by $\mathcal{V}$ to $\mathcal{P}$ are sampled uniformly at random and are independent of $\mathcal{P}$ messages. Moreover, $\mathcal{V}$ random choices are made public. In this paper, we deal with Σ-protocols that are not necessarily sHVZK and show only the interactive version of those protocols. Using the Fiat-Shamir heuristic [13], it is possible to convert an interactive proof into a *non-interactive* proof.

Inner-Product Argument relation. The *inner-product relation* follows the general definition (3). The standard IPA protocol is an interactive proof for the following relation, with $\mathbf{g}, \mathbf{h}, u, T$ public paramenters:

$$\mathcal{R}_{\mathcal{IPA}} = \{(\mathbf{g}, \mathbf{h} \in \mathbb{G}^n,\ u, T \in \mathbb{G}\ ;\ \mathbf{a}, \mathbf{b} \in \mathbb{Z}_p^n)\ :\ T = \mathbf{g}^{\mathbf{a}}\mathbf{h}^{\mathbf{b}} \cdot u^{\langle \mathbf{a}, \mathbf{b} \rangle}\} \tag{1}$$

where $\mathbf{g}$ and $\mathbf{h}$ are vectors of (independent) generators, u is group element, T a vector commitment and group element, $\mathbf{a}$ and $\mathbf{b}$ are vectors of scalar elements. The goal by the prover is to convince the verifier that he knows the two vectors $\mathbf{a}$ and $\mathbf{b}$ for the statement $T = \mathbf{g}^{\mathbf{a}}\mathbf{h}^{\mathbf{b}} \wedge c = \langle \mathbf{a}, \mathbf{b} \rangle$, where c is the resulting value from the inner-product of the two vectors $\mathbf{a}, \mathbf{b}$. Morever, this value c is given as a part of the vector commitment T by means of the additional group element u. We refer the reader to [7] for the relation where c is not given as a part of the vector commitment and call that relation $\mathcal{R}_{\mathcal{BP}}$.

4 Sigma Inner-Product with Constant Rounds

In this section we present a 3-move IPA protocol denoted by Π_1 with inefficient communication complexity in protocol 1. Before engaging in Π_1, the prover and verifier run in a 2-move protocol Π_0 where, upon the verifier samples and sends $y \xleftarrow{\$} \mathbb{Z}_p^*$, both compute

$$T' = T \cdot u^{y \cdot c}$$

with $T = \mathbf{g}^{\mathbf{a}}\mathbf{h}^{\mathbf{b}}$, $c = \langle \mathbf{a}, \mathbf{b} \rangle$ and generators $\mathbf{g}, \mathbf{h}$ public parameters. Then, prover and verifier engage in Π_1 with the inputs substitution $u^y \rightarrow u$ and $T' \rightarrow T$, as specified in [7].

Protocol 1. Σ-IPA Π_1 for relation 1

1: **input**: $(\mathbf{g}, \mathbf{h} \in \mathbb{G}^n, u, T \in \mathbb{G};\ \mathbf{a}, \mathbf{b} \in \mathbb{Z}_p^n)$
2: $\quad \mathcal{P}$'s input: $(\mathbf{g}, \mathbf{h}, u, T, \mathbf{a}, \mathbf{b})$
3: $\quad \mathcal{V}$'s input: $(\mathbf{g}, \mathbf{h}, u, T)$
4: **output**: $\mathcal{V}$ accepts or rejects

5: $\quad \mathcal{P}$ computes:
6: $\qquad L = \mathbf{g}_{lo}^{\mathbf{a}_{hi}} \cdot \mathbf{h}_{hi}^{\mathbf{b}_{lo}} \cdot u^{\langle \mathbf{a}_{hi}, \mathbf{b}_{lo} \rangle} \in \mathbb{G}$
7: $\qquad R = \mathbf{g}_{hi}^{\mathbf{a}_{lo}} \cdot \mathbf{h}_{lo}^{\mathbf{b}_{hi}} \cdot u^{\langle \mathbf{a}_{lo}, \mathbf{b}_{hi} \rangle} \in \mathbb{G}$
8: $\quad$ **end** $\mathcal{P}$
9: $\quad \mathcal{P} \rightarrow \mathcal{V} : L, R$

10: $\mathcal{V}: x \xleftarrow{\$} \mathbb{Z}_p^*$
11: $\mathcal{V} \rightarrow \mathcal{P}: x$
12: $\mathcal{P}$ computes:
13: $\mathbf{a}' = x \cdot \mathbf{a}_{lo} + \mathbf{a}_{hi} \in \mathbb{Z}_p^{n/2}$
14: $\mathbf{b}' = \mathbf{b}_{lo} + x \cdot \mathbf{b}_{hi} \in \mathbb{Z}_p^{n/2}$
15: **end** $\mathcal{P}$
16: $\mathcal{P} \rightarrow \mathcal{V}: \mathbf{a}', \mathbf{b}'$
17: $\mathcal{V}$ computes and checks:
18: $T' = L \cdot T^x \cdot R^{x^2} \in \mathbb{G}$
19: $T' \stackrel{?}{=} \mathbf{g}_{lo}^{\mathbf{a}'} \cdot \mathbf{g}_{hi}^{x\mathbf{a}'} \cdot \mathbf{h}_{lo}^{x\mathbf{b}'} \cdot \mathbf{h}_{hi}^{\mathbf{b}'} \cdot u^{\langle \mathbf{a}', \mathbf{b}' \rangle}$
20: **end** $\mathcal{V}$

The protocol Π_1 is inefficient since the communication cost is only reduced by a factor of 2, i.e., the prover sends two vectors $\mathbf{a}', \mathbf{b}'$ of size $n/2$ with respect to the witness size of n. The protocol Π_1 is not required to be zero-knowledge, but it is complete and special-sound as stated in Theorem 1.

Theorem 1 (Σ-IPA). *The Σ-protocol Π_1 is a 3-move perfectly complete and 3-special-sound argument for the relation (1).*

Proof. **Completeness.** Following the definition (5) of *perfect completeness*, if the prover follows the protocol the proof is always accepted. Hence, it is sufficient to show that the following holds: $T' = \mathbf{g}_{lo}^{\mathbf{a}'} \cdot \mathbf{g}_{hi}^{x\mathbf{a}'} \cdot \mathbf{h}_{lo}^{x\mathbf{b}'} \cdot \mathbf{h}_{hi}^{\mathbf{b}'} \cdot u^{\langle \mathbf{a}', \mathbf{b}' \rangle} = L \cdot T^x \cdot R^{x^2}$.

3-Special-Soundness. We follow the definition (6) of *k-special-soundness*. Let $((L,R), x_i, (\mathbf{a}'_i, \mathbf{b}'_i,))$ for $i = 1, \ldots, 3$, be the accepting transcripts obtained by rewinding the prover three times after the prover sends L, R. Assuming $x_1, x_2, x_3 \in \mathbb{Z}_p$ are pairwise distinct challanges, we can find three values $v_1, v_2, v_3 \in \mathbb{Z}_p$ by inverting a Vandermonde matrix with non-zero determinant. Thus, it can be shown that for each $\mathbf{a}', \mathbf{b}'$ the tuples $\bar{\mathbf{a}} := \sum_{i=1}^{3} (v_i \mathbf{a}', v_i x_i \mathbf{a}')$, $\bar{\mathbf{b}} := \sum_{i=1}^{3} (v_i x_i \mathbf{b}', v_i \mathbf{b}')$ and $c = \sum_{i=1}^{3} v_i \cdot \langle \mathbf{a}', \mathbf{b}' \rangle$ are valid extracted witnesses for relation (1). Given that the statement $\mathbf{g}^{\bar{\mathbf{a}}} \mathbf{h}^{\bar{\mathbf{b}}} \cdot u^c = T$ holds, this completes the proof.

We now need one additional rewind for the 2-move protocol Π_0. By the soundness of protocol Π_1 the extractor can obtain witnesses $\mathbf{a}$ and $\mathbf{b}$ such that $T \cdot u^{y \cdot c} = \mathbf{g}^{\mathbf{a}} \mathbf{h}^{\mathbf{b}} u^{y \cdot \langle \mathbf{a}, \mathbf{b} \rangle}$. The extractor rewinds and runs the prover with a different challenge y', thus obtaining $T \cdot u^{y' \cdot c} = \mathbf{g}^{\mathbf{a}'} \mathbf{h}^{\mathbf{b}'} u^{y' \cdot \langle \mathbf{a}', \mathbf{b}' \rangle}$. Then, by combining the two equalities we get $\mathbf{g}^{\mathbf{a}-\mathbf{a}'} \mathbf{h}^{\mathbf{b}-\mathbf{b}'} u^{y \cdot \langle \mathbf{a}, \mathbf{b} \rangle - y' \cdot \langle \mathbf{a}', \mathbf{b}' \rangle} = u^{c \cdot (y - y')}$. Hence, either we have found a non-trivial DLOG relation from definition (2), or $\mathbf{a} = \mathbf{a}'$ and $\mathbf{b} = \mathbf{b}'$. In the latter case, it fallows that $u^{(y-y') \langle \mathbf{a}, \mathbf{b} \rangle} = u^{c \cdot (y - y')}$, which implies $c = \langle \mathbf{a}, \mathbf{b} \rangle$. Hence, we have found valid witnesses $\mathbf{a}$ and $\mathbf{b}$ for the statement $T = \mathbf{g}^{\mathbf{a}} \mathbf{h}^{\mathbf{b}} \wedge c = \langle \mathbf{a}, \mathbf{b} \rangle$.

5 Compressed Sigma Inner-Product with LOG Rounds

Following the definition (8) of composable proofs, it turns out that $\Pi_0 \diamond \Pi_1$ is $(2,3)$-special-sound for the Σ-IPA. However, the size of the proof is only reduced by a factor of 2. We can further reduce the proof to a logarithmic size with respect to the witness, following the recursive composition of the *folding strategy* of BP. Furthermore, this is a similar strategy that leads the standard Σ-protocols to the *compressed* form of Attema et al. [2,3]. Thus, we can apply recursion multiple times until the two vectors $\mathbf{a}, \mathbf{b}$ have constant size. Such recursive composition is shown in protocol 2, denoted with Π_2, and considers vectors whose initial size is a power of two, i.e., $n = 2^\mu$ for some $\mu \in \mathbb{N}$.

Protocol 2. Compressed Σ-IPA Π_2 for relation 1

1: **input**: $(\mathbf{g}, \mathbf{h} \in \mathbb{G}^n, u, T \in \mathbb{G};\ \mathbf{a}, \mathbf{b} \in \mathbb{Z}_p^n)$
2: $\mathcal{P}$'s input: $(\mathbf{g}, \mathbf{h}, u, T, \mathbf{a}, \mathbf{b})$
3: $\mathcal{V}$'s input: $(\mathbf{g}, \mathbf{h}, u, T)$
4: **output**: $\mathcal{V}$ accepts or rejects

5: $\mathcal{P}$ computes:
6: $n = 2^\mu \in \mathbb{N}$
7: $L_1 = \mathbf{g}_{lo}^{\mathbf{a}_{hi}} \cdot \mathbf{h}_{hi}^{\mathbf{b}_{lo}} \cdot u^{\langle \mathbf{a}_{hi}, \mathbf{b}_{lo} \rangle} \in \mathbb{G}$
8: $R_1 = \mathbf{g}_{hi}^{\mathbf{a}_{lo}} \cdot \mathbf{h}_{lo}^{\mathbf{b}_{hi}} \cdot u^{\langle \mathbf{a}_{lo}, \mathbf{b}_{hi} \rangle} \in \mathbb{G}$
9: **end** $\mathcal{P}$
10: $\mathcal{P} \rightarrow \mathcal{V} : L_1, R_1$
11: $\mathcal{V} : x_1 \xleftarrow{\$} \mathbb{Z}_p^*$
12: $\mathcal{V} \rightarrow \mathcal{P} : x_1$
13: $\mathcal{P}$ and $\mathcal{V}$ compute:
14: $\mathbf{g}^{(2)} = \mathbf{g}_{lo} \circ \mathbf{g}_{hi}^{x_1} \in \mathbb{G}^{n/2}$
15: $\mathbf{h}^{(2)} = \mathbf{h}_{lo}^{x_1} \circ \mathbf{h}_{hi} \in \mathbb{G}^{n/2}$
16: $T_2 = L_1 \cdot T^{x_1} \cdot R_1^{x_1^2} \in \mathbb{G}$
17: **end** $\mathcal{P}$ and $\mathcal{V}$
18: $\mathcal{P}$ computes:
19: $\mathbf{a}^{(2)} = x_1 \cdot \mathbf{a}_{lo} + \mathbf{a}_{hi} \in \mathbb{Z}_p^{n/2}$
20: $\mathbf{b}^{(2)} = \mathbf{b}_{lo} + x_1 \cdot \mathbf{b}_{hi} \in \mathbb{Z}_p^{n/2}$
21: **end** $\mathcal{P}$
22: $\vdots$
23: $\mathcal{P}$ computes:
24: $\mathbf{a}' := \mathbf{a}^{(\mu-1)}$, $\mathbf{b}' := \mathbf{b}^{(\mu-1)} \in \mathbb{Z}_p^2$
25: $\mathbf{g}' := \mathbf{g}^{(\mu-1)}$, $\mathbf{h}' := \mathbf{h}^{(\mu-1)} \in \mathbb{G}^2$
26: $L_{\mu-1} = \mathbf{g}'^{\mathbf{a}'_{hi}}_{lo} \cdot \mathbf{h}'^{\mathbf{b}'_{lo}}_{hi} \cdot u^{\langle \mathbf{a}'_{hi}, \mathbf{b}'_{lo} \rangle} \in \mathbb{G}$
27: $R_{\mu-1} = \mathbf{g}'^{\mathbf{a}'_{lo}}_{hi} \cdot \mathbf{h}'^{\mathbf{b}'_{hi}}_{lo} \cdot u^{\langle \mathbf{a}'_{lo}, \mathbf{b}'_{hi} \rangle} \in \mathbb{G}$
28: **end** $\mathcal{P}$
29: $\mathcal{P} \rightarrow \mathcal{V} : L_{\mu-1}, R_{\mu-1}$
30: $\mathcal{V} : x_\mu \xleftarrow{\$} \mathbb{Z}_p^*$
31: $\mathcal{V} \rightarrow \mathcal{P} : x_\mu$
32: $\mathcal{P}$ and $\mathcal{V}$ compute:
33: $g := g^{(\mu)} = \mathbf{g}'_{lo} \circ \mathbf{g}'^{x_\mu}_{hi} \in \mathbb{G}$
34: $h := h^{(\mu)} = \mathbf{h}'^{x_\mu}_{lo} \circ \mathbf{h}'_{hi} \in \mathbb{G}$
35: $T_\mu = L_{\mu-1} \cdot T_{\mu-1}^{x_\mu} \cdot R_{\mu-1}^{x_\mu^2} \in \mathbb{G}$
36: **end** $\mathcal{P}$ and $\mathcal{V}$
37: $\mathcal{P}$ computes:

38: $\quad a := a^{(\mu)} = x_\mu \cdot \mathbf{a}'_{lo} + \mathbf{a}'_{hi} \in \mathbb{Z}_p$
39: $\quad b := b^{(\mu)} = \mathbf{b}'_{lo} + x_\mu \cdot \mathbf{b}'_{hi} \in \mathbb{Z}_p$
40: **end** $\mathcal{P}$
41: $\mathcal{P} \rightarrow \mathcal{V} : a, b$
42: $\mathcal{V}$ computes and checks:
43: $\quad c = \langle a, b \rangle$
44: $\quad T_\mu \stackrel{?}{=} g^a \cdot h^b \cdot u^c$
45: **end** $\mathcal{V}$

The protocol Π_2 is efficient in terms of communication costs, that is, it is a $(2\mu+1)$-round protocol with $\mu = log_2(n)$ and n is the witness length. This implies the proof is logarithmic-sized considering that the prover exchanges $(L_1, R_1)...(L_{\mu-1}, R_{\mu-1})$ group element and 2 elements of $\mathbb{Z}_p$ with the verifier, while the verifier sends $log_2(n)$ elements of $\mathbb{Z}_p$. The protocol Π_2 is complete $(3,...,3)$-special-sound argument, always satisfying relation (1) as stated in theorem 2, but this time with halved-size witnesses in each round.

Theorem 2 (Compressed Σ-IPA). *The Σ-protocol Π_2 is a $(2\mu+1)$-move perfectly complete and $(3,...,3)$-special-sound argument for the relation (1).*

Proof. **Completeness.** It follows directly.

$(3,...,3)$-Special-Soundness. Following the definition (7), it can be shown that Π_2 is $(k_1,...,k_\mu)$-special-sound, where $k_i = 3$ for all $i \in [1, \mu]$. That is a generalization of the extractor analysis of Π_1, where now we have $\mu = log_2(n)$. It follows that we can use the same extractor, but this time it takes in total a 3-ary tree of accepting transcripts of depth $log_2(n)$, thus it runs in polynomial time in n. This complete the proof.

We now can replace the Π_1 protocol with Π_2 and compose $\Pi_0 \diamond \Pi_2$ to obtain a $(2\mu+3)$-move complete and $(2,3,...,3)$-special-sound IPA with logarithmic-size proof. From definition (8) of composable proofs, the composition is well-defined: the transcript $(y, (\mathbf{a}, \mathbf{b}))$ for Π_0 on public input $(\mathbf{g}, \mathbf{h}, u, T, c)$, is accepting if and only if the tuple $(\mathbf{a}, \mathbf{b})$ is a witness for statement $x = \psi(\mathbf{g}, \mathbf{h}, u, T, c, y) \mapsto (T \cdot u^{y \cdot c})$ with ψ an efficient computation, and thus $(x; (\mathbf{a}, \mathbf{b})) \in \mathcal{R}_{BP}$. Similarly, Π_2 has an efficient computation ψ' and considers accepting transcripts if and only if the witnesses $(\mathbf{a}, \mathbf{b})$ belong to a new relation $\mathcal{R}_{IPA}$ (1) with halved-size witnesses at each recursive step.

6 Efficiency Analysis

We now give an analysis of the efficiency of Σ-protocols Π_1 and Π_2 (protocol 1 and 2 respectively) with regards to communication and computational complexity. The protocol Π_1 has constant rounds but inefficient communication complexity. Indeed, prover sends to the verifier: 2 elements of $\mathbb{G}$ and 2 elements of $\mathbb{Z}_p^{n/2}$ of size $n/2$ with respect to the witness size n. The verifier sends only 1 element of $\mathbb{Z}_p^*$ to the prover. The protocol Π_2 reduces the communication complexity to logarithmic. Indeed, the prover sends to the verifier: $2 \cdot (\log(n) - 1)$ elements of $\mathbb{G}$ and 2 elements of $\mathbb{Z}_p$, with respect to the witness size n. The veirifer sends $\log(n)$ elements of $\mathbb{Z}_p^*$ to the prover. Finally, protocol

Π_2 improves the computational complexity of the BP's IPA by a factor of 2. Indeed, our prover and verifier algorithms execute $\sum_{j=1}^{\log_2(n)} \frac{n}{2^{j-1}}$ exponentiations and 0 inversions for computing the new generators $\mathbf{g}, \mathbf{h}$ in each j-th round. Instead, prover and verifier algorithms of the BP's IPA execute $\sum_{j=1}^{\log_2(n)} \frac{2n}{2^{j-1}}$ exponentiations and 2 inversions in each j-th round.

7 Conclusions and Future Work

In this paper, we propose compressed Σ-IPA, an argument of knowledge of two commited vectors following the compressed Σ-protocols theory and the original IPA relation. Our Σ-IPA maintains the same logarithmic communication complexity of the original IPA, while reducing the computational complexity by a factor of 2. As a future work, we want to introduce *accumulators* to enhance the overall verification time with logarithmic efficiency.

References

1. Alonso, K.M., et al. Zero to monero (2020)
2. Attema, T.: *Compressed Σ-protocol theory*. PhD thesis, Leiden University (2023)
3. Attema, T., Cramer, R.: Compressed-protocol theory and practical application to plug & play secure algorithmics. In: Annual International Cryptology Conference, pp. 513–543. Springer (2020)
4. Bootle, J., Cerulli, A., Chaidos, P., Groth, J., Petit, C.: Efficient zero-knowledge arguments for arithmetic circuits in the discrete log setting. In: Fischlin, M., Coron, J.-S. (eds.) EUROCRYPT 2016. LNCS, vol. 9666, pp. 327–357. Springer, Heidelberg (2016). https://doi.org/10.1007/978-3-662-49896-5_12
5. Bowe, S., Grigg, J., Hopwood, D.: Recursive proof composition without a trusted setup. Cryptology ePrint Archive (2019)
6. Bünz, B., Agrawal, S., Zamani, M., Boneh, D.: Zether: towards privacy in a smart contract world. In: Bonneau, J., Heninger, N. (eds.) FC 2020. LNCS, vol. 12059, pp. 423–443. Springer, Cham (2020). https://doi.org/10.1007/978-3-030-51280-4_23
7. Bünz, B., Bootle, J., Boneh, D., Poelstra, A., Wuille, P., Maxwell, G.: Bulletproofs: Short proofs for confidential transactions and more. In: 2018 IEEE Symposium on Security and Privacy (SP), pp. 315–334. IEEE (2018)
8. Bünz, B., Chiesa, A., Mishra, P., Spooner, N.: Proof-carrying data from accumulation schemes. Cryptology ePrint Archive (2020)
9. Bünz, B., Maller, M., Mishra, P., Tyagi, N., Vesely, P.: Proofs for inner pairing products and applications. In: Tibouchi, M., Wang, H. (eds.) ASIACRYPT 2021. LNCS, vol. 13092, pp. 65–97. Springer, Cham (2021). https://doi.org/10.1007/978-3-030-92078-4_3
10. Corradini, F., Mostarda, L., Scala, E.: ZeroMT: multi-transfer protocol for enabling privacy in off-chain payments. In: Barolli, L., Hussain, F., Enokido, T. (eds.) AINA 2022. LNNS, vol. 450, pp. 611–623. Springer, Cham (2022). https://doi.org/10.1007/978-3-030-99587-4_52
11. Daza, V., Ràfols, C., Zacharakis, A.: Updateable inner product argument with logarithmic verifier and applications. In: Kiayias, A., Kohlweiss, M., Wallden, P., Zikas, V. (eds.) PKC 2020. LNCS, vol. 12110, pp. 527–557. Springer, Cham (2020). https://doi.org/10.1007/978-3-030-45374-9_18

12. Fauzi, P., Meiklejohn, S., Mercer, R., Orlandi, C.: Quisquis: a new design for anonymous cryptocurrencies. In: Galbraith, S.D., Moriai, S. (eds.) ASIACRYPT 2019. LNCS, vol. 11921, pp. 649–678. Springer, Cham (2019). https://doi.org/10.1007/978-3-030-34578-5_23
13. Fiat, A., Shamir, A.: How to prove yourself: practical solutions to identification and signature problems. In: Odlyzko, A.M. (ed.) CRYPTO 1986. LNCS, vol. 263, pp. 186–194. Springer, Heidelberg (1987). https://doi.org/10.1007/3-540-47721-7_12
14. Jivanyan, A.: Lelantus: towards confidentiality and anonymity of blockchain transactions from standard assumptions. In: IACR Cryptol. ePrint Arch., p. 373 (2019)
15. Lee, J.: Dory: efficient, transparent arguments for generalised inner products and polynomial commitments. In: Nissim, K., Waters, B. (eds.) TCC 2021. LNCS, vol. 13043, pp. 1–34. Springer, Cham (2021). https://doi.org/10.1007/978-3-030-90453-1_1
16. Lindell: Parallel coin-tossing and constant-round secure two-party computation. J. Cryptol. **16**(3), 143–184 (2003). https://doi.org/10.1007/s00145-002-0143-7
17. Scala, E., Dong, C., Corradini,F., Mostarda, L.: Zero-knowledge multi-transfer based on range proofs and homomorphic encryption. In: International Conference on Advanced Information Networking and Applications, pp. 461–472. Springer (2023)
18. Scala, E., Mostarda, L.: Range proofs with constant size and trustless setup. In: International Conference on Advanced Information Networking and Applications, pp. 301–310. Springer (2023)

A Decentralized Blockchain-Based Platform for Secure Data Sharing in Cloud Storage Model

Houaida Ghanmi[1,2](✉), Nasreddine Hajlaoui[2], Haifa Touati[2], Mohamed Hadded[3], Paul Muhlethaler[4], and Saadi Boudjit[5]

[1] National School of Computer Science (ENSI), University of Manouba, Manouba, Tunisia
houaida.ghanmi22@gmail.com

[2] Hatem Bettahar IResCoMath Research Lab, University of Gabes, 6033 Gabes, Tunisia
haifa.touati@cristal.rnu.tn

[3] Abu Dhabi University, Abu Dhabi, United Arab Emirates
mohamed.elhadad@adu.ac.ae

[4] National Institute for Research in Digital Science and Technology (INRIA), 78150 Paris, France
paul.muhlethaler@inria.fr

[5] Laboratoire de Traitement et Transport de l'Information Lab (L2TI), Institut Galilée, University of Paris 13, 99 Av J-Baptiste Clément, 93430 Villetaneuse, France
saadi@univ-paris13.fr

Abstract. In recent years, cloud-based storage systems have played a crucial role in facilitating secure user communication. These systems enable businesses to gain convenient, on-demand network access to a shared pool of configurable computing resources and effectively reduce overall costs by outsourcing the necessary services. However, it introduces new security challenges in management and control, specifically related to secure services and data sharing in distributed databases. In addition, until now there are still limits, especially in terms of security, access control, and centralization problems. In this paper, we propose a secure distributed cloud file storage and sharing system based on the Ethereum blockchain and Interplanetary File System (IPFS) to address these shortcomings. First, we encrypt data with the AES symmetric algorithm before storing it on IPFS, to resolve data leaks and tampering in centralized cloud storage. Second, we implement a distributed and reliable access control policy by sharing the list of authorized users on the blockchain network as a smart contract. We use the ECC algorithm to share encryption keys between data owners and users. Our solution is designed to ensure secure communication, data storage reliability over IPFS, and data controllability without relying on a centralized cloud storage architecture. Finally, the proposed scheme was tested and evaluated on the Ethereum test network. Performance evaluation shows that our system provides a feasible and reliable environment and could effectively resist both isolated and collaborative attackers, as well as untrusted cloud servers.

1 Introduction

Recently, with the rapid development of cloud computing and big data technology, the importance of cloud computing has become noticeable; More and more companies and

L. Barolli (Ed.): AINA 2024, LNDECT 202, pp. 338–348, 2024.
https://doi.org/10.1007/978-3-031-57916-5_29

individuals are choosing to outsource their data to the cloud service. It allows users to access the network and use storage resources, enjoys low cost, scalability, and cloud availability, reducing the cost of managing users for large amounts of data. Additionally, cloud service users can be freed from software updates, periodic maintenance, and storage infrastructure maintenance. Although it is a widely accepted technology, security and confidentiality remain the obstacles to using cloud computing [1]. In fact, in traditional cloud data-sharing systems, there are several problems. Data stored in the cloud may be lost or damaged due to damage by attackers, hardware and software failures, or unavailability of data due to a single point of failure risk [2,3]. On the one hand, traditional access control systems can suffer from issues of transaction traceability, user authorization, data ownership management, and access control preservation. On the other hand, Cloud Service Providers (CSPs) may not be entirely trustworthy, they may still hide data corruption from users to protect their interests, or deliberately delete data that users rarely access [4]. Additionally, they may hide accidental data loss from users to preserve their reputation and will not be able to determine whether the data is processed and stored securely. At present, the emergence of blockchain technology in cloud computing has attracted the attention of a large number of researchers [3], which can provide a distributed database system with the potential to create a reliable and decentralized environment for cloud storage. Additionally, once data is stored in the blockchain, no user can modify this data. In cloud technology, data owners outsource their sensitive information to the cloud to share it with authorized users and access their data from anywhere via the Internet whenever they need. Since information on the blockchain can be considered immutable and traceable, we combine the features of blockchain technology and symmetric and asymmetric encryption to secure data stored in the cloud. In this paper, we propose a data storage and access control system, for secure data sharing in the cloud to solve the problems mentioned above, such as lack of transparency and traceability, single point of failure, and high communication costs. Additionally, reliable and immutable metadata is recorded on the blockchain, so that the data owner can easily monitor user access behavior.

The main contributions of our paper are as follows:

- We harness the inherent capabilities of blockchain technology to establish robust access control within our secure cloud storage framework. Our approach introduces automatic access control managed by smart contracts, complemented by the secure storage of encrypted data in the cloud and metadata within the blockchain network. This framework ensures fine-grained access control without the need for a trusted authority.
- We leverage Ethereum accounts within our system for user authentication, associating a unique identifier with each account to validate the provenance of public keys stored on the blockchain network. This strategic implementation enhances the security and integrity of our framework.
- We leverage the Elliptic Curve Cryptography (ECC) algorithm to encrypt secret keys used during file encryption. This not only enhances the security of data sharing but also fosters secure collaboration between the Data Owner (DO) and Data User (DU). This dual-layered security approach enhances the overall integrity and confidentiality of our proposed cloud storage framework based on the Ethereum blockchain.

- We assess performance and security through the deployment of our proposed architecture on an Ethereum test network. Evaluating both performance metrics and associated costs, our analysis demonstrates the feasibility, efficiency, and effective functionality of our framework.

The rest of the document is organized as follows. Section 2 presents related work. A detailed description of the proposed framework is given in Sect. 3. Section 4 deals with the performance evaluation of the proposed solution. Section 5 presents conclusions and future research directions. The notations used throughout the paper are presented in Table 2.

2 Related Work

In recent years, many people have chosen to store their data on remote storage servers to reduce pressure on local data storage. However, for privacy and security reasons, encrypted documents are usually outsourced, which presents challenges for managing document access control and key management. Indeed, access control is an essential component of the cloud which ensures its security by checking whether a user has the necessary rights to access the services he requests. Thus, current access control systems are faced with problems of confidentiality and the presence of a trusted third party. However, features of blockchain technology can mitigate these issues and access permissions can be enforced and controlled through the use of smart contracts. In this section, we discuss the current state of cloud computing access control research combined with blockchain technology, specifically to address cloud access control issues. Table 1 provides a summary of the current state of cloud access control management research.

Table 1. Literature Review

Scheme	Tech used	Key size	Gas used	Auth	BC-Based Solutions for Cloud Access Control			
					Using TP	Access Rights Transfer	Scalability	Key management
Nabeil Eltayieba et al. [5]	Smart Contract SSS ABE	64 bytes	N/A	✗	✓	✗	✗	✓
Wang et al. [6]	CP-ABE, AES Smart contract, DH	N/A	1.272 USD	✗	✗	✓	✓	✓
Saini et al. [7]	ECDH Smart contract	N/A	1,34 USD	✓	✓	✓	✗	✓
Chen, Yi, et al. [8]	ABE Smart Contract	N/A	N/A	✗	✗	✓	✗	✓
Qin, Xua, et al. [9]	Multi-authority CP-ABE Smart contract	N/A	N/A	✗	✓	✓	✗	✓
Our contribution	AES, ECC Smart contract	128 bit		✓	✗	✓	✓	✓

Nabeil Eltayieba et al. [5] developed a mechanism that combines blockchain with the advantages of signing and encryption to provide secure data sharing to ensure access control to data stored in a cloud environment. Furthermore, smart contracts are also used to secure data-sharing capabilities between different data owners and users. They analyzed the communication cost of the attribute-based signature encryption scheme in a cloud environment, including the signing key size, decryption key, and cipher text. Although in this work there are greater savings in computational overhead and communication costs, the computational cost of data encryption and authentication still needs to be improved.

Wang et al. [6] proposed a decentralized method based on attribute encryption combined with the public Ethereum chain to improve access control policies. Data side access is monitored and logged on-chain, and an access control policy can be created through the access policy tree, allowing for more flexible access control compared to the traditional template. However, this architecture increases the storage load in the chain and the data owner load.

Saini et al. [7] designed an access control mechanism based on blockchain technology that can effectively control user behavior to ensure the privacy of shared folders. The system uses smart contracts to avoid network congestion and uses elliptic curve cryptography for secure storage and retrieval of medical records. However, this solution requires more processing time due to the involvement of proxy servers and the cost of deploying the smart contract is relatively high.

Chen et al. [8] developed a blockchain-based storage and sharing framework for medical data to ensure secure data storage. The authors introduced a medical records-sharing service framework to describe managing personal medical data. In [9], Qin et al. proposed a blockchain-based multi-authority access control (BMAC) mechanism for secure data sharing in the cloud. BMAC leverages the consortium blockchain to establish trust between multiple attribute authorities while ensuring that each attribute is jointly managed by multiple institutions, which can avoid a single point of failure and trace terminal access records. At the same time, this mechanism introduces Shamir secret sharing to realize cross-domain management through smart contracts, which can realize the cooperative operation of multiple trust institutions.

Table 2. Abbreviations

CSP	Cloud Services Providers	DO	Data Owner
DHT	Distributed Hash Table	TPA	Third Party Auditor
MHT	Merkle Hash Tree	MT	Merkle Tree
PoW	Proof of Work	PoS	Proof of Stake
P2P	Peer-to-Peer	ABE	Attribute-Based Encryption
AES	Advanced Encryption Standard	IBE	Identity Based Encryption
IPFS	Interplanetary File System	EMRs	Electronic Medical Records
ECC	Elliptic curve cryptography	CP-ABE	Ciphertext Policy Attribute Based Encryption
SHA	Secure Hash Algorithm	PrK	Private Key
PK	Public Key	SK	Secret Key
SSS	Shamir's Secret Sharing	DH	Distributed Hash

3 The Proposed Blockchain-Based Cloud Storage Framework

In this section, we describe the architecture of the proposed framework designed for data sharing and storing in a cloud computing environment based on Blockchain technology. The interaction between the data owner and the data user is realized through Ethereum smart contract technology, so each data user's access is recorded in the Ethereum blockchain network. The operational flow of our scheme is detailed below.

Figure 1 shows the overall architecture of our proposed model based on blockchain technology. The scheme we propose offers users both a secure outsourcing of their data and decentralized, efficient storage. However, a significant concern is that cloud service providers are not completely trustworthy. To mitigate this, the data owner generates a secret key and metadata for each file. Subsequently, the generated keys are used to encrypt the data files locally (at the DO) using the symmetric AES algorithm before outsourcing them. AES is a symmetric key encryption algorithm known for providing fast and secure end-to-end encryption [11]. Then, the encrypted files are stored on IPFS nodes to ensure confidentiality. To address security issues for cloud data sharing we will use Smart Contracts to store information about the encrypted files (Metadata). Therefore, information transferred between data users and data owners is neither tampered with nor repudiated because every contract call is recorded on the blockchain. Most importantly, all users must have an Ethereum account to be able to register on the networks. During the registration process, a unique ID will be assigned to each user in association with their public key. This task will be carried out based on the asymmetric ECC algorithm with smart contracts to share secret keys. To ensure secure sharing and access control of data four main entities will be included in our model as follows: Data Owner (DO), IPFS, Data User (DU), and Blockchain (BC):

- **Data Owner (DO):** DO manages information such as encrypted files and shares their data with authorized users.
- **Blockchain (BC):** Blockchain is mainly responsible for storing metadata such as a list of authorized users, a list of encrypted secret keys, addresses at which data is stored in the cloud, hash files, and public verification of data integrity. Access to the blockchain is allowed to data owners and users who have passed their authentication.
- **IPFS:** IPFS is a decentralized storage system that operates on a distributed hash table (DHT) to access files on the IPFS network (DHT locates the file via a content-based address). It is mainly responsible for storing encrypted data uploaded by the data owner.
- **Data User (DU):** As a data user, DU requests access to data with permission from the corresponding DO via blockchain. Once Du is properly authenticated and authorized, he can obtain the secret key encrypted with its corresponding public key. Subsequently, DU decrypts the encrypted file to obtain the original file. The owner of the public key will be the only user capable of decrypting the secret key.

Having identified the main entities of our current approach, let us move on to our method's structured processes, namely, the authentication the storage, and the retrieval process.

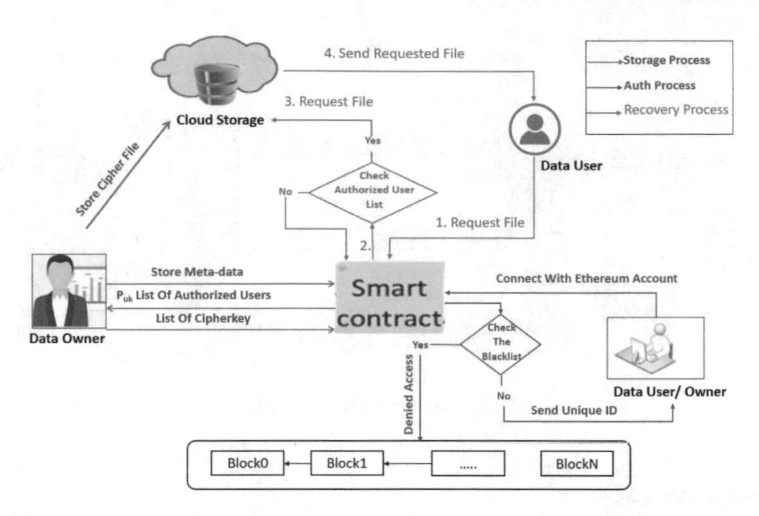

Fig. 1. Global architecture of the proposed Blockchain-based cloud storage framework

3.1 Authentication Process

When a DU requests a file, it must have an Ethereum account and a valid public key to connect to the network before any other process. As shown in Fig. 2, the proposed authentication model is described as follows.

The data owner/user connects with their Ethereum account to the Framework. Upon login, a smart contract called Check Black List (CBL) will be invoked to check whether this account already exists in the blocked user's list or not. For registration purposes, they must submit their own public key as input to the Blockchain, and the smart contract (GetID) will generate a unique identifier in association with a public key for each user as output.

Then, DO/Du can publish his PK on the blockchain network and integrate it into an Ethereum transaction to be immutable in the network. As shown in steps 2 to 4 in Fig. 2, a smart contract called "GetID" acts as a factory to produce a unique ID for each new user after they sign up. When DU/DO sends a registration request to the decentralized framework, Ethereum addresses will be used first to authenticate DU/DO's identity. Once the identity is verified, the DO/DU's Ethereum account address is added as an authorized user in the smart contract. If an attacker manages to intercept the Puk public key and modify it on the network, this malicious act will not go unnoticed, thanks to the traceability advantage of blockchain technology. Only users whose private key matches the public key stored in blockchain networks will be able to decrypt files using the shared secret key.

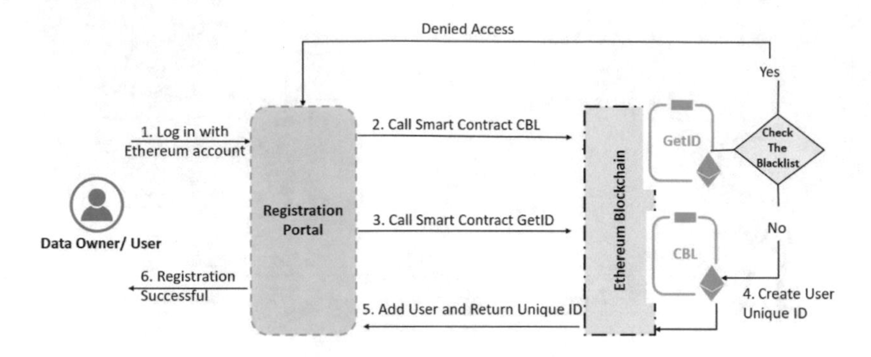

Fig. 2. Authentication process.

3.2 Storage Process

In our work, the user data storage process consists of eight steps below:

1. DO/DU generates a secret key (SK) using the AES algorithm and generates a key pair using the ECC algorithm: private key (PrK) and public key (PK). These keys will be used by the DO and the DU. The public key will be used by the DU during the registration phase and will be used by the DO to encrypt the secret key (SK).
2. DO encrypts the file with the key (SK): (F, SK) → (CF), and signs the file with the SHA-2 function to allow users to verify the integrity of the files, as shown in the Fig. 4.
3. DO saves the CF encrypted file to IPFS and records the location of the file returned by IPFS.
4. To use PK keys to encrypt (SK), DU/DO stores its public key on the Ethereum blockchain using smart contracts.
5. DO records metadata, namely the addresses of files where they are stored in the cloud, the list of users authorized to access DO data, and the list of secret keys encrypted on the blockchain.
6. The DO extracts the list of public keys from the blockchain to share its secret key.
7. DO encrypts the same secret key SK with different public keys PK of users who have access to the DO file, that is, it uses the public keys of the blockchain network to encrypt SK as CSK and embed this information encrypted in an Ethereum transaction.
8. Then the DO can store the list of encrypted secret keys on the blockchain.

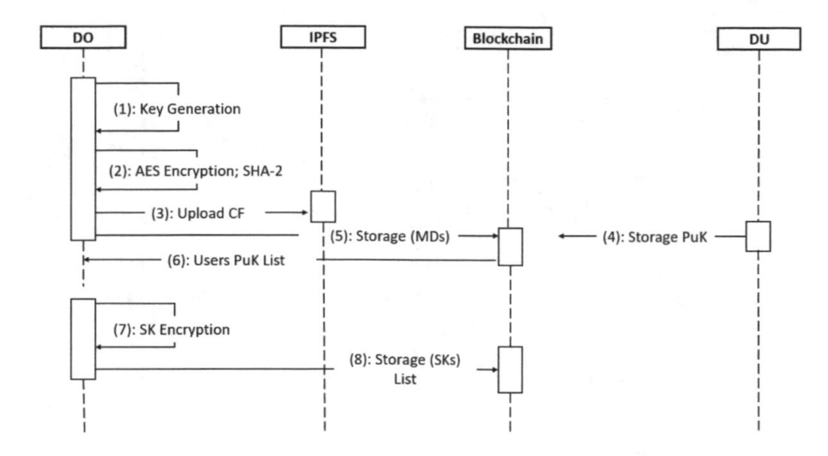

Fig. 3. Storage process.

3.3 Retrieval Process

This recovery process starts when a data requester needs to get a specific file. To retrieve the data, the user can get the information of the requested data through blockchain metadata. However, once a DU requests access, the Blockchain will check the user's authorization to access the requested file. If he is authorized, he will receive the address of the requested file in the cloud as well as the secret key encrypted with his public key. As shown in Fig. 3, our proposed framework mainly consists of five smart contracts, namely; Check Black List Contract (CBL), GetID Contract (GetID), GetAccess Contract (GAC), Grant Contract (GC), and Revocation Contract (RC), each implementing access control policies. These smart contracts are deployed on the blockchain network to manage access to shared data stored in the cloud. Each of these contracts is presented in detail as follows.

1) Check Black List Contract (CBL): checks if the user is blacklisted.
2) GetID Contract (GetID): Give a unique ID associated with the user's public key in the network.
3) GetAccessContract (GAC): It includes the access control policies. The basic principle of the GAC is to determine whether or not a person is eligible to access a requested CipherFile by checking the list of authorized users.
4) Grant Agreement (GC): After verifying that the candidate has access to the requested file, GC grants access to the candidate by sending him the CipherFile and the CipherKey.
5) Revoke Contract (RC): It revokes access control in case of validation failure, misconduct, or bombardment of multiple requests in a short period. Once the DU has obtained the CipherKey and the CipherFile, it can decrypt the secret key with its private key. The owner of the corresponding public key will be the only user able to decrypt the obtained files.

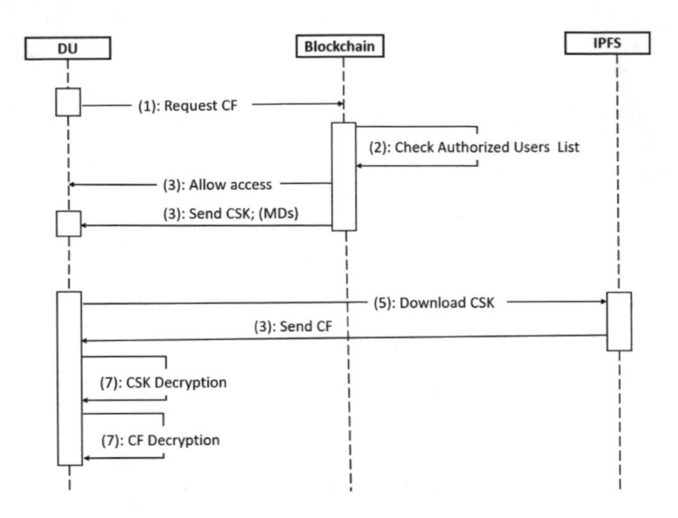

Fig. 4. Retrieval process.

4 Performance Evaluation and Analysis

In this section, we implement our proposed model on the Ethereum platform and evaluate the feasibility and performance of our decentralized application. The specific configuration of the experimental platform and experimental environment is as follows: 1.60 GHz Intel Core i5 processor, 12 GB RAM, and the system is Windows 10. We used the Turing complete scripting language, Solidity, to write smart contracts. Hardhat is a personal blockchain network with virtual accounts with unique account addresses that provides developers with 100 test Ether when linked to a Metamask online wallet, used by Hardhat to execute contracts and to pay computing costs. After successful compilation, taking literature [10] as an example, which is a traditional cloud storage framework. This experiment compares download times of files of different sizes, namely 700 KB, 2.1 MB, 10 MB, 20 MB, and 48 MB, in distributed IPFS networks and traditional centralized cloud storage, as shown in Fig. 5. Figure 5 shows that the time required for downloading will increase with the increase in file size. Compared with the traditional storage system, there is no obvious difference between the IPFS network and traditional cloud storage. Therefore, security issues that may be caused by third-party cloud storage can be resolved effectively.

Since the amount of gas required depends on the implementation of the smart contract and the number of transactions required for its execution, the user who calls functions in Ethereum smart contracts must pay transaction fees measured in units of gas.

Figure 6 lists the gas operations costs on the smart contracts used in our solution. Gas costs calculated by the hardhat network are expressed in Ethereum gas. This unit represents the processing needed to perform smart contract functions, to facilitate the cost analysis of the experimental data in this section, 1 gas price 1 Gwei = 10^{-9} ether. After the data owner executes the key generation algorithm, the ciphertext of DO's secret key is stored in the blockchain, and the AddSecretKey operation is performed, which requires a cost of approximately 58 000 Gwei; After the data owner uploads

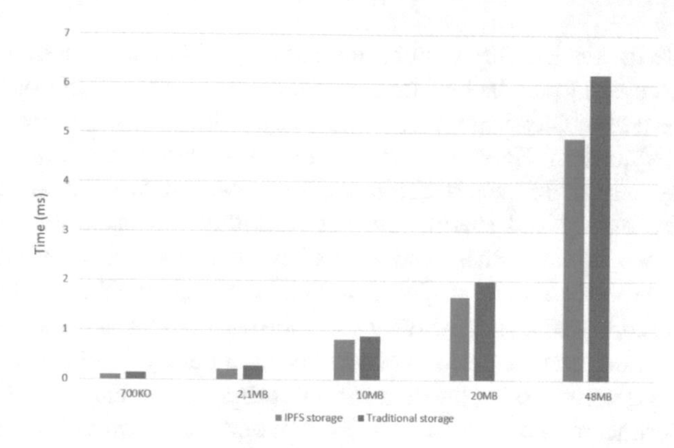

Fig. 5. Comparison of traditional download time and IPFS distributed download time based on file size.

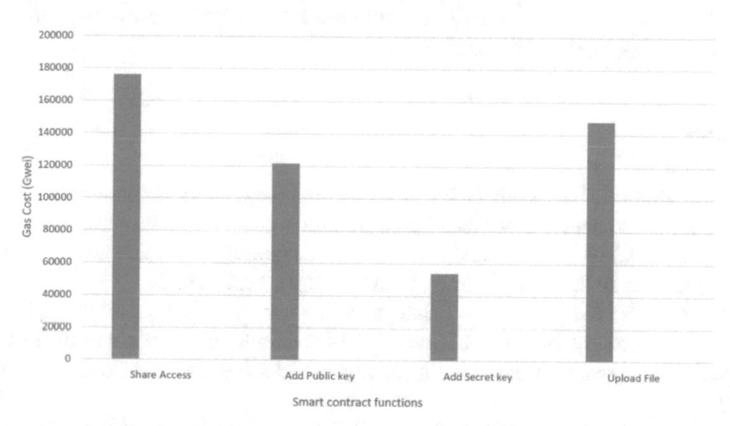

Fig. 6. Cost analysis for smart contract function execution.

the encrypted document, the cost of the UploadFile operation is approximately 144 000 Gwei; The data owner develops an access policy and stores the list of authorized users and its public key in the Ethereum blockchain. The cost of running ShareAccess and AddPublicKey operations is approximately 178 000 Gwei and 121 000 Gwei, respectively. However, when storing a secret key of size 128 bits, the gas consumption is approximately 58 000 Gwei, and when storing a public key of size 256 bits, the gas consumption is approximately 121 000 Gwei. This demonstrates that as the data size increases, the amount of gas consumed also increases. However, even though the amount of data increased, fuel consumption was not significantly different when metadata was uploaded to Ethereum using the suggested methodology.

5 Conclusion and Future Work

Since among the problems related to cloud security is the fact that the management of data access rights relies entirely on one or more trust centers, these are therefore very susceptible to attacks. Furthermore, the security of the data-sharing process cannot be guaranteed due to the lack of transparency in the management of access rights. This paper proposes a blockchain-based data-sharing and access control method using AES and ECC algorithms, to solve some problems related to traditional data-sharing schemes. Using blockchain smart contracts, we achieved distributed access control management by storing and sharing the list of authorized users on the blockchain network. Thus, we store the encrypted secret keys with the list of public keys of the corresponding authorized users, while the data is stored on IPFS (Off-chain). Furthermore, the proposed system can effectively solve the trust problem, make the whole data-sharing process more transparent, and ensure the confidentiality of outsourced data by preventing users from accessing the data without proper credentials. Performance evaluation and experimental results demonstrated that our system provides a more efficient and scalable environment for outsourced cloud data. As part of future work, we will study how to implement a data auditing process in the proposed architecture in more detail. This ensures that uploaded documents are not tampered with by malicious users and improves the security of the architecture.

References

1. Zhang, X., Grannis, J., Baggili, I., Beebe, N.L.: Frameup: an incriminatory attack on storj: a peer to peer blockchain enabled distributed storage system. Digit. Investig. **29**, 28–42 (2019)
2. Sharma, P., Jindal, R., Borah, M.D.: Blockchain technology for cloud storage: a systematic literature review. ACM Comput. Surv. (CSUR) **53**(4), 1–32 (2020)
3. Khanna, A., et al.: Blockchain-cloud integration: a survey. Sensors **22**(14), 5238 (2022)
4. Zhang, Y., Chunxiang, X., Lin, X., Shen, X.: Blockchain-based public integrity verification for cloud storage against procrastinating auditors. IEEE Trans. Cloud Comput. **9**(3), 923–937 (2019)
5. Eltayieb, N., Elhabob, R., Hassan, A., Li, F.: A blockchain-based attribute-based signcryption scheme to secure data sharing in the cloud. J. Syst. Architect. **102**, 101653 (2020)
6. Wang, S., Wang, X., Zhang, Y.: A secure cloud storage framework with access control based on blockchain. IEEE Access **7**, 112713–112725 (2019)
7. Saini, A., Zhu, Q., Singh, N., Xiang, Y., Gao, L., Zhang, Y.: A smartcontract-based access control framework for cloud smart healthcare system. IEEE Internet Things J. **8**(7), 5914–5925 (2020)
8. Chen, Y., Ding, S., Zheng, X., Zheng, H., Yang, S.: Blockchain-based medical records secure storage and medical service framework. J. Med. Syst. **43**, 1–9 (2019)
9. Qin, X., Huang, Y., Yang, Z., Li, X.: A blockchain-based access control scheme with multiple attribute authorities for secure cloud data sharing. J. Syst. Architect. **112**, 101854 (2021)
10. Ghanmi, H., Hajlaoui, N., Touati, H., Hadded, M., Muhlethaler, P.: A secure data storage in multi-cloud architecture using blowfish encryption algorithm. In: Barolli, L., Hussain, F., Enokido, T. (eds.) AINA 2022. LNNS, vol. 450, pp. 398–408. Springer, Cham (2022). https://doi.org/10.1007/978-3-030-99587-4_34
11. Bedoui, M., et al.: An improvement of both security and reliability for AES implementations. J. King Saud Univ.-Comput. Inf. Sci. **34**(10), 9844–9851 (2022)

Enhancing Security and Efficiency: A Lightweight Federated Learning Approach

Chunlu Chen[1(✉)], Kevin I-Kai Wang[2], Peng Li[3], and Kouichi Sakurai[4]

[1] Graduate School of Information Science and Electrical Engineering, Kyushu University, Fukuoka, Japan
chen.chunlu.270@s.kyushu-u.ac.jp

[2] Department of Electrical, Computer and Software Engineering, The University of Auckland, Auckland, New Zealand
kevin.wang@auckland.ac.nz

[3] School of Computer Science and Engineering, The University of Aizu, Aizuwakamatsu, Japan
pengli@u-aizu.ac.jp

[4] Faculty of Information Science and Electrical Engineering, Kyushu University, Fukuoka, Japan
sakurai@inf.kyushu-u.ac.jp

Abstract. Recently, as big data and AI technology advance, data privacy and security are increasingly critical. Federated Learning (FL) has become a key solution in machine learning to address these concerns. In this paper, we present a secure and lightweight FL scheme. It employs masking and Secret Sharing (SS) to securely aggregate data from distributed clients, thereby reducing the demands of model training on system resources. The scheme also computes data similarity among clients to evaluate each client's contribution, defending against challenges posed by malicious clients. This approach safeguards privacy, facilitates accurate model updates, and addresses the challenges of limited resources in edge computing environments. We subjected our framework to rigorous validation using MNIST datasets. Experimental outcomes unequivocally substantiate the efficacy of our proposed methodology.

1 Introduction

In recent years, the advancement of Artificial Intelligence (AI) has propelled the growth of interconnected Internet of Things (IoT), generating vast amounts of data, a significant portion of which comprises user information. However, there is a noticeable lack of adequate protection for this information data, which is a crucial concern.

Federated Learning (FL) emerges in this context, especially in sectors where user data is highly sensitive, such as healthcare, finance, IoT, and smart cities [1,2]. This approach mitigates the need to transfer raw data, thus reducing potential privacy risks. Following the training phase, these devices return the refined model parameters back to a central server for integration. This iterative process continues until the model converges, ensuring data remains localized. However, the growing concerns about privacy breaches in FL systems have become a significant topic of interest in recent academic

L. Barolli (Ed.): AINA 2024, LNDECT 202, pp. 349–359, 2024.
https://doi.org/10.1007/978-3-031-57916-5_30

and industry discussions [3]. While FL inherently aims to enhance data privacy by keeping data localized, the potential for indirect data inference or identification through model parameters remains a concern [4,5]. Furthermore, due to its distributed architecture, it may also attract attackers' attention, who could potentially influence model training through poisoning attacks. These challenges underscore the importance of developing more secure and resilient FL systems.

Moreover, with the extensive integration of Federated Learning (FL) in edge computing, the storage and computational capabilities of edge devices have emerged as potential bottlenecks in model training [6]. Therefore, to address these challenges, recent strategies are focused on developing and implementing more effective methods [7]. These methods aim to improve data management and the learning process [8,9]. However, these approaches have their limitations. For example, inappropriate or overly generalized data augmentation can lead to data distortion or misrepresentation. Hence, in environments with limited resources, it is crucial to develop learning algorithms that are both efficient and capable of overcoming these challenges.

In summary, within the FL edge computing framework, several critical aspects such as data quality, resource limitations, and privacy protection require careful consideration. This paper introduces a solution to address the challenges of limited resources in edge computing environments and to defend against potential malicious clients. Furthermore, to safeguard client-uploaded information, we employ a dual approach combining masking techniques with Secret Sharing (SS) methods, utilizing lightweight encryption algorithms to ensure privacy protection. The primary contributions of our work include:

- We have developed a lightweight FL method that integrates masking and SS scheme to safeguard data privacy. By diminishing the dependency on complex encryption algorithms, our approach significantly eases the deployment of FL models on edge devices, even those with limited computational resources.
- We address the issue of potential malicious clients poisoning attacks in edge computing by computing data similarity. By monitoring the characteristics of different data samples and assigning weights accordingly, our system enhances the system's ability to manage and utilize data more efficiently.

The structure of this paper is organized as follows: Sect. 2 provides an overview of the related work. Section 3 outlines the threat model and design goals, and details our proposed scheme. Section 4 is dedicated to analyzing the security of this scheme. Experimental evaluations are presented and discussed in Sect. 5. The paper concludes with Sect. 6.

2 Related Works

In this section, we explore the key research works that are relevant to this paper.

Encryption technologies are commonly employed in FL to safeguard data privacy, with popular methods including Homomorphic Encryption (HE) [10], SS [11], etc. However, these techniques may heighten the computational burden and system complexity [12]. Differential Privacy (DP) adds intentional ambiguity to data or specific features to prevent external entities from extracting individual raw data from the information transmitted by devices [13]. Nonetheless, excessive noise can impair model accuracy, while insufficient noise may compromise the privacy of training data [14]. Secure Multi-Party Computation (MPC) protocols, often integrated with DP techniques, aim to maintain data privacy and minimize communication overheads, but are limited by their algorithmic complexity [15]. Blockchain technology in FL offers a decentralized and secure framework, enhancing data integrity and auditability [16]. However, it also introduces increased computational and communication demands.

Currently, the common lightweight approach is implemented through the addition of masks, which diminishes the computational requirements for encryption by overlaying data with these masks. Within the FL setting, employing this lightweight masking method helps in reducing computational overhead while preserving data privacy and maintaining model accuracy [17,18]. In FL, since data is dispersed across numerous devices, masks enable efficient data processing locally on these devices without necessitating the transfer of sensitive data to a central server [19]. In practical applications, this mask-based lightweight method is extensively used in diverse areas [20–22]. In addition, training efficiency is improved through heuristic-based worker selection algorithms [23]. Higher computational efficiency is achieved through the encoder-decoder architecture [24]. By adjusting only certain parts of the model, computational demands are reduced [25].

While existing security mechanisms provide substantial protection, achieving a balance among model accuracy, complexity, and efficiency continues to be a formidable challenge in FL. We aim to construct a robust, lightweight FL system that can defend against malicious clients. This endeavor focuses not only on optimizing the system for reduced computational load but also on ensuring stringent privacy safeguards for the data involved.

3 System Design

In this section, we present the threat model, design goals, and provide a detailed exposition of our proposed scheme.

3.1 Threat Model

Our considered threat model is as follows:

Honest-but-Curious Server: In this model, we assume the server to be honest-but-curious, meaning it faithfully adheres to the protocol but might attempt to glean additional information. Specifically, the server executes computations accurately and returns genuine results. However, it may endeavor to acquire more information than necessary from the clients.

Byzantine Clients: For the clients in this model, we consider two scenarios. The first involves honest-but-curious clients, who adhere to the protocol but may attempt to access data from other clients. The second scenario pertains to malicious clients who typically launch attacks with specific objectives in mind. Such clients might try to compromise the model's performance by uploading incorrect data, thereby negatively impacting the overall integrity of the system.

3.2 Design Goals

Our aim is to establish a secure and lightweight FL framework, meticulously crafted for edge computing environments characterized by limited resources and data constraints. The primary goal of this framework is engineered to function effectively within the bounds of resource scarcity. Specifically, our scheme needs to achieve the following requirements:

- **Maintaining Model Accuracy:** Despite the reduction in complexity, the model must retain its ability to make accurate predictions and learn effectively from the distributed data.
- **Reducing Model Complexity:** This involves simplifying the model architecture or employing techniques that reduce the computational load, making the model more suitable for devices with limited resources.
- **Data Privacy Protection:** Despite the optimizations for efficiency and complexity, the system continues to uphold strict data privacy standards, ensuring that client data remains secure and private.

3.3 Our Scheme

This method emphasizes data privacy by employing masking, which avoids the requirement for complex encryption algorithms. Simultaneously, it utilizes a blend of SS algorithms and masking techniques to securely process and aggregate data from distributed clients in a FL system. Furthermore, we assess each client's contribution by calculating data similarity among clients and use this information to guide the global model aggregation. Additionally, we assume that in our system there is at least one honest but curious client, whose data serves as a reference for the computation of data similarity.

We envision a FL system comprising one server and N edge devices (clients). Each edge device engages in training the model disseminated by the server using its local dataset. At the same time, each device collects the average sample of its local dataset. The server then calculates the similarity between the average sample uploaded by clients and the sample uploaded by the honest client, retains model parameters with high similarity, and aggregates them, and disseminates the aggregated parameters back to clients, facilitating the continuation of training in subsequent rounds. The implementation approach is outlined as follows:

- **Initialization:** The server initializes the global model Θ and broadcasts it to clients C_i. lient C_j is also one of the clients, and it is honest but curious, where $i, j \in N = \{0, 1, 2, ...n\}$. This initialization by the server ensures that each client starts model training from a unified baseline.
- **Training and Masking:**
  - **Local Model Training:** Upon receiving the initialized model, each client utilizes their local dataset D_i to update the model parameters θ_i, simultaneously collecting the average data $D_{i_{avg}}$ from their local dataset.
  - **Mask Generation and Data Encryption:** In this context, our approach draws upon insights from research [20]. Our protocol employs masking to obscure the locally trained model parameters θ_i and the computed average data $D_{i_{avg}}$,

thus safeguarding their privacy from the server and other clients. Initially, each client C_i selects a random number R_i as a mask and leverages threshold SS technology to split this mask R_i into shares. The share $S_{R_{(i,i)}}$ is retained by client C_i, while the other shares $S_{R_{(i,j)}}$ are distributed to the corresponding clients C_j, where $1 \leq i, j \leq n$ and $i \neq j$. Upon receiving $n-1$ shares $S_{R_{(i,k)}}$ from other clients, where $k = 1, \ldots, i-1, i+1, \ldots, n$, client C_i computes the aggregate of the respective shares as:

$$S_{R(i)} = \sum_{k=1}^{n} S_{R(i,k)}$$

After aggregation, C_i uploads the result $S_{R(i)}$ to the server. Since these random masks are shared by the respective clients, only more than t client can collectively compute the sum of the random masks. Client C_i applies its generated random number R_i to compute the local encrypted updates:

$$Enc_i(G) = G_i + R_i$$

$$Enc_i(D_{avg}) = D_{i_{avg}} + R_i$$

and uploads them to the server. Among them, it is known that client C_j is honest but curious. This client also trains the model using the aforementioned method and computes the average sample $D_{j_{avg}}$, which is then encrypted and uploaded to the server.

- **Aggregation and Unmasking:** The server collects masked gradient updates, average data, and secret shares of masks from clients. Initially, the masks R_i are reconstructed using the Lagrange interpolation method, which can be calculated as follows:

$$R_i = \sum_{k=1}^{t} S_{R_{(i,i)}} \cdot \prod_{j=1 j \neq k}^{1} \frac{0 - x_j}{x_k - x_j}$$

Here, x_j and x_k are predefined values, typically a sequence of consecutive positive integers like $1, 2, 3, \ldots, n$, where n represents the total number of participating clients. Each x_i value must be unique to ensure that the secret share received by each client is distinct, thus preventing any confusion or duplication during the secret reconstruction process. Then, the true average data $D_{i_{avg}}$ and gradient G_i are calculated by reconstructing the mask R_i as follows:

$$D_{i_{avg}} = Enc_i(D_{avg}) - R_i$$

$$G_i = Enc_i(G) - R_i$$

Next, the server computes the weights w_i based on the obtained $D_{i_{avg}}$ and $D_{j_{avg}}$, which measure each client's contribution to the global model. The calculation method is:

$$w_i = SSIM(D_{i_{avg}}, D_{j_{avg}})$$

Here, we choose the Structural Similarity (SSIM) [26] as the metric for assessing the similarity between two samples. This choice is made because our experimental dataset is the MNIST handwritten digit dataset, which has high image structural similarity. SSIM is determined primarily by three characteristics of the images: luminance, contrast, and structure. $D_{j_{avg}}$ is the average data collected from known honest client D_j, which is used as the basis for data similarity comparison. $D_{i_{avg}}$ is the average data collected from other clients D_i. Finally, the server performs a weighted average update of the global model Θ based on the decoded gradients G_i and their respective weights w_i:

$$\Theta = \sum_{i=1}^{N} \tau_i \cdot G_i, \tau_i = \{ \begin{matrix} 1, w_i \geq 0.5 \\ 0, w_i < 0.5 \end{matrix}$$

SSIM formula calculates a value between -1 and 1, where 1 indicates perfect similarity. Here, the aggregation of the gradient G_i for each client C_i into the global model Θ only occurs if the weight w_i is greater than 0.5. If w_i is 0.5 or less, that client's gradient will not be included in the aggregation. This modification ensures that only contributions from clients with a high enough similarity score (as indicated by w_i) are considered in the model update.

Our method, combining mask and SS, offers a potential way to balance gradient contributions of clients while protecting their data privacy. Through SS, even if some information is intercepted, it is impossible to independently decipher a client's data distribution or the true weight of their contribution. Furthermore, by evaluating client contributions, it is also possible to defend against attacks from malicious clients.

4 Security Analysis

In this section, we analyze the security and demonstrate that our scheme satisfies the security goals.

Theorem 1. *Our scheme is robust against Byzantine clients, provided that at least t clients are honest-but-curious.*

Proof. Assume we have a FL system with N clients, where a subset of these clients may exhibit Byzantine behavior, characterized by either malicious actions or faulty data submissions that can potentially corrupt the learning process.

In our scheme, each client C_i, where i ranges from 1 to N, updates its local gradients G_i and uses SS to distribute these updates as shares. Byzantine clients may attempt to introduce false data, but their impact is limited by the inherent properties of the SS scheme. Without the cooperation of at least t honest-but-curious clients, Byzantine clients cannot reconstruct the individual updates or influence the aggregated model significantly. Furthermore, honest-but-curious clients, while adhering to the protocol, do not have access to other clients' specific data or contributions. They can only view their local data and the global model update, preventing them from inferring individual updates of other clients. Therefore, as long as the number of honest-but-curious clients meets or exceeds the threshold t, the scheme remains secure against Byzantine threats, maintaining the integrity and confidentiality of the participants' data. □

Theorem 2. *Our scheme is secure against the information obtained by malicious clients attacking the server.*

Proof. Consider a scenario where malicious clients attempt to compromise the server to gain access to sensitive information. In our FL framework, data privacy is primarily preserved through a combination of SS and data masking techniques.

Assume a FL environment where each client C_i (for $i = 1, 2, \ldots, N$) contributes to the learning process by updating their local models. Each client C_i masks their local model update G_i with a randomly chosen number R_i. The masking process can be represented as $G'_i = G_i + R_i$, where G'_i denotes the masked update. The random number R_i is chosen from a uniform distribution, ensuring that the masked update G'_i is statistically indistinguishable from random noise. The masked random number R_i is then split into shares using a t-threshold SS scheme. Let S_{R_i} denote the set of shares of R_i. The security of the SS scheme ensures that unless t or more shares are combined, the original masked random number R_i cannot be reconstructed. In the case of malicious clients attempting to attack the server and extract information, they face the same constraints as the server. Due to the randomness of R_i and the secure nature of the SS scheme, any information obtained by the adversary is statistically indistinguishable from random data. Thus, the adversary cannot gain meaningful insights about the original model updates. □

5 Evaluation

In this section, we present the experimental results of the our parameter secure aggregation and secure masks reusing protocols in terms of accuracy, efficiency, and security. As a baseline for comparison, we will juxtapose our proposed scheme with the Federated Averaging (FedAvg) approach, an algorithm that primarily aggregates model updates from multiple clients through averaging calculations. Our goal is to demonstrate that our approach not only matches the efficiency benefits of the FedAvg algorithm, but also enhances the system's security.

5.1 Experiment Setup

In this section, we provide detailed descriptions of our experimental setup.
Datasets: We conducted an evaluation of our training framework utilizing the MNIST dataset. The MNIST [27] dataset comprises a training set of 60,000 samples and a test set of 10,000 samples. Each image in this dataset is a 28×28 pixel grayscale representation of handwritten digits ranging from 0 to 9. Additionally, each image is accompanied by a corresponding label that denotes the digit depicted in the image.
FL Setup: We have constructed a central server system with 10 client nodes, simulating both IID and non-IID data distributions [28] within this setup.
Model: We have developed a Convolutional Neural Network (CNN) model using the PyTorch framework for the MNIST handwritten digit recognition task. This model comprises two convolutional layers, a Dropout layer and two fully connected layers.
Implementation: The experiments are implemented in PyTorch. We simulate a set of clients and a centralized server on one deep learning workstation (i.e., NVIDIA GeForce RTX 4090 GPU).

Hyperparameter: We run 50 communication rounds and perform local training on the client side for 20 epochs. We assumed that we have 10 clients. We used a learning rate $\eta = 0.01$, batch size 128 and Stochastic Gradient Descent (SGD) optimizer.

5.2 Efficiency

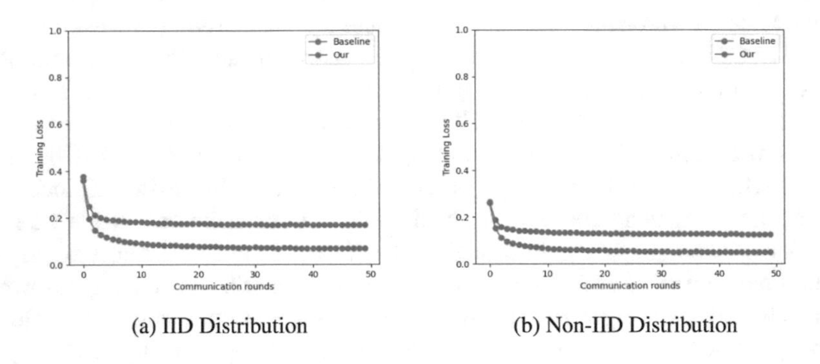

(a) IID Distribution (b) Non-IID Distribution

Fig. 1. Training Loss vs Communication rounds

In this section, our evaluation focuses on experimental efficiency. We conducted a comprehensive evaluation of the training losses incurred by both the FedAvg algorithm and our proposed method. To ensure a thorough assessment, we designed experiments under two distinct data distribution scenarios: IID and Non-IID. In the IID scenario, the data distribution across each client is uniform, ensuring consistency in the data samples. Conversely, in the non-IID scenario, each client's data distribution is varied, reflecting a more heterogeneous setting. As illustrated in Fig. 1, our findings reveal that the training loss of our method close to FedAvg.

By utilizing a mask to safeguard the data and employing SS technique to split the mask values, our method effectively prevent data leakage. This strategic implementation ensures that the additional computational overhead in our scheme, compared to FedAvg, is limited to the SS component, adding only 0.2 to 0.3 h to the training time. The masks, generated as random numbers, allow us to keep the computational demand within manageable limits.

In addition, we have added a data similarity computation step in the aggregation phase, we evaluated the training duration required for our experiments. The detail showns in Table 1, our findings indicate that the training duration of our method is comparable to that of FedAvg. This suggests that our approach maintains competitive performance in terms of training efficiency.

Table 1. Scheme Efficiency

Scheme	Training Time (IID)	Training Time (Non-IID)	Model Size	Parameter
FedAvg	3 h	3 h	0.083 MB	22000
Our Scheme	3.3 h	3.2 h	0.083 MB	22000

5.3 Security

To evaluate the robustness of our scheme, we simulated a poisoning attack, specifically a label-flipping attack [29,30]. We assumed that attackers aim to degrade model accuracy. Therefore, in our experimental setup, we introduced attackers comprising 50% of the total participants to simulate a highly adversarial environment. This approach allowed us to assess the system's resilience against malicious attempts to corrupt the learning process. Our results, shown in Table 2, indicate that in the absence of attackers, our accuracy is almost identical to the baseline. Remarkably, even with 50% attackers, our scheme maintains over 95% accuracy. This demonstrates the significant advantages of our algorithm in terms of resilience against malicious attacks.

Table 2. Scheme Security

Scheme	Without Attackers (IID)	Without Attackers (Non-IID)	50% Attackers (IID)	50% Attackers (Non-IID)
FedAvg	99.06%	99.11%	50.1%	58.8%
Our Scheme	98.3%	97.21%	**98.37%**	**97.69%**

6 Conclusions

FL currently stands as a dynamic and crucial field of research. However, this also attracts the attention of attackers, spurring the creation of sophisticated attack strategies. Consequently, the development of robust defense techniques in FL is of paramount importance. This paper aims to tackle the prevailing challenges in FL, particularly focusing on edge computing and data privacy concerns. We propose a FL methodology that integrates masking techniques and SS, aimed at bolstering data privacy. This approach reduces the reliance on complex encryption solutions, thereby facilitating the implementation of FL in devices with limited computational resources. Furthermore, to defend against potential malicious clients in edge computing scenarios, we evaluate data similarity among clients. As for the future direction of our research, we plan to explore and utilize a broader range of experimental datasets. These datasets will cover more diverse application scenarios and more complex data types to ensure that our research results have higher practicality and wider applicability. Additionally, we also plan to expand our methodology to cover more diverse federated learning scenarios, including distributed and dynamic network architectures. As the deployment of end-devices burgeons, FL emerges as a vital machine learning paradigm, necessitating a collaborative and interdisciplinary approach from the broader research community.

Acknowledgements. The research of the first author is partially supported by the Japan Science and Technology Agency, Support for Pioneering Research Initiated by the Next Generation (JST SPRING) under Grant JPMJSP2136. The research of the second author is partially supported by the International Exchange, Foreign Researcher Invitation Program of National Institute of Information and Communications Technology (NICT), Japan. The first and fourth authors are funded by JSPS international scientific exchanges between Japan and India, Bilateral Program DTS-JSP, grant number JPJSBP120227718.

References

1. Abdulrahman, S., Tout, H., Ould-Slimane, H., Mourad, A., Talhi, C., Guizani, M.: A survey on federated learning: the journey from centralized to distributed on-site learning and beyond. IEEE Internet Things J. **8**(7), 5476–5497 (2021)
2. Wen, J., Zhang, Z., Lan, Y., Cui, Z., Cai, J., Zhang, W.: A survey on federated learning: challenges and applications. Int. J. Mach. Learn. Cybern. **14**(2), 513–535 (2023)
3. Ratnayake, H., Chen, L., Ding, X.: A review of federated learning: taxonomy, privacy and future directions. J. Intell. Inf. Syst. **61**, 1–27 (2023)
4. Yin, X., Zhu, Y., Hu, J.: A taxonomy, review, and future directions, a comprehensive survey of privacy-preserving federated learning. ACM Comput. Surv. **54**, 1–36 (2021)
5. Rodríguez-Barroso, N., Jiménez-López, D., Luzón, M.V., Herrera, F., Martínez-Cámara, E.: Survey on federated learning threats: concepts, taxonomy on attacks and defences, experimental study and challenges. Inf. Fusion **90**, 148–173 (2023)
6. Sattler, F., Müller, K.-R., Samek, W.: Clustered federated learning: model-agnostic distributed multitask optimization under privacy constraints. IEEE Trans. Neural Netw. Learn. Syst. (TNNLS) **32**(8), 3710–3722 (2021)
7. Qin, Z., Deng, S., Zhao, M., Yan, X.: FedAPEN: personalized cross-silo federated learning with adaptability to statistical heterogeneity. In: Proceedings of the 29th ACM SIGKDD Conference on Knowledge Discovery and Data Mining, pp. 1954–1964 (2023)
8. Rebuffi, S.-A., Gowal, S., Calian, D.A., Stimberg, F., Wiles, O., Mann, T.A.: Data augmentation can improve robustness. Adv. Neural. Inf. Process. Syst. **34**, 29935–29948 (2021)
9. Wang, F., Li, B., Li, B.: Federated unlearning and its privacy threats. IEEE Netw., 1–7 (2023)
10. Fang, H., Qian, Q.: Privacy preserving machine learning with homomorphic encryption and federated learning. Future Internet **13**(4), 94 (2021)
11. Fazli Khojir, H., Alhadidi, D., Rouhani, S., Mohammed, N.: FedShare: secure aggregation based on additive secret sharing in federated learning. In: Proceedings of the 27th International Database Engineered Applications Symposium, pp. 25–33 (2023)
12. Ma, J., Naas, S.-A., Sigg, S., Lyu, X.: Privacy-preserving federated learning based on multi-key homomorphic encryption. Int. J. Intell. Syst. **37**, 5880–5901 (2022)
13. Xie, Y., Chen, B., Zhang, J., Wu, D.: Defending against membership inference attacks in federated learning via adversarial example. In: International Conference on Mobility, Sensing and Networking (MSN), pp. 153–160. IEEE (2021)
14. Ren, H., Deng, J., Xie, X.: GRNN: generative regression neural network-a data leakage attack for federated learning. ACM Trans. Intell. Syst. Technol. (TIST) **13**(4), 1–24 (2022)
15. Truex, S., et al.: A hybrid approach to privacy-preserving federated learning. In: ACM Workshop on Artificial Intelligence and Security, pp. 1–11 (2019)
16. Qammar, A., Karim, A., Ning, H., Ding, J.: Securing federated learning with blockchain: a systematic literature review. Artif. Intell. Rev. **56**(5), 3951–3985 (2023)
17. Zhang, Z., et al.: LSFL: a lightweight and secure federated learning scheme for edge computing. IEEE Trans. Inf. Forensics Secur. **18**, 365–379 (2022)

18. So, J., et al.: LightSecAgg: a lightweight and versatile design for secure aggregation in federated learning. Proc. Mach. Learn. Syst. **4**, 694–720 (2022)
19. Cao, Z., et al.: Privacy matters: vertical federated linear contextual bandits for privacy protected recommendation. In: Proceedings of the 29th ACM SIGKDD Conference on Knowledge Discovery and Data Mining, pp. 154–166 (2023)
20. Wei, Z., Pei, Q., Zhang, N., Liu, X., Celimuge, W., Taherkordi, A.: Lightweight federated learning for large-scale IoT devices with privacy guarantee. IEEE Internet Things J. **10**, 3179–3191 (2021)
21. Yang, C., et al.: RaftFed: a lightweight federated learning framework for vehicular crowd intelligence. *arXiv preprint* arXiv:2310.07268 (2023)
22. Guo, Y., Wu, Y., Zhu, Y., Yang, B., Han, C.: Anomaly detection using distributed log data: a lightweight federated learning approach. In: 2021 International Joint Conference on Neural Networks (IJCNN), pp. 1–8. IEEE (2021)
23. Zhu, W., Goudarzi, M., Buyya, R.: Flight: a lightweight federated learning framework in edge and fog computing. *arXiv preprint* arXiv:2308.02834 (2023)
24. Meng, D., Li, H., Zhu, F., Li, X.: FedMONN: meta operation neural network for secure federated aggregation. In: IEEE International Conference on High Performance Computing and Communications; IEEE International Conference on Smart City; IEEE International Conference on Data Science and Systems (HPCC/SmartCity/DSS), pp. 579–584 (2020)
25. Qin, Z., Yao, L., Chen, D., Li, Y., Ding, B., Cheng, M.: Revisiting personalized federated learning: Robustness against backdoor attacks. *arXiv preprint* arXiv:2302.01677 (2023)
26. Wang, Z., Bovik, A.C., Sheikh, H.R., Simoncelli, E.P.: Image quality assessment: from error visibility to structural similarity. IEEE Trans. Image Process. **13**(4), 600–612 (2004)
27. LeCun, Y., Bottou, L., Bengio, Y., Haffner, P.: Gradient-based learning applied to document recognition. Proc. IEEE **86**(11), 2278–2324 (1998)
28. Hsu, T.-M.H., Qi, H., Brown, M.: Measuring the effects of non-identical data distribution for federated visual classification. *arXiv preprint* arXiv:1909.06335 (2019)
29. Shejwalkar, V., Houmansadr, A., Kairouz, P., Ramage, D.: Back to the drawing board: a critical evaluation of poisoning attacks on production federated learning. In: 2022 IEEE Symposium on Security and Privacy (SP), pp. 1354–1371. IEEE (2022)
30. Fang, M., Cao, X., Jia, J., Gong, N.: Local model poisoning attacks to {Byzantine-Robust} federated learning. In: 29th USENIX Security Symposium (USENIX Security 2020), pp. 1605–1622 (2020)

Using Biometric Data to Authenticate Tactical Edge Network Users

Guilherme Falcão da Silva Campos[1(✉)], Jovani Dalzochio[1], Raul Ceretta Nunes[2], Luis Alvaro de Lima Silva[2], Edison Pignaton de Freitas[3], and Rafael Kunst[1]

[1] Universidade do Vale do Rio dos Sinos – UNISINOS, Av. Unisinos, 950, São Leopoldo, RS, Brazil
{guilhermefscampos,jdalzochio}@edu.unisinos.br, rafaelkunst@unisinos.br
[2] Universidade Federal de Santa Maria – UFSM, Av. Roraima, 1000, Santa Maria, RS, Brazil
{ceretta,luisalvaro}@inf.ufsm.br
[3] Universidade Federal do Rio Grande do Sul – UFRGS, Farroupilha, Porto Alegre, RS, Brazil
epfreitas@inf.ufrgs.br

Abstract. The Internet of Things (IoT) is impacting several areas. Using sensors and actuators in different contexts, such as smart cities, industry 4.0, healthcare, and agriculture, creates new interactions that can only occur due to connecting devices. As IoT devices join the military domain, new problems arise, like security and authentication in unreliable networks. This article proposes a novel approach for continuous user authentication to improve wearable device security in the Internet of Battle Things (IoBT) context. The proposed approach uses a Recurrent Neural Network (RNN) to directly optimize the embedding using the triplet loss. The created embedding represents the authorized user gait as a n-dimensional vector and enables comparison with other users' gait using a L_2 distance corresponding to a measure of gait similarity. The proposed system achieves a 92.12% rank-1 accuracy on user identification and a 12.33% equal error rate for the user validation task.

1 Introduction

Data is generated at an unprecedented rate, changing how people relate to the world. Modern technologies allow people to communicate and collaborate, making extensive amounts of information and data available through the Internet. Since data comes from heterogeneous sources, many technological trends arise, including the Internet of Things (IoT), the increase in cloud providers and services, and the spread of smart devices, including wearable devices [1].

According to Suri et al. [2], IoT is an interdisciplinary field that combines networks, embedded hardware and software, detection technologies, information management, data analysis, and visualization. Kott et al. [3] point out that

L. Barolli (Ed.): AINA 2024, LNDECT 202, pp. 360–371, 2024.
https://doi.org/10.1007/978-3-031-57916-5_31

several sectors such as health, government, and industry have used these technologies, and one of them has shown considerable interest, the military sector. Suri et al. [2] also claim that IoT concepts originated in the defense community, resulting from pioneering work on sensor networks and light low-power computing platforms.

The defense community needs to harness the benefit of IoT to improve the capabilities of battlefield combat and efficiently manage war resources. This emerging area that uses IoT technology for defense "things" is called the *Internet of Battle Things (IoBT)*. In battlefield scenarios, communications between strategic war assets, such as aircraft, warships, armored vehicles, ground stations, and soldiers, can lead to better coordination, which can be performed by IoBT [4].

The problem in battlefield coordination presents a scenario where information is crucial and cannot be available to enemies. Therefore, security measures must ensure authenticity and integrity. Security solutions typically include authentication systems involving passwords, key cards, fingerprints, face recognition, and graphical passwords. However, users can forget passwords and lose a key card, where passwords are vulnerable to brute force attacks, stain attacks, and dictionary attacks [5]. Dealing with such problems is fundamental in the military context. This concern becomes even more important considering the networked applications used by the troops in tactical edge networks.

In this scenario, the vulnerabilities can be mitigated with a new security layer using Continuous Authentication (CA), a paradigm based on constantly passively verifying the user without the need to carry another task or another device to authenticate a user [6,7]. In Continuous Authentication systems, user information is continuously acquired and processed. If a device is stolen or lost, it should block the unauthorized user. To achieve this CA systems can use behavioral biometric traits to extract information that does not require user input or action, including motion, gait, keystroke dynamics, touch gesture, voice, and multimodal methods [8]. As a result, behavioral biometrics can increase security in a mobile and rapidly changing scenario. Using behavioral biometrics for CA is useful as a form of complementary technology or second-factor authentication.

This work proposes a novel method for user identification and verification through a new layer of network authentication. Using wearable three-axis accelerometers to obtain data, this work presents an embedding to represent the user's gait behavior. This proposal uses a Recurrent Neural Network (RNN) trained with the triplet loss to generate a n-dimensional embedding. The proposed approach enables a wearable device to be continuously authenticated for short periods, even when the network is unavailable. It means that if someone not authorized tries to access the device - even with the password - the access will be denied as soon as the device detects variations in the user's usage pattern. Consequently, the system will prevent enemies from accessing sensitive information.

The main contributions of this work are:

- Implement a Recurrent Neural Network as a behavioral biometric encoder. With this approach, the RNN can be used in a continuous authentication environment and encoder and compare system users;
- The RNN can create embeddings that can be saved and compared later on. The network is trained using the triplet loss [9] to learn the separation between users. This way, new users do not trigger new training for the RNN encoder.

The proposal is evaluated considering three embedding dimensions: 32, 64, and 128. It considers user verification and user identification tasks for performance evaluation. For the verification task, it compares the performance of the different embedding sizes in comparison to related work, considering metrics like validation rate (VAL), false accept rate (FAR), false rejection rate (FRR), and equal error rate (EER). These metrics are assessed using a L_2 distance through different thresholds. This work compares the recognition rate (RR, or rank-1 accuracy) for the user identification task, attaining a rate of 92.12%.

The remainder of this paper is organized as follows. Section 2 discusses related work. Section 3 presents the proposed approach. Section 4 discusses the results, and Sect. 5 concludes the paper by presenting final remarks and directions for future work.

2 Related Work

Similar solutions in the literature employ deep learning to user authentication using inertial sensors [10,11]. This application of machine learning is part of a system that implements a continuous authentication mechanism, allowing wearable devices to have an additional security layer. Continuous authentication is an implicit user validation process that captures a user's behavior using various sensors. One method that is gaining increasing interest is gait-based authentication. Gait recognition identifies an individual by collecting data from environmental and wearable sensors [8]. This process verifies the user's authentication periodically or constantly, using biometrics without interfering with the device's operation.

Gait biometrics involves identifying a user based on their walking patterns. Generally, gait recognition can involve two types of data: (I) a sequence of images or a time series generated from inertial sensors. Accelerometers and gyroscopes are standard solutions to gather data while the subject is walking [11]. A continuous authentication system for smartphone users proposed by [12] used smartphone unlabeled accelerometer movement patterns collected in an open environment. The collected data's feature extraction and analysis process was performed to train the models, reducing the total number of features based on correlation and applying data normalization. The authors applied K-means clustering and a Random Forest classifier to identify the phone-usage context automatically. Five machine learning methods, one for each context, were trained to classify genuine users and impostors using data collected from 57 users. Random Forest performed best, with a mean equal error rate of 5.6%.

Shankar et al. [13] presents a continuous authentication scheme composed of five processes: (I) data collection from accelerometers and events such as user's data entry, (II) data standardization using min_max and z-score methods, (III) data extraction characteristic, where the solution extracts time-domain features like maximum amplitude, minimum amplitude, variance, mean, and median, (IV) the selection of characteristics using the Butterfly Optimization Algorithm, and finally (V) the identification of users using Deep Auto Encoder and Softmax Regression techniques. The authentication proposed by the authors achieved an equal error rate of 0.05% while walking and 0.03% while standing.

Abuhamad et al. [14] propose AUToSen, a deep learning based on continuous authentication. Accelerometer, gyroscope, magnetometer, elevation, and screen touch data collected in the experiments went through a process to remove noise, normalize the data, align time, and prepare to train a RNN with Long Short-Term Memory that will authenticate the user. The experiments showed that only three sensors (accelerometer, gyroscope, and magnetometer) were necessary to enable authentication with an F1-score of approximately 98%.

Zou et al. [11] study gait recognition using smartphones in the wild. They use inertial sensors equipped by most smartphones, like accelerometers and gyroscopes, to record the data. After collecting data, the authors split data between walking and non-walking sessions because only walking data is helpful for gait extraction. The authors modeled the problem as a time series segmentation problem to perform the gait data extraction and create a U-Net. After that, they clean the gait cycles to remove the phone's orientation and perform user identification and authentication using a fusion between convolutional and RNN. The experiment reports accuracies above 96% for user authentication and identification.

Another framework for user authentication using mobile devices based on gait analysis is presented by Giorgi et al. [10]. The proposed framework handles all the continuous authentication stages, including data collection, data preprocessing, classification, and policy enforcement, using a pipeline of four components. The experiment preprocesses the data by exploiting noise filters, sliding windows for sampling, and data augmentation using permutation, scaling, and jittering to improve data quality and increase the number of training samples. The user verification step implements both supervised and unsupervised learning. Supervised learning considers double sampling to decide whether or not to authenticate the user. The unsupervised learning approach implements an encoder-decoder architecture. The authors report accuracies above 96% for supervised learning and 79% for unsupervised learning and equal error rates lower than 10% and 22%, respectively.

Through those works, continuous authentication using gait analysis on devices with inertial sensors has been combined with several machine-learning techniques. It shows that checking behavioral attributes implicitly and periodically enables learning algorithms to perform user verification and validation.

This study proposes a way to avoid retraining the model every time a new user is added to the database by learning embeddings capable of separating users.

Another contribution is by minimizing the pre-processing requirements for time series data, the RNN can model the relation in the time series and create an embedding using the full context without the need to summarize the data. The following section will detail the proposed solution in this context.

3 Proposed Solution: GRU Encoder

The proposed solution involves authenticating communicating users using a Tactical Edge Network (TEN) [15]. Scenarios in which TENs can be applied include (but are not limited to) borderline surveillance and the exploration of enemy-occupied areas. Environmental access may be limited or restricted, and no communication infrastructure frequently exists. The communication network in these scenarios is inherently dynamic due to mobile nodes such as small UAVs, ground vehicles, and the devices carried by military personnel [16], meaning that nodes are constantly connecting and disconnecting from the network. Therefore, continuous authentication is crucial to guarantee that the communication remains safe as military operatives move across several network partitions that occasionally and opportunistically interconnect, as seen in Fig. 1.

The machine learning-based proposal to implement continuous authentication is presented in Fig. 2 and detailed in the following subsection. Subsection 3.1 presents the dataset that feeds the RNN, which we detail in Subsect. 3.2. Subsection 3.3 presents the proposed embedding to authenticate the users.

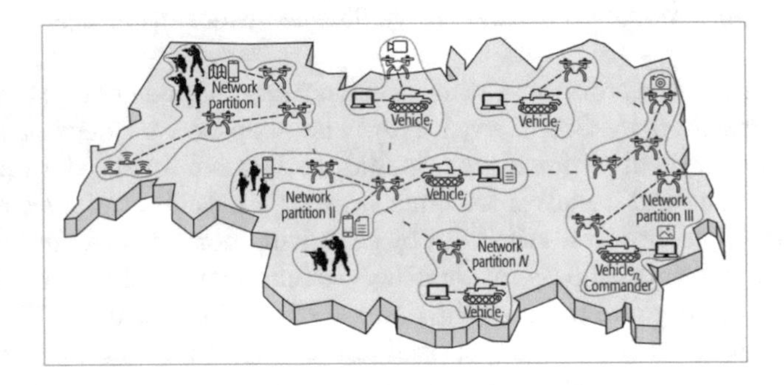

Fig. 1. Network Scenario [16]

3.1 Dataset

The used dataset is the ZJU-GaitAcc detailed in [17]. This dataset measures gait acceleration using Wii Remotes fastened at five body locations. According to the authors, they are placed to cover the essential articulation structures of the body: the left upper arm, the right wrist, the right side of the pelvis, the left thigh, and the right ankle. The time series generated by those sensors are manually annotated with the user segments and motion cycles. The data is sampled at 100 Hz, and the collection process was divided into two sessions with a time interval varying from 1 week to 6 months.

There are 175 subjects in the dataset, and 153 of them are present in two sessions. The 22 remaining volunteers were only present in one session. This behavior of the subjects led the authors to divide the records into three sessions. Session 0 includes data from the 22 volunteers whose data is unavailable in both sessions, while sessions 1 and 2 contain 153 subjects. In each session, the participant was recorded straightly walking six times, so each record is composed of five-time series, one for each Wii Remote. This work uses sessions 1 and 2 for the neural network training and session 0 to evaluate the embedding performance.

3.2 Recurrent Neural Network

The proposed solution relies on a recurrent neural network (RNN) architecture to create an embedding from a time series collected from an accelerometer sensor. The goal is to create an embedding $f(x)$ from a time series x into a feature space $\mathbb{R}^d$ where measures from the same person are grouped. This approach adapts the method described by [9], which uses a convolutional neural network (CNN) to create facial embedding, and the network is trained using the triplet loss. This work uses an RNN to create embedding for gait measures with the triplet loss and evaluate user authentication and verification task results.

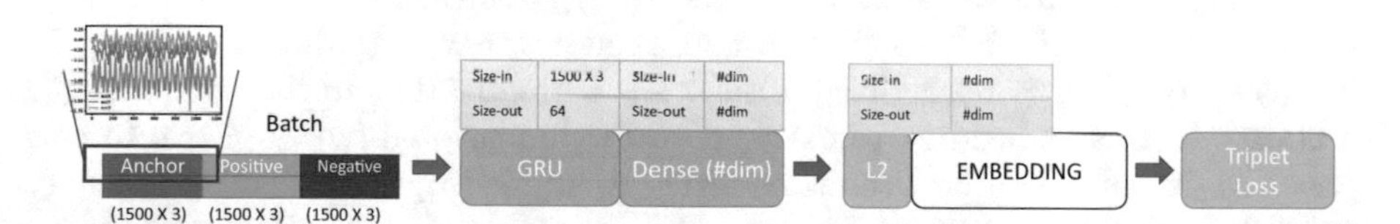

Fig. 2. Model structure. The network consists of an input layer, a GRU layer followed by a dense layer, and a L_2 normalization. The dense layer defines the embedding dimensions. In this work, we use 32, 64, and 128 output units; in the image, it is the #dim. The model uses the triplet loss during training. The input layer receives three recordings: an anchor, a positive, and a negative. The anchor and the positive belong to the same person, while the negative belongs to someone else.

In the performed experiments, the embedding creates a RNN based on a Gated Recurrent Unit (GRU) introduced by [18]. Figure 2 presents the proposed model structure. The input data comprises a 2D vector of dimension 3 X 1500, corresponding to the three recorded axes and 1500 time steps. The network uses a GRU with 64 output units and a dense layer with the embedding size as output units. Three embedding dimensions are used in the experiments: 32, 64, and 128. This vector is constrained to live a d-dimensional hypersphere, so a l_2 normalization is used to help with training stability as seen in [9,19].

The network training considers the triplet loss defined by Schroff et al. [9]. This loss ensures that an anchor time series of a specific person is closer to all other series belonging to the same person than any series belonging to someone else. So, during training, a triplet is created containing an anchor (x_i^a), a positive example (x_i^p), and a negative example (x_i^n). The triplet loss function follows Eq. 1.

$$L = \sum_{i}^{N} \left[\|f(x_i^a) - f(x_j^p)\|_2^2 - \|f(x_i^a) - f(x_j^n)\|_2^2 + \alpha \right]_+ \quad (1)$$

This study randomly selects the positive and negative samples for each batch. Moreover, we train our neural network using the Adam optimizer [20], embedding dimensions 32 and 64 and a batch size of 512 samples. This work considers 10% dropout on the GRU layer during training and a leaky rectified linear unit (Leaky ReLU) as the activation function on the dense layer.

3.3 Embedding

The embedding in this work is based on learning an Euclidean embedding for time series so that the Euclidean distance between the series corresponds to gait similarity. Measuring the distance between time series is helpful for various analysis algorithms [21]. The proposal consists of a learning approach using a deep learning method to create an embedding space where distance metrics cluster time series from the same individual. The embedding dimension is defined by the number d of outputs from the dense layer. So, it is possible to define an embedding of a data point $x_i \in \mathbb{R}^N$ as $f(x_i)$, where x_i is the time series to be embedded and $f : \mathbb{R}^N \rightarrow \mathbb{R}^d$ is a neural network with parameter Θ [19]. In this method, $f(x_i)$ is normalized with a L_2 normalization to have unit length to constrain this embedding to exist on the d-dimensional hypersphere. In this space the distance $D_{i,j}$ between time series x_i and x_j represents their similarity and is defined as the euclidean distance $D_{ij} = ||f(x_i) - f(x_j)||_2$. This embedding is created using the loss function defined in (1).

This work experiments with different embedding dimensions to verify which is best for the proposed tasks. So, the proposed embedding is a 1D vector with d-positions representing a time series (Fig. 2). In the verification task, the L_2 distance between the embedding of two-time series is used to classify them as belonging to the *same* person or to *different* people. As in [9], all pairs (i, j) of the same identity are denoted $\mathcal{P}_{\text{same}}$, where all pairs of different identity are $\mathcal{P}_{\text{diff}}$.

3.4 Evaluation Metrics

We evaluate the proposed method on the user identification task on the ZJU-GaitAcc dataset [17]. Given a pair of two time series from accelerometer data we calculate a L_2 distance and use a threshold to determine if they are from the same or different users.

A *true accept* is defined as the distance $D(i, j)$ is less or equal then given a threshold d as:

$$\text{TA}(d) = \{(i, j) \in \mathcal{P}_{\text{same}}, \text{ with } D(x_i, x_j) \leq d\} \quad (2)$$

and the pairs that were incorrectly classified as *same* as:

$$\mathrm{FA}(d) = \{(i, j) \in \mathcal{P}_{\mathrm{diff}}, \text{ with } D(x_i, x_j) \leq d\} \tag{3}$$

these are called *false accept*. Here, a set of pairs that should have been classified as *different* is added:

$$\mathrm{FN}(d) = \{(i, j) \in \mathcal{P}_{\mathrm{same}}, \text{ with } D(x_i, x_j) > d\} \tag{4}$$

but should have been classified as *same* (*false negative*).

Those sets are used to calculate the validation rate $\mathrm{VAL}(d)$ as:

$$\mathrm{VAL}(d) = \frac{|\mathrm{TA}(d)|}{|\mathcal{P}_{\mathrm{same}}|} \tag{5}$$

it measures the likelihood that the authentication system will accept an authorized user correctly. The false accept rate $\mathrm{FAR}(d)$ as:

$$\mathrm{FAR}(d) = \frac{|\mathrm{FA}(d)|}{|\mathcal{P}_{\mathrm{diff}}|} \tag{6}$$

it measures the likelihood that the authentication system will incorrectly accept access by an unauthorized user. The false rejection rate $\mathrm{FRR}(d)$ as:

$$\mathrm{FRR}(d) = \frac{|\mathrm{FN}(d)|}{|\mathcal{P}_{\mathrm{same}}|} \tag{7}$$

it measures the likelihood that the authentication system will reject an authorized user incorrectly.

The FAR and the FRR are also used to calculate the equal error rate (EER). The (EER) is the point in the plot where the FRR and FAR are equal. It describes the overall accuracy of a biometric system. As seen in Fig. 4.

For the identification task, a recognition rate (RR, or rank-1 accuracy) is used, as seen in [17], so the results are comparable. The RR is calculated for the holdout set and is defined as the closest embedding using a L_2 distance as

$$\mathrm{RR} = \frac{\text{No of correct classified}}{\text{Total Samples}} \tag{8}$$

where correct means the closest embedding belongs to the user and total samples.

4 Results

This section presents the results of applying the proposed solution to the ZJU-GaitAcc. Data from sessions 1 and 2 are used to train the neural network and leave session 0 as a hold-out test set. This set has 22 people with six records, each containing five sensors, resulting in 660 examples in this validation. The method is evaluated on the user verification and the user identification.

4.1 Results Evaluation

The results present how the embedding size influences the network's user verification and identification performance.

In Fig. 3, it is possible to observe the detailed evaluation of the user verification metrics on the test set using three embedding dimensions. It shows that the embedding with 64 dimensions works better for the user verification task.

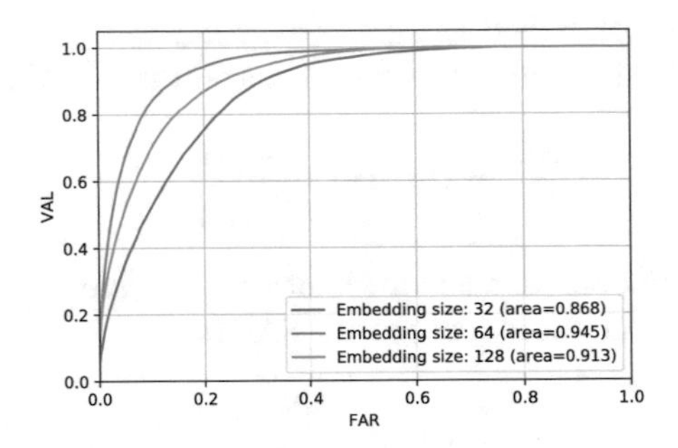

Fig. 3. This plot shows the complete ROC for the three embedding dimensions on the test set.

Figure 4 displays the curves FAR and FRR that each embedding dimension produced and where they cross each other. This point is the EER. One can see that the result follows the ROC curve seen before (Fig. 4). The embedding with 64 dimensions has the lowest EER at 12.33%, while the worst embedding is the one with 32 dimensions with a EER at 21.63%. Table 1 compares the acquired results with those presented in the original paper [17].

The RR in the hold-out set is used for the user identification task. The embedding with dimension 32 has RR = 61.21% the 64 dimension has RR = 87.88% and the 128 units embedding has RR = 92.12%. These results are reported in Table 1. The embedding dimension size directly impacts the model's ability to identify the user, and the embedding with 128 units achieved the best performance.

Different embedding dimensions were explored and compared to the performance of user verification and user identification as reported in Table 1. One would expect the larger embeddings to perform at least as well as the smaller ones; however, the larger one performed better in the user identification task, and the one with 64 dimensions performed better in the user verification task.

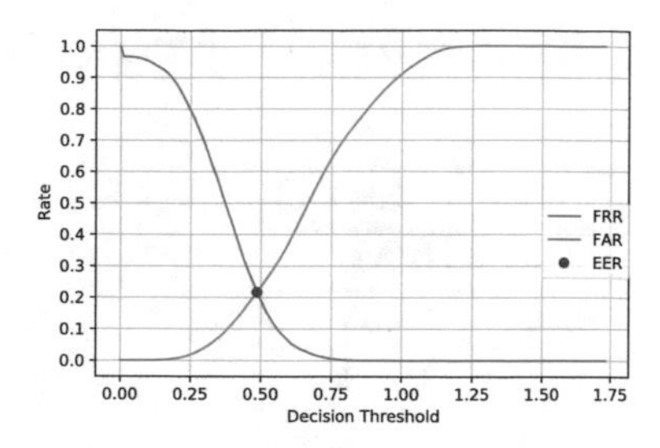

(a) FAR and FRR curves for the embedding with 32 dimensions (EER = 0.2163).

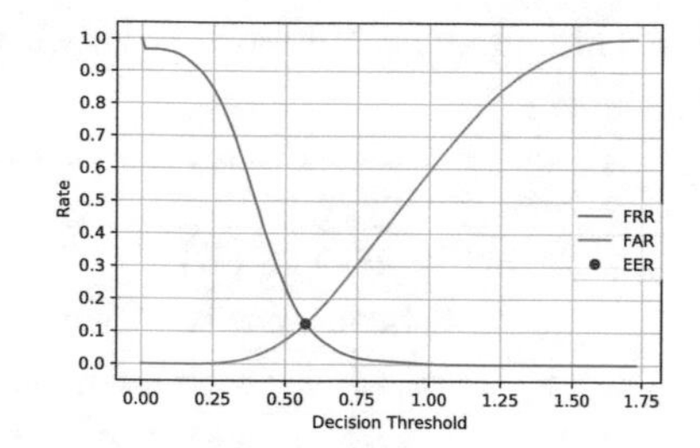

(b) FAR and FRR curves for the embedding with 64 dimensions (EER = 0.1233).

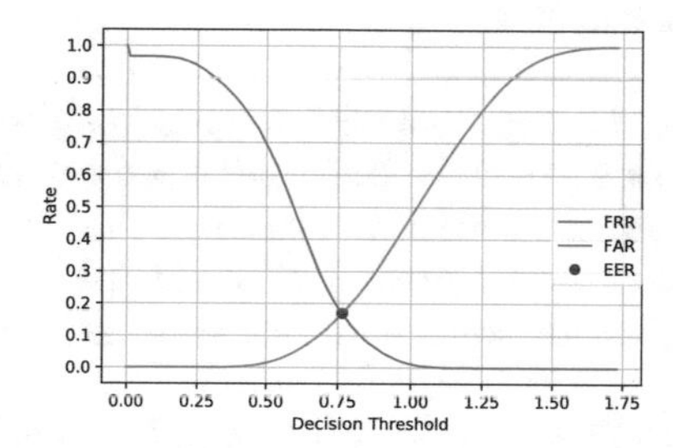

(c) FAR and FRR curves for the embedding with 128 dimensions (EER = 0.1691).

Fig. 4. Equal error rate (EER) for each embedding dimension.

4.2 Results Comparison

The proposed method was compared with the results in the original work that introduced the ZJU-GaitAcc [17]. The proposed embedding was trained to be independent of the body location of the sensor, so each time series stands as a user sample. The original work uses a method to group more than one body location to increase performance. Due to this difference, their single-body location was compared with the single-location performance of this work. The average performance for each body location in their work was calculated so that it was possible to have a better comparison.

Table 1 reports all results of the proposed solution regarding both tasks, verification with the EER and identification with RR. It also reports the performance of the original ZJU-GaitAcc work. The original work performs better at the verification task, whereas the model proposed performs better at the identification task.

Table 1. Comparison between the acquired result and those reported by Zhang et al. [17]

Methods	RR (%)	EER (%)
Zhang et al. [17][a]	66.16	**9.92**
Embedding with 32 dimensions	61.21	21.63
Embedding with 64 dimensions	87.88	12.33
Embedding with 128 dimensions	**92.12**	16.91

[a] Average from their single body locations results.

5 Concluding Remarks

As the amount of mobile and wearable devices grows on the battlefield, solutions like the one discussed in this paper are essential to protect sensitive data from the enemy. Authentication systems have flaws, so research to increase device security is always necessary. The proposal presented in this paper can be implemented along with other measures to increase the overall performance of the security system.

This proposal consists of a method using neural networks to map time series to a compact Euclidean space where distances correspond to a similarity measure between user gait. This is based on data generated from inertial sensors, and it is done with the end goal of deploying a new security layer that can work offline as well with authentication computing done at the edge. As generating the embedding compresses the original dataset, less storage is needed to identify the user gait.

With this method, it was possible to achieve a recognition rate of 92.12% for the largest embedding and an equal error rate of 12.33% for the embedding with 64 dimensions. So, deciding which embedding dimension is better depends on the task that needs to be solved.

For future work, there is a need to apply this method to other datasets and experiment with larger networks. The rationale for this need is that the model is expected to cover situations not covered in the ZJU-GaitAcc. This addition can make the model more robust and cover other types of devices.

Acknowledgments. We thank the Brazilian Army Strategic Program ASTROS for financial support through the SIS-ASTROS GMF project (898347/2020) - TED 20-EME-003-00.

References

1. Oussous, A., Benjelloun, F.Z., Lahcen, A.A., Belfkih, S.: J. King Saud Univ. Comput. Inf. Sci. **30**(4), 431 (2018). https://doi.org/10.1016/j.jksuci.2017.06.001. http://www.sciencedirect.com/science/article/pii/S1319157817300034
2. Suri, N., et al.: In: 2016 International Conference on Military Communications and Information Systems (ICMCIS), pp. 1–8 (2016)

3. Kott, A., Swami, A., West, B.J.: Computer **49**(12), 70 (2016)
4. Farooq, M.J., Zhu, Q.: In: 2017 15th International Symposium on Modeling and Optimization in Mobile, Ad Hoc, and Wireless Networks (WiOpt), pp. 1–8 (2017)
5. Shafique, U., et al. (2017)
6. Patel, V.M., Chellappa, R., Chandra, D., Barbello, B.: IEEE Signal Process. Mag. **33**(4), 49 (2016). https://doi.org/10.1109/MSP.2016.2555335. https://ieeexplore.ieee.org/document/7503170. Conference Name: IEEE Signal Processing Magazine
7. Stragapede, G., Vera-Rodriguez, R., Tolosana, R., Morales, A., Acien, A., Le Lan, G.: Pattern Recognit. Lett. **157**, 35 (2022). https://doi.org/10.1016/j.patrec.2022.03.014. https://www.sciencedirect.com/science/article/pii/S016786552200071X
8. Abuhamad, M., Abusnaina, A., Nyang, D., Mohaisen, D.: arXiv preprint arXiv:2001.08578 (2020)
9. Schroff, F., Kalenichenko, D., Philbin, J.: CoRR **abs/1503.03832** (2015). http://arxiv.org/abs/1503.03832
10. Giorgi, G., Saracino, A., Martinelli, F.: Pattern Recogn. Lett. **147**, 157 (2021)
11. Zou, Q., Wang, Y., Wang, Q., Zhao, Y., Li, Q.: Deep learning-based gait recognition using smartphones in the wild (2020)
12. Kumar, R., Kundu, P.P., Shukla, D., Phoha, V.V.: In: 2017 IEEE International Joint Conference on Biometrics (IJCB), pp. 177–184. IEEE (2017)
13. Shankar, V., Singh, K.: IEEE Access **7**, 48645 (2019)
14. Abuhamad, M., Abuhmed, T., Mohaisen, D., Nyang, D.H.: IEEE Internet Things J. (2020)
15. Tortonesi, M., Morelli, A., Stefanelli, C., Kohler, R., Suri, N., Watson, S.: IEEE Commun. Mag. **51**(10), 66 (2013). https://doi.org/10.1109/MCOM.2013.6619567
16. Zacarias, I., Gaspary, L.P., Kohl, A., Fernandes, R.Q.A., Stocchero, J.M., de Freitas, E.P.: IEEE Commun. Mag. **55**(10), 22 (2017). https://doi.org/10.1109/MCOM.2017.1700239
17. Zhang, Y., Pan, G., Jia, K., Lu, M., Wang, Y., Wu, Z.: IEEE Trans. Cybernet. **45**(9), 1864 (2015). https://doi.org/10.1109/TCYB.2014.2361287. Conference Name: IEEE Transactions on Cybernetics
18. Cho, K., van Merrienboer, B., Gülçehre, Ç., Bougares, F., Schwenk, H., Bengio, Y.: CoRR **abs/1406.1078** (2014). http://arxiv.org/abs/1406.1078
19. Wu, C.Y., Manmatha, R., Smola, A.J., Krahenbuhl, P.: In: Proceedings of the IEEE International Conference on Computer Vision (ICCV), pp. 2840–2848 (2017)
20. Kingma, D.P., Ba, J.: (2017). http://arxiv.org/abs/1412.6980
21. Shanmugam, D.: A tale of two time series methods: representation learning for improved distance and risk metrics. Thesis, Massachusetts Institute of Technology (2018). https://dspace.mit.edu/handle/1721.1/119575

SovereignRx: An Electronic Prescription System Based on High Privacy, Blockchain, and Self-Sovereign Identity

Maurício de Vasconcelos Barros(✉) and Jean Everson Martina

Programa de Pós-Graduação em Ciência da Computação, Departamento de Informática e Estatística, Universidade Federal de Santa Catarina, Florianópolis, SC 88040-900, Brazil
mauricio.barros@posgrad.ufsc.br, jean.martina@ufsc.br

Abstract. This work provides a comprehensive review and critical analysis of blockchain-based electronic medical prescription systems, with a focus on the Self-Sovereign Identity (SSI) paradigm. The comparative analysis will look at major implementations, covering features such as identification methods, data storage techniques, privacy measures, access controls, and interoperability. Risks due to the permanence of data on a blockchain and possible improvements in computing in terms of scalability and speed are presented, along with a discussion on stronger data protection measures. We introduce "SovereignRx", a novel system that integrates SSI and blockchain technologies to address privacy concerns in healthcare prescriptions. SovereignRx uses Decentralized Identifiers (DIDs) for party identification and uses distributed nodes that reduce the maintenance costs between healthcare entities. The proposed architecture, based on Hyperledger projects (Indy, Aries, and Fabric), supports secure data transportation, tracing of dispensation, and interoperability based on the FHIR standard. It utilizes data minimization, selective disclosure and biometric authentication to enable the protection of prescription information, with dispensation control and audit functionalities. Furthermore, challenges of decentralized identity adoption are addressed, along with future improvements in the usability of systems for digital health care.

Keywords: Self-Sovereign Identity · Blockchain in Healthcare · Digital Medical Prescriptions

1 Introduction

The recent COVID-19 pandemic [44] has accelerated digitalization in healthcare, including a significant increase in telemedicine [43]. However, this rapid transition has presented challenges, particularly in privacy. Health data, being sensitive, requires compliance with laws such as the GDPR in Europe [1], or HIPAA in the United States [2].

L. Barolli (Ed.): AINA 2024, LNDECT 202, pp. 372–383, 2024.
https://doi.org/10.1007/978-3-031-57916-5_32

In this scenario, classic identities and documents are becoming digital [3,38], including medical prescriptions. Paper prescriptions, subject to tampering and errors, may be responsible for up to 30% of the problems in hospitalizations [6]. Electronic prescribing, defined as the exchange of prescription information digitally [39], reduces errors and manipulations.

Many current electronic prescribing systems are centralized [19], susceptible to single points of failure and privacy issues. Centralized systems can be expensive and opaque [24]. In contrast, an implementation with Blockchain [9] can offer greater independence from single controllers, as well as high security, integrity, transparency, and shared costs [28], in addition to efficient auditing, vital for medication monitoring.

Self-Sovereign Identity [7], where the user controls their data, is suitable for sensitive data such as health information. A system that combines Blockchain and Self-Sovereign Identity focuses on privacy and ownership of health data. With data minimization, the user shares only what is necessary, maintaining control over their information.

This work is structured as follows: Concepts of Self-Sovereign Identity and Blockchain are discussed in Sects. 2 and 3, respectively. A systematic review of Electronic Prescriptions and Blockchain is presented in Sect. 4. A proposed model for Electronic Prescription is detailed in Sect. 5, followed by conclusions in Sect. 6 and challenges and future work in Sect. 7.

2 Digital Identity

As characterized in the ISO 24760-1 document, focusing on security and privacy in identity management, the concept of Digital Identity is defined as "a set of attributes related to an entity" [4]. On the other hand, [5] conceptualizes Digital Identity as an entity's representation in a particular context. In this way, Digital Identity can be formulated as a set of attributes of an entity correlated to a particular context.

The Digital Identity concepts are usually managed within centralized identity management models, putting them under the control of Identity Providers, who own the data associated with these identities, leaving the users without the direct ownership of their digital identity.

2.1 Self-Sovereign Identity

In the Self-Sovereign Identity approach, user-centric, the identity is independent of services or Identity Providers [8]. The user holds their data and digital sovereignty, with usual implementation via Blockchain. It includes three participants: the issuer, the holder, and the verifier. The Distributed Ledger serves as a trust validator.

Entities are identified by Decentralized Identifiers (DID), autonomous and controlled by the owner for a verifiable digital identity [11]. A DID, standardized by W3C, can identify various entities, associated with a DID Document, a JSON detailing public keys and service endpoints, among other data.

DIDs and DID Documents on the Blockchain, with an appropriate governance architecture, ensure persistence and resistance to censorship. Verifiable Credentials, linked to the DID, are verified against the DID Document. These credentials include data about the subject, issuer, type of credential, and alleged attributes [10], digitally representing physical credentials, with metadata for cryptographic authentication of the issuer.

3 Blockchain

Satoshi Nakamoto, the pseudonym of the creator of Blockchain, introduced Bitcoin in 2008 as a peer-to-peer electronic cash system [9]. He criticized the reliance of digital commerce on large financial institutions and presented Blockchain as an alternative, using a distributed timestamp server to chronologically validate transactions.

Blockchain, is an interconnected sequence of blocks containing transactions, the hash of the previous block, and a nonce. Transactions are stored in a Hash Tree, allowing for quick verifications without downloading all transactions. The validity of blocks is ensured across the network through consensus algorithms like Proof of Work or Proof of Stake.

Following Nakamoto's Bitcoin, new Blockchain formats emerged, focusing on scalability and energy efficiency, and the introduction of smart contracts on the Ethereum network [12].

4 Blockchain and Electronic Medical Prescriptions

A systematic review was conducted to assess the use of Blockchain in Electronic Medical Prescriptions, emphasizing the paradigm of Self-Sovereign Identities. The investigation focused on improving patient data privacy compared to traditional methods.

The research was guided by the following question: "How are Self-Sovereign Identities and Blockchain being applied in Electronic Medical Prescriptions?" The search was conducted in October 2022 using the *string*:

("E-Prescription" OR "Electronic Prescription" OR "Digital Prescription") AND ("Blockchain" OR "Verifiable Credentials" OR "Verifiable Credential" OR "SSI" OR "Self Sovereign Identity")

The databases consulted included IEEE Xplore [15], ACM Digital Library [16], ScienceDirect [17], and Springer Link [18]. The initial search resulted in 60 articles, as shown in Table 1. Inclusion and exclusion criteria were applied, as detailed in Table 2.

After applying these criteria, 11 articles remained. A Snowballing technique was used, resulting in an additional 275 articles. After the exclusion of duplicates and re-application of the criteria, 14 articles were included in the final review.

Table 1. Search results by database.

Database	Results
IEEE Xplore	28
ACM Digital Library	1
Science Direct	11
Springer Link	20
Total	60

Table 2. Inclusion and exclusion criteria.

Inclusion Criteria	
CI-1	Works that propose Electronic Prescriptions using Blockchain.
Exclusion Criteria	
CE-1	Work not available
CE-2	Secondary study
CE-3	Electronic Prescription is not the central theme

4.1 Data Extraction

To facilitate future references, Table 3 assigns an ID to each selected work. These works were analyzed to answer the research question. In Table 4, various aspects are compared: storage of the patient's personal data, location, storage method and authentication for accessing this data, type of Blockchain used, use of the Self-Sovereign Identity paradigm, treatment of medication dispensing and its audit, and use of interoperability methods such as HL7 [41] or FHIR [40].

Many implementations use similar approaches. Works A1, A2, A3, A4, A6, A7, A13, and A14 store encrypted prescription data in a smart contract, accessed by a private key.

Works A4, A6, A7, A13, and A14 do not store patients' personal data, increasing privacy. On the other hand, A11 and A9 do not focus on privacy, storing data in plain text, with A9 allowing access without authentication. A5 and A8 are exceptions, not storing prescription data on the Blockchain, but using the distributed ledger for search and access control.

Explicit patient consent is required in A1, A2, A5, A6, A7, A8, A13, and A14. However, only A1, A2, A6, A7, and A14 use encryption with the patient's public key, ensuring greater control. Three types of Blockchain were used: Hyperledger Fabric [13], Ethereum [33], and a combination of CosmWasm [34] and Tendermint [35], with A4 using a public Ethereum Blockchain.

Most address the dispensation of medication, except A6 and A10. Only A5 and A6 adopt the FHIR interoperability model. No work cites or uses the Self-Sovereign Identity paradigm.

4.2 Discussion

This review analyzed various implementations of electronic medical prescriptions based on Blockchain, highlighting an emerging pattern in the management of

Table 3. Selected Works

ID	Article Name	Year	Reference
A1	A Blockchain-based Data Governance with Privacy and Provenance: a case study for e-Prescription	2022	[19]
A2	A New Blockchain-based Electronic Medical Record Transferring System with Data Privacy	2020	[20]
A3	A Secure Blockchain-based Prescription Drug Supply in Health-care Systems	2019	[21]
A4	Authentic Drug Usage and Tracking with Blockchain Using Mobile Apps	2020	[22]
A5	Blockchain Technology Use Cases in Healthcare	2018	[23]
A6	DMMS: A Decentralized Blockchain Ledger for the Management of Medication Histories	2019	[24]
A7	Exploiting smart contracts in PBFT-based blockchains: A case study in medical prescription system	2022	[25]
A8	OpTrak: Tracking Opioid Prescriptions via Distributed Ledger Technology	2020	[26]
A9	Pharmaceutical uses of Blockchain Technology	2018	[27]
A10	Providing Electronic Health Care Services Through A Private Permissioned Blockchain	2020	[28]
A11	RxBlock: Towards the design of a distributed immutable electronic prescription system	2020	[29]
A12	SecureRx: A blockchain-based framework for an electronic prescription system with opioids tracking	2021	[30]
A13	Towards a decentralized e-prescription system using smart contracts	2021	[31]
A14	VigilRx: A Scalable and Interoperable Prescription Management System Using Blockchain	2022	[32]

patient data. Distributed ledgers and encryption are used to enhance the privacy and security of data, with access conditioned on patient consent.

The studies explore the decentralized and secure features of distributed ledgers, but there is room for improvement. Many systems utilize smart contracts and on-chain storage, resulting in a large volume of operations on the blockchain, accompanied by various forms of encryption to store and share data.

There are risks to the immutability of on-chain data, especially personal and sensitive data, even when encrypted. Current protections may not be sufficient in the future, with advancements in conventional computing and the potential development of quantum computing [36].

There is also limited use of interoperability standards, such as HL7 and FHIR, indicating opportunities to improve integration with other healthcare systems.

Notably, none of the studies employed the Self-Sovereign Identity paradigm. A Self-Sovereign approach, avoiding on-chain storage, would reduce processing and storage consumption on the Blockchain, as well as eliminate the risks of future decryption of sensitive records. It would facilitate the portability of data across different healthcare providers and borders, giving patients absolute control over their data.

5 SovereignRx

Leveraging the autonomy and decentralization of Blockchain and seeking improvements from the analysis of the state of the art in digital medical prescriptions, a system named SovereignRx is proposed, based on the paradigm of Self-Sovereign Identity and with a focus on privacy.

In this system, a doctor issues a medical prescription as a Verifiable Credential, stored in the patient's mobile digital wallet. When presenting this Verifiable

Table 4. Comparative Table

ID	Identification	Where	Privacy	Access	Blockchain	SSI	Dispensing	Audit	FHIR
A1	Yes	Smart Contract	Encrypted Data	Private Key	CosmWasm + Tendermint	No	Yes	Yes	No
A2	Yes	Smart Contract	Encrypted Data	Private Key	Generic	No	Yes	Yes	No
A3	Yes	Smart Contract	Encrypted Data	Private Key	Generic	No	Yes	Yes	No
A4	No	Smart Contract	Encrypted Data	Private Key	Public Ethereum	No	Yes	Yes	No
A5	Yes	Off-chain Database	Secure Database	Private Key	Private Ethereum	No	Yes	Yes	Yes
A6	No	Smart Contract	Encrypted Data	Private Key	Hyperledger Fabric	No	No	No	Yes
A7	No	Smart Contract	Encrypted Data	Private Key	CosmWasm + Tendermint / Private Ethereum	No	Yes	Yes	No
A8	Not Specified	Off-chain Database	Not Specified	Biometrics/ Private Key	Private Ethereum	No	Yes	Yes	No
A9	Yes	Smart Contract	Plain Text Data	No Authentication	Private Ethereum	No	Yes	Yes	No
A10	Yes	Smart Contract	Not Specified	Private Key	Hyperledger Fabric	No	No	No	No
A11	Yes	Smart Contract	Plain Text Data	Role-Based Authentication	Private Ethereum	No	Yes	Yes	No
A12	Yes	Smart Contract	Encrypted Data	Role-Based Authentication	Private Ethereum	No	Yes	Yes	No
A13	No	Smart Contract	Encrypted Data	Private Key	CosmWasm + Tendermint / Hyperledger Fabric	No	Yes	No	No
A14	No	Smart Contract	Encrypted Data	Private Key	Private Ethereum	No	Yes	No	No

Credential to the pharmacist, the authenticity of the prescription is confirmed by the doctor's digital signature, verifiable through the doctor's DID registered in the Distributed Ledger.

SovereignRx uses data minimization and selective disclosure to protect prescription information, while biometric authentication can allow patient verification without exposing personal data.

5.1 System Overview

To implement the SovereignRx system, three Hyperledger projects are used: Hyperledger Indy, Hyperledger Aries, and Hyperledger Fabric, all open source.

Hyperledger Indy, a Blockchain specialized in Self-Sovereign Identities, is an appropriate choice for the prototype. In its Distributed Ledger, credential schemas, credential definitions, public DIDs, public keys, and cryptographic accumulators for credential revocation are stored, essential for verifying issuer signatures.

Hyperledger Aries, originally part of Indy, was separated into an independent project. Compatible with Indy, Aries manages the issuance of Verifiable Credentials and the peer-to-peer DIDComm communication protocol.

Finally, Hyperledger Fabric, a multipurpose private Blockchain, processes smart contracts (or chaincode) and allows the creation of private communication channels [14]. In the context of SovereignRx, Fabric is used for control and audit by regulatory entities, where private channels enable customization of data access, restricting it to auditors and inspectors.

5.2 Archiecture

The architecture of SovereignRx is detailed in Fig. 1. Doctors and pharmacists register their DIDs in Hyperledger Indy (1), obtaining a unique identity in the system.

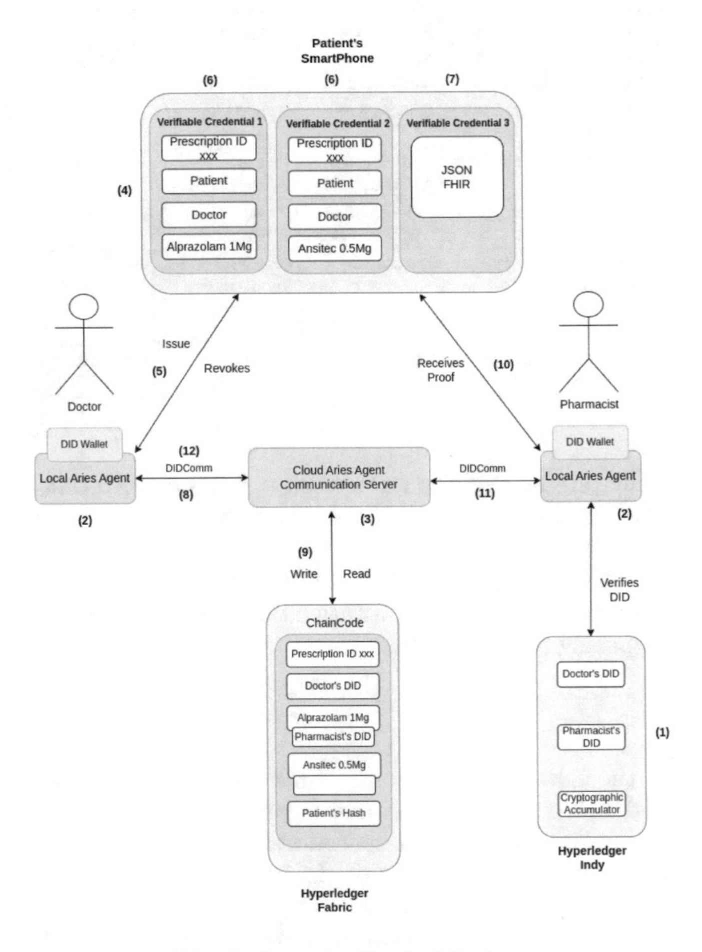

Fig. 1. SovereignRx Architecture

They operate an Aries Agent via ACA-Py framework [37], with doctors represented by Health Units and pharmacists by pharmacies (2). An Aries Agent server (3) acts as an intermediary, managing communications and operations in Hyperledger Fabric.

The doctor issues an electronic prescription as Verifiable Credentials, sent to the patient's digital wallet (5). Each medication in the prescription is a separate Verifiable Credential (6), allowing dispensations at different pharmacies. An additional credential (7) contains the prescription in FHIR format [40], ensuring interoperability with compatible systems.

The doctor's Aries Agent also sends data to the Aries Agent Server (8), including medication, the doctor's DID, prescription ID, and a patient identification hash, generated from the patient's personal data. A new smart contract is created in Fabric (9), as shown in code 1.1, controlling the dispensation and facilitating the audit.

```
1 #Call initiatePrescription()
2 Function: "initiatePrescription" Args: ["001", "5
      abXqmLc87oSRPbejbXAnQ", "6493751eea3097b571bde0fc2a14fca87...
      b25efb1cdcd15ad9f27f28ed8db", "Tylenol 500mg"]
3 #Response
4 2023-06-29 22:51:33.377 -03 0001 INFO [chaincodeCmd]
      chaincodeInvokeOrQuery -> Chaincode invoke successful. result:
       status:200
```

Code 1.1. *chaincode* is initiated with the prescription ID, doctor's DID, patient's *hash*, and medication.

At the pharmacy, the patient presents the prescription via Aries Agent (10). The pharmacist verifies the Verifiable Credentials and authenticates the patient. When dispensing a medication, the pharmacist's Aries Agent informs the Aries Agent Server (11), which requests the revocation of the corresponding credential (12). The pharmacist's DID is associated with the dispensed medication in the Smart Contract on Fabric, as seen in the Code 1.2. The process can be repeated for other medications.

```
1 #Call ConsultPrescription()
2 "function":"ConsultPrescription()", "Args":["001"]
3 #Response
4 2023-06-29 22:57:53,198 -03 0001 INFO [chaincodeCmd]
      chaincodeInvokeOrQuery -> Chaincode invoke successful. result:
       status:200 payload: {"prescriptionId":"001","doctorDID":"
      SabXqmLc87oSRPbejbXAnQ","patientHash":"6493751
      eea3097b571bde0fc2a14fca87...b25efb1cdcd15ad9f27f28ed8db","
      medication":"Tylenol 500mg","pharmacistDID":"
      jH6gFC8LU8HfgEb0QjTfd"}
```

Code 1.2. The dispensed medication is associated with the Pharmacist's DID.

Repeated dispensation attempts generate errors (Code 1.3). The record in Fabric assists in dispensation and auditing, maintaining control of potential abuses, although patients are pseudonymously identified.

```
1 #Call DispenseMedication()
2 "function":"DispenseMedication", "Args":["001", "
      jH6gFC8LU8HfgEb0QjTfd"]
3 #Response
4 Error: endorsement failure during invoke. response status:500
      message:"The medication has already been dispensed for the
      prescription 001"
```

Code 1.3. The pharmacist cannot dispense medication that has already been dispensed.

This architecture with Hyperledger Fabric also allows for expansion into the management of the medication supply chain, with separate communication channels in Fabric allowing distinct functions such as dispensation auditing and supply chain control.

6 Conclusion

In conclusion, this study provides a thorough examination of blockchain-based electronic medical prescription systems, especially focusing on the Self-Sovereign Identity (SSI) paradigm. The comparative review explores the different identification methods, data storage, privacy safeguards, and interoperability standards. The proposed system, "SovereignRx", introduces a novel approach that harnesses the potential synergy of SSI and blockchain.

SovereignRx uses decentralized identifiers (DIDs) strategically to unique identification, while also ensuring secure data transport, dispensation tracking, and interoperability through adherence to the FHIR standard. The design uses Hyperledger projects, specifically Indy, Aries, and Fabric in the building, prioritizing user privacy within healthcare prescriptions and facilitating data ownership.

The system focuses on dealing with challenges and presenting a forward view on healthcare data management. It fosters a focus on user-centricity, data security, and interoperability, laying the groundwork for a patient-centric and privacy-respecting paradigm in electronic medical prescriptions.

Future iterations of this endeavor will require widespread adoption of Decentralized Identities, standardization efforts, and most importantly, usability improvements to truly realize the full potential of this transformative approach. The proposed system serves as a promising model for the development of secure, privacy-preserving, and patient-controlled healthcare systems.

7 Challenges and Future Work

The widespread adoption of Decentralized Identities, such as Self-Sovereign Identity, faces significant challenges. A crucial aspect is the development of standards, with the W3C already establishing the Verifiable Credentials Data Model 1.0 [10] and Decentralized Identifiers (DIDs) v1.0 [11]. However, additional standards for interoperability and portability are still needed, currently under development by W3C and the Hyperledger Project [42].

Another concern is the ability of users to manage their own data. Not everyone may be prepared for such a responsibility. Issues like data loss on smartphones require backup and redundancy solutions. Furthermore, cases involving minors or incapable individuals require mechanisms for identity delegation.

While Decentralized Identity technology is promising for increasing privacy and granting users ownership of their data, especially in sensitive contexts such as health data, further improvements are needed, particularly in usability. End-user acceptance depends on good usability, a factor that should be prioritized in future systems like the one proposed in this work.

References

1. European Parliament, Council of the European Union: Regulation (EU) 2016/679. http://data.europa.eu/eli/reg/2016/679/oj. Accessed 13 Sep 2023
2. Centers for Medicare & Medicaid Services: The Health Insurance Portability and Accountability Act of 1996 (HIPAA). https://aspe.hhs.gov/reports/health-insurance-portability-accountability-act-1996. Accessed 13 Sep 2023
3. Yong, J., Tiwari, S., Huang, X., Jin, Q.: Constructing robust digital identity infrastructure for future networked society. In: Proceedings of the 2011 15th International Conference on Computer Supported Cooperative Work in Design (CSCWD), pp. 570-576 (2011). https://doi.org/10.1109/CSCWD.2011.5960129
4. ISO: ISO/IEC 24760-1:2019(en) IT Security and Privacy - A framework for identity management - Part 1: Terminology and concepts. https://www.iso.org/obp/ui/#iso:std:iso-iec:24760:-1:ed-2:v1:en. Accessed 13 Sep 2023
5. El Maliki, T., Seigneur, J.-M.: A survey of user-centric identity management technologies. In: The International Conference on Emerging Security Information, Systems, and Technologies (SECUREWARE 2007), pp. 12-17 (2007). https://doi.org/10.1109/SECUREWARE.2007.4385303
6. Santos, A.M.: Inpatient's medical prescription errors. Einstein (São Paulo) **7**(3) (2009). https://pesquisa.bvsalud.org/portal/resource/pt/lil-530793
7. Allen, C.: The path to self-sovereign identity. (2016). http://www.lifewithalacrity.com/2016/04/the-path-to-self-soverereign-identity.html. Accessed 13 Sep 2023
8. Mühle, A., Grüner, A., Gayvoronskaya, T., Meinel, C.: A survey on essential components of a self-sovereign identity. CoRR, abs/1807.06346 (2018). http://arxiv.org/abs/1807.06346. Accessed 13 Sep 2023
9. Nakamoto, S.: Bitcoin: a peer-to-peer electronic cash system. (2008). https://bitcoin.org/bitcoin.pdf. Accessed 13 Sep 2023
10. Sporny, M., Longley, D., Chadwick, D.: Verifiable credentials data model v1.1. (2019). https://www.w3.org/TR/vc-data-model/. Accessed 13 Sep 2023
11. Sporny, M., Longley, D., Sabadello, M., Reed, D., Steele, O., Allen, C.: Decentralized identifiers (DIDs) v1.0. (2021). https://www.w3.org/TR/did-core/. Accessed 13 Sep 2023
12. Buterin, V.: Ethereum whitepaper. (2014). https://ethereum.org/content/whitepaper/whitepaper-pdf/Ethereum_Whitepaper_-_Buterin_2014.pdf. Accessed 13 Sep 2023
13. Hyperledger fabric: a blockchain platform for the enterprise (2023). https://hyperledger-fabric.readthedocs.io/en/release-2.5/. Accessed 13 Sep 2023

14. Shalaby, S, Abdellatif, A.A., Al-Ali, A., Mohamed, A., Erbad, A., Guizani, M.: Performance evaluation of hyperledger fabric. In: 2020 IEEE International Conference on Informatics, IoT, and Enabling Technologies (ICIoT), pp. 608-613 (2020). https://doi.org/10.1109/ICIoT48696.2020.9089614
15. Institute of electrical and electronics engineers: IEEE Xplore (2023). https://ieeexplore.ieee.org/. Accessed 13 Sep 2023
16. Association for computing machinery: ACM digital library (2023). https://dl.acm.org/. Accessed 13 Sep 2023
17. Elsevier: ScienceDirect (2023). https://www.sciencedirect.com/. Accessed 13 Sep 2023
18. Springer Nature: Springer Link (2023). https://link.springer.com/. Accessed 13 Sep 2023
19. Garcia, R.D., Ramachandran, S.G., Jurdak, R., Ueyama, J.: A blockchain-based data governance with privacy and provenance: a case study for e-Prescription. In: 2022 IEEE International Conference on Blockchain and Cryptocurrency (ICBC), pp. 1-5 (2022). https://doi.org/10.1109/ICBC54727.2022.9805545
20. Li, J.: A new blockchain-based electronic medical record transferring system with data privacy. In: 2020 5th International Conference on Information Science, Computer Technology and Transportation (ISCTT), pp. 141-147 (2020). https://doi.org/10.1109/ISCTT51595.2020.00032
21. Ying, B., Sun, W., Radwan, M.N., Nayak, A.: A secure blockchain-based prescription drug supply in health-care systems. In: 2019 International Conference on Smart Applications, Communications and Networking (SmartNets), pp. 1-6 (2019). https://doi.org/10.1109/SmartNets48225.2019.9069798
22. Benita, R., Kumar, S., Ganesh, Murugamantham, B., Murugan A: Authentic drug usage and tracking with blockchain using mobile apps. Int. J. Interact. Mob. Technol. (iJIM) **14**(17), 20-32 (2020). https://doi.org/10.3991/ijim.v14i17.16561
23. Zhang, P., Schmidt, D.C., White, J., Lenz, G.: Casos de uso da tecnologia Blockchain na saúde. In: Raj, P., Deka, G.C. (eds.) Blockchain Technology: Platforms, Tools and Use Cases, Advances in Computers, vol. 111, pp. 1-41. Elsevier (2018). https://doi.org/10.1016/bs.adcom.2018.03.006
24. Li, P., Nelson, S.D., Malin, B.A., Chen, Y.: DMMS: a decentralized blockchain ledger for the management of medication histories. Blockchain Healthc. Today **2**(38) (2019). https://doi.org/10.30953/bhty.v2.38
25. Garcia, R.D., Ramachandran, G., Ueyama, J.: Exploiting smart contracts in PBFT-based blockchains: a case study in medical prescription system. Comput. Netw. **211**, 109003 (2022). https://doi.org/10.1016/j.comnet.2022.109003
26. Zhang, P., et al: OpTrak: tracking opioid prescriptions via distributed ledger technology. Int. J. Inf. Syst. Soc. Change (IJISSC) **10**(2), 45-61 (2019). https://ideas.repec.org/a/igg/jissc0/v10y2019i2p45-61.html
27. Thatcher, C., Acharya, S.: Pharmaceutical uses of blockchain technology. In: 2018 IEEE International Conference on Advanced Networks and Telecommunications Systems (ANTS), pp. 1-6 (2018). https://doi.org/10.1109/ANTS.2018.8710154
28. Navaratna, L., Wijesinghe, N., Pilapitiya, U.: Providing electronic health care services through a private permissioned blockchain. In: 2020 2nd International Conference on Advancements in Computing (ICAC), vol. 1, pp. 144-149 (2020). https://doi.org/10.1109/ICAC51239.2020.9357135
29. Thatcher, C., Acharya, S.: RxBlock: towards the design of a distributed immutable electronic prescription system. Netw. Model. Anal. Health Inf. Bioinf. **9**(58) (2020). https://doi.org/10.1007/s13721-020-00264-5

30. Alnafrani, M., Acharya, S.: SecureRx: a blockchain-based framework for an electronic prescription system with opioids tracking. Health Policy Technol. **10**(2), 100510 (2021). https://doi.org/10.1016/j.hlpt.2021.100510
31. Garcia, R.D., Zutião, G.A., Ramachandran, G., Ueyama, J.: Towards a decentralized e-prescription system using smart contracts. In: 2021 IEEE 34th International Symposium on Computer-Based Medical Systems (CBMS), pp. 556-561 (2021). https://doi.org/10.1109/CBMS52027.2021.00037
32. Taylor, A., Kugler, A., Marella, P.B., Dagher, G.G.: VigilRx: a scalable and interoperable prescription management system using blockchain. IEEE Access **10**, 25973–25986 (2022). https://doi.org/10.1109/ACCESS.2022.3156015
33. Ethereum: Welcome to Ethereum (2023). https://ethereum.org/, last accessed 2023/09/13
34. CosmWasm: smart contract platform expanding beyond Cosmos (2023). https://cosmwasm.com/. Accessed 13 Sep 2023
35. Tendermint: building the most powerful tools for distributed networks (2023). https://tendermint.com/. Accessed 13 Sep 2023
36. Grote, O., Ahrens, A., Benavente-Peces, C.: A review of post-quantum cryptography and crypto-agility strategies. In: 2019 International Interdisciplinary PhD Workshop (IIPhDW), pp. 115-120 (2019). https://doi.org/10.1109/IIPHDW.2019.8755433
37. Hyperledger Aries: Hyperledger Aries Cloud Agent - Python (2023). https://github.com/hyperledger/aries-cloudagent-python. Accessed 15 Sep 2023
38. Conselho Nacional de Arquivos: Glossário - Documentos Arquivísticos Digitais. (2016). https://www.gov.br/conarq/pt-br/assuntos/camaras-tecnicas-setoriais-inativas/camara-tecnica-de-documentos-eletronicos-ctde/2016_CTDE_Glossario_V7.pdf. Accessed 25 Sep 2023
39. Aldughayfiq, B., Sampalli, S.: Digital health in physicians' and pharmacists' office: a comparative study of e-prescription systems' architecture and digital security in eight countries. Omics: J. Integr. Biol. **25**(2), 102-122 (2021). https://doi.org/10.1089/omi.2020.0085
40. Vorisek, C.N., et al.: Fast healthcare interoperability resources (FHIR) for interoperability in health research: systematic review. JMIR Med. Inform. **10**(7), e35724 (2022). https://doi.org/10.2196/35724
41. Hammond, W.E.: HL7-more than a communications standard. Stud. Health Technol. Inform. **96**, 266–71 (2003). PMID: 15061555
42. Avellaneda, O., et al.: Decentralized identity: where did it come from and where is it going?. IEEE Commun. Stand. Mag. **3**(4), 10-13 (2019). https://ieeexplore.ieee.org/document/9031542. Accessed 13 Sep 2023
43. McKinsey & Company: Telehealth: A quarter-trillion-dollar post-COVID-19 reality? https://www.mckinsey.com/industries/healthcare/our-insights/telehealth-a-quarter-trillion-dollar-post-covid-19-reality. Accessed 13 Sep 2023
44. UN News: WHO chief declares end to COVID-19 as a global health emergency. https://news.un.org/en/story/2023/05/1136367. Accessed 13 Sep 2023

Privacy-Preserving Location-Based Services: A DQN Algorithmic Perspective

Manish Pandey[1], Harkeerat Kaur[1(✉)], Sudipta Basak[1], and Isao Echizen[2]

[1] Indian Institute of Technology Jammu, Jammu and Kashmir, India
{2022pds0024,harkeerat.kaur,2022pis0030}@iitjammu.ac.in
[2] National Institute of Informatics Tokyo, Tokyo, Japan
iechizen@nii.ac.jp

Abstract. With the increasing prevalence of location-based services, preserving user privacy has become a paramount concern. In this research, we propose a comprehensive privacy-preserving approach using Deep-Q networks which uses the best property of both reinforcement learning and deep learning models. The model intelligently gathers contextual information, such as app category, user frequency, and context, to accurately predict user's privacy behavior towards location data access requests. To further enhance privacy protection, we incorporate an obfuscated region technique. The combination of these techniques empowers users with personalized privacy preferences while safeguarding sensitive data. Additionally, we address potential adversarial attacks through adversarial training and differential privacy. Our approach contributes to a more privacy-centric data landscape, allowing users to make informed decisions about their data sharing in today's data-driven world.

1 Introduction

Amidst the ubiquitous digital tapestry woven by the Internet of Things (IoT) and web-based applications, convenience and connectivity reign supreme. From smart homes whispering our preferences to wearable devices chronicling our every step, we readily share troves of personal data - preferences, behaviors, even location trails - to access the bounty of online services. Yet, this boundless data collection, particularly by location-based services and other mobile apps, casts a long shadow of privacy concerns [1,3].

The continuous, granular mapping of our movements and preferences necessitates innovative solutions that uphold *privacy by design* principles within the very fabric of IoT frameworks [15,19,20]. Users deserve to revel in the benefits of these services while retaining absolute control over the dissemination of their digital footprints [2,14]. To bridge this critical gap, we propose a transformative concept: *smart data* - data imbued with an inherent ability to *think for itself.*

In this research, we unveil a groundbreaking model built upon the revolutionary notion of an *intelligent web-based agent* acting as a virtual sentinel for users in the online realm. This dynamic entity will autonomously govern the

L. Barolli (Ed.): AINA 2024, LNDECT 202, pp. 384–399, 2024.
https://doi.org/10.1007/978-3-031-57916-5_33

release of data based on individual preferences and directives, striking a delicate equilibrium between privacy and service functionality.

The crux of our approach lies in the development of an adaptive and automated *smart data agent model* empowered by reinforcement learning (RL). Unlike traditional learning paradigms, RL thrives on dynamic adaptation and learning through real-time interaction with the environment. By mimicking human-like learning processes, RL empowers our proposed agent to make informed decisions that maximize potential rewards - ultimately safeguarding user privacy with unparalleled efficacy.

To our knowledge, this research represents transition towards personalized *smart data agents* [12] tailored to each user's unique privacy calculus, sculpted through the intricate dance of reinforcement learning. This paradigm shift promises to redefine the user-technology relationship, empowering individuals to reclaim control over their digital selves in the ever-evolving IoT landscape.

The paper unfolds as follows: Sect. 2 reviews existing approaches to the location privacy problem. Section 3 details our proposed smart data agent model using Reinforcement Learning, encompassing context-aware sensing, privacy prediction, and memory replay. Section 4 analyzes simulation results for USER A and USER B. Section 5 explores obfuscated regions for enhanced privacy. Section 6 summarizes key findings, concluding the paper on the effectiveness of our model in fortifying location privacy for smart data agents.

2 Location Privacy and Related Works

The preservation of location privacy has emerged as a critical concern, encompassing the control over the dissemination, storage, and sharing of individuals' location information. Numerous mechanisms have been proposed to manage users' location data in the realms of Machine Learning, Deep Learning, and Cryptography. One such approach involves a centralized third-party mechanism, where a trusted anonymizer or third-party entity acts as an intermediary between the user and Location- Based Services (LBS) [9]. The anonymizer conceals the user's actual location within a set of locations, forming a *cloaking region*. Several enhancements to this method, such as K-anonymity [11] and L-diversity, distributed Kanonymity [8,13], and (K, T)-anonymity [7], have been suggested . However, concerns persist regarding the reliability of the anonymizer and its vulnerability to potential attacks. Additionally, the need to maintain an adequate number of users might lead to the expansion of cloaking regions, negatively impacting service quality and time. Another approach is the collaborative mechanism, wherein users collaborate to conceal their location information. This method requires location-aware devices connected over wireless infrastructures (e.g., Bluetooth, Wi-Fi) to communicate with each other. Peers in the network can share location-specific information obtained from LBS, aiding others in seeking similar data without revealing their own actual location [4]. Challenges with this approach include neighbor presence, the likelihood of similar queries among peers, cooperation willingness, and network viability for such operations, which can be influenced by geographical location and population density. Alternatively, the user-centric mechanism empowers individuals to safeguard their location by distorting, enlarging, or reducing the target region. This approach involves various techniques, such as geometric obfuscation, submission of fake queries directly to LBS, historical proximity, and kernel transformation.

Despite the diversity of proposed methods, a fundamental concern remains unaddressed: the limited choice between using the original location or a distorted version. For instance, a map app that requires precise location for navigation might misuse this information to record a user's location even when not in use. Furthermore, individuals' privacy preferences are influenced by their personality and profession. While a teenager might not mind revealing their exact location to a social networking LBS, it would raise significant concerns for a medical practitioner or military personnel. Moreover, a user's behavior on an app can influence their preference for location privacy. Thus, there is a pressing need for an intelligent model that assesses the environment, predicts the required level of privacy, and adapts to changes over time. Amidst these evolving strategies for preserving location privacy, it is essential to acknowledge the dynamic nature of this challenge. Location-based technologies and the data privacy landscape continue to advance. As new technologies emerge and threats evolve, the need for adaptable and intelligent privacy-preserving mechanisms becomes increasingly apparent [14,15]. The quest for striking a balance between location-based service utility and individual privacy remains a multifaceted journey. Our ongoing research in Reinforcement Learning aims to contribute to this dynamic field by providing a robust and responsive solution that can navigate the intricate landscape of location privacy.

Recently in 2020, we proposed a preliminary model of such a data agent using supervised learning on static input-output training data with a simple neural network architecture [5]. However, this approach proved cumbersome and lacked scalability for all possible scenarios. Thereafter in 2021 we proposed an advanced, reconfigurable, self-evolving model using Reinforcement Learning (RL) [21] to simulate a proxy agent that accommodates individual preferences, app behaviors, and context [5]. While initial Q-learning models for location privacy showed promise, their limited scalability(with respect to the state-action space) and potential for suboptimal decisions due to overestimation bias and slow convergence motivated our exploration of deep Q-networks for a more robust and adaptable solution.In light of these shortcomings, our research delves into deep Q-networks (DQN), a powerful fusion of reinforcement learning (RL) and deep learning, as a promising avenue for personalized location privacy. Unlike traditional Q-table-based RL approaches, DQNs offer enhanced scalability and adaptability, making them well-suited for the complex and dynamic nature of location privacy challenges.

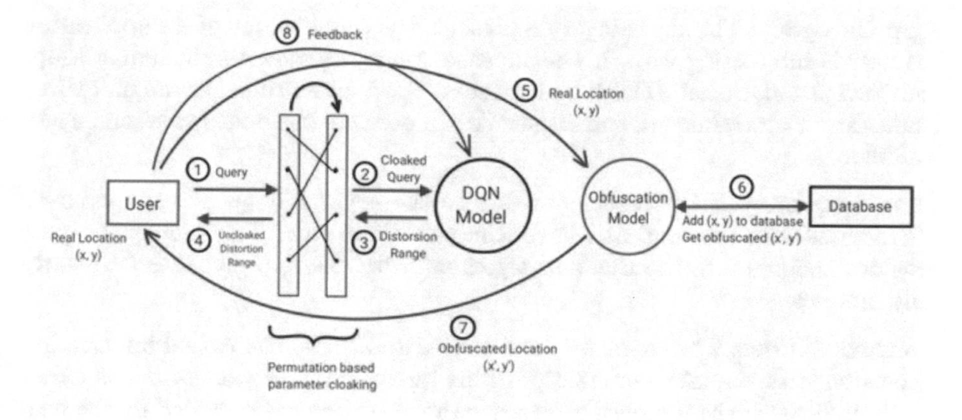

Fig. 1. Proposed model architecture

3 Model Architecture and Working

Overview: The overview of proposed system, as shown in Fig. 1, introduces a dynamic and user-centric location privacy framework for Location-Based Services (LBSs) by leveraging a Deep Reinforcement Learning (DRL)-based architecture. The system consists of interconnected components designed to collaboratively achieve privacy-preserving location sharing. Users initiate location queries through an LBS app, providing context-specific information, including app category, user frequency, and relevant keywords. An intelligent agent intercepts and forwards user queries to the DQN model, a Deep Q-learning model residing on a server equipped with a replay memory. This model analyzes received queries and user preferences to select an optimal distortion range for cloaking the user's location.

The proposed system also incorporates an Obfuscation Model, refining the cloaked location generated by the DQN model using Algorithm 2, detailed in the full research paper, to enhance anonymity and prevent re-identification. The Database stores both real and obfuscated user locations for potential reference and model optimization. Optionally, a Feedback Loop allows users to provide feedback on privacy protection and LBS service quality, enriching the DQN model's knowledge for refined decision-making through its replay memory.

At the core of the model lies Privacy by Design, a fundamental concept emphasizing context and control. To abide by these principles, the model performs four pivotal operations as described below.

3.1 Context-Aware Sensing

Here the proposed model collects three key input context features: app category, user frequency, and context values.

App Category: The app category represents the classification of the application the user is interacting with. It encompasses various categories, including Maps and Navigation, Social, Health and Fitness, Food and Drinks, Communication, Games and Entertainment, and Others [10]. These are basic categories and again scalable.

User frequency: User frequency refers to the average frequency of the user's interactions with the application over the past three days. This temporal aspect provides insights into the user's engagement with the app and how frequently they interact.

Context Values: The context values denote additional contextual information associated with the user's interaction. This binary input is represented as either 0 or 1, indicating the presence or absence of context words provided by the user during the interaction. The context words can be derived form query request and are usually headings of selected options like, book, ride, tag, order, etc.

3.2 Deep Q-Network Model

The Privacy Prediction model features a meticulously designed neural network with layers tailored for predicting user behaviors in location data access. With an input layer of n neurons, it comprehends user inputs and implications on privacy decisions. The subsequent layers, including a critical feature extractor with 24 neurons (tanh activation) and two layers addressing the vanishing gradient problem (256 neurons each, ReLU activation), capture nuanced patterns in user behavior. A fourth layer with 100 neurons (ReLU activation) adds complexity, culminating in the output layer with n neurons (sigmoid activation) producing probabilities for user decisions on location data access. This architecture empowers precise and informed predictions, unraveling individual privacy choices with depth.

The Deep Q-Learning (DQN) algorithm, as outlined in Algorithm 1, serves as the foundation for our model architecture, aimed at enhancing privacy predictions in location-based services. The model features two interconnected neural networks, namely the Main Network and the Target Network, both initialized with random weights. These networks share an identical architecture, and each plays a distinctive role in the reinforcement learning process.

The input to the DQN model comprises state variables ($S = \{App_{\text{cat}}, uf_{\text{cat}}, \text{context}\}$), reflecting the categorical information of the mobile application, user feedback, and contextual data. The output consists of a set of Q values corresponding to different actions in a particular state ($S, A_0, S, A_1, \ldots, S, A_5$), computed using the sigmoid activation function. Figure 5 visually represents the DQN model architecture, showcasing both the Main Network and the Target Network. The model utilizes a replay memory buffer (M) to store experiences (s, a, r, s') during each episode. This replay memory serves as a reservoir of past interactions, facilitating network training by randomly sampling batches of experiences.

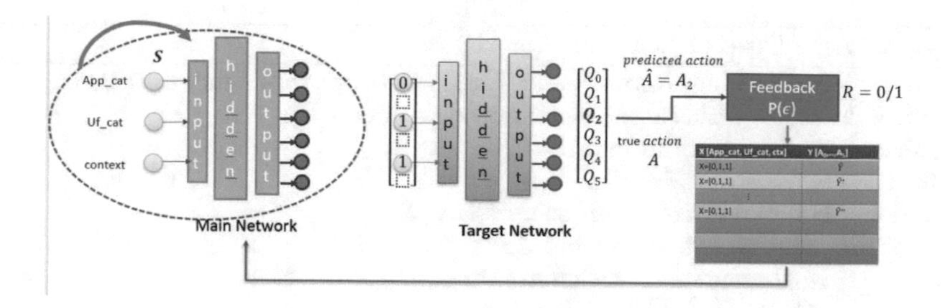

Fig. 2. DQN Model Architecture with Main and Target Networks, Replay Memory, and Feedback.

During the training process, the Main Network predicts Q values (Y) for the current state (s), selects an action (a) based on these predictions, and executes the action. The resulting reward (r) and subsequent state (s') are then stored in the replay memory. Random batches of experiences are sampled from the replay memory, and the Target Network computes target Q-values ($\hat{Y}$) for the next states.

The temporal difference error (td), representing the discrepancy between predicted Q values and actual rewards, is calculated. The Q-values of the batch in the Main Network are updated using a learning rate (α) and the temporal difference error. Additionally, the exploration probability (ϵ) is updated to balance exploration and exploitation.

If the replay memory is full, a training iteration is initiated. Batches of experiences are sampled, and the Target Network computes target Q-values. The temporal difference error is used to update the Q-values in the Main Network. This process is iterated for multiple training iterations, and the weights of the Main Network are periodically copied to the Target Network for stability.

This architecture effectively leverages the DQN algorithm, enabling the model to learn from past experiences, balance exploration-exploitation, and make informed predictions regarding users' privacy choices in location data access requests (Fig. 2).

3.3 Memory Replay for Reinforcement Learning

In the realm of Memory Replay for reinforcement learning, the pivotal process of updating output values, specifically action probabilities, using rewards and feedback constitutes a fundamental facet of the learning paradigm. This update is orchestrated through the formula at line 14 of Algorithm 1, that integrates the reward received from the environment and the feedback on the model's predictions. Representing the action probabilities predicted by the model as $\hat{y} = [\hat{y}_1, \hat{y}_2, \ldots, \hat{y}_n]$,with n denoting the number of possible actions, each $\hat{y}_i$ embodies the probability of undertaking action i. The reward can manifest as binary values (e.g., 1 for a correct action, 0 for an incorrect action).

The Memory Replay for reinforcement learning update formula is articulated as follows:$\hat{y}_i = \hat{y}_i + \alpha \cdot (\text{reward} - \hat{y}_i)$, where $\hat{y}_i$ denotes the current predicted

Algorithm 1. Deep Q-Learning Algorithm

Require: Initialize Main Network and Target Network with random weights.
Require: Initialize replay memory buffer M with capacity N.
Require: Initialize exploration probability ϵ to 1.0.
1: **for** each episode **do**
2: Sample initial state s from query space X.
3: **while** episode not terminated **do**
4: **if** random number $< \epsilon$ **then**
5: Choose a random action a from the action space.
6: **else**
7: Pass state s through the Main Network to get Y with Q-value estimates.
8: Choose action $a = \hat{Y}$.argmax()
9: **end if**
10: Execute action a, get the reward r, and the next state s'.
11: Store the experience (s, a, r, s') in replay memory M.
12: Sample a random batch of experiences (s, a, r, s') from M.
13: Compute the target Q-values $\hat{Y}$ for the next states using the Target Network.
14: Calculate the temporal difference error:

$$t_d = r + \gamma \cdot \max(\hat{Y}) - Y[a]$$

15: Update the Q-values of the batch in the Main Network:

$$Y[a] = Y[a] + \alpha \cdot t_d$$

16: Update the exploration probability ϵ (e.g., $\epsilon = \max(0.01, 0.99 \cdot \epsilon)$).
17: **end while**
18: **end for**
19: **if** replay memory M is full **then**
20: **for** each training iteration from 1 to S **do**
21: Sample a random batch of experiences (s, a, r, s') from M.
22: Compute the target Q-values $\hat{Y}$ for the next states using the Target Network.
23: Calculate the temporal difference error for the batch.
24: Update the Q-values of the batch in the Main Network:

$$Y[a] = Y[a] + \alpha \cdot t_d.$$

25: Update the exploration probability ϵ (e.g., $\epsilon = \max(0.01, 0.99 \cdot \epsilon)$).
26: **end for**
27: Copy the weights of the Main Network to the Target Network.
28: **end if**
return the trained Main Network and Target Network.

probability for action i, alpha is the learning rate governing the extent of the update, reward encapsulates the feedback received after undertaking action i, and (reward - $\hat{y}_i$) encapsulates the prediction error. Implementation of this Memory Replay-based Q-learning enables the model to dynamically adapt its action probabilities.

When the replay memory is full we randomly sample a batch of experiences from it. Random sampling breaks the temporal correlations between the consecutive experiences.It also allows the network to learn from a broader range of situations and help in generalizing better to unseen states.

3.4 Customization for Individual Privacy

Customization for Individual Privacy is a key aspect of the above reinforcement deep learning model, catering to users' unique privacy preferences which is mange using epsilon greedy strategy. This shift is facilitated by epsilon decay. As epsilon decreases, the model becomes less reliant on exploration and focuses more on exploiting the learned action probabilities to make accurate predictions. For users who prioritize privacy over utility, a higher initial epsilon value and slower decay rate can encourage more exploration, allowing the model to thoroughly understand privacy preferences. Conversely, users who prioritize utility over privacy may opt for a lower initial epsilon value and faster decay rate, enabling the model to make more confident predictions based on past experiences.

4 Experimental Study and Results

The research endeavor encompassed a comprehensive examination of the intricacies surrounding the deliberate manipulation of location data, specifically focusing on the behavior of two distinct individuals, denoted as User A and User B. These two individuals were chosen for their notable divergence in both personality traits and professional affiliations, thereby representing a diverse spectrum of potential users within the context of location-based services and their associated privacy concerns.

In the experimental setup, app categories, user frequency, and context collectively define the queries processed by the Deep Q-Network (DQN) model. App categories are numerically represented as integers from 0 to 6, including *Maps & Navigation* (0), *Social* (1), *Food & Drinks* (2), *Home Services* (3), *Communication* (4), *Entertainment* (5), and *Others* (6). The user frequency categories are simplified to three levels: *Low* (0.5 - 1.5), *Medium* (1.5 - 2.5), and *High* (2.5 - 3.5), mapped to numerical values 1, 2, and 3, respectively. The binary context variable distinguishes between *present* (context = 1) and *absent* (context = 0) states.

Table 1. State-Action Mappings for User A

App Category	Context Word	User Frequency (UF)	Preferred Action
Maps & Navigation	Present	Any of [low,mid,high]	A0
Maps & Navigation	Absent	Low	A1
Maps & Navigation	Absent	Any of [mid,high]	A1
Social	Present	Any of [low,mid,high]	A0
Social	Absent	Low	A1
Social	Absent	Mid	A2
Social	Absent	High	A3
Entertainment	Present	Any of [low,mid]	A2
Entertainment	Absent	Any of [low,mid,high]	A4
Food & Drinks	Present	Mid	A0
Food & Drinks	Absent	Mid	A2
Food & Drinks	Absent	High	A3
Home Services	Present	Mid	A0
Home Services	Absent	Mid	A2
Home Services	Absent	High	A3
Communication	Present	Any of [low,mid]	A0
Communication	Present	High	A2
Others	Present	Any of [low,mid]	A2
Others	Absent	Any of [low,mid,high]	A4

The chosen actions (A0 to A4) correspond to specific distortion ranges: A0 (0–10 m), A1 (100–200 m), A2 (200–500 m), A3 (500–1000 m), and A4 (1000–2000 m). These distortion ranges determine the degree of privacy protection applied to the user's location information. For example, if the DQN model selects A0 for a given query, it indicates a preference for minimal distortion, preserving location accuracy within the 0–10 meter range.

This categorical mapping enables a nuanced investigation into the impact of different app contexts, user frequencies, and app categories on the privacy-preserving actions recommended by the DQN model during the experimental evaluation.

User A: User A is characterized as a male university graduate who exhibits a relaxed attitude towards sharing his location. He doesn't express much concern about providing his precise location to various applications. However, there are two conditions that need to be met for him to be comfortable with location sharing. Firstly, the User Fear (UF), which likely represents his apprehension or discomfort, must be low. Secondly, the context parameter must be equal to 1, implying that the specific circumstances or context in which his location is shared align with his preferences.

The state-action mapping for User A across various app categories, context word scenarios, and user frequency levels is presented in the Table 1. For the *Maps & Navigation* category, when the context word is present, and user frequency can be low, medium, or high, the preferred action is denoted as A0. In the absence of a context word, a low user frequency corresponds to the preferred action A1, while any of medium or high user frequencies leads to A1 as well. The Table 1 outlines similar mappings for other app categories, such as *Social,* where the preferred action varies based on the presence or absence of a context word and the user frequency level. Notably, the mappings capture nuanced preferences, reflecting the complex interplay between app category, contextual information, and user frequency in determining the user's preferred privacy action.

Figure 3(a) depicts a graph that showcases the initial distortion for first 100 simulations predicted by the model during the exploration phase. As the agent is in the early stages of exploration, the distortion values fluctuate significantly, varying by large amounts. To improve its predictions and rewards, the agent seeks more frequent feedback from User A, which helps it better understand the user's preferences and refine its location distortion behavior accordingly. In first 50 simulation the model asked just 17 feedback from the user. Figure 3(b) illustrates the distortion predicted by the model during simulation 100 to 150. As the exploration rate decreases during this phase, the agent becomes more certain about its predictions and needs less frequent feedback from User A, i.e. just 8 feedback. The reduced feedback asking frequency suggests that the agent is becoming more personalized in its location distortion behavior, aligning it more closely with User A's preferences. Figure 3(c) displays the distortion predicted by the model for the final 150–200th simulation. At this point, the

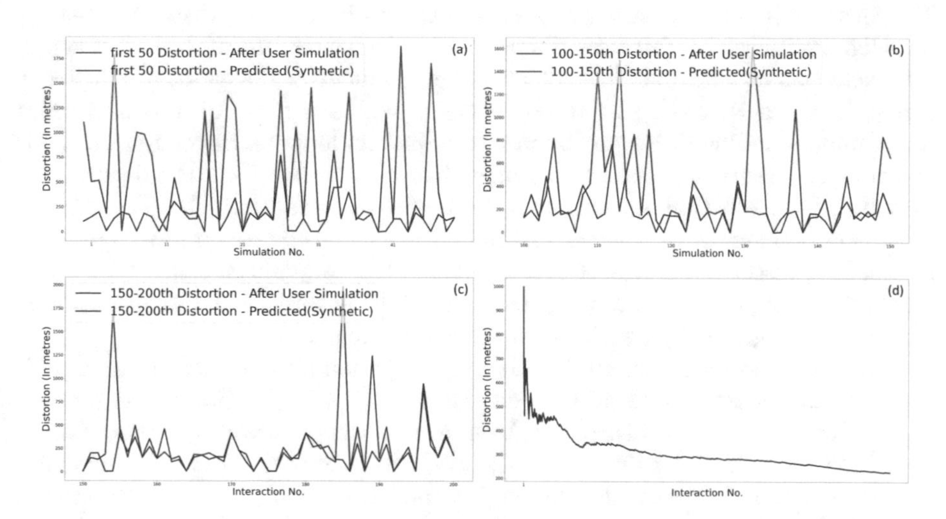

Fig. 3. User A. (a) Initial distortion comparison, (b) Distortion for 100–150th interactions approx, (c) distortion for 150–200th interactions and (d) Average distortion preference over time.

agent rarely asks User A for feedback, as it has gained sufficient knowledge about the user's preferences.At last 50 simulation the agent just asked 4 feedback from the user.The agent tends to exploit this knowledge to a greater extent, aligning its location distortion behavior with User A's preferences as much as possible. Figure 3(d) provides insights into how the model has analyzed User A's average distortion preferences over time. The model has deduced that, on average, User A desires a distortion of around 200 to 300 m. This average distortion preference aligns with the observation that User A frequently uses various applications and doesn't prioritize privacy concerns highly, likely due to his status as a university student.

USER B: User B is described as a government agency that places significant importance on maintaining location privacy. Unlike User A, User B is highly concerned about revealing his precise location to applications. For User B to consider sharing his location, certain conditions need to be met to ensure stringent privacy measures are in place. The average distortion range preferred by User B, as shown in Fig. 5(d), is between 500 to 600 m.

User B's state-action mapping, as shown in Table 2, delineated for distinct app categories, context word scenarios, and user frequency tiers, provides insight into nuanced privacy preferences. For instance, when the context word is present with a low user frequency, A0 is the favored action. However, in the absence of a context word, low user frequency corresponds to A2, while medium or high frequencies lead to the selection between A3 or A4.

Similar to User A, the graph depicted in Fig. 4(a) represents the initial distortion predicted by the model during the exploration phase. Figure 4(b), the distortion predicted by the model during interactions 100 to 150 is illustrated. The agent strives to become more personalized in its location distortion behavior, aligning it closely with User B's privacy preference in Fig. 4(c). At this stage, the agent has obtained sufficient knowledge about User B's strict privacy concerns. As a result, the agent rarely asks for feedback from User B and relies on its understanding of his preferences to adjust its location distortion behavior accordingly, ensuring a distortion range that falls within User B's desired 500 to 600 m. Figure 5(d) showcases how the model has analyzed User B's average distortion preferences over time. The model has deduced that, on average, User B prefers a distortion range of 500 to 600 m. This preference reflects User B's emphasis on high location privacy, and the model ensures that the distortion is within this range to align with his strict requirements.

Our proposed model exhibits scalability in action and learning from experience, enhancing its adaptability to diverse scenarios in location-based services (LBS). For instance, consider a user, User A, initiating a navigation query falling under the *Maps & Navigation* app category, with a medium user frequency and a present context. The model, having learned from a myriad of interactions, dynamically selects an action, such as A0, representing a distortion range of

Table 2. State-Action Mappings for User B.

App Category	Context Word	User Frequency (UF)	Preferred Action
Maps & Navigation	Present	low	A0
Maps & Navigation	Present	Any of [mid,high]	A1
Maps & Navigation	Absent	Low	A2
Maps & Navigation	Absent	Any of [mid,high]	A3 or A4
Social	Present	Any of [low,mid,high]	A2
Social	Absent	Low	A3 or A4
Entertainment	Present	low	A2
Entertainment	Present	Any of [mid,high]	A3 or A4
Entertainment	Absent	Any of [mid,high]	A3 or A4
Food & Drinks	Present	Any of [low,mid,high]	A2
Food & Drinks	Absent	low	A3
Food & Drinks	Absent	Any of [mid,high]	A3 or A4
Home Services	Present	Any of [low,mid,high]	A2
Home Services	Absent	low	A3
Home Services	Absent	Any of [mid,high]	A3 or A4
Communication	Present	low	A2
Communication	Present	Any of [mid,high]	A3 or A4
Communication	Absent	Any of [mid,high]	A3 or A4
Others	Present	low	A2
Others	Present	Any of [mid,high]	A3 or A4
Others	Absent	Any of [mid,high]	A3 or A4

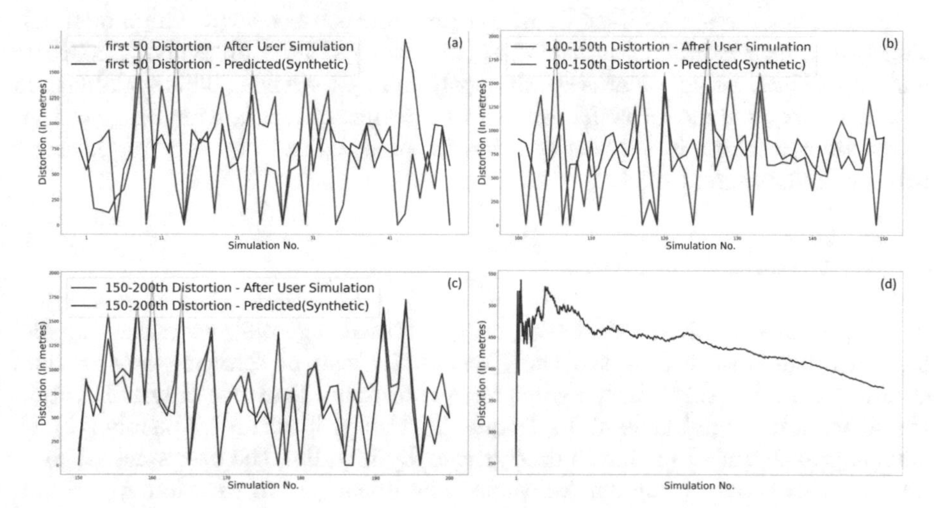

Fig. 4. User B. (a) Initial distortion comparison, (b) Distortion for 100–150th interactions approx, (c) distortion for 150–200th interactions and (d) Average distortion preference over time.

Algorithm 2. Location Obfuscation Algorithm

```
1: Initialize worldMap as a 2D matrix representing the world map, with each cell as
   a region.
2: Initialize cell as the world map at position [i][j], where each cell contains a list of
   hotspot coordinates (initially empty).
3: function GETCELL(realCoord)
4:     return row and col index of realCoord in worldMap
5: end function
6: function GETHOTSPOTINDISTORTIONRANGE(realCoord, distortionRange, cell)
7:     hotspotsInRange ← []
8:     for each coord in cell do
9:         if realCoord.distance(coord) < distortionRange.max then
10:            if distortionRange.min ≤ realCoord.distance(coord) then
11:                return coord
12:            end if
13:        end if
14:    end for
15:    return randomCoordinate(distortionRange)
16: end function
17: function OBFUSCATELOCATION(realCoord, distortionRange, worldMap)
18:    cell ← getCell(realCoord)
19:    hotspotInRange ← getHotspotsInDistortionRange(realCoord,distortionRange,
   worldMap)
20:    return hotspotInRange
21: end function
```

0–10 m, tailored to User A's preferences. This showcases the model's adaptability to specific user behaviors. Importantly, the model's scalability shines when faced with different users, like User B, whose preferences may vary. The model efficiently categorizes inputs, adjusts distortion ranges based on evolving contexts and user behaviors, and learns continuously from experience. This scalability in action ensures the model's effectiveness in accommodating a wide range of user interactions, making it a robust solution for personalized and evolving location privacy preferences in LBS.

5 Obfuscated Region

In the proposed method for creating an obfuscated region[8], we utilize the distortion range provided by the DQN model. Instead of selecting any random coordinate within this range, we opt for a coordinate that not only falls within the distortion range but is also a frequently visited location by people [17,18]. This approach aims to enhance user privacy by blending the user's actual coordinates with those of popular locations. The Location Obfuscation Algorithm, outlined in Algorithm 2, introduces a systematic approach to enhance location privacy within a world map.To implement this, we divide the world map into a grid of 2D matrix cells, with each cell representing a specific region. Within each

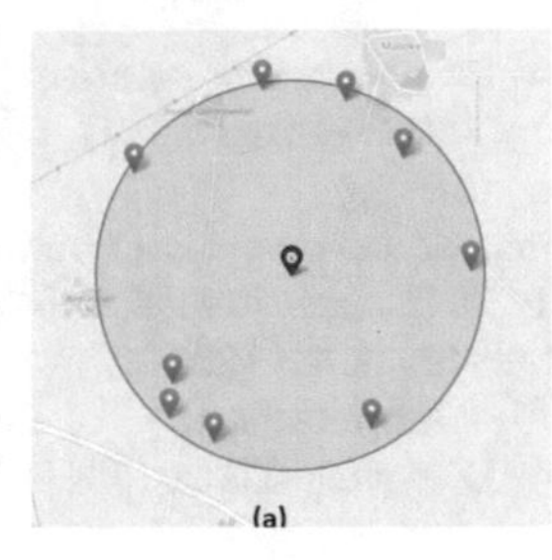

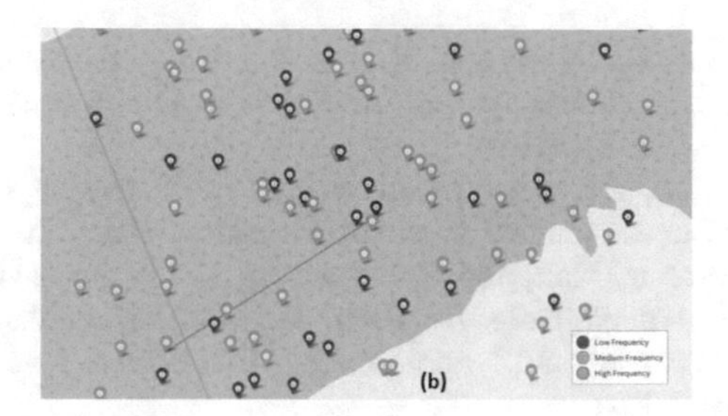

Fig. 5. (a) Predicted distortions in circular pattern and (b) the distorted location with Obfuscation.

cell, we maintain a list of hotspot coordinates, sorted based on the frequency of visitors in descending order. By leveraging this method, we can offer a more effective obfuscation of the user's location, making it harder for potential adversaries to distinguish,as shown by Fig. 6(a), the user's actual position amidst the crowd of frequently visited areas [16, 17].

6 Conclusion

In conclusion, this research paper demonstrates the effectiveness of the Privacy Prediction model, employing the Deep Q-Learning (DQN) algorithm for informed privacy predictions in location-based services. The model exhibits adaptability by dynamically adjusting action probabilities based on user interactions, refining predictions over time to align closely with user preferences. Additionally, the Location Obfuscation Algorithm introduces an innovative strategy for bolstering location privacy through a grid-based structure with hotspots and distortion ranges, contributing to heightened security in location-based services.

Acknowledgments. This work was partially supported by C3Ihub IIT Kanpur sanction 1103, JSPS KAKENHI Grant JP21H04907, and by JST CREST Grants JPMJCR18A6 and JPMJCR20D3, Japan.

References

1. Bettini, C., Jajodia, S., Samarati, P., Wang, S.X.: Privacy in Location-Based Applications: Research Issues and Emerging Trends, vol. 5599. Springer, Cham (2009)
2. Gedik, B., Liu, L.: Protecting location privacy with personalized k-anonymity: architecture and algorithms. IEEE Trans. Mob. Comput. **7**(1), 118 (2007)
3. Guo, X., Wang, W., Huang, H., Li, Q., Malekian, R.: Location privacy-preserving method based on historical proximity location. Wirel. Commun. Mob. Comput. **2020**, 1–16 (2020)
4. Hashem, T., Kulik, L.: Safeguarding location privacy in wireless Ad-Hoc networks. In: Krumm, J., Abowd, G.D., Seneviratne, A., Strang, T. (eds.) UbiComp 2007: Ubiquitous Computing. Lecture Notes in Computer Science, vol. 4717, pp. 372–390. Springer, Berlin (2007). https://doi.org/10.1007/978-3-540-74853-3_22

5. Kaur, H., Echizen, I., Kumar, R.: Smart data agent for preserving location privacy. In: 2020 IEEE Symposium Series on Computational Intelligence (SSCI), pp. 2567–2575. IEEE (2020)
6. Le, T., Echizen, I.: Lightweight collaborative semantic scheme for generating an obfuscated region to ensure location privacy. In: IEEE International Conference on Systems, Man, and Cybernetics (SMC), pp. 2844–2849. IEEE (2018)
7. Machanavajjhala, A., Kifer, D., Gehrke, J., Venkitasubramaniam, M.: l-diversity: privacy beyond k-anonymity. ACM Trans. Knowl. Discovery Data (TKDD) 1(1), 3es (2007)
8. Masoumzadeh, A., Joshi, J.: An alternative approach to k-anonymity for location-based services. Procedia Comput. Sci. **5**, 522530 (2011)
9. Mokbel, M.F.: Privacy in location-based services: state-of-the-art and research directions. In: International Conference on Mobile Data Management, pp. 228. IEEE (2007)
10. Scrapper https://pypi.org/project/google-play-scraper/. Accessed 14 May 2020
11. Takabi, H., Joshi, J.B., Karimi, H.A.: A collaborative k-anonymity approach for location privacy in location based services. In: 2009 5th International Conference on Collaborative Computing: Networking, Applications and Work sharing, pp. 19. IEEE (2009)
12. Tomko, G.J., Borrett, D.S., Kwan, H.C., Steffan, G.: SmartData: make the data think for itself. Identity Inf. Soc. **3**(2), 343362 (2010)
13. Zhong, G., Hengartner, U.: A distributed kanonymity protocol for location privacy. In: IEEE International Conference on Pervasive Computing and Communications, pp. 110. IEEE (2009)
14. Beresford, R., Stajano, F.: Location privacy in pervasive computing. IEEE Pervasive Comput. **2**(1), 4655 (2003)
15. GDPR. https://gdpr-info.eu/issues/privacy-by-design/. Accessed 14 May 2020
16. Damiani, M.L., Bertino, E., Silvestri, C.: Protecting location privacy through semantics-aware obfuscation techniques. In: Karabulut, Y., Mitchell, J., Herrmann, P., Jensen, C.D. (eds.) Trust Management II. IFIP – The International Federation for Information Processing, vol. 263, pp. 231–245. Springer, Boston (2008). https://doi.org/10.1007/978-0-387-09428-1_15
17. Ardagna, C.A., Cremonini, M., Damiani, E., De Capitani di Vimercati, S., Samarati, P.: Location privacy protection through obfuscation-based techniques. In: Barker, S., Ahn, GJ. (eds.) Data and Applications Security XXI. Lecture Notes in Computer Science, vol. 4602, pp. 47–60. Springer, Berlin (2007). https://doi.org/10.1007/978-3-540-73538-0_4
18. Ardagna, C.A., Crcmonini, M., Damiani, E., De Vimercati, S., Samarati, P.: A middleware architecture for integrating privacy preferences and location accuracy. In: Venter, H., Eloff, M., Labuschagne, L., Eloff, J., von Solms, R. (eds.) New Approaches for Security, Privacy and Trust in Complex Environments. IFIP International Federation for Information Processing, vol. 232, pp. 313–324. Springer, Boston (2007). https://doi.org/10.1007/978-0-387-72367-9_27
19. Amiri-Zarandi, M., Dara, R.A., Fraser, E.: A survey of machine learning-based solutions to protect privacy in the internet of things. Comput. Secur. **96**, 101921 (2020)

20. Bilogrevic, I., Huguenin, K., Agir, B., Jadliwala, M., Gazaki, M., Hubaux, J.-P.: A machine-learning based approach to privacy-aware information-sharing in mobile social networks. Pervasive Mob. Comput. **25**, 125142 (2016)
21. Kaur, H., Kumar, R., Echizen, I.: Reinforcement learning based smart data agent for location privacy. In: Barolli, L., Woungang, I., Enokido, T. (eds.) Advanced Information Networking and Applications. Lecture Notes in Networks and Systems, vol. 226, pp. 657–671. Springer, Cham (2021). https://doi.org/10.1007/978-3-030-75075-6_54

Knowledge Empowered Deep Reinforcement Learning to Prioritize Alerts Generated by Intrusion Detection Systems

Lalitha Chavali(✉), Paresh Saxena, and Barsha Mitra

Department of Computer Science and Information System, BITS Pilani, Hyderabad Campus, Pilani, India
{p20190423,psaxena,barsha.mitra}@hyderabad.bits-pilani.ac.in

Abstract. Intrusion detection systems (IDS) produce a vast number of alerts, many of which are false positives. Hence prioritizing these alerts for investigation is essential. It is also crucial to consider the different levels of knowledge of the defender (who prioritize alerts) about the system for developing effective defense strategies in the context of cybersecurity. Recognizing the importance of varying knowledge levels for defenders, this paper introduces **KN**owledge empowered deep reinforcement learning for **A**lert **P**rioritization (KNAP). We propose three novel knowledge empowered DRL approaches by integrating KNAP with three actor-critic methods: (i) deep deterministic policy gradient based KNAP (D-KNAP), (ii) soft actor-critic based KNAP (S-KNAP), and (iii) twin delayed deep deterministic policy gradient based KNAP (T-KNAP). The interaction between DRL based attacker and defender is framed as a zero-sum game and the double oracle approach is used to obtain the mixed strategy Nash equilibrium (MSNE). As a key performance metric, we consider the defender's loss i.e., the incapability of the defender in investigating the alerts generated by attacks. We present the performance of the proposed approaches across different levels of defender's knowledge utilizing MQTT-IoT-IDS2020 and CSE-CIC-IDS2018 dataset along with Snort IDS. Our results show that T-KNAP reduces defender's loss by 19.59% and 36.84% compared to S-KNAP and D-KNAP, respectively for the MQTT-IoT-IDS2020 dataset. Furthermore, for the CSE-CIC-IDS2018 dataset, it reduces defender's loss by 33.87% compared to S-KNAP and by 35.17% compared to D-KNAP. Moreover, there is a substantial improvement in the results when compared with traditional alert prioritization techniques like Uniform and Snort.

Keywords: Intrusion detection system · Alert prioritization · Complete and partial knowledge · Deep reinforcement learning · Mixed strategy Nash equilibrium

L. Barolli (Ed.): AINA 2024, LNDECT 202, pp. 400–411, 2024.
https://doi.org/10.1007/978-3-031-57916-5_34

1 Introduction

Starting from $3 trillion in 2015, the cost associated with cybercrime has been growing by 15% annually due to sophisticated cyberattacks [1]. Attacks such as advanced persistent threats (APTs) are known to execute precise and covert attacks making them challenging to identify [2]. To detect and mitigate these attacks, security analysts employ intrusion detection systems (IDS) [3]. IDSs generate a huge volume of alerts, many of which turn out to be false positives. According to [4], up to 99% of the alerts generated can be false positives. Due to this, network operators experience alert fatigue, where they are unable to look into each of these alerts. According to Cisco's findings, 44% of the alerts generated go unnoticed by operators because of the overwhelming volume of alerts [5]. As a result, the prioritization of these alerts for investigation emerges as a critical concern for cybersecurity professionals.

Furthermore, a defender (who protects the system) or an attacker (illegitimate user) can gather various levels of information about the overall system. We refer to such information as knowledge in this paper. This knowledge can be used by both defender and attacker to achieve their objectives [6]. The knowledge, for instance, can include internal system details or any other technical information. Partial knowledge in the context of attackers and defenders signifies limited awareness of the target system's vulnerabilities and defenses. Attackers with partial knowledge may not be able to exploit the system's weaknesses and the corresponding attacks can be easily detected by the defender[7]. Defenders with partial knowledge face challenges in accurately assessing the significance of security alerts which can lead to the risk of missing critical attacks and difficulties in developing effective response strategies. Therefore, considering defenders with varying levels of knowledge about the system is crucial for developing effective defense strategies in the context of cybersecurity. If the defender can effectively counter a strong attacker i.e., attacker with complete knowledge, it indeed implies competence in defending against weak attacker i.e., attacker with partial knowledge. The security defense systems need to be robust and reliable to counteract dynamic and sophisticated cyber attacks. Machine learning (ML) and reinforcement learning (RL) have been used in the design of security mechanisms to handle complex attacks [8]. Furthermore, deep reinforcement learning (DRL), which combines deep learning (DL) with RL, provides efficient cyber security solutions and defend against increasingly complex cyberattacks [9]. Actor-critic methods based on DRL [10] are considered as effective approaches for handling large state and action spaces and thus can efficiently handle dynamic cybersecurity environment.

In this paper, partial and complete **KN**owledge empowered deep reinforcement learning for **A**lert **P**rioritization (KNAP) has been proposed to investigate alerts generated by the IDS. This work is an extension of our prior work [11], where we proposed soft actor-critic based deep reinforcement learning for alert prioritization (SAC-AP). The main contribution of the current work is to develop DRL based robust alert investigation policies by considering varying levels of knowledge of the defender, both partial as well as complete, against the strongest attacker equipped with complete knowledge. By integrating KNAP with three actor-critic methods, we propose three novel knowledge empowered DRL approaches: (i) deep deterministic policy gradient (DDPG) [12] based KNAP (D-KNAP), (ii) soft actor-critic (SAC) [13] based KNAP (S-KNAP), and (iii) twin delayed deep deterministic policy gradient (TD3) [14] based KNAP (T-KNAP). We represent the overall design of D-KNAP, S-KNAP and T-KNAP and show its efficiency by comparing it with the existing alert prioritization methods. We illustrate the benefits of the proposed approach with MQTT-IoT-IDS2020 dataset [15] and CSE-CIC-IDS2018 dataset [16] along with Snort IDS [17]. We consider the defender's loss i.e., the defender's inability to investigate alerts triggered by attacks as the performance metric. Our results illustrate that T-KNAP outperforms S-KNAP and D-KNAP. Defender possessing complete knowledge has lower loss compared to one with partial knowledge. Our results indicate that for the MQTT-IoT-IDS2020 dataset, T-KNAP reduces defender's loss by 19.59% and 36.84% compared to S-KNAP and D-KNAP, respectively. In the case of the CSE-CIC-IDS2018 dataset, T-KNAP reduces defender's loss by 33.87% compared to S-KNAP and by 35.17% compared to D-KNAP.

The remainder of the paper is structured as follows. Section 2 illustrates the system architecture. The overall design of the proposed approaches is presented in Sect. 3. The details on the experimental setup, datasets and results are discussed in Sect. 4. Conclusion and future research directions are discussed in Sect. 5.

2 System Architecture

Figure 1 presents the system architecture. It consists of five modules: benign users, DRL based defender (denoted by v), DRL based attacker (denoted by $-v$), IDS and state. The benign users are registered users of the system whereas the attacker is an unauthorized user who intends to conduct a range of attacks. IDS receives network packets both from the attacker and the benign users and generates alerts. The alerts generated due to the benign data are known as false alerts and the alerts generated due to the attack data are known as true alerts. These alerts are sent to the defender which prioritizes the alerts. The attacker and the defender use reward based RL to generate policies. Specifically, RL includes an agent and its environment. At each time step t, the agent interacts with the environment, observes a state $s^t \in S$, where S is the set of states, and performs an action $a^t \in A$, where A is the set of actions. The selection of the action is determined by the policy $\pi(a^t|s^t)$, which is a mapping between the state s^t and the action a^t to be performed in state s^t. The agent then receives a reward of r^{t+1} and proceeds to the next state s^{t+1} [10]. In RL, the objective is to maximize the total discounted reward and generate the optimal policy. Let us denote $V_\pi(s^t)$, as a state-value function that specifies the value of being in a specific state s. It is given by, $V_\pi(s) = E_\pi\left[\sum_{f=0}^{\infty} \gamma^f r_{t+f+1} | s^t = s\right]$, where $\gamma \in [0, 1]$ is a discount factor and $E_\pi[.]$ is the expected value of a random variable in state s^t and subsequently following a policy π. Similarly, let us denote

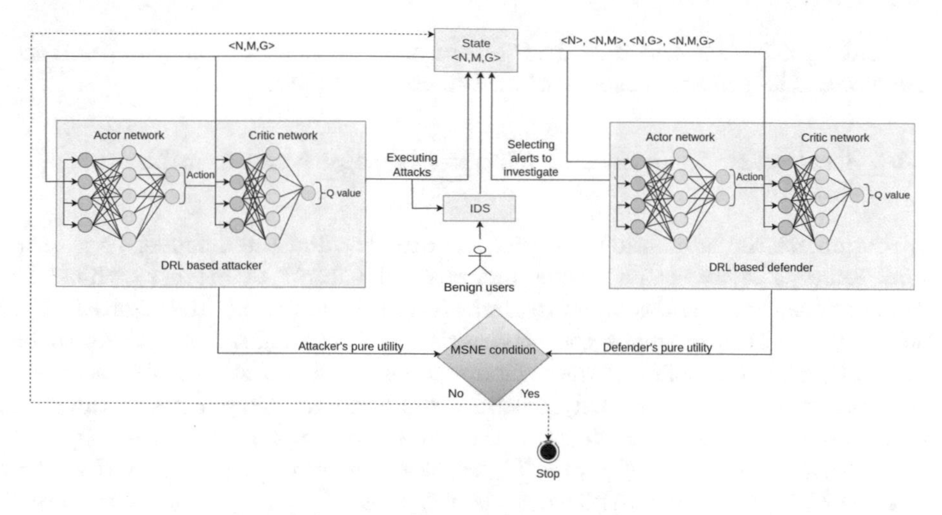

Fig. 1. System Architecture

$Q_\pi(s^t, a^t)$ as an action-value function that specifies the value of taking action a^t in a state s^t under policy π. It is given by, $Q_\pi(s^t, a^t) = E_\pi\left[\sum_{f=0}^{\infty} \gamma^f r_{t+f+1}\right]$. The optimal policy is given by, $\pi^*(s^t) = \arg\max_{a^t \in A}(Q^*(s^t, a^t))$, where, the optimal $Q^*(s^t, a^t)$ is calculated as, $Q^*(s^t, a^t) = \max_{\pi \in \Phi} Q_\pi(s^t, a^t)$ with Φ as a set of all policies.

To be able to utilize RL with the system architecture, we formulate, states, actions and rewards as follows. Three parameters are used to represent state S at time step t, i.e., $S =< N, M, G >$. The first parameter $N = N_l^t$, represents, the number of alerts of type $l \in L$ that the defender needs to investigate. The second parameter $M = M_k^t$, indicates whether an attack $k \in K$ was executed at time t. The third parameter $G = G_{k,l}^t$, represents the number of alerts of type l that are generated as a result of an attack k. The state comprises of two parts: the attacker's state and the defender's state. Based on the attacker's state, it selects an action that includes choosing a subset of attacks to execute based on the following constraint: $\sum_{k \in K} a_{-v,k}^t E_k \leq D$, where a_{-v} is a player's action, E_k is the cost of launching attack k, K is the set of attacks and D is attacker's budget. Based on the defender's state, it selects an action that includes choosing a subset of alerts to investigate based on the following constraint: $\sum_{l \in L} a_{v,l}^t C_l \leq B$, where, a_v is a player's action, C_l is a cost of investigating alert type l, L is the set of alert types and B is defender's budget. The defender's reward at time period t is given by r_v^t where, $r_v^t = -\sum_{k \in K} O_k \hat{M}_k^t$, with O_k as the loss of the defender when attack k is not detected and $\hat{M}_k^t$ is a binary indicator that represents whether the executed attack has been investigated or not. Based on the attacker's state/knowledge i.e., $< N, M, G >$, the attacker selects a policy to generate attacks whereas based on the defender's state/knowledge i.e., $< N >$,

$< N, M >, < N, G >$ *and* $< N, M, G >$, the defender selects a policy to prioritize the alerts. The process is repeated until it obtains MSNE.

2.1 Strategies, Utilities and Mixed Strategy Nash Equilibrium (MSNE)

We formulate the interaction between the attacker and the defender as a zero-sum game. The attacker's action includes all subsets of attacks, while the defender's action includes all subsets of alerts that are to be investigated. The policy of the attacker maps the attacker's state to a subset of attacks to be executed, while the policy of the defender maps the defender's state to a subset of alerts to be investigated. Let us denote the deterministic policies as the pure strategies and stochastic policies as the mixed strategies. Let Π_v represent a set of player's pure strategies and $\sum_v$ represent a set of player's mixed strategies. Let $U_v(\pi_v, \pi_{-v})$ denote player v's utility where $\sum_{x \in (-v,+v)} U_x(\pi_x, \pi_{-x}) = 0$ since it is a zero-sum game. The expected utility of a player v when it uses pure strategy, $\pi_v \in \Pi_v$, and its opponent uses mixed strategy, $\sigma_{-v} \in \sum_{-v}$, is given by [18]: $U_v(\pi_v, \sigma_{-v}) = \sum_{\pi_{-v} \in \Pi_{-v}} \sigma_{-v}(\pi_{-v}) U_v(\pi_v, \pi_{-v})$.

Furthermore, the expected utility of a player v when it uses mixed strategy, $\sigma_v \in \sum_v$, and its opponent also uses mixed strategy, $\sigma_{-v} \in \sum_{-v}$, is given by: $U_v(\sigma_v, \sigma_{-v}) = \sum_{\pi_v \in \Pi_v} \sigma_v(\pi_v) U_v(\pi_v, \sigma_{-v})$. Therefore, the defender's utility when it uses pure strategy, $\pi_v \in \Pi_v$, and its opponent (attacker) also uses pure strategy, $\pi_{-v} \in \Pi_{-v}$ is given by: $U_v(\pi_v, \pi_{-v}) = E\left[\sum_{p=0}^{\infty} \gamma^p r_v^p\right]$. The attacker's utility is given by, $U_{-v}(\pi_v, \pi_{-v}) = -U_v(\pi_v, \pi_{-v})$.

To obtain the MSNE of the zero-sum game, we employ an extension of the double oracle approach [19]. The MSNE for the two players, that use mixed strategies $(\sigma_v^*, \sigma_{-v}^*)$, is given by: $U_v(\sigma_v^*, \sigma_{-v}^*) \geq U_v(\sigma_v, \sigma_{-v}^*) \quad \forall \sigma_v \in \Sigma_v$ and it is calculated as,

$$\begin{aligned} & max\ U_v^* \\ s.t.\ U_v(\sigma_v, \pi_{-v}) & \geq U_v^*, \forall \pi_{-v} \in \Pi_{-v} \\ \sum_{\pi_v \in \Pi_v} & \sigma_v(\pi_v) = 1 \\ \sigma_v(\pi_v) & \geq 0, \forall \pi_v \in \Pi_v \end{aligned} \tag{1}$$

The DRL based attack oracle computes its best response, $\pi_{-v}(\sigma_v)$, to the mixed strategy, σ_v, of the defender. Similarly, the defense oracle computes its best response, $\pi_v(\sigma_{-v})$, to the mixed strategy, σ_{-v}, of the attacker. In addition, the policy sets (Π_v, Π_{-v}) are updated with the optimal responses. The process is repeated until there is no improvement in the policies.

3 Proposed Solutions

In this section, we present our proposed knowledge based DRL algorithms for alert prioritization: D-KNAP, S-KNAP and T-KNAP. The proposed solutions compute the optimal pure strategy (π_v) of player v against its opponent's mixed strategy (σ_{-v}), as represented by the following equation: $\pi_v = \arg\max_{\pi_v} U_v(\pi_v, \sigma_{-v})$.

We consider different levels of knowledge obtained by the attacker and the defender. Our objective in this study is to construct a robust defender by concentrating solely on various levels of the defender's knowledge.

- **Attacker's knowledge:** We consider a strong attacker capable of observing the complete state, i.e., $< N, M, G >$. This eliminates the need to formulate precise (and potentially erroneous) hypotheses concerning the attacker's knowledge of the system's current state.
- **Defender's knowledge:** We consider four varying levels of defender's knowledge about the system: (i) $\mathbf{S} =< \mathbf{N} >$ i.e., the defender's knowledge includes only N (number of alerts to investigate), (ii) $\mathbf{S} =< \mathbf{N}, \mathbf{M} >$ i.e., with the knowledge of both N and M (where M is the indicator of attack execution), the defender gains insights into the occurrence of attacks, enabling a more precise approach to investigations, (iii) $\mathbf{S} =< \mathbf{N}, \mathbf{G} >$ i.e., possessing information about N and G, the defender obtains an understanding of the total alert count and the number specifically related to potential attacks, and (iv) $\mathbf{S} =< \mathbf{N}, \mathbf{M}, \mathbf{G} >$ i.e., with complete knowledge of N, M and G, the defender has a comprehensive perspective of the system state.

3.1 Overall Design of D-KNAP, S-KNAP and T-KNAP

In this section, we present the overall design and framework for the proposed approaches.

- **D-KNAP:** The D-KNAP integrates the DDPG actor-critic method with the proposed knowledge based attacker and defender. D-KNAP contains two neural networks (1 actor and 1 critic) for each of the attacker and defender. Algorithm 1 provides details of the D-KNAP algorithm. First, both the neural networks are initialized randomly (line 1). The system state is initialized according to the player's knowledge (lines 3–7). Player v then executes the action α_v using policy π_v and player -v executes an action α_{-v} using policy π_{-v} (lines 9–10). Player v receives a reward based on the state. It stores the tth transition $< s_v^t, a_v^t, r_v^{t+1}, s_v^{t+1} >$ in a memory buffer, D (line 11). Player v then samples a minibatch, from the memory buffer (line 12). The critic and actor network's loss functions are represented in lines 15–16 where $Q'_{\theta'_v}$ and $\mu'_{\phi'_v}$ are the target critic and target actor networks, respectively. Each of these networks weights are then updated. After a fixed number of episodes, the resulting policy network μ_{ϕ_v} is returned as the parameterized optimal response to an opponent with mixed strategy σ_{-v}.
- **S-KNAP:** S-KNAP integrates the SAC method with the proposed knowledge based attacker and defender. S-KNAP contains four neural networks (1 actor, 2 critics and 1 value) for each of the attacker and defender. Algorithm 2 provides details of the S-KNAP algorithm. First, all four neural networks

Algorithm 1. D-KNAP Algorithm

1: Randomly initialize actor network, μ_{ϕ_v} and critic network, Q_{θ_v};
2: **for** *episode* $= 0,\ J-1$ **do**:
3: **if** defender is the player **then**
4: Initialize System state (i) $S =< N >$, (ii) $S =< N, M >$, (iii) $S =< N, G >$, and (iv) $S =< N, M, G >$;
5: **else if** attacker is the player **then**
6: Initialize System state $S =< N, M, G >$;
7: **end if**
8: **for** t=0, t-1 **do**:
9: $a_v^{(t)} = \mu_{\phi_v}(s_v^t)$;
10: Execute $a_v^{(t)}$ and $a_{-v}^{(t)} = \mu_{\phi_{-v}}(s_{-v}^t)$, observe reward r_v^{t+1} and transit the system state to s^{t+1};
11: Store transition $< s_v^t, a_v^t, r_v^{t+1}, s_v^{t+1} >$ in D;
12: Sample a random minibatch of N transitions $< s_v^t, a_v^t, r_v^{t+1}, s_v^{t+1} >$ from D;
13: $a_v^{t+1} = \mu_{\phi_v'}(s_v^{t+1})$;
14: $\hat{y} = r^t + \gamma Q'_{\theta_v'}(s_v^{t+1}, a_v^{t+1})$;
15: $L(\theta_v) = E\big[(Q_{\theta_v}(s_v^t, a_v^t) - \hat{y})\big]$;
16: $L(\phi_v) = -E\big[Q_{\theta_v}\big(s_v, \mu_{\phi_v}(s_v)\big)\big]$;
17: Update network weights;
18: **end for**
19: **end for**
20: **return** $\mu_{\phi_v}(s_v)$ of player v;

are initialized randomly (line 1). The system state is initialized according to the player's knowledge (lines 3–7). Player v then executes the action α_v using policy π_v and player -v executes an action α_{-v} using policy π_{-v} (lines 9–10). The player v receives a reward based on the state . It stores the tth transition $< s_v^t, a_v^t, r_v^{t+1}, s_v^{t+1} >$ into a memory buffer (line 11). Player v then samples a minibatch, from the memory buffer, D (line 12). The actor network loss function is represented in line 13 where, β is the entropy regularization factor. The value function is trained to minimize the loss (line 14). The two critic network loss functions are represented in lines 15–16 where, $V'_{\psi_v'}$ is the target value network. Each of these network's are then updated. After a fixed number of episodes, the resulting policy network μ_{ϕ_v} is returned as the parameterized optimal response to an opponent with mixed strategy σ_{-v}.

- **T-KNAP:** T-KNAP integrates the TD3 actor-critic method with the proposed knowledge based attacker and defender. T-KNAP contains three neural networks (1 actor and 2 critics) for each of the attacker and defender. Algorithm 3 provides details of the T-KNAP algorithm. First, all three neural networks are initialized randomly (line 1). The system state is initialized according to the player's knowledge (lines 3–7). Player v then executes the action α_v using policy π_v and player -v executes an action α_{-v} using policy π_{-v} (lines 9–10). Player v receives a reward based on the state. It stores the tth transition $< s_v^t, a_v^t, r_v^{t+1}, s_v^{t+1} >$ into a memory buffer, D (line 11). Player

Algorithm 2. S-KNAP Algorithm

1: Randomly initialize actor network, μ_{ϕ_v}, critic networks, $Q_{\theta_{v1}}$, $Q_{\theta_{v2}}$ and value network, V_{ψ_v} ;
2: **for** *episode* $= 0,\ J-1$ **do**:
3: **if** defender is the player **then**
4: Initialize System state (i) $S =< N >$, (ii) $S =< N, M >$, (iii) $S =< N, G >$, and (iv) $S =< N, M, G >$;
5: **else if** attacker is the player **then**
6: Initialize System state $S =< N, M, G >$;
7: **end if**
8: **for** t=0, t-1 **do**:
9: $a_v^{(t)} = \mu_{\phi_v}(s_v^t)$;
10: Execute $a_v^{(t)}$ and $a_{-v}^{(t)} = \mu_{\phi_{-v}}(s_{-v}^t)$, observe reward r_v^{t+1} and transit the system state to s^{t+1};
11: Store transition $< s_v^t, a_v^t, r_v^{t+1}, s_v^{t+1} >$ in D;
12: Sample a random minibatch of N transitions $< s_v^t, a_v^t, r_v^{t+1}, s_v^{t+1} >$ from D;
13: $L(\mu_{\phi_v}) = E\big[-\big[\min_{i=1,2} Q_{\theta_{vi}}(s_v^t, \mu_{\phi_v}(s_v^t))\big] + \beta log(\mu_{\phi_v})\big)\big]$;
14: $L(\psi_v) = E\big[\big[\min_{i=1,2} Q_{\theta_{vi}}(s_v^t, \mu_{\phi_v}(s_v^t)) - \beta log(\mu_{\phi_v}(s_v^t))\big] - V_{\psi_v}(s_v^t)\big]$;
15: $L(\theta_{v1}) = Q_{\theta_{v1}}(s_v^t, a_v^t) - \big[r_{t+1} + \gamma V'_{\psi_v'}(s_v^{t+1})\big]$;
16: $L(\theta_{v2}) = Q_{\theta_{v2}}(s_v^t, a_v^t) - \big[r_{t+1} + \gamma V'_{\psi_v'}(s_v^{t+1})\big]$;
17: Update network weights;
18: **end for**
19: **end for**
20: **return** $\mu_{\phi_v}(s_v)$ of player v;

v then samples a minibatch, from the memory buffer (line 12). The critic and actor network's loss functions are represented in lines 15–16 where $Q'_{\theta'_{vi}}$ and $\mu'_{\phi'_v}$ are the target critic and target actor networks, respectively. Each of these networks weights are then updated. After a fixed number of episodes, the resulting policy network μ_{ϕ_v} is returned as the parameterized optimal response to an opponent with mixed strategy σ_{-v}.

4 Results

This section outlines the experimental setup, including the dataset and implementation details. Our experiments are conducted on a system running Ubuntu 18.04.6 LTS with Intel(R) Core (TM) i7-10700 CPU @2.90 GHz, 16 cores and 16 GB memory. Our proposed methods are implemented using the open-source neural network training library Tensorflow. The values of various hyperparameters is presented in Table 1. The experiment is conducted 20 times with 20 distinct random seeds. Each experiment runs for 2,00,000 instances. We consider MQTT-IoT-IDS2020 [15] and CSE-CIC-IDS2018 [16] datasets with Snort IDS for performance evaluation of the proposed approaches. The MQTT-IoT-IDS2020 dataset contains network data from IoT devices that support message queuing telemetry

Algorithm 3. D-KNAP Algorithm

1: Randomly initialize actor network, μ_{ϕ_v} and critic network, Q_{θ_v};
2: **for** *episode* $= 0,\ J-1$ **do**:
3: **if** defender is the player **then**
4: Initialize System state (i) $S =< N >$, (ii) $S =< N, M >$, (iii) $S =< N, G >$, and (iv) $S =< N, M, G >$;
5: **else if** attacker is the player **then**
6: Initialize System state $S =< N, M, G >$;
7: **end if**
8: **for** t=0, t-1 **do**:
9: $a_v^{(t)} = \mu_{\phi_v}(s_v^t)$;
10: Execute $a_v^{(t)}$ and $a_{-v}^{(t)} = \mu_{\phi_{-v}}(s_{-v}^t)$, observe reward r_v^{t+1} and transit the system state to s^{t+1};
11: Store transition $< s_v^t, a_v^t, r_v^{t+1}, s_v^{t+1} >$ in D;
12: Sample a random minibatch of N transitions $< s_v^t, a_v^t, r_v^{t+1}, s_v^{t+1} >$ from D;
13: $a_v^{t+1} = \mu_{\phi_v'}(s_v^{t+1})$;
14: $\hat{y} = r^t + \gamma Q_{\theta_v'}^{'}(s_v^{t+1}, a_v^{t+1})$;
15: $L(\theta_v) = E\big[(Q_{\theta_v}(s_v^t, a_v^t) - \hat{y})\big]$;
16: $L(\phi_v) = -E\big[Q_{\theta_v}\big(s_v, \mu_{\phi_v}(s_v)\big)\big]$;
17: Update network weights;
18: **end for**
19: **end for**
20: **return** $\mu_{\phi_v}(s_v)$ of player v;

transport (MQTT) protocol. The dataset comprises of 2,284,994 benign packets and 19,792,003 attack packets with 44 features. In this dataset, there are two attack types: bruteforce and portscan. Further, Snort IDS generates two alert types: attempted-recon and misc-activity. The CSE-CIC-IDS2018 dataset contains 16,232,943 flows including 13,484,708 benign flows and 2,748,235 attack flows with 75 features. In this dataset there are six attack types: botnet, bruteforce, DoS, DDoS, infiltration and webattack. Further, Snort IDS generates nine alert types: bad-unknown, attempted-recon, misc-activity, network-scan, rpc-portmap-decode, attempted-admin, trojan-activity, protocol-command-decode and web-application-attack. As a key performance metric, we consider defender's loss $(U_v(\pi_v, \sigma_{-v}))$, i.e., the failure of the defender to investigate alerts generated due to the attacks. We have compared the proposed approaches with the following state-of-the-art alert prioritization methods: Uniform and Snort [17].

4.1 Results with MQTT-IoT-IDS2020 Dataset

Figure 2 compares the performance of D-KNAP, S-KNAP and T-KNAP with other state-of-the-art alert prioritization methods for varying levels of defender's knowledge. The defender's budget is fixed at 1000 and the attacker's budget is 125. We do not consider the scenario in which the attacker's budget is greater than the defender's budget because attackers are assumed to work alone or

Table 1. Hyperparameter configurations for experiments.

Hyperparameter	Description	Value
α_z	Critic network's learning rate	0.002
α_i	Actor network's learning rate	0.001
α_y	Value network's learning rate	0.002
γ	Discount factor	0.95
β	Entropy regularization factor	0.5
τ	Smoothing constant	0.01
ϵ_{max}	Maximum epsilon value	1
$\epsilon_{discount}$	Epsilon discount factor	0.99
max- episodes	Number of episodes	500
max- epsteps	Number of steps per episode	400

in small groups, whereas the defender can consist of a large organization or group, such as an security operation centres (SoC), with substantial resources to implement security mechanisms. As expected, the defender's loss is minimal for the defender's complete knowledge i.e., for $< N, M, G >$ case. In addition, T-KNAP outperforms S-KNAP and D-KNAP by 19.59% and 36.84%, respectively. The benefits are significantly higher when T-KNAP is compared with other baselines, Uniform and Snort.

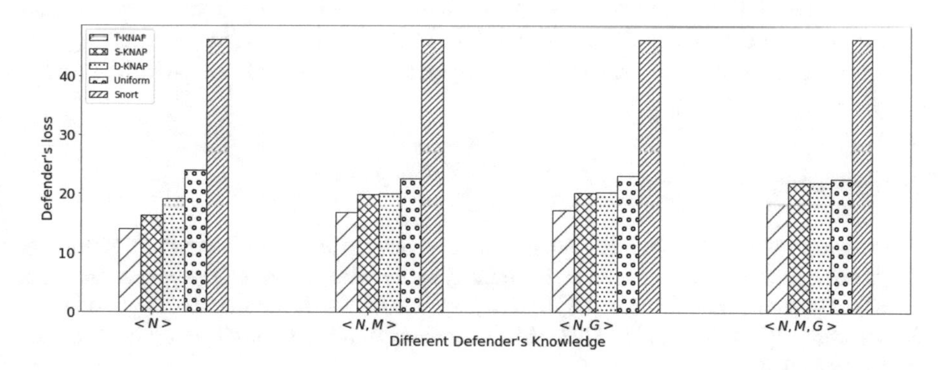

Fig. 2. Performance comparison of alert prioritization methods with MQTT-IoT-IDS2020 dataset when the defender's budget is 1000 and attacker's budget is 125.

4.2 Results with CSE-CIC-IDS2018 Dataset

We also evaluated the performance of the proposed approaches with CSE-CIC-IDS2018 which contain many more complex attacks as compared to MQTT-IoT-IDS2020 dataset. Our results in Fig. 3 show that T-KNAP outperforms S-KNAP and D-KNAP by 33.87% and 35.17%, respectively. The benefits are significantly higher when T-KNAP is compared with other baselines.

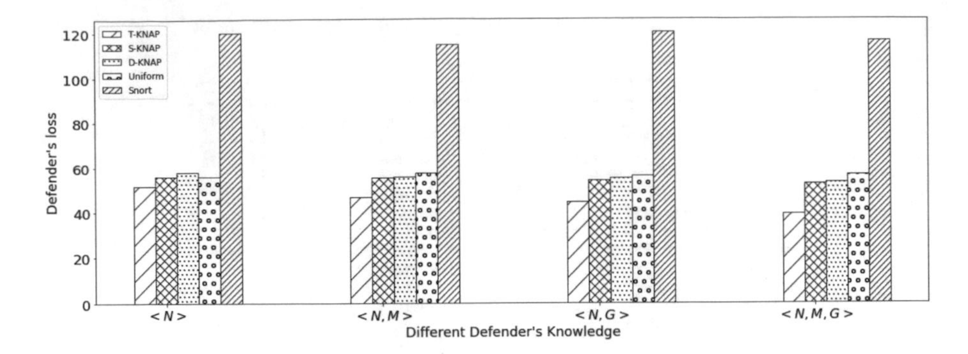

Fig. 3. Performance comparison of alert prioritization methods with CSE-CIC-IDS2018 dataset when the defender's budget is 1000 and attacker's budget is 125.

5 Conclusion

In this paper, we propose three novel knowledge empowered deep reinforcement learning based alert priortization approaches, D-KNAP, S-KNAP and T-KNAP. We illustrate the benefits of the proposed approaches with MQTT-IoT-IDS2020 dataset and CSE-CIC-IDS2018 dataset along with Snort IDS. Our results show that T-KNAP outperforms S-KNAP and D-KNAP by 19.59%, 33.87% and 36.84%, 35.17%, respectively for the MQTT-IoT-IDS2020 and CSE-CIC-IDS2018 datasets. Future work includes investigating the performance of KNAP in the following scenarios: (i) multiple attackers with different budget values, (ii) using advanced ML and DRL based IDS, and (iii) evaluating the practical applicability of the proposed approaches through real-world experiments.

References

1. Mclean, M.: Must-know cyber attack statistics and trends. Embroker (2023) https://www.embroker.com/blog/cyber-attack-statistics/#:~:text=cyber%20attacks%20have%20been%20rated,expected%20to%20double%20by%202025
2. Mandiant 2023. m-trends (2023). https://www.mandiant.com/m-trends. Accessed 10 Aug 2023
3. Liao, H.-J., Lin, C.-H.R., Lin, Y.-C., Tung, K.-Y.: Intrusion detection system: a comprehensive review. J. Netw. Comput. Appl. **36**(1), 16–24 (2013)
4. Alahmadi, B.A., Axon, L., Martinovic, I.: 99% false positives: a qualitative study of {SOC} analysts' perspectives on security alarms. In: 31st USENIX Security Symposium (USENIX Security 22), pp. 2783–2800 (2022)
5. Ulevitch, D.: Cisco 2017 annual cybersecurity report: the hidden danger of uninvestigated threats (2017). https://blogs.cisco.com/security/cisco-2017-annualcybersecurity-report-the-hidden-danger-of-uninvestigated-threats. Accessed 17 Apr 2019
6. Li, Z., Das, A.: The utility of partial knowledge in behavior models: an evaluation for intrusion detection. Int. J. Netw. Secur. **1**(3), 138–146 (2005)

7. Miller, D.J., Xiang, Z., Kesidis, G.: Adversarial learning targeting deep neural network classification: a comprehensive review of defenses against attacks. Proc. IEEE **108**(3), 402–433 (2020)
8. Nguyen, T.T., Reddi, V.J.: Deep reinforcement learning for cyber security. IEEE Trans. Neural Netw. Learn. Syst. (2019)
9. François-Lavet, V., Henderson, P., Islam, R., Bellemare, M.G., Pineau, J.: An introduction to deep reinforcement learning. arXiv preprint: arXiv:1811.12560 (2018)
10. Sutton, R.S., Barto, A.G.: Reinforcement Learning: An Introduction. MIT press, Cambridge (2018)
11. Chavali, L., Gupta, T., Saxena, P.: SAC-AP: soft actor critic based deep reinforcement learning for alert prioritization. In: 2022 IEEE Congress on Evolutionary Computation (CEC), pp. 1–8, IEEE (2022)
12. Lillicrap, T.P., et al.: Continuous control with deep reinforcement learning. arXiv preprint: arXiv:1509.02971 (2015)
13. Haarnoja, T., Zhou, A., Abbeel, P., Levine, S.: Soft actor-critic: off-policy maximum entropy deep reinforcement learning with a stochastic actor. In: International Conference on Machine Learning, pp. 1861–1870. PMLR (2018)
14. Fujimoto, S., Hoof, H., Meger, D.: Addressing function approximation error in actor-critic methods. In: International Conference on Machine Learning, pp. 1587–1596. PMLR (2018)
15. Hindy, H., Bayne, E., Bures, M., Atkinson, R., Tachtatzis, C., Bellekens, X.: Machine learning based IoT intrusion detection system: an MQTT case study (mqtt-iot-ids2020 dataset). In: Ghita, B., Shiaeles, S. (eds.) Selected Papers from the 12th International Networking Conference. Lecture Notes in Networks and Systems, vol. 180, pp. 73–84. Springer, Cham (2020). https://doi.org/10.1007/978-3-030-64758-2_6
16. CSE-CIC-IDS2018 on AWS, a collaborative project between the communications security establishment (CSE) and the Canadian institute for cybersecurity (CIC) (2018). https://registry.opendata.aws/cse-cic-ids2018
17. Network based intrusion detection system, snort. https://www.snort.org/
18. Tong, L., Laszka, A., Yan, C., Zhang, N., Vorobeychik, Y.: Finding needles in a moving haystack: Prioritizing alerts with adversarial reinforcement learning. In: Proceedings of the AAAI Conference on Artificial Intelligence, vol. 34, pp. 946–953 (2020)
19. Tsai, J., Nguyen, T.H., Tambe, M. : Security games for controlling contagion. In: Twenty-Sixth AAAI Conference on Artificial Intelligence (2012)

Container-Level Auditing in Container Orchestrators with eBPF

Fábio Junior Bertinatto, Daniel Arioza(✉), Jéferson Campos Nobre, and Lisandro Zambenedetti Granville

Institute of Informatics - Federal University of Rio Grande do Sul, Porto Alegre, Brazil
{fabio.bertinatto,daniel.almeida,jcnobre,granville}@inf.ufrgs.br

Abstract. This paper examines the application of eBPF (extended Berkeley Packet Filter) for achieving more precise auditing at the container level in container orchestrators such as Kubernetes. We address the challenges associated with auditing container behavior and highlight the advantages of leveraging eBPF to monitor container activities at the kernel level. We propose an eBPF-based solution that enhances transparency with respect to operations performed within containers. Overall, this study suggests that the use of eBPF for container-level auditing can provide valuable insights into container behavior and improve the security of containerized applications.

1 Introduction

The use of Linux containers has experienced a significant surge in popularity in recent years. This popularity comes from their ability to package and isolate applications in a portable manner. A survey conducted by the Cloud Native Computing Foundation (CNCF) in 2022 [1] revealed that 44% of the respondents reported using containers for most applications within their organizations. Furthermore, 35% of the respondents used containers for a few production systems, while 9% were actively evaluating the technology. According to recent data [2], Kubernetes[1], the leading container orchestration platform, is used by nearly half of all organizations as their primary tool for deploying and managing applications.

This surge in containerized applications and Kubernetes usage has introduced challenges, especially in real-time auditing for cluster administrators and network operators. Traditional methods, like executing a shell interpreter within the container, do not persist commands, leaving no audit trail post-termination. Kubernetes offers an *Events* mechanism for cluster events but lacks the capability to record shell interpreter commands.

Addressing this gap, our paper introduces an eBPF-based auditing solution. We employ eBPF for kernel instrumentation to capture commands executed within Bash shell interpreters by targeting the *readline* function. Our approach integrates an eBPF program loader service within containers, creating Kubernetes *Events* resources. This service provides cluster administrators with a detailed record of executed commands, enhancing the ability to audit and troubleshoot containerized applications.

[1] https://kubernetes.io.

L. Barolli (Ed.): AINA 2024, LNDECT 202, pp. 412–423, 2024.
https://doi.org/10.1007/978-3-031-57916-5_35

The paper is organized as follows: Sect. 2 covers the fundamentals of Linux containers, orchestrators, and eBPF. Section 3 reviews existing approaches using eBPF for container monitoring and auditing. Our proposed methodology and its assessment are detailed in Sect. 4. Section 5 concludes with final thoughts and potential future research directions.

2 Background

Containers have gained prominence as a viable option for deploying applications at scale, primarily due to their isolation capabilities and the ease with which applications can be packaged and deployed. As the adoption of containers increased, container orchestrators emerged as a solution for managing large-scale deployments of containerized applications. Simultaneously, eBPF has gained popularity as an efficient and flexible system monitoring solution. In this section, we explore these technologies and examine their key features to better understand how they can complement each other.

2.1 Containerization

Containers are isolated processes on a host machine, crucial in Linux systems for achieving process-level isolation through *namespaces*, *cGroups*, and *seccomp*. These kernel features isolate resources, control resource access, and restrict system calls, enhancing security and efficiency [3,4].

Figure 1 illustrates the need for an additional management interface to incorporate the aforementioned Linux kernel concepts into containers. The subsequent subsections will provide further details regarding this interface.

Container runtimes, like *runc* and *crun*, are command-line tools that manage containers in accordance with Open Container Initiative (OCI) specifications. They interact with container engines that handle additional functionalities like image retrieval and mounting [5].

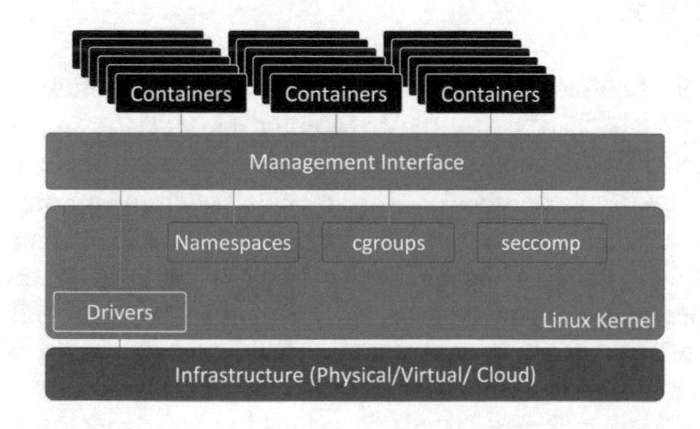

Fig. 1. Architecture of Linux Containers.

2.2 Container Orchestrators

Automated management of containerized services is vital due to the complexity of handling numerous containers. Orchestration frameworks, particularly Kubernetes, have become essential in managing container ecosystems at scale, addressing resource heterogeneity and environmental constraints [6].

2.2.1 Kubernetes

Kubernetes, a project initially developed by Google and now maintained by the CNCF, is a prominent orchestrator. It interfaces with various container engines via the Container Runtime Interface (CRI), ensuring compatibility across different container management tools [2,7].

Kubernetes objects like *Pods* represent the cluster's intended state, with each object playing a specific role in the system's overall functionality. The orchestration process involves managing these objects to maintain desired states [8].

2.3 eBPF

Extended BPF (eBPF), evolving from BPF, is a critical Linux kernel technology for system monitoring and tracing. It facilitates kernel-level program execution in response to events, making it a powerful tool for container auditing and security [9–13].

eBPF's versatility for applications like program tracing and performance analysis is particularly useful for container monitoring, providing deep insights into operations. Programs written in C-like syntax are compiled, loaded, and executed in the kernel, offering real-time system operation insights [14,15].

The programmability of eBPF allows for innovative container auditing solutions, enabling tracking of container actions for enhanced security and observability. This adaptability makes eBPF crucial in modern Linux environments for applications requiring detailed observability, such as container-level auditing [16].

3 Related Work

In the domain of container auditing and monitoring, recent studies have primarily focused on various aspects of performance monitoring, security, and network analysis, often leveraging eBPF technology. Our work aims to extend these concepts to the specific challenge of auditing commands issued by cluster administrators in containerized environments, an area that has not been thoroughly explored in existing literature.

Cassagnes et al. [6] discuss the application of eBPF for system performance monitoring in computer systems, underscoring the limitations of traditional methods. Their work highlights eBPF's ability to collect kernel-level data efficiently, which is relevant for our approach to monitor administrative actions in containers.

Nam et al. [17] explore eBPF's role in enhancing inter-container communication security. Their focus on security and unauthorized access prevention aligns with our goal of ensuring that administrative actions within containers are secure and traceable.

Liu et al. [18] present a system that uses eBPF for analyzing network traffic in dynamic containerized networks. Their comprehensive approach to network observability demonstrates the feasibility of using eBPF for in-depth monitoring in complex container environments, which is analogous to our focus on detailed command auditing.

While Burns [19] introduces the sidecar pattern for observability in Kubernetes, this method presents limitations such as resource overhead and the need for manual instrumentation. Our work aims to overcome these limitations by proposing a more integrated and automated approach to auditing within container orchestration environments.

Rice [20] suggests using a single eBPF-based agent per node, moving away from the sidecar pattern. The Tracee forensics tool, based on this concept, uses eBPF to trace activities on the host OS, including containerized applications. This approach to collecting and analyzing runtime events offers insights into implementing an efficient container-level auditing system.

To address the gap in existing literature, our work focuses on developing a method for accurately identifying and auditing specific commands executed by cluster administrators inside containers, leveraging the strengths of eBPF for real-time monitoring and security enhancement.

4 Evaluation

In this section, we assess the viability of employing eBPF programs within containers for monitoring and auditing objectives. In Sect. 4.1, we expound upon the eBPF program employed to capture executed commands within a container and elucidate our methodology for integrating it with Kubernetes. Lastly, we present the outcomes of our experiments in Sect. 4.3.

4.1 Propose and Implementation

In our implementation, we utilized an eBPF program to capture the commands executed within the Bash shell interpreter. This was achieved by instrumenting the *readline*[2] function, which is used by Bash to read user-provided commands. Our focus on Bash stems from its status as the default shell in most contemporary Linux distributions.

The eBPF program we employed monitored commands executed by any user on the system. While this may pose challenges in systems with multiple users, it becomes advantageous within the context of Linux containers. Containers have a limited perception of the system, and the program running inside a container perceives the container itself as the entire system. Consequently, our eBPF program captures commands executed by any process running within the same container.

However, running eBPF programs in containers introduces additional complexities. Notably, one major challenge is ensuring that the eBPF program runs within the same namespace as the process being monitored. Specifically, our eBPF program needs to run within the same mount and *PID* namespaces as the shell interpreter. To address this, we employed the *nsenter*[3] tool, which allows the execution of a program within specified

[2] https://git.savannah.gnu.org/cgit/readline.git/tree/readline.c?h=readline-8.2#n351.

[3] https://man7.org/linux/man-pages/man1/nsenter.1.html.

namespaces. We ran the eBPF program using nsenter to ensure it runs in the appropriate namespaces.

To initiate the eBPF program as soon as a shell interpreter starts running in a container, we introduced two additional components to our solution: a runc wrapper and a service running on the worker node alongside the *kubelet*. The *runc* wrapper detects the PID of the Bash process within any container and sends this information to the service running on the worker node. The service accepts incoming PIDs via an HTTP interface and runs the eBPF program within the specified PID's namespace using the nsenter tool.

Once the eBPF program is running within a container, it captures all commands executed within any Bash process running in that container. It communicates this information back to our service running on the host machine. This communication is facilitated by a Unix Domain Socket, which is mapped within the container and exposes the service's HTTP interface. When the service receives information from the eBPF program, it creates corresponding Events resources within the Kubernetes API server. As a result, Kubernetes cluster administrators gain a comprehensive view of the actions executed within the container.

Figure 2 illustrates the workflow of our implementation.

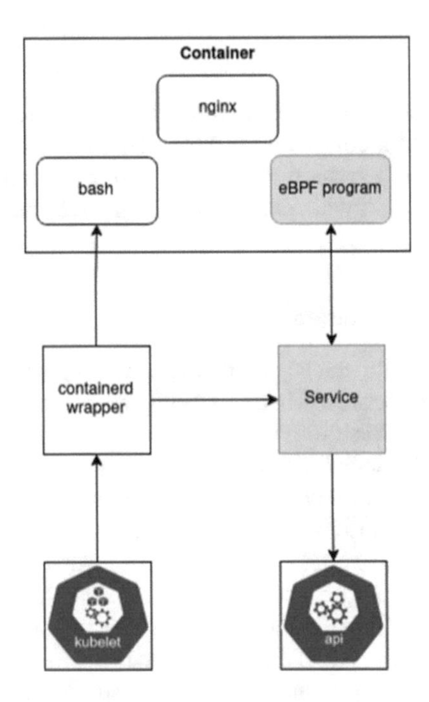

Fig. 2. Architecture of our implementation.

Our solution requires manual configuration. Specifically, a configuration change is needed in Containerd to invoke our wrapper instead of directly calling Runc. The excerpt in Listing 1 provides additional configuration details that should be added to the */etc/containerd/config.toml* file on the worker node, specifically under the *[plugins."io.containerd.grpc.v1.cri".containerd.runtimes]* section.

```
[plugins."io.containerd.grpc.v1.cri".containerd.
↪   runtimes.wrapper]
  runtime_type = "io.containerd.runc.v1"
  pod_annotations = ["*"]
  container_annotations = ["*"]

[plugins."io.containerd.grpc.v1.cri".containerd.
↪   runtimes.wrapper.options]
  BinaryName="/usr/bin/wrapper"
```

Listing 1: Containerd configuration.

To enable applications running in Kubernetes to utilize our runc wrapper and, as a result, leverage our entire solution, it is necessary to create a RuntimeClass and reference it in the workloads that require monitoring. Listing 2 illustrates the YAML file that can be used to create this resource in Kubernetes.

```
apiVersion: node.k8s.io/v1
kind: RuntimeClass
metadata:
  name: my-wrapper-name
handler: wrapper
```

Listing 2: RuntimeClass that uses our runc wrapper.

Lastly, it is necessary to specify the designated *RuntimeClass* in the deployment object of the containerized application that requires monitoring. For example, Listing 3 demonstrates a *Deployment* object that utilizes the *RuntimeClass* defined in Listing 2. Additionally, this object includes a mapping that allows the eBPF program to be accessed from the host's directory and executed by our service on the worker node via *nsenter*.

4.2 Environment

In order to assess the viability of using eBPF programs within containers, we conducted experiments in a test environment that emulates a typical deployment of containerized applications. Our test environment comprised a Kubernetes 1.26 cluster with a single node, running on a virtual machine. Opting for a single-node Kubernetes setup allowed us to concentrate on the feasibility of our approach rather than the scalability or performance of the system. The Kubernetes cluster was set up using the *hack/local-up-cluster.sh* script, which is available in the Kubernetes source code.

The virtual machine was provisioned using Vagrant version 2.2.9 and utilized the Kernel Virtual Machine (KVM) virtualization technology on a Linux-based system.

```yaml
apiVersion: apps/v1
kind: Deployment
metadata:
  name: nginx-deployment
spec:
  selector:
    matchLabels:
      app: nginx
  replicas: 1
  template:
    metadata:
      labels:
        app: nginx
    spec:
      runtimeClassName: wrapper
      containers:
      - name: nginx
        image: nginx:latest
        volumeMounts:
        - mountPath: /ebpf
          name: ebpf-program-mount-point
      volumes:
      - name: ebpf-program-mount-point
        hostPath:
          # Location on the host where
          # the eBPF program is located
          path: /path-on-the-host
```

Listing 3: Deployment Object using the custom wrapper RuntimeClass

The virtual machine was equipped with 4 GiB of Random Access Memory (RAM) and an Intel Skylake CPU with 4 cores. It ran Fedora Linux version 37 with a Linux kernel version 6.2.8.

The host machine employed for running the virtual machine was a ThinkPad P1 Gen 3 laptop, equipped with 32 GiB of RAM and an Intel i7-10850H processor featuring 12 cores. The host machine ran Fedora Linux version 37 with a Linux kernel version 6.2.8.

4.3 Experiments

After completing the manual steps outlined in Subsect. 4.1, we proceeded with a series of experiments to evaluate the effectiveness of our solution. To simulate a real-world troubleshooting scenario, we initiated a Bash shell interpreter within the running container *nginx* from Listing 3 and executed multiple commands.

These commands were designed to replicate a troubleshooting scenario where an administrator investigates the underlying cause of stalled requests within an *nginx* worker process. During this process, the administrator examines the *nginx* logs, monitors active processes, and utilizes the strace tool[4] to trace the specific system call where the issue arises.

During the experimentation process, our service successfully generated Kubernetes *Event* objects for each command executed within the container. Table 1 provides a comprehensive list of the *Events* created in our Kubernetes cluster throughout the course of our experiments. The data presented in the table was obtained using the *kubectl* command, filtering out any details unrelated to our specific experiments.

Table 1. Kubernetes Events.

Object	Message
pod/nginx-deployment-89c6ff86b-92ndx	ls
pod/nginx-deployment-89c6ff86b-92ndx	ip a
pod/nginx-deployment-89c6ff86b-92ndx	vim /etc/nginx/nginx.conf
pod/nginx-deployment-89c6ff86b-92ndx	cat /var/log/nginx/access.log
pod/nginx-deployment-89c6ff86b-92ndx	cat /var/log/nginx/error.log
pod/nginx-deployment-89c6ff86b-92ndx	ps
pod/nginx-deployment-89c6ff86b-92ndx	apt update
pod/nginx-deployment-89c6ff86b-92ndx	apt install procps
pod/nginx-deployment-89c6ff86b-92ndx	ps
pod/nginx-deployment-89c6ff86b-92ndx	strace -p 1
pod/nginx-deployment-89c6ff86b-92ndx	apt install strace
pod/nginx-deployment-89c6ff86b-92ndx	strace -p 1
pod/nginx-deployment-89c6ff86b-92ndx	strace -p 28

4.4 Comparison of Container-Level Monitoring and Auditing Solutions

In order to validate our solution, we conducted a comprehensive comparison with established tools in the market, adjusting their configurations to encompass a scope similar to ours. In this subsection we describe each of the compared solutions and provide details on the methodology and evaluation criteria used.

Falco is a real-time security monitoring tool designed to identify anomalous behaviors in applications and infrastructure. Tracee, another contender in container monitoring and security, excels at capturing and analyzing system events within containers, providing insights into container-level activities. Finally, Auditd, an established auditing solution, was included in our analysis. Auditd offers detailed system-level auditing capabilities. We compare our eBPF solution in terms of resource usage and responsiveness, justifying each comparison.

Specific rules were defined for each solution to capture and evaluate relevant container events. For Falco and Trace, we configured rules like "Terminal in Container" and

[4] https://strace.io/.

"Write below etc." to detect suspicious activities. Auditd, being an established solution, required less rule customization, showcasing the adaptability of the monitoring solutions.

Performance tests were conducted using a custom scrip that executed two specific test cases: 'Terminal in Container' and 'Write below etc.' For the 'Terminal in Container' test case, it simulated the use of a shell in a container. For the 'Write below etc.' test case, it detected attempts to write to any file below the '/etc' directory. The script was executed with the following parameters:

- **Repetitions:** 100,000 times
- **Resource Sampling Interval:** 0.1 s

The performance results are displayed in Table 2, which presents a comparison of three distinct metrics. The first column concerning the comparison of the response times, the second column regarding the CPU usage, and the third column pertaining to the assessment of memory usage. There are also graphs representing some level of the performance of the tools hence making it easier to analyze in very clear and comprehensive graphs. The evaluation comprised three rounds of tests: specifically, Test Type A was executed in a Terminal inside a Container and Test Type B that consumed many Write operations and others.

We provide a concise analysis of memory, CPU utilization, and response time from our comparative study:

- **Our Solution:** Exhibited excellent efficiency in memory and CPU usage, leveraging eBPF's event-specific capture. It outperformed Auditd and matched Tracee and Falco in resource utilization and response time, offering real-time monitoring and immediate alerts for suspicious activities.
- **Tracee:** Showed low memory and CPU impact, suitable for constrained environments, with real-time response capabilities.
- **Auditd:** Demonstrated heavier resource usage under complex rules or high event volume. It introduces latency due to log writing to disk.
- **Falco:** Provided efficient container behavior monitoring, though resource usage can escalate with extensive rules.

We propose a solution to this challenge based on eBPF that strikes the balance between resource efficiency and immediate responsiveness, offering a cost-efficient alternative for auditing and monitoring of containers. It enhances security without really adding significant overheads, making it well suited for efficient paradigms preferred by container environments.

Table 2. Comparative Analysis of Response Time, CPU Usage, and Memory Usage

Test	Tool	Response Time (s)	CPU Usage (%)	Memory Usage (MB)
Test 1A	Falco	49.16	−0.60	6.43
	Tracee	50.12	0.20	4.77
	eBPF	49.12	−0.50	6.30
	Auditd	49.34	−0.40	6.20
Test 2B	Falco	49.44	0.40	1.55
	Tracee	49.89	0.80	3.98
	eBPF	49.33	0.30	1.50
	Auditd	49.42	0.20	1.40
Test 3A	Falco	49.35	0.00	2.34
	Tracee	50.02	0.80	4.26
	eBPF	49.53	−0.10	2.20
	Auditd	49.63	0.00	2.10
Test 4B	Falco	49.54	0.00	0.76
	Tracee	49.83	0.60	3.55
	eBPF	49.74	−0.10	0.70
	Auditd	49.79	0.00	0.60
Test 5A	Falco	49.71	0.00	2.34
	Tracee	49.52	0.80	4.26
	eBPF	49.42	−0.10	2.20
	Auditd	49.44	0.00	2.10
Test 6B	Falco	49.84	0.00	0.76
	Tracee	49.93	1.60	3.55
	eBPF	49.90	−0.10	0.70
	Auditd	50.00	0.00	0.60

5 Conclusion

This solution combines security and efficacy to allow organizations in reinforcing their strategies for container security. The paper exhibits how effectively eBPF can effectively be employed in auditing tasks in Kubernetes orchestrators through capturing, as well as analyzing, the commands at a container level, basically the administrator instructions. Our eBPF-based solution flawlessly integrated within Kubernetes increases visibility, security, and observability over containerized systems delighting administrators with fine-grained insights. Experimental results carried out over synthetic Kubernetes clusters validate the efficacy of our approach in resource utilization while keeping approximately responsive the system.

Focusing on security and with a minimum system performance impact, our solution is perfect for any organizational setup looking for bulletproof container security solutions without having to compromise the system's performance. The comparison indicates the importance of memory, CPU utilization, and response time in ensuring that the security within the container environment is guaranteed. Our eBPF solution seems an effective option pointing towards its capability to optimize resources as well as its high potential to identify any future threat faster. The container-level monitoring tools offer the comparative analysis for memory, cpu usage and response with a goal of providing security for the container environment.

Looking ahead, enhancing the eBPF solution involves broadening support for various shell interpreters, evaluating its effectiveness in larger, multi-node Kubernetes clusters, and ensuring scalability under diverse workloads. Further, streamlining deployment through automation techniques like Kubernetes *DaemonSet* will improve efficiency.

Acknowledgements. This work was supported by The São Paulo Research Foundation (FAPESP) under the grant number 2020/05152-7, the PROFISSA project.

References

1. C. N. C. Foundation, "CNCF annual survey 2022" (2023). https://www.cncf.io/reports/cncf-annual-survey-2022
2. I. CDatadog, "9 insights on real-world container usage" (2023). https://www.datadoghq.com/container-report
3. Zhan, M., Li, Y., Yang, H., Yu, G., Li, G., Wang, W.: Runtime detection of application-layer CPU-exhaustion dos attacks in containers. IEEE Trans. Serv. Comput. 1–12 (2022)
4. Simonsson, J., Zhang, L., Morin, B., Baudry, B., Monperrus, M.: Observability and chaos engineering on system calls for containerized applications in docker. Future Gener. Comput. Syst. **122**, 117–129 (2021). https://www.sciencedirect.com/science/article/pii/S0167739X21001163
5. O. Specification, "About the open container initiative - open container initiative." https://opencontainers.org/about/overview/
6. Cassagnes, C., Trestioreanu, L., Joly, C., State, R.: The rise of eBPF for non-intrusive performance monitoring. In: Proceedings of IEEE/IFIP Network Operations and Management Symposium 2020: Management in the Age of Softwarization and Artificial Intelligence, NOMS 2020, no. August 2019 (2020)
7. Authors, T.K.: Container runtime interface (cri) (2023). https://kubernetes.io/docs/concepts/architecture/cri/
8. Authors, T.K.: Pods (2023). https://kubernetes.io/docs/concepts/workloads/pods/
9. McCanne, S., Jacobson, V.: The BSD packet filter: a new architecture for user-level packet capture. In: Proceedings of the Winter 1993 USENIX Conference, pp. 259–269 (1993)
10. Høiland-Jørgensen, T., et al.: The eXpress data path: fast programmable packet processing in the operating system kernel. In: CoNEXT 2018 - Proceedings of the 14th International Conference on Emerging Networking EXperiments and Technologies, pp. 54–66 (2018)
11. Abranches, M., Michel, O., Keller, E., Schmid, S.: Efficient network monitoring applications in the kernel with ebpf and xdp. In: 2021 IEEE Conference on Network Function Virtualization and Software Defined Networks (NFV-SDN), pp. 28–34 (2021)
12. Schulist, J., Borkmann, D., Starovoitov, A.: Linux socket filtering aka Berkeley packet filter (BPF) (2023). https://www.kernel.org/doc/Documentation/networking/filter.txt
13. Gebai, M., Dagenais, M.R.: Survey and analysis of kernel and userspace tracers on linux: design, implementation, and overhead. ACM Comput. Surv. **51**(2), 1–33 (2018)

14. eBPF.io Authors, "ebpf - introduction, tutorials & community resources," (2023). https://ebpf.io
15. Vieira, M.A.M., Castanho, M.S., Pacífico, R.D.G., Santos, E.R.S., Júnior, E.P.M.C., Vieira, L.F.M.: Fast packet processing with ebpf and xdp: concepts, code, challenges, and applications. ACM Comput. Surv. 53(1), 1–36 (2020). https://doi.org/10.1145/3371038
16. Abranches, R., Tuma, F., Guimarães, R., Vieira, M.: ebpf: a game-changer for network monitoring and security. IEEE Commun. Surv. Tutorials **23**(2), 1815–1842 (2021)
17. Nam, J., Lee, S., Porras, P., Yegneswaran, V., Shin, S.: Secure inter-container communications using xdp/ebpf. IEEE/ACM Trans. Netw. **31**, 1–14 (2022)
18. Liu, C., Cai, Z., Wang, B., Tang, Z., Liu, J.: A protocol-independent container network observability analysis system based on eBPF. In: Proceedings of the International Conference on Parallel and Distributed Systems - ICPADS, vol. 2020-Decem, pp. 697–702 (2020)
19. Burns, B.: Designing Distributed Systems. O'Reilly Media, Inc. (2018)
20. Rice, L.: What Is eBPF? O'Reilly Media, Inc. (2022)

A Study on Privacy-Preserving Transformer Model for Cross-Domain Recommendation

Jing Ning and Kin Fun Li(✉)

University of Victoria, Victoria, Canada
{jingning,kinli}@uvic.ca

Abstract. As customer relationship management becomes increasingly data-driven, cross-domain recommendation (CDR) systems are critical in leveraging insights from different domains to enhance customer experience. However, the aggregation of cross-domain user data raises significant privacy concerns. We propose a transformer-based CDR model that shows improved performance on key metrics such as Mean Reciprocal Ranking (MRR) and hit rate. In this model, we introduce a privacy-preserving method using embedded mask and differential privacy to protect user information. Our contribution is twofold: we propose ways to protect user privacy in CDR and analyze the balance between accuracy and privacy.

1 Introduction

Cross-domain recommendation (CDR) systems improve the quality of recommendations by utilizing information from other domains to overcome the limitations of single-domain data. In scenarios where user-item interaction is sparse, CDR systems provide a solution by transferring knowledge across domains, such as recommending movies based on a user's book preferences. Current research at CDR is focused on developing more mature algorithms that can accurately map user preferences in similar domains to enrich the recommendation process and gain a more comprehensive understanding of user interests.

As recommendation systems become more connected with our digital lives, privacy technologies are becoming increasingly important. In a cross-domain scenario, the challenge is twofold: protect sensitive information that may be more exposed in a CDR without compromising recommendation quality. With increased regulatory scrutiny and consumer awareness of data privacy, adopting privacy protection mechanisms in CDR systems is not a technical requirement, but a necessity to maintain user trust and comply with legal frameworks such as the General Data Protection Regulation (GDPR).

This paper makes several contributions to the privacy-preserving cross-domain recommendations. First, we propose a privacy-preserving method that uses embedded masking and differential privacy (DP) techniques. Second, we

L. Barolli (Ed.): AINA 2024, LNDECT 202, pp. 424–435, 2024.
https://doi.org/10.1007/978-3-031-57916-5_36

apply an advanced transformer-based model to the CDR, which has demonstrated good performance across multiple evaluation metrics while guaranteeing user privacy. Finally, we provide an analysis of the trade-off between accuracy and privacy, offering insights into the optimal configuration of DP parameters that achieve a balance between recommended quality and privacy concerns.

Section 2 examines existing literature, providing a background on CDR systems and privacy-preserving methodologies. Section 3 describes the methodologies employed, including the specifics of embedding masking and DP mechanisms. Section 4 presents the experimental setup, evaluation metrics, and results, showcasing the effectiveness of our proposed model. Section 5 discusses the experiment results, particularly the trade-off between privacy and accuracy, and Sect. 6 concludes the paper with a summary of our contributions and potential direction for future research.

2 Background and Related Works

Cross-Domain Recommendation. CDR has emerged as a solution to the limitations of single-domain recommenders, including data sparsity and the cold-start problem. By utilizing data from interrelated domains, CDR systems benefit from relational learning through collective matrix factorization, and shared representations that enhance recommendation tasks [1]. The DTCDR framework advances this concept by improving recommendation quality across dual targets [2], while collaborative cross networks like CoNet refine outputs through user-item interaction data across domains [3]. Furthermore, bi-directional transfer learning models employ transfer graph collaborative filtering to facilitate bidirectional knowledge transfer [4], and deep learning adaptations like the Deepapf model use deep attentive probabilistic factorization for nuanced recommendations [5]. The embedding and mapping approaches project interactions into a shared latent space, enabling cross-domain recommendations [6]. Tackling the cold-start problem, semi-supervised learning approaches use both labeled and unlabeled data for robust recommendations [7], while comprehensive frameworks like RecBole standardize research and deployment in the field [8]. These advancements and the field's development, including future directions, are recorded in extensive surveys [9,10].

Privacy-Preserving. Privacy-preserving is a critical concern in cross-domain recommendations to ensure the confidentiality of sensitive user data during knowledge transfer. A variety of techniques have been developed to address this, such as differential private knowledge transfer [11], which maintains data privacy while enabling recommendations across domains. Special attention has been given to protecting location data, as seen in privacy-preserving systems that secure users' location preferences [12]. Privacy-preserving matrix factorization methods further protect user data across various domains [13], with local differential privacy enhancing the security of individual user data within the recommendation process [14]. The PPGenCDR framework represents a step towards robust privacy-preserving mechanisms, ensuring system stability along-side user privacy in cross-domain recommendations [15]. Similarly, the DPLCF model introduces a differentially private approach to local collaborative filtering, emphasizing privacy at the user-item interaction level [16]. The extensive analysis of security and privacy status in recommendation systems by Himeur et al. emphasizes the urgency of the privacy-preserving mechanism [17].

Transformer Model. The Transformer, initially developed for natural language processing, has gained acceptance for the implementation in recommendation systems, introducing advanced modeling techniques to the field. The BERT4Rec system by Sun et al. represents a significant shift, employing a transformer-based architecture with bidirectional encoder representations to capture sequential user interactions with enhanced nuance [18]. This approach differs from the traditional unidirectional models, leading to a more complex understanding of user behavior [4]. The Behavior Sequence Transformer by Chen et al. further refines this technology for e-commerce, predicting user actions by analyzing behavior data [19], while the SSE-PT model by Wu et al. personalizes sequential recommendations by incorporating individual user preferences into the transformer model [20]. These innovations are built on the foundational work of Vaswani et al., who introduced the self-attention mechanism, a critical element of transformer models that have significantly impacted the capacity for accurately predicting user preferences in recommendation systems [21].

Motivation and Prospective Direction. While transformers have revolutionized the recommendation systems by adeptly capturing the complexities of user-item interactions, they still have potential in the specific area of privacy-preserving cross-domain recommendation. The inherent capability of transformers to handle sequential data and their powerful self-attention mechanisms can significantly benefit CDR by providing a nuanced understanding of user preferences across domains. However, there is still room for improvement in the effectiveness of existing privacy protection methods in CDR, as they struggle to strike a balance between recommendation accuracy and privacy. To address these shortcomings, there is a need for innovative methods that can use the strengths of transformers within a privacy-preserving framework for CDR. By introducing a privacy-preserving method designed for transformer-based CDR systems, we can conduct an analysis of the trade-offs between accuracy and privacy. This analysis would not only enhance the recommendation quality by taking full advantage of the transformer architecture but also ensures privacy controls.

3 Preliminary

This section introduces some basic concepts that are essential to our model. We introduced the models used in our study in the following aspects: neural collaborative filtering; transformer, the machine learning model used;and the privacy-preserving method used: . This section introduces some basic concepts that are essential to our model. We introduced the models used in our study in the following aspects: neural collaborative filtering; transformer, the Machine learning model used; and differential privacy, the privacy-preserving method used.

3.1 Neural Collaborative Filtering

Neural Collaborative Filtering (NCF) is an advanced approach in recommendation systems that moves beyond traditional matrix factorization (MF) methods, as introduced by He et al. It employs neural networks to handle complex user-item interactions, transforming the basic MF equation $r_{ui} = p_u^T q_i$ into a more dynamic, non-linear framework. In NCF, user-item interactions are approached as a binary classification problem, focusing on predicting the likelihood of these interactions.

The architecture of NCF combines a dual-pathway approach, integrating Generalized Matrix Factorization (GMF) and Multi-Layer Perceptron (MLP). The GMF component is responsible for modeling linear interactions using the equation $\hat{y}_{ui}^{GMF} = \sigma(p_u^T q_i)$, where p_u and q_i are latent vectors representing the user and item, respectively. The MLP, on the other hand, captures non-linear interactions through $\hat{y}_{ui}^{MLP} = \sigma(MLP(p_u, q_i|\Theta_{MLP}))$, where $MLP(p_u, q_i|\Theta_{MLP})$ is a multi-layer perceptron that processes the concatenated user and item vectors, p_u and q_i, parameterized by Θ_{MLP}.

The final prediction in NCF is a fusion of these two pathways, aiming to balance linear and non-linear elements for improved recommendation accuracy:

$$\hat{y}_{ui} = \sigma(h^T[\alpha \cdot \hat{y}_{ui}^{GMF} \oplus (1-\alpha) \cdot \hat{y}_{ui}^{MLP}])$$

In this equation, h is a vector projecting the combined outputs to a prediction score, and α is a hyperparameter that balances the contributions of the GMF and MLP components.

To effectively implement NCF, especially in cross-domain recommendations, their neural network parameters need to be carefully designed and optimized, as shown in Table 1.

Table 1. Parameters of the Neural Collaborative Filtering Model

Parameter	Description
p_u and q_i	Latent vectors representing the user and item, respectively, in the GMF part
σ	A non-linear activation function, such as sigmoid, applied to both the GMF and MLP outputs
Θ_{MLP}	The set of parameters (weights and biases) of the MLP, which are learned during training
h	A vector in the final prediction layer that projects the combined GMF and MLP outputs to a prediction score
α	A hyperparameter that balances the contributions of the GMF and MLP components, determining their influence on the final prediction

3.2 Transformer Model

The transformer model provides weighting of input data through its self-focusing mechanism, which is particularly beneficial for cross-domain recommendation systems that need to integrate user interactions from multiple domains [4,21]. The scale dot product attention of the transformer is given by the following formula:

$$\text{Attention}(Q, K, V) = \text{softmax}\left(\frac{QK^T}{\sqrt{d_k}}\right)V$$

In this equation, Q (Query) represents the set of queries, which could be likened to user or item representations in recommendation systems, K (Key) corresponds to keys associated with items or users, and V (Value) is linked to items or users. The term d_k denotes the dimension of the key vectors, employed to scale the dot product and maintain manageable values within high-dimensional models.

This allows all locations to be processed in parallel, significantly more efficiently than traditional RNNS and LSTMS. The Transformer further utilizes multi-head attention to focus on different parts of the data, enhancing its representational capabilities:

$$\text{MultiHead}(Q, K, V) = \text{Concat}(\text{head}_1, ..., \text{head}_h)W^O$$
$$\text{where head}_i = \text{Attention}(QW_i^Q, KW_i^K, VW_i^V)$$

This mechanism allows the model to simultaneously attend to information from various representation subspaces at different positions. The weight matrices W_i^Q, W_i^K, and W_i^V for each head are learned during training, enabling the model to capture different facets of the input data. The output of these concatenated heads is then transformed by another weight matrix, W^O. The number of heads, denoted by h, signifies the capacity of the model to focus on varied aspects of the input.

Each position is independently processed through a feed-forward network:

$$\mathrm{FFN}(x) = \max(0, xW_1 + b_1)W_2 + b_2$$

This network consists of two linear transformations with a ReLU activation in the intermediate stage, where W_1 and W_2 are the respective weight matrices, and b_1 and b_2 are bias vectors.

Incorporating the transformer into cross-domain recommendation offers the promise of improved recommendation quality by capturing complex user behaviors across diverse domains, showcasing the model's scalability and flexibility in learning long-range user-item interactions.

3.3 Differential Privacy

DP provides a framework for quantifying privacy risks, essential for privacy-preserving mechanisms in cross-domain recommendation systems. Its implementation usually involves a Gaussian mechanism, which adds noise to the analytical output to mask the effects of any single data point, guaranteeing personal privacy [4].

The Gaussian mechanism is mathematically expressed as:

$$O(x) = f(x) + \mathcal{N}(0, \sigma^2 \Delta^2)$$

Here, $O(x)$ denotes the algorithm's output, $f(x)$ is the original function, and $\mathcal{N}(0, \sigma^2 \Delta^2)$ represents Gaussian noise with a variance dependent on the function's sensitivity Δ and the desired privacy level σ.

In practice, the sensitivity Δ is managed through a 'clip_norm' parameter that limits individual data contributions, while σ determines the noise scale. The careful adjustment of these parameters is crucial in cross-domain systems to protect user privacy without overly diminishing the utility of the recommendations. Integrating DP into cross-domain recommendations requires striking a trade-off between the accuracy of recommendations and user privacy, ensuring that insights from one domain do not compromise the privacy of individuals in another.

4 Proposed Method

In this section, we will introduce the model in detail from the following parts. First, the overall flow of the model is introduced. The following describes the three main parts that make up the model: the embedding, the recommended model, and the training flow.

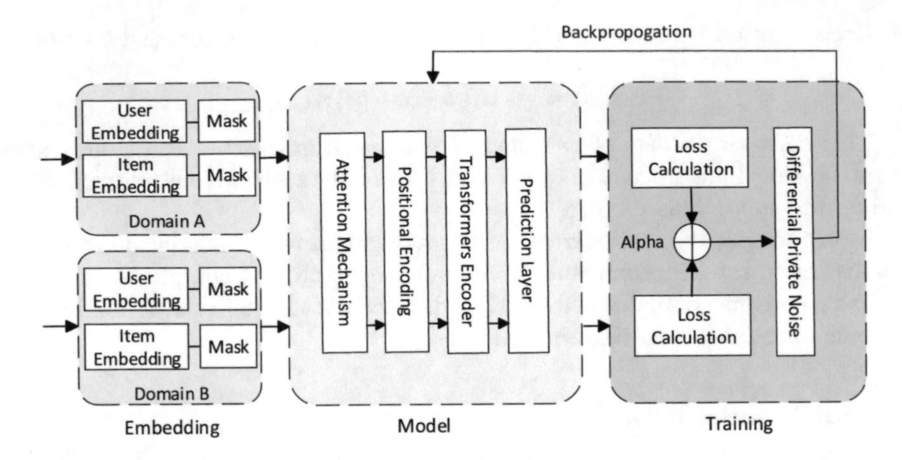

Fig. 1. Architecture Overview

4.1 Architecture Overview

Figure 1 presents a comprehensive framework for privacy-preserving cross-domain recommendation, including two parallel domains, source and target. Each domain contains user and project embeddings that represent potential characteristics of users and projects. After the mask mechanism, these embeddings are fed into a central embedding model whose task is to generate shared potential spaces for cross-domain knowledge transfer. The loss calculation module within each domain calculates the error between the prediction and the actual interaction, guides the iterative training process, and adjusts the embedding to minimize losses. This training follows privacy constraints and uses privacy technologies to protect sensitive data. As the system evolves through training iterations, the embeddings improve, ensuring that the model's recommendations improve accuracy without compromising user privacy, thus maintaining a delicate balance between practicality and confidentiality.

4.2 Embedding

In cross-domain recommendation systems, the embedding process is important for transforming raw user and item data into a dense, lower-dimensional space. This process is particularly sensitive as it involves direct interaction with user and item identifiers which could potentially expose private data. To address this, embeddings are masked, especially for users and items that do not exist in the overlapping domain space, effectively protecting their privacy. In the NeuMF-based model, separate user and item embeddings are created for both the source and target domains. To ensure privacy, the weights for non-overlapping users and items are set to negative infinity, which masks those embeddings. This process allows for the secure integration of domain-specific knowledge while ensuring that private data remains protected.

4.3 Recommendation Model

The cross-domain recommendation model uses the transformer architecture to enhance recommendation performance across different domains. It utilizes the attention mechanism to focus on specific aspects of user-item interactions. The transformer is combined with positional encoding, which provides information about the order of items in the sequence data, which is important for capturing the context within user activity sequences. At the center of the model is the transformer encoder, which processes the embedded representations of users and items. That is where the model learns complex patterns and dependencies, a task made possible by the transformer's ability to handle long-range dependencies within the data. Following the encoding process, the representations are fed into a prediction layer, which is tasked with outputting the final recommendation scores. This layer typically consists of a fully connected network, and its parameters are fine-tuned to optimize recommendation accuracy.

4.4 Training

The training process in cross-domain recommendation includes several key steps. The core process is the loss calculation, where individual losses from the source and target domains are computed using a predefined loss function. These losses are then combined into a single composite loss by weighting them with a parameter, α, which represents the relative importance of each domain in the training process.

DP plays a crucial role in training. To maintain privacy, DP noise is added to the gradients during backpropagation. This is achieved by first clipping the gradients to a specified norm, defined by 'clip_norm', and then adding Gaussian noise with a standard deviation of 'sigma' to the clipped gradients. This ensures that the model updates do not reveal sensitive information.

The composite loss, a weighted sum of the source and target domain losses adjusted by the parameter α, guides the model updates. Backpropagation is used to calculate gradients based on this composite loss, which is then modified by the DP mechanism before updating the model parameters. This privacy-preserving training ensures that the model benefits from the combined data of both domains, enhancing recommendation performance while maintaining privacy-preserving. The balance between loss from both domains, adjusted by α, and the combination of DP noise is the key to training privacy-preserving cross-domain recommendation model.

5 Experiment and Result Analysis

This section verifies our model from three aspects: experimental setup, trade-off study, and analysis of experimental results.

5.1 Experiments Settings

As shown in Table 2, we used MovieLens-1M with 6039 users and 3308 items, and MovieLens-100K with 944 users and 1197 items, to test a cross-domain recommendation system. The sparsity of these datasets is 95.82% and 92.77%, respectively.

We used the most common performance metrics in the recommendation valuations, which are described below:

- MRR (Mean Reciprocal Rank): Reflects the rank of the first correct recommendation, with higher values indicating better performance.
- NDCG (Normalized Discounted Cumulative Gain): Measures the ranking quality, emphasizing the top of the list.
- HIT: Checks if the correct recommendation is within the top-n list without considering the rank order.

Table 2. Statistics of the MovieLens Datasets

Dataset	Users	Items	Interactions	Sparsity
MovieLens-1M	6039	3308	835789	95.82%
MovieLens-100K	944	1197	81705	92.77%

5.2 Privacy Trade-Off

In our evaluation of the differential privacy (DP) parameters on the recommendation system's performance, we observed a trade-off between privacy preservation and accuracy. The DP mechanism incorporates two key parameters: 'sigma', representing the noise level, and 'clip_norm', the clipping threshold. Our findings indicate that a lower 'sigma' tends to improve accuracy as it introduces less noise into the system but at the expense of user privacy. Conversely, a higher 'sigma' value enhances privacy by adding more noise, although it may negatively impact the system's accuracy. The 'clip_norm' parameter controls the maximum contribution of any individual user, with lower values enhancing privacy by reducing the influence of outliers and higher values allowing for greater individual contribution, thereby increasing accuracy but potentially compromising privacy.

The experimental results as shown in Fig. 2, demonstrate that the best performance is achieved with a 'sigma' of 0.05 and a 'clip_norm' of 100, striking an optimal balance between privacy and accuracy. This configuration outperformed others in terms of Mean Reciprocal Rank (MRR@10), Normalized Discounted Cumulative Gain (NDCG@10), and Hit Rate (HIT@10), where the @10 suffix indicates that these metrics are calculated considering only the top 10 recommendations, suggesting its effectiveness in maintaining a high recommendation quality while ensuring user privacy. As 'sigma' decreases, there is a clear trend of improved performance, especially when coupled with higher 'clip_norm' values. However, the performance gains tend to plateau or slightly decline at the highest 'clip_norm' levels, underscoring a diminishing return on accuracy gains relative to privacy loss.

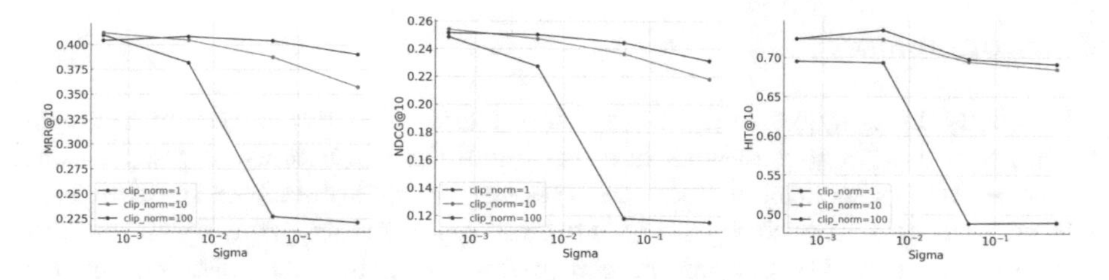

Fig. 2. Impact of different Sigma and Clip_norm

5.3 Result Analysis

In a comparative analysis of cross-domain recommendation models, our DP-Transformers model exhibited better performance across multiple metrics, indicating its effectiveness in producing relevant recommendations. The model demonstrated the highest scores in terms of MRR, NDCG, and HIT. Particularly, at MRR@10, the DP-Transformers achieved a score of 0.4084, outperforming the next best model, EMCDR, which scored 0.3808. In terms of NDCG@10 and HIT@10, the model again excelled with scores of 0.2504 and 0.7348 respectively, suggesting that the DP-Transformers not only recommend relevant items but also rank these items higher in the recommendation list, enhancing the user experience.

The prioritization of the DP-Transformers model also performs well, with the model maintaining its superiority at MRR@20, and getting best results to EMCDR at NDCG@20, HIT@20. This trend continues at the top-50 recommendation list, where the model's scores in MRR@50, NDCG@50, and HIT@50 are 0.4244, 0.348, and 0.9276 respectively. The consistently good performance across all metrics and list lengths highlights the model's robustness and its ability to scale effectively, catering to diverse user preferences while maintaining the quality of recommendations (Table 3).

Table 3. Evaluation of Various Models

Model	mrr@10	ndcg@10	hit@10	mrr@20	ndcg@20	hit@20	mrr@50	ndcg@50	hit@50
CLFM	0.3771	0.2325	0.685	0.4042	0.2602	0.8094	0.4078	0.3284	0.9148
EMCDR	0.3808	0.2345	0.6882	0.401	**0.2776**	**0.836**	0.404	0.3433	0.9244
Co-net	0.3578	0.2199	0.6713	0.4018	0.2592	0.8104	0.4051	0.3189	0.9095
DCDCSR	0.3389	0.2091	0.6766	0.4083	0.2717	0.8243	0.4114	0.3362	0.9169
DeepAPF	0.3129	0.2004	0.6691	0.3565	0.2383	0.7913	0.3597	0.2977	0.8892
DTCDR	0.3751	0.2315	0.6903	0.4041	0.2624	0.7955	0.4081	0.3278	0.9148
NATR	0.3278	0.1957	0.6628	0.4081	0.2645	0.8168	0.4112	0.3297	0.9084
DP-Transformers	**0.4084**	**0.2504**	**0.7348**	**0.4104**	0.2731	0.8307	**0.4244**	**0.348**	**0.9276**

6 Conclusion

This study investigates and analyzes the trade-off between recommendation accuracy and privacy-preserving in cross-domain recommendation systems. Our proposed DP-Transformers model achieves better performance on multiple metrics without compromising privacy. Through experiments, we have shown that carefully adjusting DP parameters can produce a satisfactory balance between user privacy and recommendation accuracy. These results not only demonstrate the practicality of the model but also illustrate its adaptability in real-world applications where privacy cannot be ignored. Our study also provides a reference for further improvement of these methods.

References

1. Singh, A.P., Gordon, G.J.: Relational learning via collective matrix factorization. In: Proceedings of the 14th ACM SIGKDD International Conference on Knowledge Discovery and Data Mining (KDD 2008), pp. 650–658 (2008). https://doi.org/10.1145/1401890.1401969
2. Zhu, F., Chen, C., Wang, Y., Liu, G., Zheng, X.: DTCDR: a framework for dual-target cross-domain recommendation. In: Proceedings of the 28th ACM International Conference on Information and Knowledge Management (CIKM 2019), pp. 1533–1542 (2019). https://doi.org/10.1145/3357384.3357992
3. Hu, G., Zhang, Y., Yang, Q.: oNet: collaborative cross networks for cross-domain recommendation. In: Proceedings of the 27th ACM International Conference on Information and Knowledge Management (CIKM 2018), pp. 667–676 (2018). https://doi.org/10.1145/3269206.3271684
4. Liu, M., Li, J., Li, G., Pan, P.: Cross domain recommendation via bi-directional transfer graph collaborative filtering networks. In: Proceedings of the 29th ACM International Conference on Information & Knowledge Management (CIKM 2020), pp. 885–894 (2020). https://doi.org/10.1145/3340531.3412012
5. Yan, H., Chen, X., Gao, C., Li, Y., Jin, D.: Deepapf: deep attentive probabilistic factorization for multi-site video recommendation. In: Presented at IJCAI, TC (2019). https://www.ijcai.org/
6. Man, T., Shen, H., Jin, X., Cheng, X.: Cross-domain recommendation: an embedding and mapping approach. In: IJCAI (2017). https://www.ijcai.org/
7. Kang, S.K., Hwang, J., Lee, D., Yu, H.: Semi-supervised learning for cross-domain recommendation to cold-start users. In: Proceedings of the 28th ACM International Conference on Information and Knowledge Management (2019). https://dl.acm.org/
8. Zhao, W.X., et al.: Recbole: towards a unified, comprehensive and efficient framework for recommendation algorithms. In: Proceedings of the 30th ACM International Conference on Information and Knowledge Management (2021). https://dl.acm.org/
9. Zang, T., Zhu, Y., Liu, H., Zhang, R., Yu, J.: A survey on cross-domain recommendation: taxonomies, methods, and future directions. ACM Trans. Inf. Syst. (2022). https://dl.acm.org/
10. Zhu, F., Wang, Y., Chen, C., Zhou, J., Li, L., Liu, G.: Cross-domain recommendation: challenges, progress, and prospects. arXiv preprint arXiv:2103.01696 (2021)

11. Chen, C., Wu, H., Su, J., Lyu, L., Zheng, X., Wang, L.: Differential private knowledge transfer for privacy-preserving cross-domain recommendation. In: Proceedings of the ACM Web Conference 2022 (2022). https://dl.acm.org/
12. Gao, C., Huang, C., Yu, Y., Wang, H., Li, Y., Jin, D.: Privacy-preserving cross-domain location recommendation. In: Proceedings of the ACM on Interactive, Mobile, Wearable and Ubiquitous Technologies (2019). https://dl.acm.org/
13. Ogunseyi, T.B., Avoussoukpo, C.B., Jiang, Y.: Privacy-preserving matrix factorization for cross-domain recommendation. IEEE Access (2021). https://ieeexplore.ieee.org/
14. Shin, H., Kim, S., Shin, J., Xiao, X.: Privacy enhanced matrix factorization for recommendation with local differential privacy. IEEE Trans. Knowl. Data Eng. (2018). https://ieeexplore.ieee.org/
15. Liao, X., Liu, W., Zheng, X., Yao, B., Chen, C.: PPGenCDR: a stable and robust framework for privacy-preserving cross-domain recommendation. arXiv preprint arXiv:2305.16163 (2023)
16. Gao, C., Huang, C., Lin, D., Jin, D., Li, Y.: DPLCF: differentially private local collaborative filtering. In: Proceedings of the 43rd International ACM SIGIR Conference on Research and Development in Information Retrieval (2020). https://dl.acm.org/
17. Himeur, Y., Sohail, S.S., Bensaali, F., Amira, A.: Latest trends of security and privacy in recommender systems: a comprehensive review and future perspectives. Comput. Secur. (2022). https://www.sciencedirect.com/journal/computers-and-security
18. Sun, F., et al.: BERT4Rec: Sequential recommendation with bidirectional encoder representations from transformer. In: Proceedings of the 28th ACM International Conference on Information and Knowledge Management (2019). https://dl.acm.org/
19. Chen, Q., Zhao, H., Li, W., Huang, P., Ou, W.: Behavior sequence transformer for e-commerce recommendation in Alibaba. In: Proceedings of the 1st International Conference on (Conference Name) (2019). https://dl.acm.org/
20. Wu, L., Li, S., Hsieh, C.J., Sharpnack, J.: SSE-PT: sequential recommendation via personalized transformer. In: Proceedings of the 14th ACM Conference on Recommender Systems (2020). https://dl.acm.org/
21. Vaswani, A., Shazeer, N., Parmar, N., et al.: Attention is all you need. In: Advances in Neural Information Processing Systems (NeurIPS) (2017). https://proceedings.neurips.cc/

Exploring SDN Based Firewall and NAPT: A Comparative Analysis with Iptables and OVS in Mininet

Md Fahad Monir(✉) and Azwad Fawad Hasan(✉)

Department of Computer Science and Engineering, Independent University, Dhaka, Bangladesh
{fahad.monir,2020222}@iub.edu.bd

Abstract. In recent years, Software-Defined Networking (SDN) has emerged as a transformative paradigm for managing and controlling computer networks while offering enhanced flexibility and scalability compared to traditional networking models. SDN-driven network modules enhance business network efficiency, aid in experimental network studies, simplify network management and automation, and improve the reliability and speed of the internet in daily use. In this study, we implemented Open vSwitch (OVS) controller-based SDN networking modules in Mininet and assessed their performance, with particular emphasis on packet loss and jitter. Network Address Port Translation (NAPT) middleware, OVS controller-based firewalls, and a mix of both in a single middleware were implemented in networking modules and tested with different networking policies. A comparative analysis was conducted between iptables and OVS policies to understand their different effects on packet loss and jitter. The objective of this research was to investigate the performance differences between various SDN-based network module types-in this case, OVS-based modules. The results indicate that multiple OVS policies increase packet loss and jitter, whereas iptables exhibit better performance. This study also provides insights into the trade-offs between OVS and iptables in SDN middleware, highlighting the scope for optimization in future research.

Keywords: Open vSwitch · OVS Firewall · NAPT · iptables · Mininet · OpenFlow · Jitter · Packet Loss

1 Introduction

Software Defined Networking (SDN) is a revolutionary concept for next-generation networks that offers programmability and scalability while providing an open networking framework that allows dynamic orchestration. In the dynamic landscape of SDN, Open vSwitch (OVS) is a pivotal component that offers essential capabilities for virtualized network environments. As the realm of SDN evolves, understanding the performance characteristics of OVS-based

L. Barolli (Ed.): AINA 2024, LNDECT 202, pp. 436–447, 2024.
https://doi.org/10.1007/978-3-031-57916-5_37

middleware becomes important. This study delves into the evaluation of OVS-controller-based network modules, such as firewalls and network address port translators, in a simulated environment using Mininet, with a specific focus on packet loss and jitter as performance indicators.

Our investigation encompasses several distinct network modules, each representing a unique configuration like NAPT and firewalls-an imperative component for network security and intrusion detection. These configurations also include iptables-based network module, OVS controller-based firewalls such as OVS policy-based NAPT and a combination of both within a network module. The research explores the influence of different OVS policies, particularly those applied in firewalls. The investigation scrutinizes the effects of these policies on packet loss and jitter, drawing comparisons with the outcomes observed when employing iptables.

Prior research has extensively covered SDN-based network modules and controllers. For example, in a paper by Keerthana *et al.* titled "Performance Comparison of Various Controllers in Different SDN Topologies", the focus was on measuring round-trip time with controllers like OVS, POX, and Ryu [1]. Furthermore, Badotra *et al.* developed an SDN-based firewall operating at the application and transport layers of the TCP/IP model using the Python-based Ryu controller [2]. The implemented firewall successfully blocked traffic from both transport and application. But, no performance testing like packet loss and jitter were done.

Moreover, the author in [3] demonstrated latency performance in an SDN using a POX controller on an OVS switch in Mininet, comparing scenarios with and without a firewall. Similarly, authors *et al.* in [4] analyzed jitter and bandwidth tests in an SDN environment with a POX controller on an OVS switch, incorporating a firewall. In a comparative study by Gupta *et al.* [5], various SDN controllers, including Floodlight, Ryu, OpenDayLight, were evaluated in Mininet. Similarly, authors *et al.* in [6–8] implemented and carried out different kinds of tests on SDN-based network modules such as NAPT, POX firewall, click-based IDS. However, the OVS controller was not the primary focus of any of these studies.

To the best of our knowledge, no work has evaluated the performance of OVS controller-based middleware in Mininet, which is required as baseline data for SDN-based middleware by evaluating the jitter and packet loss for different network configurations. This study aims to fill a notable gap in the systematic evaluation of OVS controller-based middleware within simulated environments by providing valuable insights into the trade-offs and performance implications associated with OVS policies and iptables in SDN middleware.

Summary of Novel Contributions

- This study provides baseline performance data for SDN-based middleware using Mininet's OVS controller.
- Research shows that more OVS firewall policies generally lead to higher packet loss and jitter.

- The research examines and highlights the distinctions between iptables and OVS policies in firewall and NAPT functionalities. It discloses that, on average, there is a 0.91% variance in packet loss and a 0.003 ms difference in jitter values between iptables and OVS policies.
- This paper reveals that switch-level OVS rules increase overhead compared with kernel-level iptables, offering vital baseline data for future SDN middleware optimization.

2 Methodology

The primary objective of this study was to analyze the impact of diverse network modules on packet loss and jitter over a range of bandwidths and time intervals in a simulated environment using Mininet. Python scripts were developed to create these networking modules, simulating various scenarios involving Open vSwitch (OVS) configurations, firewall functionalities, and Network Address Port Translation (NAPT). Subsequently, iPerf tests were conducted to evaluate packet loss and jitter under different conditions. This study involved systematic variations in network configurations, enabling a comprehensive comparison of the implemented network modules. Oracle VirtualBox version 7.0.6 was used to run Mininet on Ubuntu 22.04.3-LTS 64-bit. The host device, equipped with a Ryzen 5800H processor and 32 GB of RAM, allocated 23750 MB of RAM and eight cores to the virtual machine. Furthermore, a Linux-based SDN emulator called Mininet (version 2.3.1b4) was used. It can create virtual networks that closely resemble real-world physical networks and produce accurate results [9]. It simplifies programming with a responsive, topology-aware CLI that makes adding different network components easily [10]. In addition, a multi-layer, open-source virtual switch called an Open vSwitch was used in this project. This software switch offers virtual machine network services and is compatible with all popular hypervisor platforms. It is designed for networking in virtual environments and can be programmed using OpenFlow or other protocols [11].

2.1 Network Simulation

Our network topology was developed such that we placed our designed OVS network modules between three distinct network zones: public, demilitarized, and private. This network topology was chosen to demonstrate a scenario with varying security levels across different zones, ranging from a minimally secure public zone to a more protected demilitarized zone through the OVS firewall, culminating in a highly secure private zone equipped with NAPT.

In the simulations depicted in Fig. 1, these zones are separated using an Open vSwitch (OVS) based NAPT network module and an OVS-firewall network module. The NAPT is placed between the private and demilitarized zones, whereas the firewall is positioned between the public and demilitarized zones. Within the demilitarized zone, there are four entities: two hosts (H5 and H6), a web server, and a DNS server. The public and private zones collectively contain seven hosts. A Python script using Mininet was used to create the illustrated network topology, with the OVS controller managing network operations. In addition, a log file records specific activities such as ping and HTTP requests each time the Mininet topology is activated. We also executed HTTP POST and PUT requests to transmit data to the web server from host H1 in the demilitarized zone. After sending these requests, we confirmed that the web server successfully received and updated these POST messages, verifying the functionality and connectivity of the web server. Subsequently, upon successfully verifying this, we used the 'ovs-ofctl' program to add a policy to our switch, allowing only DNS service for UDP. HTTP was allowed for all ports except port 4000 for TCP from the public to the demilitarized zone.

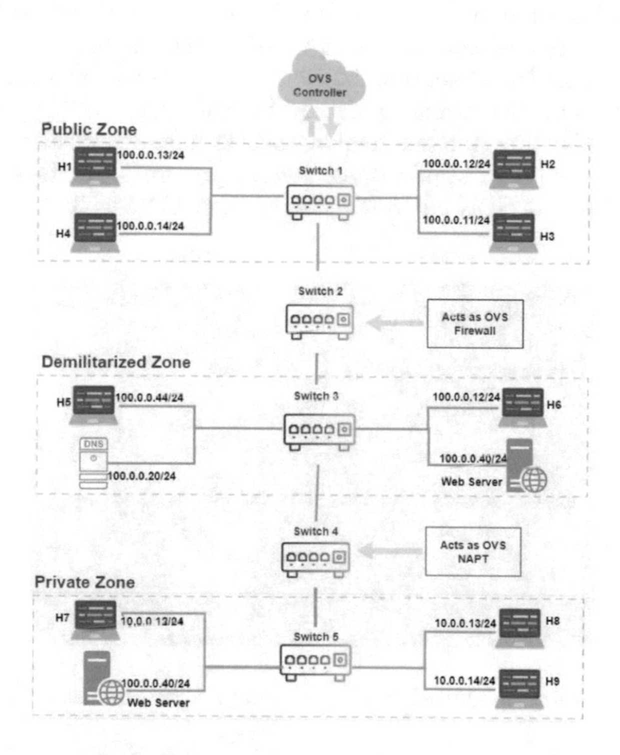

Fig. 1. Network topology showing how the NAPT and OVS firewall separate the public and private zones from the demilitarized zone.

Our experiment was conducted in three stages. The initial phase involved setting up an Open vSwitch connected to an OVS controller within our network framework. After this, we performed multiple iPerf tests to evaluate various network configurations, particularly focusing on analyzing packet loss and jitter. In these tests, iPerf UDP/TCP data packets were transmitted and received on a client-server basis. For instance, H1 in the public zone was used as a client, while H5 or the Webserver in the demilitarized zone acted as the server. The purpose of sending iPerf UDP and TCP packets was to observe the impact on the packet delivery ratio between a client and a server due to the presence of an intermediary device, such as a firewall. Figure 2 illustrates the outcomes of these iPerf tests, specifically showing the UDP throughput results through an Open vSwitch.

For packet loss evaluation, UDP packets were sent through port 4000 at various bandwidths (500 MB, 600 MB, 700 MB, 800 MB, 900 MB, and 1000 MB). Each UDP session had a fixed time interval of 10 s and used a UDP buffer size of 208 Kbytes (default setting). Notably, each test was repeated ten times to enhance reliability and accuracy. The recorded data were averaged to ensure precision. In the case of jitter measurement, similar iPerf tests were conducted, maintaining a fixed bandwidth of 1000 MB and varying the time interval from 10 to 50 s. Later, graphs were drawn to visualize the results.

This meant that the OpenFlow L2 learning switch was now acting as a firewall and was responsible for managing flow table entries, enforcing firewall rules, dynamically configuring rule sets, making packet handling decisions, facilitating communication with SDN applications, handling network events, and coordinating firewall policies in SDN environments, as shown in Fig. 3.

```
(base) kali@Ubuntu:~/Desktop/Final$ sudo python3 3_oneFirewall_manyPolicies.py
*** Adding controller
*** Add switches
*** Add hosts
*** Add links
*** Starting network
*** Configuring hosts
h1 h2
*** Starting iperf test between h1 and h2
*** with switch 2
Performing iperf UDP test from h1 to h2 on port 4000
Bandwidth sent by h1: 1.05 Gbits/sec
Bandwidth received by h2: 1.05 Gbits/sec
------------------------------------------------------------
Client connecting to 10.0.0.2, UDP port 4000
Sending 1470 byte datagrams, IPG target: 11.22 us (kalman adjust)
UDP buffer size:  208 KByte (default)
------------------------------------------------------------
[  1] local 10.0.0.1 port 4000 connected with 10.0.0.2 port 4000
[ ID] Interval       Transfer     Bandwidth
[  1] 0.0000-10.0071 sec  1.22 GBytes  1.05 Gbits/sec
[  1] Sent 891572 datagrams
[  1] Server Report:
[ ID] Interval       Transfer     Bandwidth        Jitter   Lost/Total Datagrams
[  1] 0.0000-10.0058 sec  1.22 GBytes  1.05 Gbits/sec   0.003 ms 1256/891571 (0.14%)
[  1] 0.0000-10.0058 sec  67 datagrams received out-of-order
```

Fig. 2. Output of a server report from Mininet.

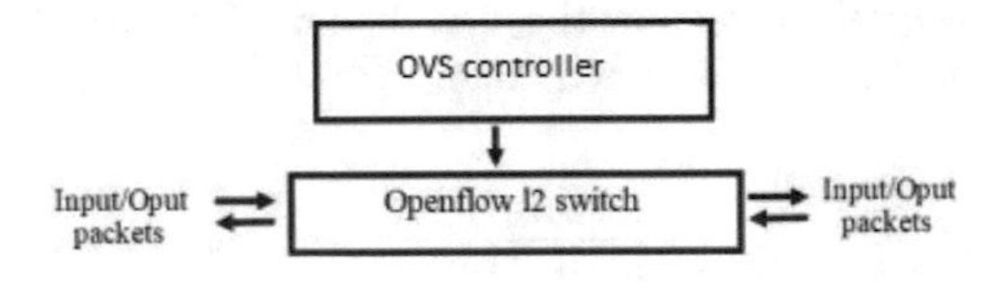

Fig. 3. OVS controller attached to an Openflow L2 learning switch.

Afterward, additional policies were implemented on the switch to restrict the passage of TCP packets through additional ports, for instance, ports 4000 and 4002. In addition, the entry of UDP packets was prohibited through ports

4001 and 4003. A simplified explanation of the switch's operation is given by the flowchart in Fig. 4, which focuses on the permitted service policy for DNS (UDP) and HTTP (TCP).

In the third phase, the firewall policies were replaced with policies that enabled the switch to act like a NAPT network module. The policies implemented ensured the correct translation of the IP and port numbers of the packets. For example, the NAPT policies would translate packets from 10.0.0.0:10001 to 100.0.0.0:4000. This means that any packet sent from port 10001 on the host with the IP address 10.0.0.0 will be re-written to have the IP address 100.0.0.0 and port 4000 before it is sent to the other host.

The policies were modified again. The same NAPT policies are applied along with the same multiple policies as before with OVS policies. The Open vSwitch now acts as a firewall and NAPT network module.

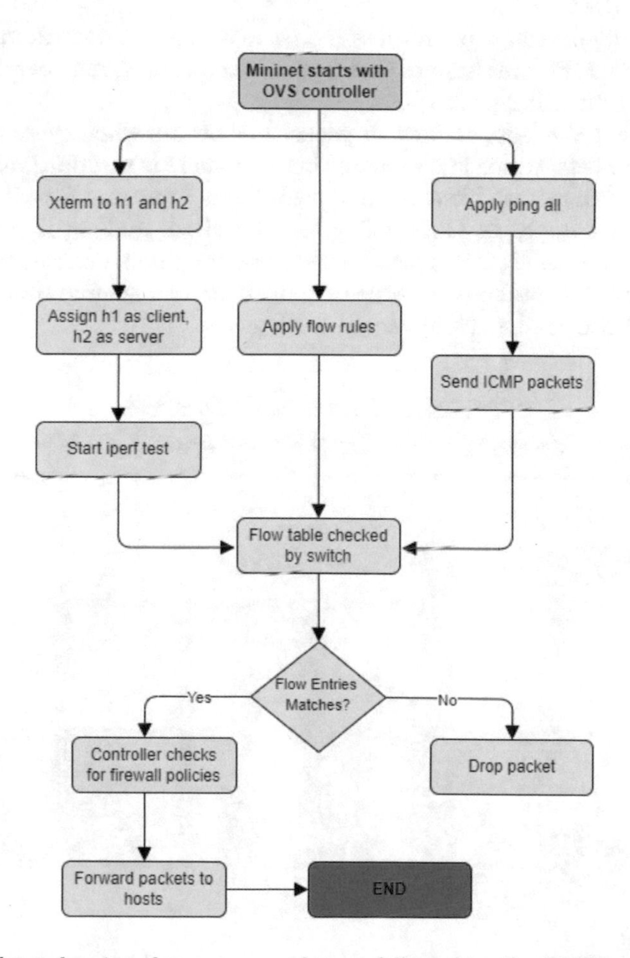

Fig. 4. Flowchart showing the sequence of steps followed by the OVS firewall (middleware).

In the last phase, the middleware was replaced with a central node, configured to enforce firewall and Network Address Port Translation (NAPT) functionalities using iptables. In this setup, iptables policies on the intermediary node replicate the previously described tasks, incorporating identical NAPT features and firewall policies.

3 Results and Analysis

The graph depicted in Fig. 5 is a clustered bar chart that compares packet loss to bandwidth for five distinct network setups. Megabytes (MB), which range from 500 MB to 1000 MB in steps of 100 MB, are represented by the x-axis as bandwidth. The packet loss percentage is displayed on the y-axis, with values ranging from 0% to 1.5%.

For all configurations, packet loss increases as the bandwidth increases from 500 MB to 1000 MB. This implies that increased data rates may overload network components, increasing packet loss.

We find that the least amount of packet loss occurs when there is no middleware between hosts, which is reasonable because there is no additional processing overhead. In general, we observe that packet loss increases for all input bandwidths when we apply an OVS policy to the switch, making it function like a firewall. This implies that increased packet loss may be a consequence of higher data transfer rates, perhaps because of congestion or resource constraints when handling and processing the increased traffic.

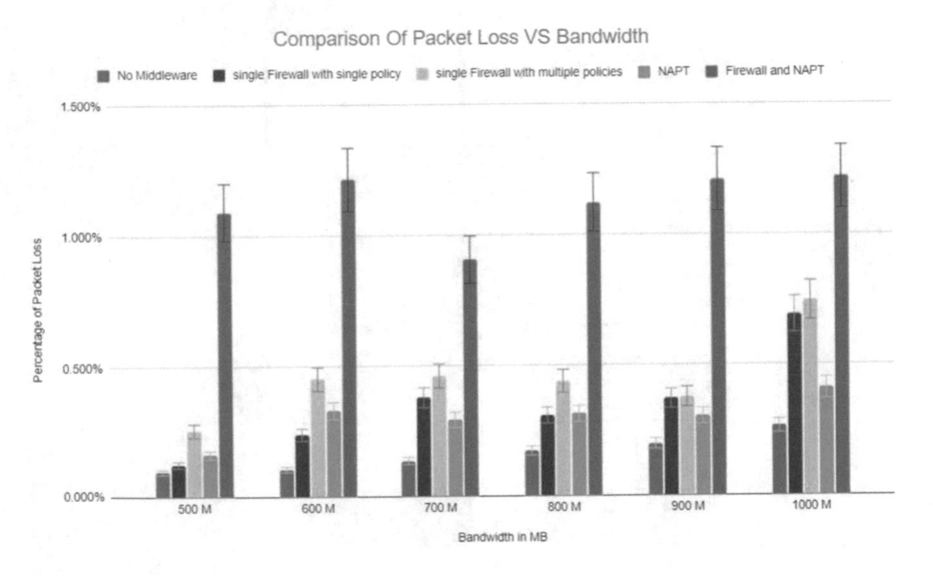

Fig. 5. Measurement of packet loss for OVS-based network modules.

The same pattern recurs when the firewall is subjected to multiple policies. The packet loss increases with the number of policies applied on a single firewall.

We observe an intermediate level of packet loss-that is, less packet loss than firewalls but more packet loss than the topology without middleware-when the firewall rules are removed and policies are applied to make the switch function as a NAPT device.

Packet loss was highest when both the firewall rules and the NAPT policies were applied simultaneously. We observe a similar trend where packet loss rises with increasing input bandwidth.

Figure 6 summarizes packet jitter for UDP traffic with a fixed bandwidth of 1000 Mbits and a variable time interval. Jitter appears to rise over time in all configurations, albeit with some variations. According to the results, the network configuration without middleware has the least overall jitter when compared to other network configurations. Similarly, a network configuration with a single firewall and multiple policies typically has more jitter than a network configuration with just one firewall and one policy. However, in certain instances, a firewall with a single policy exhibits higher jitter values than one with multiple policies, indicating that the jitter is not linearly increased by the additional policies. The NAPT configuration exhibits the second-lowest jitter values for the majority of time intervals. The trend generally resembles the previous packet loss graph, meaning that NAPT jitter is generally higher when compared to a network configuration without middleware but lower when compared to firewalls with single and multiple policies. This implies that the firewall policies have a greater effect

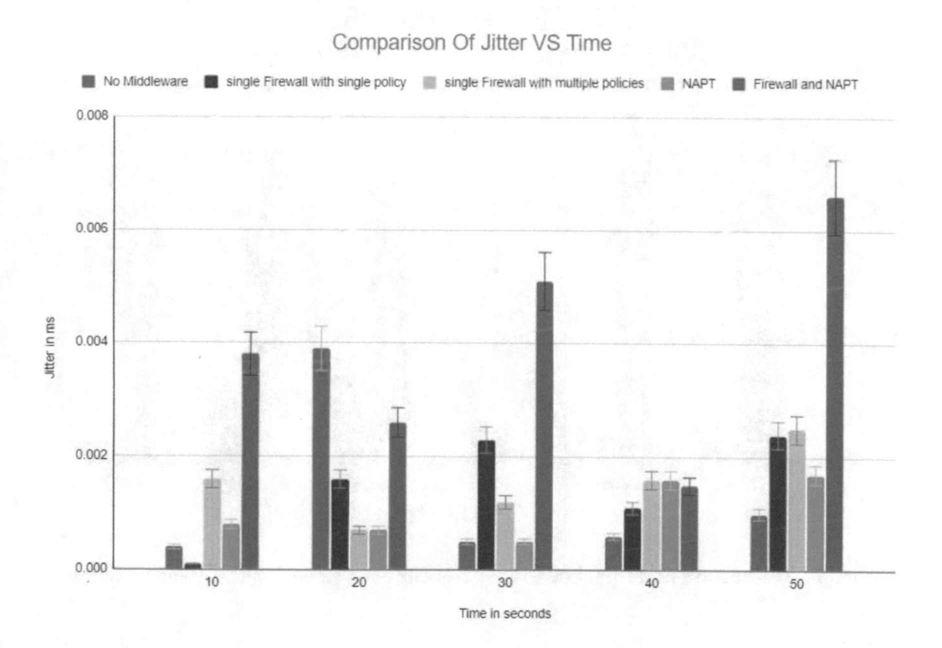

Fig. 6. Measurement of jitter variation over time for OVS-based network modules.

on jitter than the NAPT policy. The network configuration with both firewall and NAPT generally had the highest jitter values, indicating that it has a major impact on packet timing consistency.

The performance of NAPT with a Firewall using Open vSwitch (OVS) policies is contrasted with that of NAPT with a Firewall using iptables policies in Fig. 7, which shows a comparison of packet loss percentages across various bandwidths. The percentage of packet loss tends to increase with increased bandwidth for both setups, although the rate of increase varies. With increased bandwidth, the packet loss is not strictly linear; rather, it varies. The graph indicates that OVS policies result in higher packet loss. When compared to the iptables policy scenario, this setup consistently displays a higher percentage of packet loss across all bandwidths (500 MB to 1000 MB).

Figure 8 shows the performance of a firewall and network address port translation (NAPT) setup using either Open vSwitch (OVS) policies or iptables policies. It is a bar chart that compares jitter in milliseconds over time in seconds. As time goes on, the jitter increases for both the OVS and iptables policies. At the 10-s mark, the jitter is lower and usually worsens every time interval thereafter. When compared to iptables policies, the OVS policies consistently exhibit higher jitter at each time interval. This implies that there may be increased packet delay variability due to the implementation of the OVS policy. In addition, the graph shows that although the overall amount of jitter is fairly low, there is a noticeable relative difference between the two types of policies.

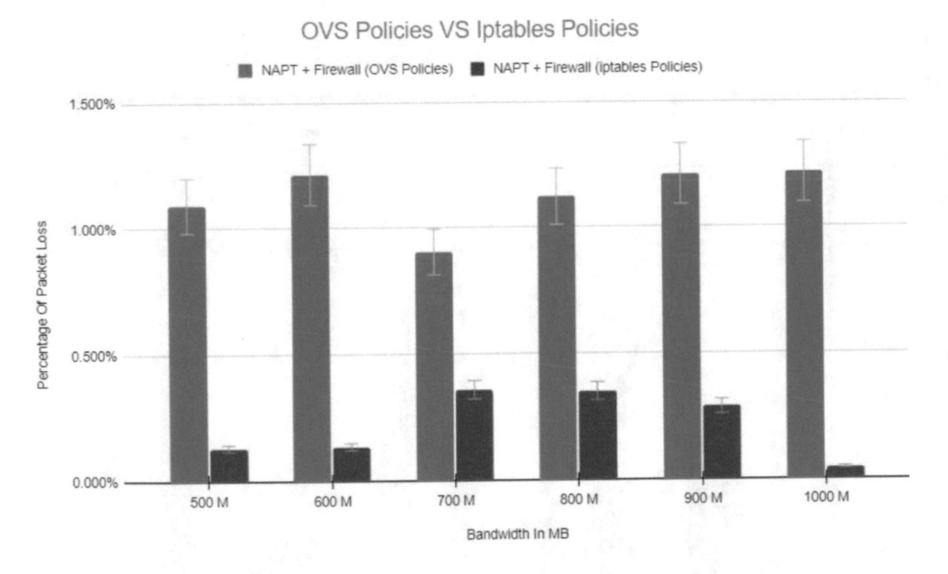

Fig. 7. Measurement of packet loss variation over time for different policies.

To summarize, we discovered that when we implemented multiple OVS policies in our firewall, the overall jitter and packet loss increased. Additionally, we discovered that the network configuration with multiple firewall policies and NAPT rules implemented in the middleware showed the highest overall packet loss and jitter. To understand this difference, it is essential to understand how flow rules operate. These rules, which are sent to network switches, dictate packet management based on header details such as source and destination IPs and TCP/UDP ports and are customized for specific network needs. Upon reaching a switch, packets are evaluated for actions such as forwarding, dropping, or modification based on these rules. Simple rules might direct packets from one point to another, whereas complex rules could involve inspecting packet contents, monitoring connection states, applying QoS policies, or executing transformations such as NAPT. The level of simplicity or complexity of these rules directly affects packet processing, potentially explaining the observed variations in packet loss and jitter in different network configurations in our study.

As part of our work, we also tested the performance of iptables and OVS policies to implement middleware that can perform firewall and NAPT functions. We discovered that when iptables were used in the network configuration, there was a significant reduction in packet loss and jitter.

One reason behind this difference lies in the abstraction level: OVS rules function at the switch level, while iptables control packet flow to and from virtual hosts within Mininet at the kernel level (host level). In a Mininet network, each host can have its IP table configured, but OVS rules utilize flow tables instead of the IP tables present in the Linux kernel. Mininet relies on the Linux

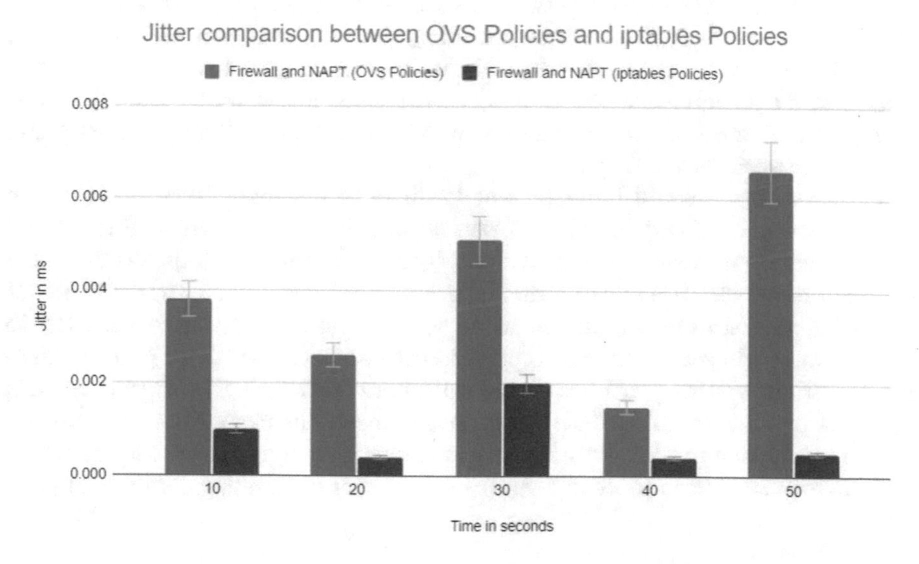

Fig. 8. Measurement of jitter variation over time for different policies.

networking stack for its hosts, causing OVS to operate partially in kernel space and partially in user space. When the flow table experiences frequent misses, packets need to be pushed to user space for processing, leading to significant overhead and potential packet loss, especially under high loads. In contrast, iptables operate solely within the kernel space (also known as Netfilter [12]), enabling more efficient packet handling with fewer context switches. Furthermore, iptables are stateful, allowing them to monitor connection states and generally perform basic operations like accepting and rejecting packets. However, OVS may be less adept at connection tracking, potentially affecting the handling of NAT sessions and resulting in more dropped packets during system overload. The performance of OVS is significantly influenced by both the size of the flow table and the complexity of the flow rules, as increased packet loss can stem from sub-optimally optimized flow rules. Conversely, iptables rules, processed linearly, can operate very effectively with a well-optimized set of rules [13].

Alternatively, network congestion may cause this discrepancy due to buffer overflows, hardware constraints (such as CPU and RAM), or inherent limitations in the software.

4 Conclusion and Future Work

In this study, OVS-based SDN network modules were set up and their performance tests were performed to evaluate packet loss and jitter during data transmission. The results indicate that an increase in OVS policies on an open vSwitch correlates with higher packet loss, which is an important baseline for future research in SDN based network modules. Additionally, it was found that OVS policies are less effective than iptable policies in reducing packet loss and jitter. However, it is crucial to acknowledge that our research was conducted in a basic Mininet simulation environment with only a few hosts; therefore, our findings might not fully reflect the complexities present in more advanced and intricate network setups [14].

Future research could build on our findings to optimize OVS flow rules in order to lower overhead and boost packet handling effectiveness. Subsequent research endeavors may concentrate on algorithmic enhancements to flow rule administration, with the aim of reducing user space-kernel space context switches and permitting stateful functions. Moreover, this study could be expanded to include real-world scenarios and applications to observe how these policies function in various settings with more intricate network topologies. By doing this, it may be possible to gain a better understanding of the capabilities and limitations of the current implementations and to investigate hybrid approaches, which combine the use of both tools to achieve even greater performance improvements.

References

1. Keerthana, B., Balachandra, M., Hebbar, H., Muniyal, B.: Performance comparison of various controllers in different SDN topologies. In: Jeena Jacob, I., Gonzalez-Longatt, F.M., Kolandapalayam Shanmugam, S., Izonin, I. (eds.) Expert Clouds and Applications. LNNS, vol. 209, pp. 297–309. Springer, Singapore (2022). https://doi.org/10.1007/978-981-16-2126-0_26
2. Badotra, S., Singh, J.: Creating firewall in transport layer and application layer using software defined networking. In: Saini, H.S., Sayal, R., Govardhan, A., Buyya, R. (eds.) Innovations in Computer Science and Engineering. LNNS, vol. 32, pp. 95–103. Springer, Singapore (2019). https://doi.org/10.1007/978-981-10-8201-6_11
3. AbdulRidha Hussein, M.: A proposed multi-layer firewall to improve the security of software defined networks. Int. J. Interact. Mobile Technol. (iJIM) **17**(02), 153–165 (2023). https://doi.org/10.3991/ijim.v17i02.36387
4. Othman, W.M., Chen, H., Al-Moalmi, A., Hadi, A.N.: Implementation and performance analysis of sdn firewall on POX controller. In: Proceedings of the 2017 IEEE 9th International Conference on Communication Software and Networks (ICCSN), Guangzhou, China, pp. 1461–1466 (2017). https://doi.org/10.1109/ICCSN.2017.8230351
5. Gupta, N., Maashi, M.S., Tanwar, S., Badotra, S., Aljebreen, M., Bharany, S.: A comparative study of software defined networking controllers using mininet. Electronics **11**, 2715 (2022). https://doi.org/10.3390/electronics11172715
6. Monir, M.F., Uddin, R., Pan, D.: Behavior of NAPT middleware in an SDN environment. In: 2019 4th International Conference on Electrical Information and Communication Technology (EICT), Khulna, Bangladesh, pp. 1–5 (2019). https://doi.org/10.1109/EICT48899.2019.9068752.
7. Monir, M.F., Pan, D.: Application and assessment of click modular firewall vs POX firewall in SDN/NFV framework. In: 2020 IEEE Region 10 Conference (TENCON), pp. 991–996 (2020). https://doi.org/10.1109/TENCON50793.2020.9293713.
8. Monir, M.F., Uddin, R., Pan, D.: Implementation of a click based IDS on SDN-NFV architecture and performance evaluation. In: 2021 IEEE International Black Sea Conference on Communications and Networking (BlackSeaCom), Bucharest, Romania, pp. 1–6 (2021). https://doi.org/10.1109/BlackSeaCom52164.2021.9527751.
9. Flauzac, O., Robledo, E.M.G., Nolot, F.: Is mininet the right solution for an SDN testbed?. In: Proceedings of the 2019 IEEE Global Communications Conference (GLOBECOM), Waikoloa, HI, USA, pp. 1–6 (2019). https://doi.org/10.1109/GLOBECOM38437.2019.9013145.
10. Mininet: an instant virtual network on your laptop (or other PC) - Mininet. Mininet. https://mininet.org/. Accessed 12 Nov 2023
11. Open vSwitch Project, Open vSwitch controller. https://manpages.ubuntu.com/manpages/trusty/en/man8/ovs-controller.8.html. Accessed 25 Nov 2023
12. Purdy, G.N.: Linux iptables Pocket Reference: Firewalls, NAT & Accounting. O'Reilly Media, Inc. pp. 1–21 (2004)
13. Xuan, L.-F., Wu, P.-F.: The optimization and implementation of iptables rules set on linux. In: Proceedings of the 2015 2nd International Conference on Information Science and Control Engineering, Shanghai, China, pp. 988–991 (2015). https://doi.org/10.1109/ICISCE.2015.223.
14. Hardin, B., Comer, D., Rastegarnia, A.: On the unreliability of network simulation results FROM Mininet and iPerf. Int. J. Future Comput. Commun. **12**(1) (2023). https://www.ijfcc.org/vol12/596-NC01.pdf

SDN Supported Network State Aware Command and Control Application Framework

Carlos André Rodrigues da Silva[1], Lauro de Souza Silva[1], Julio César Santos dos Anjos[2], Jorgito Matiuzzi Stocchero[1], Juliano Wickboldt[1], and Edison Pignaton de Freitas[1](✉)

[1] Institute of Informatics, Federal University of Rio Grande do Sul (UFRGS), Porto Alegre, RS, Brazil
jstocchero@ael.com.br, {jwickboldt,epfreitas}@inf.ufrgs.br

[2] Graduate Program on Teleinformatics Engineering, Federal University of Ceara (UFC), Fortaleza, CE, Brazil
jcsanjos@ufc.br

Abstract. Command and Control (C2) applications have particular performance requirements depending on their criticality. In a military operation scenario, network parameters frequently vary due to the harsh conditions of this environment. Thus, meeting application requirements is a challenge. This paper proposes a framework to address this challenge by employing a network state-aware approach in which the C2 applications tune the network to meet their requirements using Software Defined Networking orchestration combined with Information-Centric Networking mechanisms. The acquired results provide evidence of the success of the proposed approach in meeting the requirements to accomplish the mission.

1 Introduction

The 21st century saw an increased complexity of military operations, in which events happen and change very fast, and there is a need for accurate and timely information. The speed at which events unfold on the battlefield is influenced by various factors, including the nature of the conflict, the employed technologies, and the strategies adopted by the commanders. Having accurate and timely information on the battlefield enhances the situational awareness of the commanders and is, consequently, essential for strategic planning, making the best decisions, and for the overall success of the mission. [1]

In parallel with these factors, systems have been designed to assist in the management and decision-making of commanders, known as Command and Control Systems (C2) ([2]). They stand out for their ability to provide situation awareness to the decision-makers. Thus, C2 becomes an indispensable tool for decision-making ([3]). A crucial factor in this process is timely information transmission. However, how can this goal be achieved with multiple network nodes

L. Barolli (Ed.): AINA 2024, LNDECT 202, pp. 448–459, 2024.
https://doi.org/10.1007/978-3-031-57916-5_38

communicating with insufficient bandwidth? This is a complex issue requiring smart network management.

In this context, an important concept is Command and Control Agility [2]. C2 Agility applied to military operations can be defined as a strategy that provides flexibility to change the behavior on time by observing relevant opportunities and changes in the battlefield. In the context of the modern battlefield, this flexibility is reflected in the operational capacities of the networked C2 systems used by the troops, thus demanding resources of the underlining network that supports them [4].

Following the characteristics necessary to obtain C2 Agility, observing the intrinsic dependence of the underlining network, this work implements a framework called C2-APP, which integrates a software-defined networking (SDN) network orchestration with the high-level needs of C2 applications, exploring the capabilities of the Information-centric networking (ICN) paradigm. To achieve this goal, a proof-of-concept is implemented using Open Network Operating System (ONOS), OpenFlow, Open vSwitch, and MiniNDN to simulate an Internet of Battle Things (IoBT) [5] comprising of an illustrative application scenario with ground sensors, drones, soldiers, and armored vehicles.

The main contributions of this work are:

- The implementation of a framework that enables the application to adjust the network to meet the performance requirements in military operations;
- The combination of SDN and ICN paradigms to tune the network according to the application needs.

This work is organized as follows: Sect. 2 discusses related work. Section 3 offers an overview of the proposed approach. Section 4 demonstrates the design of the network-aware application approach. Section 5 presents the network dimension that supports the proposed approach. Section 6 showcases the experiments and results. Finally, Sect. 7 presents the conclusion and directions for future work.

2 Related Work

The work presented in [6] aimed to investigate technological resources for providing QoS guarantee in SDN networks. It implemented a software called SDQ (Software-Defined Queuing), which contains two main modules, namely queue management and traffic engineering. The first one uses ONOS CLI command line interface and Open Virtual Switch Database Management (OVSDB) protocol to manage queues in the switches, while the latter works on the flow rules to communicate with the switches. The experiments performed resulted in: stable, limited and very low latency for high and medium priority traffic packets. The work of Abbou bears similarities to this study, as it employs comparable technologies such as Mininet, ONOS, OpenFlow, and OVS. Both studies have implemented applications designed to work in conjunction with the ONOS controller. The primary objective of Abbou's work was to enhance the controller's performance by creating Quality of Service (QoS) and queues, a feature that is still not available in version 3.0 of ONOS.

The study by [7] explores the application of ICN in the Internet of Things (IoT), discussing its associated challenges, uses, and problems. The paper highlights the inefficiency of mobility support and the ineffectiveness of cache utilization in the network within the IoT context. It proposes the integration of IoT with ICN as a strategy to reduce data and conserve energy in IoT devices.

In an earlier study [8], an integration of SDN with ICN in the IoBT is proposed to meet the high-level requirements of C2 systems, emphasizing their network-centric nature. Since network dynamics are essential in such a situation, heterogeneous nodes were used in the study to validate the benefits of the proposed architecture. A proof of concept of the architecture, consisting of building a tactically realistic scenario for evaluation on Mininet wifi using the Ryu SDN controller, compared IP-based, ICN-SDN and pure SDN implementations using metrics such as latency and network load. The results were significant, showing improvements in the metrics compared to traditional approaches. This current paper improved on this earlier design and used the ONOS [9] instead of Ryu. ONOS is a fully distributed network operating system ideal for multi-domain wide area networks (WAN). ONOS provides scalability and fault tolerance with a distributed core that maintains a global view of each controller instance running on different servers (vehicles), optimizes data flows, and provides network services to the application, which in turn provides C2 services to users in the form of operational planning and monitoring.

Despite other relevant related work, for the lack of space, this section focused on the most relevant ones, summarized in the comparison provided in Table 1.

Table 1. Comparison between Solutions

Reference	Technologies	Objective
C2-APP (This work)	SDN, ICN, and IoBT	Network state aware application and dynamic configuration
[8]	SDN, ICN, and IoBT	Combined employment of ICN/SDN in IoBT Context
[7]	ICN and IoT	Employment of ICN in IoT
[6]	SDN	Creation of QoS rules and queues in ONOS

3 Approach Overview

This work proposes using QoS to prioritize specific flows on tactical battlefield networks. At certain moments of combat, some network nodes may obtain a bandwidth prioritization in relation to other nodes that are on the same network. However, due to the dynamics of the combat unfolding, a complete network configuration is not feasible, but the flows need to be adjusted according to the needs in runtime.

3.1 Physical Model

Figure 1 shows a conceptual view of the network assets on the battlefield scenario. This model has the following components:

- Armored personnel carrier vehicle (VBTP): the vehicle belonging to the commander's platoon. This vehicle carries the SDN controller, the C2-App, and a wireless access point;
- Soldiers (soldier1, soldier2, soldier3, and soldier4): military personnel belonging to the platoon that provides security to the vehicle surrounding. They are equipped with tablet-type devices that are network nodes with the function of access points and workstations;
- Drones (drone1 and drone2): a network node that has the function of access point and workstation;
- Sensors: wireless devices positioned on strategic positions on the battlefield. They communicate with each other through an ad-hoc network. The platoon drones collect the information acquired by the sensors through a delay-tolerant network (DTN) connection;
- Access point (AP1): an entry device in the SDN network. The nodes within the Wi-Fi zone are connected to the network (soldier1, soldier2, soldier3, soldier4, drone1, drone1, and VBTP);
- NDN connection: used to connect and receive information of interest between the network nodes;
- IP connection: used for communications between the soldiers of the platoon;
- DTN connection: used for communications between the sensors and drones that come within range for communication.

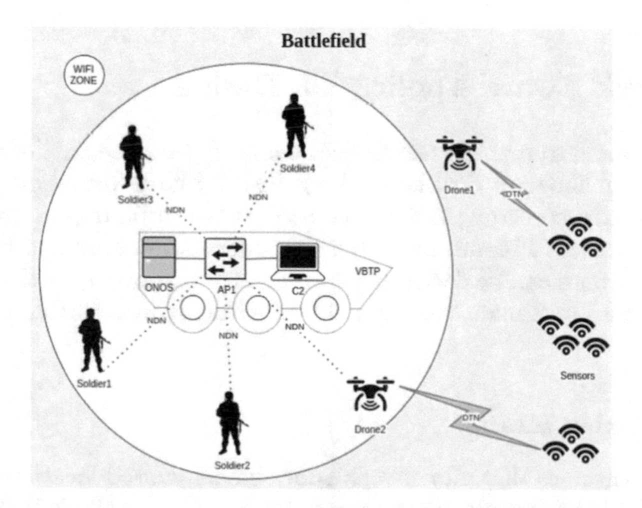

Fig. 1. Overview of the network nodes on the battlefield scenario.

3.2 Logical Model

In this work, each node represents a pair of devices composed of 1 switch and 1 host, as shown in Fig. 2. The network architecture is generated by a Python script in the MiniNDN emulator. The network also uses OVS 3.1, OpenFlow 1.4, and ONOS 3.0 as the SDN controller.

Although in a real deployment wireless communication is a must, this work uses a wired network as if it were Wi-Fi, i.e., wired communication with the parameters of a wireless one. This is due to the characteristics of the emulation environment, which is more stable in the wired version. However, this aspect does not affect the characterization of the network as it were a wireless one, as all parameters are set so that the network performs as wireless.

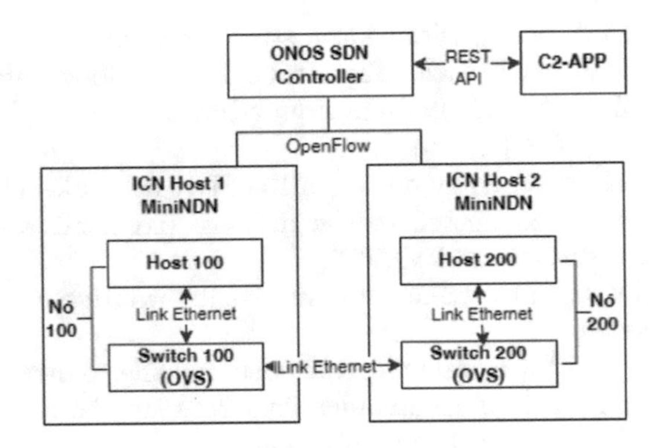

Fig. 2. Internal view of the nodes' configuration and placement in the network architecture.

4 Network Aware Application Design

The C2-App aims to manage the devices connected on the battlefield and thus assist the commander in decision-making tasks. The information acquired by the devices scattered throughout the combat area can be important for accomplishing the mission. If transmitted quickly, information obtained from sensors, soldiers, and drones can be decisive in gaining an advantage over the enemy. The C2-App system has 2 main modules: the Monitoring interface and the Area of Interest.

4.1 Monitoring Interface

The C2-App assumes that all network nodes have updated location information and are available for consultation through the ONOS controller's REST API. C2-App uses a geographic information system (GIS) from Google Maps and Open Street Maps tiles. Tiles are arranged images that form the base of a map. The tiles used in this work are road maps, satellite photos, and topography. Figure 3 shows the use of the Google Maps road map tile. The monitoring module is of great importance to the system because it is possible to follow the positioning of the friendly troop (blue color) and the enemy troop (red color), as shown in Fig. 3.

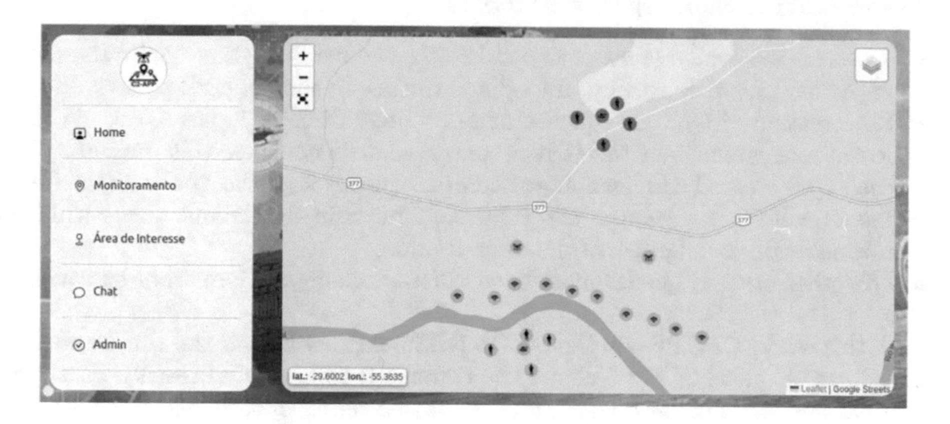

Fig. 3. C2-APP Monitoring Interface.

4.2 Area of Interest Interface

It is a region of the battlefield with the interest of the operation command, according to the C2-APP system screen shown in Fig. 4. A new window is opened when accessing the Area of Interest menu item. On the left-hand side of the map, a drawing toolbar is enabled, which can be used to make markings on the map.

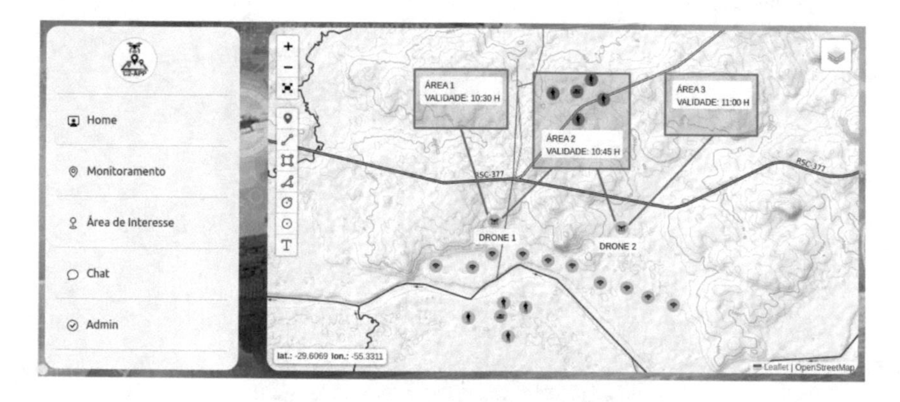

Fig. 4. Area of Interest Interface.

4.3 Network Node Reconfiguration via REST API

C2-APP uses the ONOS services through the REST API (Representational State Transfer) to obtain network information and establish flow priorities. The network node reorganization is a procedure that the C2-APP system can perform. This procedure requires specific conditions:

- Based on the available network bandwidth, the bandwidth usage limits must be planned for the creation of QoS and queues in the network nodes;
- The creation of QoS and queues must be done directly in the OVS, via the operating system since the ONOS controller does not have this function;
- Each QoS created can have more than two queues, and the queue 0 (the first queue) will be the default. The value configured in the default queue will be automatically configured to the desired node;
- The other created queues must have different configurations from the queue 0;
- In this work, C2-APP via the ONOS REST API, will send the command for the use of queue 3; Fig. 5 shows the command to create an intent, with the MAC addresses of the source and destination nodes and the use of queue 3.
- Figure 6 shows the command used to create QoS in node s200-eth1. A bandwidth of 500 kbps was configured for the network. The command also shows the creation of queues 0, 1, 2 and 3. Queue 0 was created with a bandwidth of at least 100 kbps and at most 200 kbps, queue 1 with a bandwidth of at least 200 kbps and at most 300 kbps, queue 2 with a bandwidth of at least 300 kbps and at most 400 kbps and queue 3 was created with a minimum bandwidth of 400 kbps. All nodes that do not have QoS configurations are under the service of best effort. In the best effort service, the network tries to forward the packets as fast as possible and does not provide any guarantee of delay, throughput or any QoS requirement.

```
{ "type": "HostToHostIntent",
 "appId": "org.onosproject.fwd",
 "priority": 100,
 "treatment": {
       "instructions": [
         {
           "type": "QUEUE",
           "queueId": 3
         }
       ],
       "deferred": []
       },
 "one": "CA:93:C4:1E:F6:B3/None",
 "two": "AA:A3:41:2E:94:AB/None"
}
```

Fig. 5. Command to create intent in ONOS

```
ovs-vsctl --\
 -- set port s200-eth1 qos=@newqos --\
 --id=@newqos create qos type=linux-htb other-config:max-rate=500000
  queues=0=@0,1=@1,2=@2,3=@3 --\
 --id=@0 create queue other-config:min-rate=100000 other-config:max-rate=200000 --\
 --id=@1 create queue other-config:min-rate=200000 other-config:max-rate=300000 --\
 --id=@2 create queue other-config:min-rate=300000 other-config:max-rate=400000 --\
 --id=@3 create queue other-config:min-rate=400000
```

Fig. 6. Command to create QoS and queues

5 Network Dimension

The ONOS controller is at the heart of the software-based network featured in this work. It natively integrates the Intent Framework, which is the service that allows prioritizing data flows for the SDN network. The ONOS controller is responsible for translating these policies into a network configuration. One of the advantages of ONOS is its ability to install intents, a key feature.

Intent-based networking (IBN) can abstract the mechanisms of how networks are configured to focus on specifying the goals that the network should achieve. As ONOS has its own Intent Framework, it gains functionalities compared to other controllers [10]. The installations of intents [11] in ONOS can be done through the REST API and the Command-line Interface (CLI).

A REST API uses the JavaScript Object Notation (JSON) format, and the calls it receives are made via URL. The main options available for access via REST API in the ONOS controller are: devices, flows, hosts, intents, links, and meters [9].

The simplest way to interact with ONOS is using the Graphical User Interface (GUI). Accessing it requires authentication using a username and password via the URL on the IP address of the ONOS server.

6 Experiments and Results

The performed experiments aim to assess the proposed approach's efficacy to dynamically adjust the network according to the needs.

6.1 Experiments Setup

Figure 7 shows two situations on a battlefield, represented by letters A and B. In both situations, the elements represent network nodes (Commander's Vehicle, Drone 1, Drone 2, and Soldier), arrows (exchange of information between nodes), and IoBT (represents all other network nodes not previously mentioned). In situation "A" the Commander's Vehicle requires information from the drones about possible threats, having the priority to receive it. Situation "B" represents the case in which Drone 2 detects a threat, and the priority is dynamically changed to prioritize the data delivery to the Soldier, as it is closer to the threat than the Commander's Vehicle.

Figure 8 represents the network model that maps a snapshot of the application scenario depicted in Fig. 7 in a network with 36 nodes. The nodes in yellow represent the NDN network configured for the experiments (the nodes in the IoBT cloud). The network segment comprising nodes 200, 24, 25, 18, 9, 10, 3, 2, and 1 represents the sequence of the nodes that forward data to node 100 (Commander's Vehicle). The segment comprising nodes 200, 24, 25, 18, 9, 10, 5, and 6 represents the sequence that forwards data to node 300 (Soldier). The nodes are mobile, but for this experiment, mobility is not a relevant aspect.

Figure 8 shows the position of nodes 100, 200, 300, and 400 in the network topology. Figure 7-A presents four network nodes, Commander, Drone 1, Drone 2, and Soldier, with priority for node 100 (Commander's Vehicle), without any priority for the other nodes (best effort). Figure 7-B, presents three network nodes, Commander's Vehicle, Drone 2, and Soldier, with priority for node 300 (Drone 2). Drone 2 is the one closest to the threat. After the Soldier receives the task of engaging with the enemy, all information acquired by Drone 2 will be prioritized towards him.

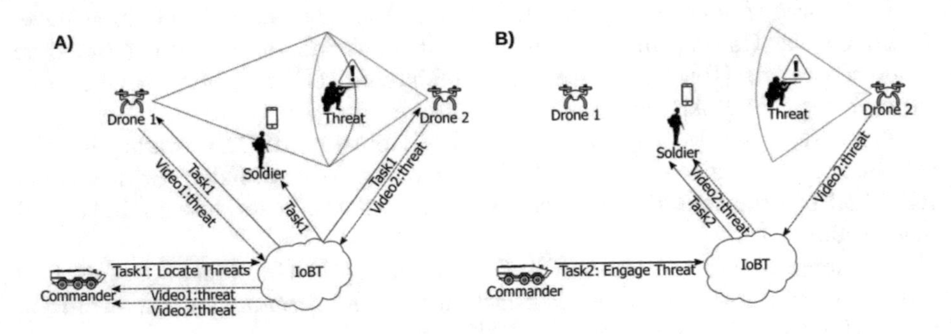

Fig. 7. Node prioritization application scenario. [8]

Table 2. Emulation details

Operational System	Ubuntu 22.04
Software and protocols	Docker, ONOS 3.0, OpenVSwitch 3.1, and OpenFlow 1.4
Topology	Mesh
Number of nodes	36
Bandwidth per link	500 kbps
Delay per link	10ms
Queues	0 (100 - 200), 1 (200 - 300), 2 (300 - 400), 3 ($\geq$ 400)
Best effort	nodes 100, 300, 400
IoBT	nodes 1 - 32
Priority 3	node 200
Queue mechanism	Hierarchical Token Bucket (HTB)

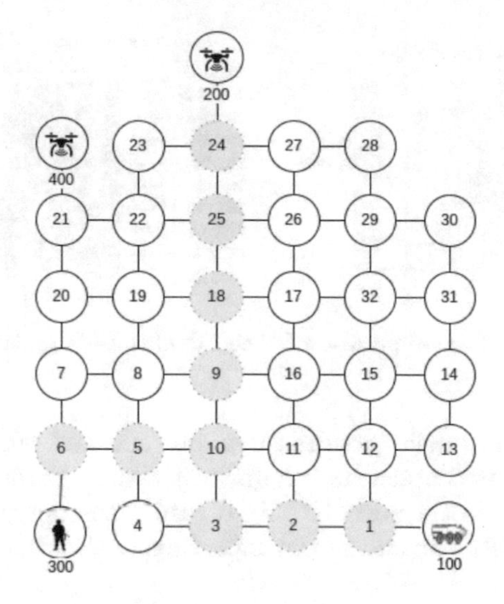

Fig. 8. Emulated network implementing the application scenario.

To assess the efficacy in performing the network configuration to achieve the objective illustrated in Fig. 7, the network presented in Fig. 8 was implemented in MiniNDN using the parameters depicted in Table 2.

Background traffic is generated to stress the network to represent realistic usage conditions. For this purpose, the command "ping -f -w 100 hx" was used, where "x" is the target node, to create network congestion and thus test the efficiency of the proposed approach in overloaded situations. 36 pings were sent from nodes 7, 12, 14, 20, 27, 29, and 31 targeting another 15 nodes: 4, 8, 11, 13, 15, 16, 17, 19, 21, 22, 26, 28, 30, and 32.

6.2 Results and Discussion

Experiments were conducted to obtain the performance of node 200 (queue 3) compared to node 100 (best effort). Delay and packet loss percentages were used as comparison metrics.

Figure 9 shows the packet loss results between nodes 100 and 300. In the first graph, none of the compared nodes have prioritization. The result shows a loss of 48% of the packets from 300 and 45% from 100. In the second graph, node 300 is configured with QoS queue 3 and node 100 has no prioritization. The result shows a loss of 34% of the packets from 300 and 43% from 100. The comparative analysis between the two presented graphs showed that the prioritization of node 300 was efficient as it reduced packet loss by 14%.

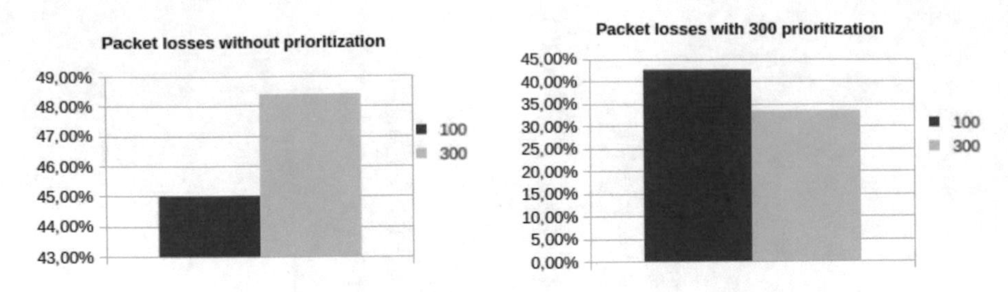

Fig. 9. Comparation for the Packet Loss Metric.

Figure 10 shows the delay results for nodes 100 and 300. From second 0 to second 40, the compared nodes have no prioritization. From second 41 onwards, node 300 is prioritized over node 100. As a result, the delay for node 300 is less than for node 100 until the end of the experiment.

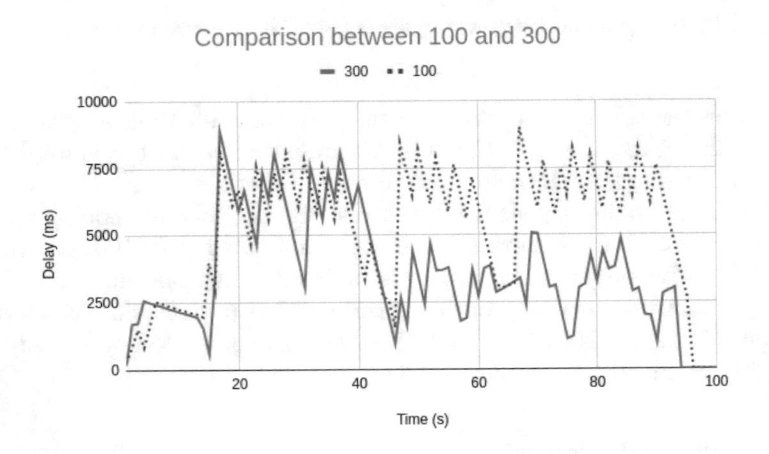

Fig. 10. Comparation for the Delay Metric.

7 Conclusion

This work presented a proposal of a network-aware command and control application that can enforce adjustments to the network to meet the application goals. This was possible by a reconfigurable network supported by SDN and ICN paradigms. The acquired results provide evidence of the success of the proposed approach with the decrease in packet loss and delay by the prioritized node.

Future works point to deploying the proposed approach in real-world devices coping with practical implementation challenges and handling network partitioning.

Acknowledgments. This work was partly sponsored by CAPES, - Finance Code 001 and CNPq, projects 309505/2020-8 and 311773/2023-0, and partly by the Brazilian Army (S2C2 project, ref. 2904/20).

References

1. Tucholski, H.M.: Future command and control: closing the knowledge gaps. Air Space Power J. **06**, 5–17 (2021)
2. Alberts, D.S., et al.: Command and control (c2) agility, Tech. rep. (2014)
3. Zhou, W., Bao, W., Sun, X., Wan, J., Xu, Y., Gao, Y.: A reconciliation model of agile c2 organization based on converged networks. In: 2020 6th International Conference on Big Data and Information Analytics (BigDIA), pp. 58–65 (2020)
4. Eisenberg, D.A., Alderson, D.L., Kitsak, M., Ganin, A., Linkov, I.: Network foundation for command and control (c2) systems: literature review. IEEE Access **6**, 68782–68794 (2018)
5. Kott, A., Swami, A., West, B.J.: The internet of battle things. Computer **49**(12), 70–75 (2016)
6. Taleb Abbou and Song: A software-defined queuing framework for QoS provisioning in 5G and beyond mobile systems. IEEE Network **35**(2), 168–173 (2021)
7. Zhang, Z., Lung, C.-H., Wei, X., Chen, M., Chatterjee, S., Zhang, Z.: In-network caching for ICN-based IoT (ICN-IoT): a comprehensive survey. IEEE Internet Things J. **10**(16), 14595–14620 (2023)
8. Matiuzzi Stocchero, J., Dexheimer Carneiro, A., Zacarias, I., et al.: Combining information centric and software defined networking to support command and control agility in military mobile networks. Peer-to-Peer Netw. Appl. **16**, 765–784 (2023)
9. ONOS Wiki. Open network operating system (2022)
10. Ujcich, B.E., Sanders, W.H.: Data protection intents for software-defined networking. In: 2019 IEEE Conference on Network Softwarization, pp. 271–275 (2019)
11. Irfan, T., Hakimi, R., Risdianto, A.C., Mulyana, E.: ONOS intent path forwarding using dijkstra algorithm. In: 2019 International Conference on Electrical Engineering and Informatics (ICEEI), pp. 549–554 (2019)

Reliability-Aware SFC Protection by Using Nodes with Spare Resources

Hung-You Chen and Pi-Chung Wang(✉)

Department of Computer Science and Engineering, National Chung Hsing University, No. 250, Kuokuang Road, Taichung, Taiwan
pcwang@nchu.edu.tw

Abstract. Network Function Virtualization (NFV) can reduce expenses in network services deployment while enhancing flexibility, adaptivity, and reliability of network services. To optimize resource allocation, each service request will be expressed as a service function chain (SFC), where each SFC may include multiple virtualized network functions (VNFs). Then, the SFC paths are calculated to improve the overall performance of network services. Any VNF failure within an SFC may result in service interruption. Installing backup VNFs is an effective approach to improve the reliability of SFCs, but previous approaches may consume substantial amount of resources to achieve only limited reliability. In this work, we propose a novel approach to improve the reliability of SFCs without excessive backup resource consumption. Our approach generates both primary and backup paths for each SFC simultaneously to meet the requirements of reliability, where the backup node of an SFC can be the primary nodes of other SFCs. The simulation results show that the proposed approach can achieve high reliability with high acceptance ratios. Moreover, the backup resource can be reduced as compared to previous algorithms.

1 Introduction

NFV enables networking software to be independent of network appliance hardware. They provide fundamental units of network services and form SFCs that originates from the source, passes through various VNFs and ultimately reaches the destination. SFCs are deployed on the substrate network of physical devices, including servers, paths, switches, and other networking equipment, as well as physical links that connect them. Numerous research studies of NFV have been conducted. Alwakeel et al. described the main entities of NFV with UML diagrams [1]. Software functions, such as load balancers and firewalls, are implemented with the concept of NFV and evaluated in terms of packet loss and throughput [6,7]. As compared to traditional network deployment, NFV has higher flexibility and scalability so users can receive good network services. An SFC may suffer from failures when one or more VNFs fails.

The previous research has addressed the issue of VNF failures. To ensure uninterrupted network service, a solution must address the VNF failures by allocating backup nodes. For example, Huang et al. [3], provided a proactive failure recovery mechanism for stateful NFV. Tomassilli et al. [9] considered an approach to protect paths by global

L. Barolli (Ed.): AINA 2024, LNDECT 202, pp. 460–470, 2024.
https://doi.org/10.1007/978-3-031-57916-5_39

rerouting. Another work by Thorat et al. [8] presents SSP to efficiently recover failure while avoiding traffic overhead. Xu et al. [12] presented a dynamic heterogeneous backup strategy, which considers an attack scenario in SFC. Although the previous recovery mechanisms can improve reliability, they usually consume considerable network resources.

There are research attempting to ensure the SFC reliability with low resource consumption. For example, Kanizo et al. [4] suggested a new plan on deploying backup schemes for network functions that guarantee high survivability with a low resource consumption. Galdamez et al. [2] proposed a region-disjoint mapping algorithm that can map the virtual networks into non-overlapping geographical areas and prevent it from large-scale regional failures. Liu et al. [5] presented a deployment mechanism to reuse VNF instances. Accordingly, a well-designed backup scheme should achieve a balance between network service reliability and resource consumption. Wang et al. [10] revolved around instance sharing and resource allocation. Instance sharing implies that SFCs can share a physical node if it can provide adequate resources. The backup arrangement is based on the resources of the primary nodes. Moreover, both primary and backup nodes are determined simultaneously. Woldeyohannes et al. [11] also proposed another algorithm, CoShare, reduce resource consumption for VNF backup.

In this work, we propose a method to maintain the reliability of SFC with reasonable backup resources. We present an algorithm based on service function graph (SFG) that considers resource allocation and backup arrangement during the VNFs forwarding-graph construction phase prior to the phase of mapping VNFs to physical devices. The simulations are conducted on a medium-scale topology. The experiment results demonstrate that the proposed method can improve acceptance ratio and enhanced reliability of SFCs. It also reduces the overall consumption for backup resources in SFCs.

The subsequent sections of this article are: Sect. 2 outlines the system model, and Sect. 3 introduces the proposed algorithms. In Sect. 4, the experiment results are presented. Finally, Sect. 5 concludes this work.

2 System Model

Several components are introduced as follows: the substrate network, VNFs, and service requests (SFCs). The substrate network is composed of nodes and links connecting these nodes. The symbol $G(N, L)$ represents the substrate network, where N represents the set of nodes and L indicates the set of links. An SFC consists of several VNFs, where each VNF represents a specific network function. By specifying the number and types of VNFs in an SFC, multiple sequences of SFCs can be created. We deploy SFCs on the substrate network by mapping VNFs to the physical nodes.

Each VNF has its own reliability, which is typically represented as a decimal value below 1. The reliability value signifies the probability of successful execution for the particular network function. An SFC can be expressed as $S = \{s_j | j = 1, 2, ..., |S|\}$. Each VNF, represented as $v_i = (c^{v_i}, r^{v_i}, e^{v_i})$, where v_i is a specific network function, c^{v_i} indicates the resource requirement of v_i, r^{v_i} denotes the reliability of v_i, and e^{v_i} is the maximum allowable number of SFCs sharing v_i.

Service requests are created and mapped onto the substrate network based on their corresponding paths. The primary paths are designed to pass through the primary VNFs, while the backup paths are designed to pass through the backup VNFs. Both the primary and backup VNFs have the same type, but they are arranged on different physical nodes. In the event of a failure of the primary VNF, the traffic will be rerouted through the backup path in order to bypass the failed node and traverse the backup VNF instead.

The reliability of a VNF involves two main factors: the value of the VNF reliability and the value of the substrate node that accommodates the VNF. The reliability value is determined by using Eq. (1), where x_{ik} is a 0/1 variable to indicate v_i is mapped to node n_k or not. r^{v_i} represents the reliability of v_i and r^{n_k} is the reliability of n_k.

$$r_{ik} = x_{ik} \cdot r^{v_i} \cdot r^{n_k} \tag{1}$$

Consequently, the reliability of an SFC, s_j, can be yielded by calculating the product value of all r_{ik} on the SFC, as shown in Eq. (2).

$$r^{s_j} = \prod_{\forall v_i \in s_j, \forall n_k \in N} r_{ik} \tag{2}$$

Equation (3) further expresses the reliability of v_i by allocating a backup instance in node l.

$$r^{v_i}_{backup} = 1 - (1 - r_{ik})(1 - x_{il} \cdot r^{v_i} \cdot r^{n_l}) \tag{3}$$

Apparently, the reliability of a VNF can be improved by allocating a backup instance in a different node.

3 Proposed Method

In this section, we present the proposed approach for allocating backup resources. Our approach mixes the backup instances with the primary instances of SFCs. When a node is selected to accommodate NFVs of a type, the spare instances is also allocated to serve as the backup instances of different NFVs. When an event of a node failure occurs, the traffic to primary instances of the failed node are redirected to the spare node for service restoration. We attempt to ensure that all instances have spare instances to significantly improve the reliability of network services. The detailed process is outlined in the following subsections.

Specifically, the resources of each physical node are divided into N parts, where each part can accommodate one VNF. Then, the first node is allocated by allocating a VNF of a SFC. The other $N-1$ parts are remained as unused for spare instances. Since this node has the most spare instances, it is also known as a spare node. For the other VNFs of the same type, their instances are allocated in different nodes, where these nodes have only one spare instance, as shown in Fig. 1.

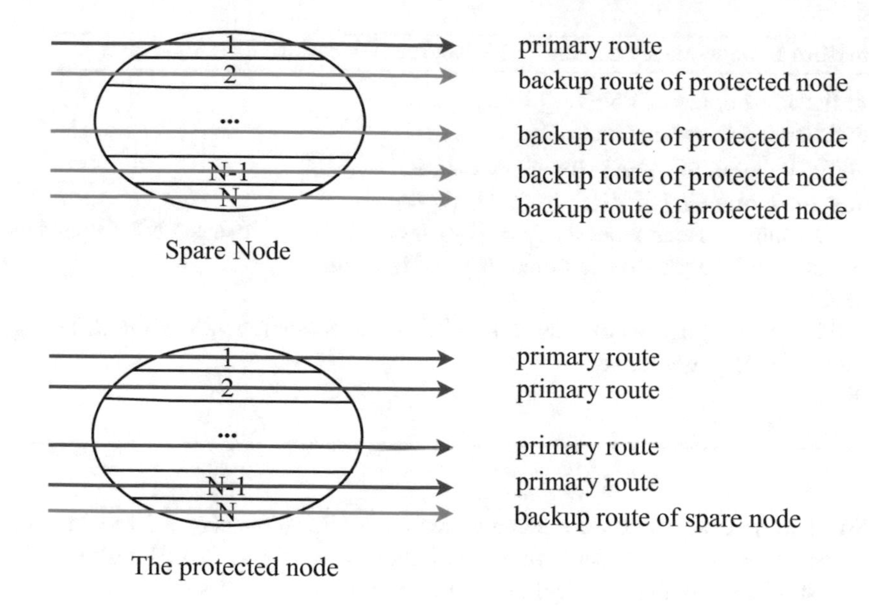

Fig. 1. Backup concept

We use an example to illustrate further. In Fig. 2, we have thirteen SFCs with VNF1. In the procedure of allocating During the backup process, the spare node is selected to accommodate VNF1 of the first SFC, where a node can support up to four instances of VNF1. The VNF1 instances of the other twelve SFCs are arranged in another four physical nodes, where each node accommodate only three VNF1 instances. The resource allocation scheme ensures that the spare resources of the spare node can serve as the backup instances for the VNF1 instances of the other $N - 1$ nodes and the VNF1 instance of the first node can also be protected by the other $N - 1$ nodes since these nodes have spare resources.

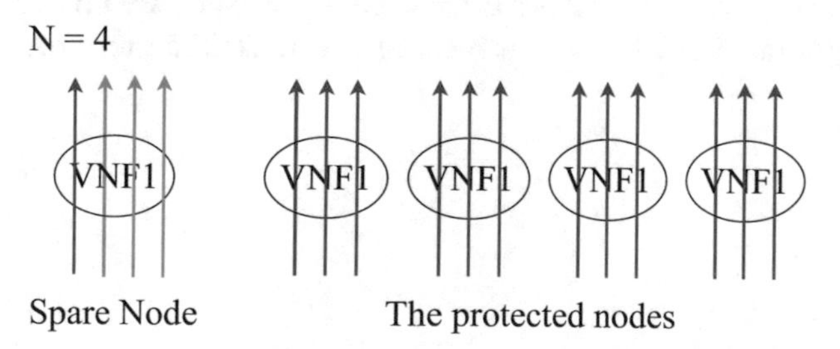

Fig. 2. Protected nodes and Spare node

Algorithm 1. Dependency Sorting and Resource Allocation SFG Algorithm

Input: list of grouping of VNFs : $Q = \{q_j\}$
Output: Service Function Graph $T'(.)$
1: Initial a Graph for service functions **Result**;
2: **for** the grouping j of VNFs in Q, $j \in |Q|$ **do**
3: Arrange service requests $S_j = \{s_k \mid k = 1,2,...,|S_j|\}$ using VNF Dependency Sorting SFC Sequence Algorithm; // See Algorithm 2
4: **end for**
5: Add one of sequences of request to $T'(.)$ using Resource Allocation SFG Algorithm; // See Algorithm 3
6: **Result** $\leftarrow T'(.)$
7: **return Result**

We note that the proposed scheme of resource allocation may reserve additional spare instances. In the previous example, if there are only 12 VNF1 instances, our scheme still requires five physical nodes for both primary and spare instances. It is also possible to allocate three nodes for primary VNF1 instances and one spare node without allocating any VNF1 instance. For the latter case, the spare node is used to protect the failure of one of three nodes with primary VNF1 instances. However, with the proposed scheme, we can allocate the spare node to the node with the highest reliability in order to improve the reliability of all SFCs.

Next, we construct an SFG from the first SFC to the last SFC by sharing their instances. The procedure of SFG construction and resource allocation is depicted in Algorithm 1. The algorithm employs another two algorithms for the generation of feasible VNF sequences (Algorithm 2) and resource allocation (Algorithm 3). Initially, an empty SFG is generated. Algorithm 2 is then used to generate all feasible sequences of VNFs in each SFC. It employs sorting and recursion to systematically evaluate where a sequence can meet the requirements of VNF dependency. The acceptable sequences undergo continuous sorting and recursive processing until the final VNF is reached to generate in a comprehensive list of acceptable VNF sequences. Once the list of feasible SFC sequences is available, Algorithm 3 selects a single sequence from the list. The selected sequence serves as a basis for constructing an SFG. By utilizing the selected sequence, Algorithm 3 efficiently constructs an SFG to facilitate the process of resource allocation.

Algorithm 2. VNF Dependency Sorting for feasible SFC Sequences

Input: $begin = 0$, $\overrightarrow{D}_j^Q = \{d_i \mid v_i \in I_j^Q\}$
Output: $S_j = \{s_k \mid k = 1,2,...,|S_j|\}$

```
1: if begin = |I_j^Q| then
2:     for each v_i do
3:         if d_{i-1} < d_i then
4:             this sequence not accepted;
5:             return
6:         else
7:             continue sorting;
8:         end if
9:     end for
10:    Sorting service request with D_j^Q;
11:    Next, append service request to S_j;
12: else
13:    Recursion and Swap;
14:    for i ∈ [begin, |I_j^Q|] do
15:        i is the order of VNF in the sequence;
16:        swap begin and i according the VNF type;
17:        recursion (I_j^Q, begin + 1);
18:        swap begin and i according the VNF type;
19:    end for
20: end if
21: return S_j
```

The process of checking the sequence for acceptance in Algorithm 2 entails two fundamental steps: sorting and swapping. In the first step, the VNFs of a SFC is sorted based on the predefined dependency order. Then, the VNFs with the same dependency order are swapped to generate all feasible sequences. Upon completion, Algorithm 2 generates an output comprising of all feasible sequences of an SFC.

After examining the list of feasible sequences, Algorithm 3 proceeds to construct an SFG. The construction process begins by considering the first SFC request. For an incoming VNF, it checks whether the spare node has already been incorporated into the SFG. If the answer is negative, it allocates a new node and designates it as a spare node. However, if the spare node already exists in the SFG, Algorithm 3 simply inserts a new instance to accommodate the request. The process is repeated for subsequent requests of the same type until a node is assigned $N - 1$ requests, as indicated in lines 2 to 11 of Algorithm 3, where N denotes the maximum number of allowable instances for a VNF in a node. This step ensures the appropriate allocation of resources for a SFG.

Algorithm 3. Resource Allocation SFG Algorithm

Input: $T(.)$, N
Output: $T'(.)$
1: Set pre-arranged instance v_p
2: **for** each *reqs* $\in$ T(.) **do**
3: **for** each $v_i \in r_k$ **do**
4: check the spare instance v_s with same VNF type is in T(.);
5: or the VNF type first appeared;
6: **if** v_s is exists **then**
7: $v_c \leftarrow$ choose a instance of T(.) with enough resources (N - 1) and arrange v_c on it;
8: **else**
9: Initial the new instance v_s
10: **end if**
11: **end for**
12: Connect one edge from v_c to v_p;
13: $v_p \leftarrow v_c$;
14: Replace pre-arranged instance v_p with current instance v_c;
15: **end for**
16: SFG $T'(.)$ construction finished;
17: **return** $T'(.)$

Since the SFG contains information of all requests, it utilizes the mapping strategy (Algorithm 4) to arrange instances in the SFG and map them to the corresponding physical nodes. First, we group all instances of SFG into several small SFGs according to the connectivity, as depicted in lines 1 to 6 of Algorithm 4. Second, the small SFG with the most instances is processed first, as depicted in lines 7 to 11 of Algorithm 4.

Algorithm 4. SFG Mapping Algorithm

Input: $T'(.)$
Output: *placement*
1: Divide each instances in $T'(.)$ into the small SFG based on the connectivity;
2: **for** each *instance* $\in$ T(.) **do**
3: **for** each *req* pass through *instance* **do**
4: add *req* into the *small SFG group*;
5: **end for**
6: **end for**
7: Sorting small SFG groups by the number of instances of them;
8: **for** each *small SFG group* $\in$ T(.) **do**
9: Backtracking the mapping composition of physical nodes in the iteration times;
10: Find the minimum SFC delay of *placement* in the iteration times;
11: **end for**
12: **return** *placement*

4 Performance Evaluation

In this section, we conduct the proposed method and its algorithm. We evaluate the performance of the proposed method. We also compare our scheme with two previous algorithms, CoShare and ISRCA. All executions of three algorithms are conducted on a four core and 32 GB of RAM computer. For the service requests, each request consisted of two or three different-type VNFs. There are nine VNF types, and the corresponding resource requirements for VNF nodes are randomly assigned from 60 to 100. The reliability values of VNF instances are ranged from 0.99 to 0.999. The reliability requirement of a request is set at 0.975.

We employ the condensed European topology with 278 nodes for our experiments, where each node had 300 units of resources for deploying VNF instances. Each node is also randomly assigned a value of hardware reliability from 0.9 to 0.999. We assume that physical nodes on the substrate network can host only one type of VNF. For the condensed European topology, there are 10 to 110 SFC requests. Each experiment is repeated by 1000 times to calculate the average values.

The two parameters are used to estimate the values of the proposed algorithms:

- *Acceptance ratio*: The ratio formula is counting the accepted SFCs and then dividing by the number of all SFCs in a delivery.
- *Backup resources*: The efficiencies of backup resource allocation of all methods are evaluated, which meeting the requirements of reliability to consider the number of instances used for backup and backup resource consumption.
- *Backup resources*: The efficiencies of backup resource allocation of all methods are evaluated, which meeting the requirements of reliability to consider the number of instances used for backup.

We first show the reliability performance of different algorithms with different number of requests in Fig. 3. It has been observed that as the number of requests increases, the SFC reliability tends to decrease. This trend is particularly evident in the context of ISRCA, where an increasing number of requests leads to a higher proportion of nodes without any backup. To address this issue, the CoShare method has been proposed, which enables nodes to share their resources with multiple SFCs for backup purposes to increase the number of backup nodes. As compared to CoShare, our algorithm demonstrates higher reliability due to the enhanced stability of the spare node. In our approach, we ensure that the whole node is under protection. Moreover, when multiple backup nodes are available, the reliability approaches a value close to 1 to achieve a significant improvement in reliability.

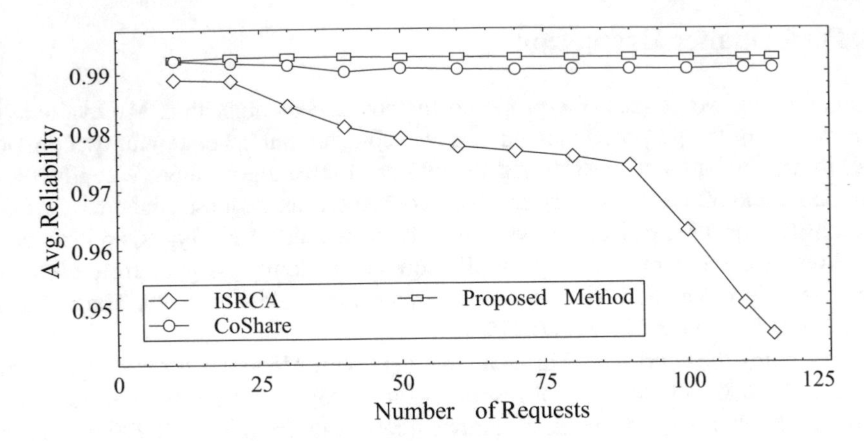

Fig. 3. Average reliability

Next, we show the average acceptance ratio of different algorithms, as illustrated in Fig. 4. This metric provides insights to the efficiency and effectiveness of our approach in accepting and processing a significant number of requests. By considering the impact of the number of requests on reliability based on instance sharing and improving upon it with our proposed approach, we achieve higher reliability and stability. This is demonstrated by the increased reliability of the spare node and the superior reliability achieved with multiple backup nodes. The average acceptance ratio further indicates the effectiveness of our method in handling a large number of requests.

In the last results, we show the total number of instances in Fig. 5. The results show that our algorithm may require more instances than CoShare owing to the less efficient backup instance allocation. However, this is reasonable tradeoff for better reliability and acceptance ratio.

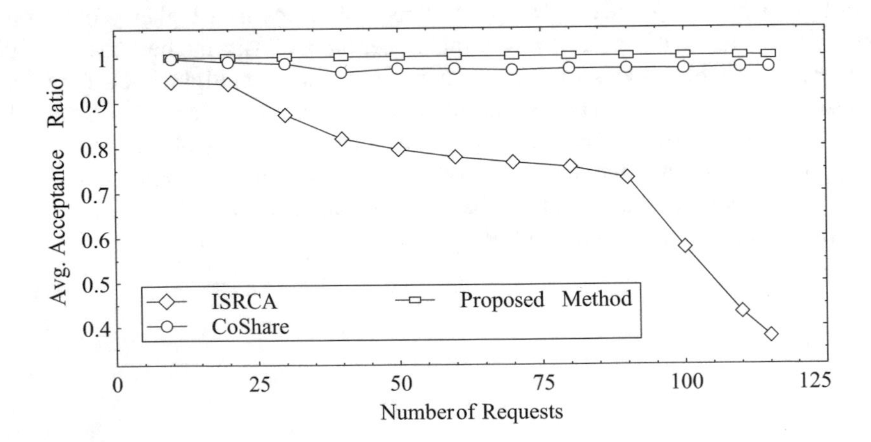

Fig. 4. Average accept ratio

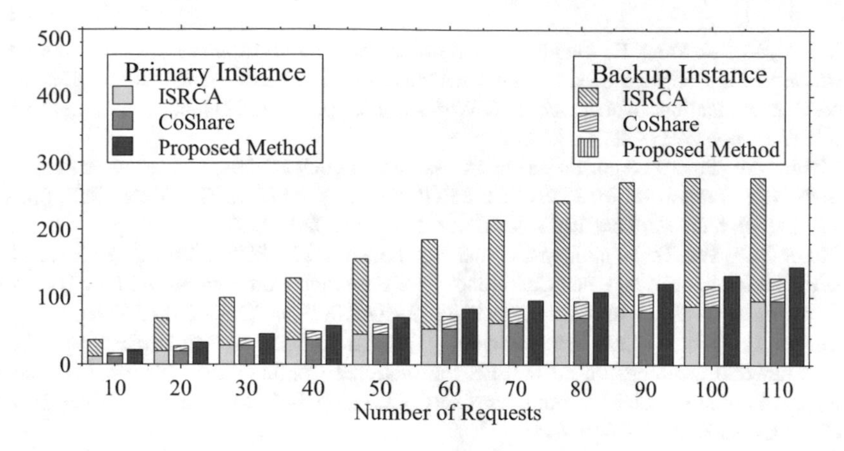

Fig. 5. Average number of instances

5 Conclusion

NFV enables SFC to offer users adaptability by deploying network devices through software. The reliability of SFC is influenced by both hardware devices and software. In this work, we propose a method to maintain the reliability of SFC through backup. When considering backup, it is essential to allocate backup resources effectively. We present an SFG construction algorithm that considers resource allocation and backup arrangement during the VNFs forwarding-graph construction phase prior to the mapping phase. We also provide a mapping strategy for SFG that achieves good performance in terms of latency. The experiment results demonstrate that the proposed method achieves superior performance of reliability and acceptance ratio for different number of SFC requests. Our future work would focus on minimizing the number of backup instances while maintaining the reliability performance.

Acknowledgements. This work was supported by NSTC Taiwan under Grant NSTC 112-2221-E-005-059-MY2 and NSTC 111-2221-E-005-045.

References

1. Alwakeel, A.M., Alnaim, A.K., Fernandez, E.B.: Toward a reference architecture for NFV. In: 2019 2nd International Conference on Computer Applications Information Security (ICCAIS), pp. 1–6 (2019). https://doi.org/10.1109/CAIS.2019.8769449
2. Galdamez, C., Ye, Z.: Resilient virtual network mapping against large-scale regional failures. In: 2017 26th Wireless and Optical Communication Conference (WOCC), pp. 1–4 (2017). https://doi.org/10.1109/WOCC.2017.7928978
3. Huang, Z., Huang, H.: Proactive failure recovery for stateful NFV. In: 2020 IEEE 26th International Conference on Parallel and Distributed Systems (ICPADS), pp. 536–543 (2020). https://doi.org/10.1109/ICPADS51040.2020.00075
4. Kanizo, Y., Rottenstreich, O., Segall, I., Yallouz, J.: Optimizing virtual backup allocation for middleboxes. In: 2016 IEEE 24th International Conference on Network Protocols (ICNP), pp. 1–10 (2016). https://doi.org/10.1109/ICNP.2016.7784411

5. Liu, Y., Xu, Z., Yang, F., Kuang, L.: Node-resource- and user-demand-aware resource allocation in NFV-enabled elastic optical networks. In: 2021 IEEE International Conference on Communications Workshops (ICC Workshops), pp. 1–6 (2021). https://doi.org/10.1109/ICCWorkshops50388.2021.9473602
6. Monir, M.F., Pan, D.: Application and assessment of click modular firewall vs pox firewall in SDN/NFV framework. In: 2020 IEEE REGION 10 CONFERENCE (TENCON), pp. 991–996 (2020). https://doi.org/10.1109/TENCON50793.2020.9293713
7. Monir, M.F., Pan, D.: Exploiting a virtual load balancer with SDN-NFV framework. In: 2021 IEEE International Black Sea Conference on Communications and Networking (BlackSeaCom), pp. 1–6 (2021). https://doi.org/10.1109/BlackSeaCom52164.2021.9527807
8. Thorat, P., Dubey, N.K.: Pre-provisioning protection for faster failure recovery in service function chaining. In: 2020 IEEE International Conference on Electronics, Computing and Communication Technologies (CONECCT), pp. 1–6 (2020). https://doi.org/10.1109/CONECCT50063.2020.9198654
9. Tomassilli, A., et al.: Poster: design of survivable SDN/NFV-enabled networks with bandwidth-optimal failure recovery. In: 2019 IFIP Networking Conference (IFIP Networking), pp. 1–2 (2019). https://doi.org/10.23919/IFIPNetworking46909.2019.8999461
10. Wang, Y., et al.: Reliability-oriented and resource-efficient service function chain construction and backup. IEEE Trans. Netw. Serv. Manage. **18**(1), 240–257 (2021). https://doi.org/10.1109/TNSM.2020.3045174
11. Woldeyohannes, Y.T., Tola, B., Jiang, Y., Ramakrishnan, K.K.: CoShare: an efficient approach for redundancy allocation in NFV. IEEE/ACM Trans. Netw. **30**(3), 1014–1028 (2022). https://doi.org/10.1109/TNET.2021.3132279
12. Xu, S., Ji, X., Liu, W.: Enhancing the reliability of NFV with heterogeneous backup. In: 2019 IEEE 3rd Information Technology, Networking, Electronic and Automation Control Conference (ITNEC), pp. 923–927 (2019). https://doi.org/10.1109/ITNEC.2019.8729059

Highly Reliable Communication Using Multipath Slices with Alternating Transmission

Italo Tiago da Cunha[1(✉)], Eduardo Castilho Rosa[2], Rodrigo Moreira[3], and Flávio de Oliveira Silva[4,5]

[1] Federal University of Jataí (UFJ), Jataí, GO, Brazil
italo@ufj.edu.br
[2] Goiano Federal Institute (IFGoiano), Catalão, GO, Brazil
eduardo.rosa@ifgoiano.edu.br
[3] Federal University of Viçosa (UFV), Rio Paranaíba, MG, Brazil
rodrigo@ufv.br
[4] Federal University of Uberlândia (UFU), Uberlândia, MG, Brazil
flavio@ufu.br
[5] Centro Algoritmi, University of Minho, Braga, Portugal
flavio@di.uminho.pt

Abstract. The development of mobile networks with network-slicing technologies has facilitated the delivery of increasingly customized applications to customers in various business verticals. Network slicing technology has made it possible to tailor applications to meet specific and rigorous requirements. However, Ultra-Reliable Low-Latency Communications (URLLC) remain challenging due to deploying network slices over an unreliable physical substrate. Traditionally, reliability has been enhanced through multipath, but this approach is limited to delivering URLLC. To address this issue, we propose and evaluate a method based on P4 that utilizes slices to enable reliability using the Alternating Transmission (AT) technique without Packet Duplication (PD). Our approach guarantees slice reliability while saving bandwidth compared to existing state-of-the-art methods.

1 Introduction

Mobile networks significantly facilitate the world's digital transformation by revolutionizing the performance of everyday tasks [1]. 5G technology has enabled innovations such as real-time video transmission, integration with cloud computing, and compatibility with the Internet of Things (IoT). These advancements have enabled the successful design and deployment of various technologies, including technology-enhanced clothing, health monitors, autonomous vehicles, smart cities, and Industry 4.0 [1].

Towards digital transformation, Network Slices (NS) has received attention and efforts in developing and standardizing the mobile network 5G. NS provides deployment of isolated logical networks, self-managed with heterogeneous requirements over a general purpose network [2,3]. NS enables on-demand personalized services in networks with limited resources, enabling optimal utilization in static and mobile environments [4]. That is, NS allows several customized services on demand to be served by

L. Barolli (Ed.): AINA 2024, LNDECT 202, pp. 471–482, 2024.
https://doi.org/10.1007/978-3-031-57916-5_40

the same physical network, in which resources can be dynamically allocated through logical slices according to Quality of Service (QoS) requirements.

Improving the dependability of connections across a network that is likely to experience issues is a significant challenge for new network architectures. These systems are frequently utilized in critical applications, such as smart grids, industrial Internet, remote surgeries, intelligent transportation systems, vehicle communications, high-speed trains, drones, and industrial robots. There is no universal standard for consuming these applications due to their unique characteristics [3,5].

Recent advances in literature focused on tackling the challenge of ultra-reliable network slicing by utilizing PD in Multi-Connectivity (MC) scenario [6]. These efforts do not rely on alternating transmission. This study presents a novel, highly reliable communication approach using multiple slices bringing together AT with multipath support. Using data-plane programmability, we prototype our approach using Mininet and a programmable P4-based software switch (*bmv2*) [7,8]. This prototype uses a dispatcher supporting multipath together AT between peers. In this work, multipath support means more than two paths between the communicating peers. The AT capacity uses the multiple slices to provide reliability by optimizing bandwidth consumption.

The main contribution of this work is to use the NS concept to increase reliability by using multipath support with AT. Our work avoids the downsides of PD with an optimized bandwidth consumption that increases reliability by increasing the number of NS.

The remainder of this work is organized as follows: Sect. 2 presents a short state-of-art survey and related work to our rationale. Section 3 describes the proposed packet multipath alternating transmission and usage scenarios. Section 4 presents the experimental scenario we used to evaluate our approach and discuss experimental results. Finally, in Sect. 5, we conclude the paper while highlighting the future research directions.

2 Related Work

This section discusses related studies aimed at deploying ultra-reliable network slicing. Finally, we contrast our work in Table 1 by highlighting important features and their differences.

To achieve high reliability and low latency, Islambouli et al. 2019 [9] make use of more than one network interface, network, and path per User Entity or, User Equipment - UE, for sending packets, with emphasis on duplicates of fragments of packets. In this scenario, there is not only an over-allocation of resources, but there can also be a considerable consumption of resources, such as batteries and data.

The work of Mahmood et al. 2019 [10], provides, according to the authors, a complete analysis of MC. In particular, the improvement of reliability with MC, measured in terms of interruption probability gain, which is obtained analytically. In addition, the operational cost of the MC in terms of resource utilization is also analyzed. Overall, [10] emphasizes that the price for increased reliability with MC is almost double that of single connectivity regarding resources and additional signaling overhead. However, it was considered that resource efficiency is not the main performance indicator for many applications that require high reliability. However, this arouses a strong motivation to investigate more resource-efficient MC schemes. This time, the present work aims to optimize resources to provide reliability, as initially shown in [11].

According to Gebert and Wich 2020 [12], they used *survival time* to devise a new communication scheme “AT”, which enables the alternate sending of packets, either through DC or carrier aggregation. It should be noted that both in the alternating transmission of packets via DC and carrier aggregation, there is no duplication of packets and that the work does not address URLLC and NS. And through the results obtained, it may be interesting to apply AT in the context of URLLC, combined with multipath slices - what we did here.

Something interesting at Centenaro et al. 2020 [13] is that they address DC in downlink by proposing two mechanisms that prevent unnecessary duplicate transmissions to optimize the spectrum. It should also be noted that the mechanisms seek to provide transmission redundancy only at the end of a communication.

To provide reliability, Shahriar et al. 2020 [14] propose the incorporation of two techniques to reduce the resources of backup: (i) bandwidth elasticity, i.e., the adjustment of the amount of bandwidth guaranteed in case of failures to a minimum value; and (ii) provisioning of multiple alternate paths - restricted to the EON scope. Thus, the simulated work presents interesting contributions, especially in formalism, but it does not understand E2E communication. Based on this, the present research will expand and implement this in E2E communication via multipath to increase reliability.

Concerning the simulated work of Sweidan et al. 2020 [15], which is an extension of the work of Islambouli et al. 2019 [9], the authors discuss a more general and practical network model that serves multiple applications to identify specific paths in the network as part of a dedicated slice that meets stringent performance requirements. In a more general scope, the objective is to maximize the number of flows allowed by applications, considering different requirements, both in terms of delay and reliability, inherent to ongoing URLLC applications. Differently from the work proposed here, [15] did not use slices containing replicas of packages and even elasticity in the referred ones. These points are essential to guarantee reliability and low latency in E2E communication - even more so if it crosses CN.

For short, the works of Mishra et al. [16,17] and Kar et al. 2023 [18], via Simu5G develop a novel NR-DC architecture, a multi-connectivity with Packet duplication via Simu5G and Reduces frequent handovers and allows high UE mobility with no additional complexity, respectively.

Showing the relevance of MC, Susloparov et al. 2022 [19] considered a scenario in which AR/VR traffic is served in a 5G system that supports MC. Another research from Paropkari and Beard 2023 [20] presents a technique that duplicates enough packets across multiple connections to meet the outage criteria.

Majamaa et al. 2023 [21] confirm that packet duplication (PD) through multi-connectivity is a promising solution to ensure reliable communication in such networks. With this in mind, they propose a dynamic PD activation scheme for Non-Terrestrial Networks (NTNs) based on hybrid automatic repeat request feedback.

Elias et al. 2023 [22] analyzes the problem of admission control and resource allocation in MC scenarios, considering different requirements and 5G NR features. To solve these problems, they developed an optimization framework that decides which users to admit in the system, whether to activate multiple connections to satisfy user requirements and how to allocate radio resources.

Kamboj et al. 2023 [23] punctuated that software-defined multipath routing is a viable approach to fulfill such QoS requirements by improving the data delivery performance through multipath. Opportunity to improve the data delivery performance through multipath to achieve QoS. Another research that caught our attention came from Großmann and Homeyer 2023 [24], which presented a solution based on P4 to duplicate packets to send them to their destination via multiple routes. This differs from what we propose in the number of paths that can be used and the Alternating Transmission technique that we use.

Interestingly, the idea of incoming packets in an MC way is extended to wireless technology. Levitsky et al. 2022 [25] confirm that when they wrote that Multi-link Operation (MLO) is a key feature of Wi-Fi 7, which is currently under development. MLO improves throughput and latency by allowing two devices to establish multiple links between them. The maximal gains are achieved when the links do not interfere, and the devices can use them independently. And we know, as shown in the Evaluation section, that this is true in the context of what we're proposing.

Alsakati et al. 2023 [26] evaluated the performance of MLO, using different policies, in serving Augmented Reality (AR) applications compared to Single-Link (SL). They support the idea that AR applications require high throughput, low latency, and high reliability to ensure a high-quality user experience. Related to the idea of MLO, Ali and Bellalta 2023 [27] proposed a collaborative ML approach to training models across multiple distributed agents without exchanging data to learn the best MLO-Link Allocation (LA) collaboratively, what can be appreciated to the adoption of new paths.

Carrascosa-Zamacois et al. 2023 [28] presented the first work studying the performance of MLO using real spectrum occupancy measurements. All this work related to MLO reinforces that the idea of alternating transmission of packets in a multipath way - guaranteeing reliability- is a promising area of research relevant to our work.

We summarize our brief literature review in Table 1. The rationale behind this table relies on the contrast we employed, considering some features. The column "DC/MC" means: Dual Connectivity and/or Multi-Connectivity. It's important to highlight that the multipath capacity means more than two paths between the communicating peers in the work.

Table 1. Short State-of-the-Art Survey.

Approach	Addresses 5G/B5G	DC/MC	Reliability	Multipath Support	Alternating Transmission
Islambouli *et* al. 2019 [9]	●	●	●	●	○
Mahmood *et* al. 2019 [10]	●	●	●	○	○
Gebert; Wich 2020 [12]	●	●	○	○	●
Centenaro *et* al. 2020 [13]	●	●	●	○	○
Shahriar *et* al. 2020 [14]	●	●	●	●	○
Sweidan *et* al. 2020 [15]	●	●	●	●	○
Mishra *et* al. 2021 [16]	●	●	●	○	○
Mishra *et* al. 2022 [17]	●	●	●	○	○
Levitsky *et* al. 2022 [25]	○	●	○	○	○
Susploparov *et* al. 2022 [19]	●	●	●	○	○
Paropkari; Beard 2023 [20]	●	●	○	○	○
Kar *et* al. 2023 [18]	●	●	●	○	○
Elias *et* al. 2023 [22]	●	●	●	○	○
Ali; Bellalta 2023 [27]	○	●	●	○	○
Alsakati *et* al. 2023 [26]	○	●	●	○	○
Carrascosa-Zamacois *et* al. 2023 [28]	○	●	○	○	○
Majamaa *et* al. 2023 [21]	●	●	●	○	○
Kamboj *et* al. 2023 [23]	●	●	○	●	●
Großmann; Homeyer 2023 [24]	○	●	●	○	○
Our Work	●	●	●	●	●

3 Multipath Slices with Alternating Transmission

This section details the reliability of using multiple NS to provide multipath and AT. First, we assumed that a network interface does not always send packets because it depends on the upper layer (application layer) for this to occur. Thus, there is a small timeframe in which the "link" medium is idle. This idle time is presented in Fig. 1 by time distance between two packets. Figure 1 also abstracts three slices with distinct paths between two communicating peers. If the packet within the transmission uses different slices, the picture represents AT without PD. Thus, we have no reliability in the event of failure.

As shown in Fig. 1, the sender transmits packets in multipath mode to the destination, which may occur through three different slices. No guarantee mechanism is available for the retransmission of a packet in the event of failure of any of the slices, which is the responsibility of higher layers and most likely requires a retransmission request from the recipient for this to occur. This causes a delay in the acknowledgment of packets, as they may have been corrupted or even lost. Therefore, it compromises real-time applications not only in terms of reliability but also in terms of delay.

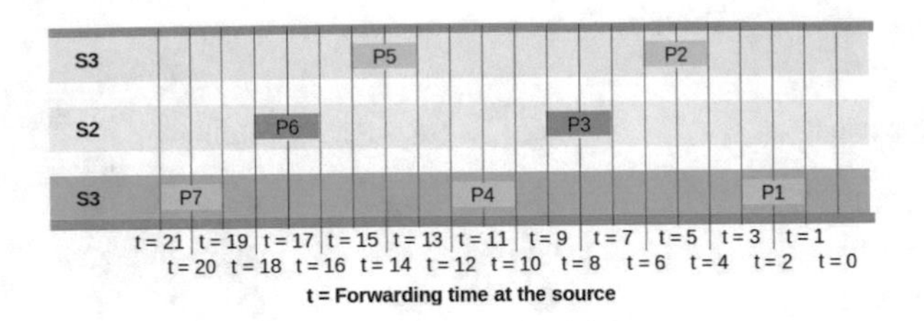

Fig. 1. Scenario of packets forwarding with no reliability using Multipath Slices with AT.

The rationale behind our method is to fit duplicates in this time interval when slices are idle, as shown in Fig. 2. That is, the copy of a packet that has just been passed through one slice is merged into another. In this way, slice bandwidth usage will be optimized, and there will be a significant increase in reliability, which is essential for URLLC applications.

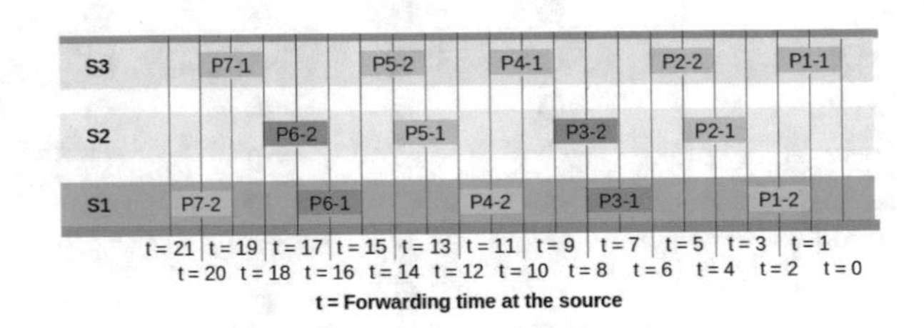

Fig. 2. Approach to get reliability via Multipath Slices using three slices with AT.

Figures 2 and 3 illustrate the behavior of our method in an ideal scenario with no packet loss and three and four active slices, respectively. Idle slices are used to send both original and redundant packets in an interleaved manner, thus ensuring the reliability of the communication. Notably, using PD is not indistinct, flooding all slices with replicas. As shown in Figs. 2 and 3, the PD occurs in a controlled way with a single copy alternating the transmissions between the available slices during communication.

From the analysis of Fig. 3, it can be observed that using our technique of alternating packet forwarding between slices, the number of duplicates in the network as a whole dropped to 50%.

Figure 4 shows the proposed approach for a failure scenario. According to Fig. 4, communication is served by three distinct slices in which duplicates of packets that have just been transmitted are interspersed. For example, Packet 2 "P2-1" was initially forwarded by the sender through Slice S2 at time instant t3 and, being an instant of time later, forwarded a copy of it through Slice S3 "P2-2". This optimizes the use of slices and increases the reliability. If there is a loss or corruption of a packet, communication will be maintained without the need for retransmission.

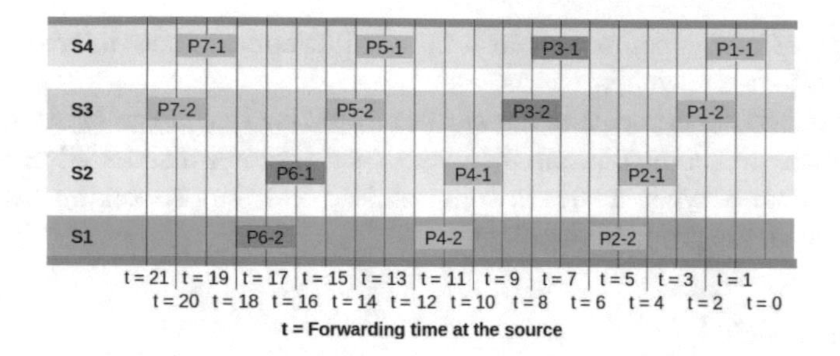

Fig. 3. Approach to get reliability via Multipath Slices using four slices with AT.

Alternatively, multipath slices can handle and divide traffic loads among other slices. As shown in Fig. 4, these slices are already active and inserted in the communication without the need for additional time to instantiate a new slice [14, 29, 30].

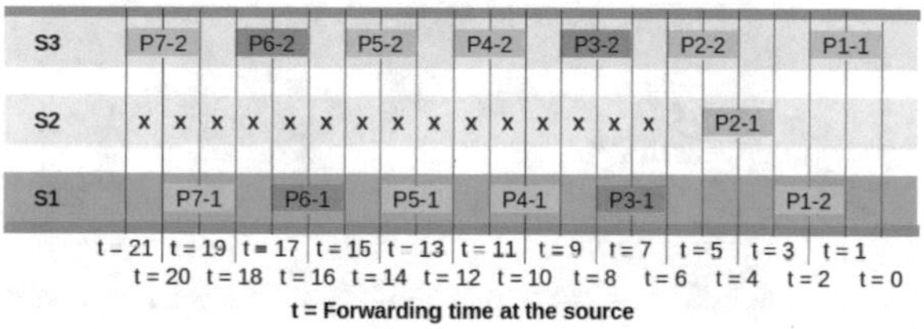

Fig. 4. Fail in slice 2 and communication working without loss or delay.

According to Fig. 4, at time frame 6, Slice S2 no longer forwards packets, either because of intermittent or permanent failure. However, communication is not disrupted since slices S1 and S3 are still active, absorbing the traffic load of the one that failed, the packets (P5-1) and (P3-2, P6-2), respectively. It should be noted that such an approach changes the sender slice for some packets, as happened with P4-1, which in Fig. 2 arrived through Slice 3, and now, through Slice 1 (Fig. 4).

NS support is a standard 5G/B5G network. Our method explores the native capacity of these networks to provide reliable communication. So, instead of considering that each slice offers different capabilities according to the application on top, the approach here is to use multiple slices deployed in multiple paths with alternating transmission to provide to the User Equipments (UEs) a highly reliable scenario natively. We argue that a single NS will not support applications that require highly reliable communication.

4 Experimental Evaluation

In this section, we describe the experimental setup and the preliminary results. We conducted two experiments based on a mininet to simulate a multi-slice environment.

The topology of both experiments consisted of multi-homed nodes interconnected with BMv2 [31] software switch instances.

We conducted experiments to answer the following questions: (1) Can communication between two communicants be maintained through distinct slices? (2) In the worst-case scenario, when a slice is compromised and "stops working," what impact does this have on the proposed topology?

4.1 Experiment 1

This experiment evaluated the effect of link failure on the End-to-End (E2E) throughput when using our dispatcher, in contrast to the traditional approach based on single-channel communication. This experimental scenario was similar to the Dual Connectivity (DC) with the PD rationale. As shown in Fig. 5, the end hosts are connected through two slices, each instantiated as an overlay tunnel on top of switches sw1, sw2, and sw4.

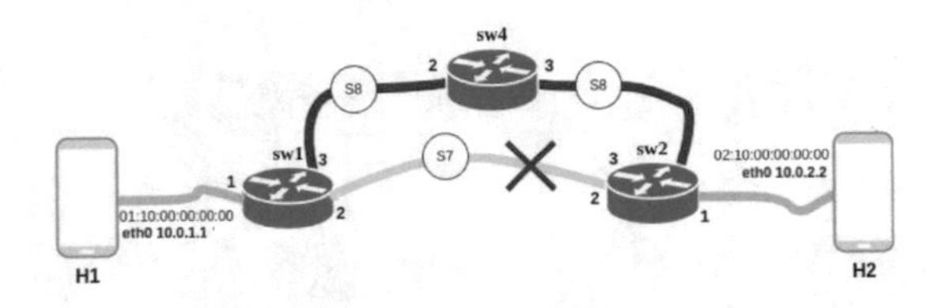

Fig. 5. Experimental scenario to use Multipath Slices in experiment 1.

We used the Python Scapy library to send 2000 packets from H1 to H2. The packets are sent continually with zero interpacket time. Thus, the maximum rate achieved was based on the speed at which the script library could send packets over the virtual interface. To simulate a link failure, we disabled the output face in switch 1 when the packet 1000th was sent.

The unexpected behavior of throughput up to 20 s in the simulation is due to the nondeterministic latency of `bmv2`, which is normal for software switches. However, as shown in Fig. 6, the throughput stabilizes from 20 s of simulation onward, and we can observe that the throughput is similar in both the multi-slice and traditional approaches. This behavior is expected in scenarios where link failures do not occur.

However, just after the second slice failure, we can see that the throughput goes to zero in the traditional method, whereas in our multi-slice approach, it is possible to maintain communication with 50% of the initial throughput. This is because, in our approach, we intelligently split the traffic between multiple slices to ensure reliability. We must note that the loss will not affect the destination because the user received another copy of the data.

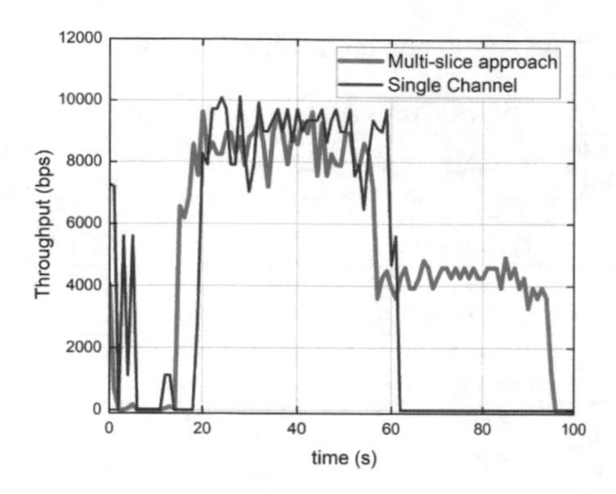

Fig. 6. Throughput when using a multi-slice approach.

4.2 Experiment 2

In the second experiment, we worked on three active slices, S8, S7, and S5, in a scenario with four switches and two hosts, as illustrated in Fig. 7. It should be noted that this scenario did not result in packet loss or even path interruption.

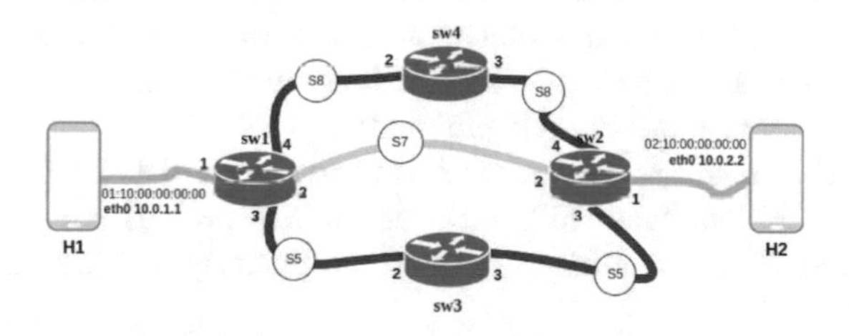

Fig. 7. Experimental scenario to use multipath slices.

As shown in Fig. 7, with three functional slices, the traffic in all of them was 66,66% of the total traffic for a single connection or in the PD scenario. In other words, a significant gain in bandwidth and, consequently, in reliability, especially cause, is not necessary to initialize a new slice in case of failure. Table 2 summarizes and extrapolates the results of Experiment 2 by considering four different slices.

Our method increased the reliability while reducing bandwidth wasting compared with naive PD. The best results were obtained when we used four slices for incoming packets, as shown in Table 2. In other words, the bandwidth consumption decreases by increasing the number of paths ("slices"). Therefore, we can aggregate more paths with smaller bandwidths to provide elasticity.

Table 2. Reliability and number of packets crossing the network

# of packets generated by the simulation	Source Transmission		Achieved Reliabiliy	Per Flow	
	Using	# of Packets		Bandwidth Consumption	# of Packets
1200	Single		○	100%	1200
3000	Connectivity		○	100%	3000
1200	Packet		●	100%	1200
3200	Duplication		●	100%	3000
1200	3 Slices		●	66.6%	800
3000	(Our Method)		●	66.6%	2000
1200	4 Slices		●	50%	600
3000	(Our Method)		●	50%	1500

5 Concluding Remarks

In this paper, we proposed and evaluated a new method for ultra-reliable network slicing by leveraging idle slices and network time slots. Existing approaches do not consider alternating transmissions in communication that involves more than two paths. On the other hand, our method can forward traffic with some redundant packets through multiple slices interleaved. Our solution enables new slices to be predesigned, supporting connectivity reliability when there is a problem between the communicants.

We introduced and assessed a mechanism that offers enhanced network-slicing dependability. Our evaluations demonstrated that, as the number of slices increased, the level of reliability also increased significantly.

In future work, we will refine our methods to ensure reliability in asymmetric network slices with different bandwidth allocations. Our findings provide fresh perspectives on achieving ultra-reliability in NS while efficiently utilizing unused resources from other slices.

Acknowledgements. The authors thanks the National Council for Scientific and Technological Development (CNPq) under grant number 421944/2021-8 (call CNPq/MCTI/FNDCT 18/2021), the Research Support Foundation of the State of São Paulo (FAPESP) grant number 2018/23097-3, for the thematic project Slicing Future Internet Infrastructures (SFI2) and Centro Algoritmi, funded by Fundação para a Ciência e Tecnologia (FCT) within the RD Units Project Scope 2020–2023 (UIDB/00319/2020) for partially support this work.

References

1. O'Connell, E., Moore, D., Newe, T.: Challenges associated with implementing 5G in manufacturing. Telecom **1**(1), 48–67 (2020)
2. Moreira, R., Rosa, P.F., Aguiar, R.L.A., Silva, F.O.: NASOR: a network slicing approach for multiple autonomous systems. Comput. Commun. **179**, 131–144 (2021)
3. Martins, J.S.B., et al.: Enhancing network slicing architectures with machine learning, security, sustainability and experimental networks integration. IEEE Access **11**, 69144–69163 (2023)
4. Chang, C.-Y., et al.: Performance isolation for network slices in industry 4.0: the 5Growth approach. IEEE Access **9**, 166990–167003 (2021)

5. Liu, Y., Clerckx, B., Popovski, P.: Network slicing for eMBB, URLLC, and mMTC: an uplink rate-splitting multiple access approach. IEEE Trans. Wireless Commun. **23**, 2140–2152 (2023)
6. Silva, M., Santos, J., Curado, M.: The path towards virtualized wireless communications: a survey and research challenges. J. Netw. Syst. Manage. **32**, 12 (2023)
7. Bosshart, P., et al.: P4: programming protocol-independent packet processors. ACM SIGCOMM Comput. Commun. Rev. **44**, 87–95 (2014)
8. Lantz, B., O'Connor, B.: A mininet-based virtual testbed for distributed SDN development. ACM SIGCOMM Comput. Commun. Rev. **45**, 365–366 (2015)
9. Islambouli, R., Sweidan, Z., Sharafeddine, S.: Dynamic multipath resource management for ultra reliable low latency services. In: 2019 IEEE Symposium on Computers and Communications (ISCC), pp. 987–992, June 2019. ISSN 2642-7389
10. Mahmood, N.H., Karimi, A., Berardinelli, G., Pedersen, K.I., Laselva, D.: On the resource utilization of multi-connectivity transmission for URLLC services in 5G new radio. In: 2019 IEEE Wireless Communications and Networking Conference Workshop (WCNCW), pp. 1–6, April 2019
11. Cunha, I., Rosa, E., Silva, F.: Comunicação Fim-a-Fim Altamente Confiável e de Baixa Latência entre UEs Móveis no Contexto de 5G/B5G via Multipath Slices Elásticas. In: Anais do XIII Workshop de Pesquisa Experimental da Internet do Futuro, Porto Alegre, RS, Brasil, pp. 41–46. SBC (2022). ISSN 2595-2692. Event-place: Fortaleza/CE
12. Gebert, J., Wich, A.: Alternating transmission of packets in dual connectivity for periodic deterministic communication utilising survival time. In: 2020 European Conference on Networks and Communications (EuCNC), June 2020. ISSN 2575-4912
13. Centenaro, M., Laselva, D., Steiner, J., Pedersen, K., Mogensen, P.: Resource-efficient dual connectivity for ultra-reliable low-latency communication. In: 2020 IEEE 91st Vehicular Technology Conference (VTC2020-Spring), May 2020. ISSN 2577-2465
14. Shahriar, N., et al.: Reliable slicing of 5G transport networks with bandwidth squeezing and multi-path provisioning. IEEE Trans. Netw. Serv. Manag. **17**, 1418–1431 (2020)
15. Sweidan, Z., Islambouli, R., Sharafeddine, S.: Optimized flow assignment for applications with strict reliability and latency constraints using path diversity. J. Comput. Sci. **44**, 101163 (2020)
16. Mishra, P., Kar, S., Bollapragada, V., Wang, K.-C.: Multi-connectivity using NR-DC for high throughput and ultra-reliable low latency communication in 5G networks. In: 2021 IEEE 4th 5G World Forum (5GWF), pp. 36–40, October 2021
17. Mishra, P., Kar, S., Wang, K.-C.: Performance evaluation of 5G multi-connectivity with packet duplication for reliable low latency communication in mobility scenarios. In: 2022 IEEE 95th Vehicular Technology Conference: (VTC2022-Spring), pp. 1–6, June 2022. ISSN 2577-2465
18. Kar, S., Mishra, P., Wang, K.-C.: A novel single grant-based uplink scheme for high throughput and reliable low latency communication. In: 2023 19th International Conference on Wireless and Mobile Computing, Networking and Communications (WiMob), pp. 169–174, June 2023. ISSN 2160-4894
19. Susloparov, M., Krasilov, A., Khorov, E.: Providing high capacity for AR/VR traffic in 5G systems with multi-connectivity. In: 2022 IEEE International Black Sea Conference on Communications and Networking (BlackSeaCom), pp. 385–390, June 2022
20. Paropkari, R.A., Beard, C.: Multi-connectivity-based adaptive fractional packet duplication in cellular networks. Signals **4**(1), 251–273 (2023)
21. Majamaa, M., Martikainen, H., Puttonen, J., Hämäläinen, T.: On enhancing reliability in B5G NTNs with packet duplication via multi-connectivity. In: 2023 IEEE International Conference on Wireless for Space and Extreme Environments (WiSEE), pp. 154–158, September 2023. ISSN 2380-7636

22. Elias, J., Martignon, F., Paris, S.: Multi-connectivity in 5G new radio: optimal resource allocation for split bearer and data duplication. Comput. Commun. **204**, 52–65 (2023)
23. Kamboj, P., Pal, S., Bera, S., Misra, S.: QoS-aware multipath routing in software-defined networks. IEEE Trans. Netw. Sci. Eng. **10**, 723–732 (2023)
24. Großmann, M., Homeyer, T.: Emulation of multipath transmissions in P4 networks with Kathará. In: KuVS Fachgespräch - Würzburg Workshop on Modeling, Analysis and Simulation of Next-Generation Communication Networks 2023 (WueWoWAS 2023), pp. 1–4. Universität Würzburg (2023)
25. Levitsky, I., Okatev, Y., Khorov, E.: Feasibility of simultaneous transmit and receive in Wi-Fi 7 multi-link devices: demo. In: Proceedings of the Twenty-Third International Symposium on Theory, Algorithmic Foundations, and Protocol Design for Mobile Networks and Mobile Computing, MobiHoc 2022, New York, NY, USA, pp. 293–294. Association for Computing Machinery, October 2022
26. Alsakati, M., Pettersson, C., Max, S., Moothedath, V.N., Gross, J.: Performance of 802.11be Wi-Fi 7 with multi-link operation on AR applications. In: 2023 IEEE Wireless Communications and Networking Conference (WCNC), pp. 1–6, March 2023. ISSN 1558-2612
27. Ali, R., Bellalta, B.: A federated reinforcement learning framework for link activation in multi-link Wi-Fi networks. In: 2023 IEEE International Black Sea Conference on Communications and Networking (BlackSeaCom), pp. 360–365, July 2023
28. Carrascosa-Zamacois, M., Geraci, G., Knightly, E., Bellalta, B.: Wi-Fi multi-link operation: an experimental study of latency and throughput. IEEE/ACM Trans. Netw. **32**, 308–322 (2023)
29. Zhang, W., Lei, W., Zhang, S.: A multipath transport scheme for real-time multimedia services based on software-defined networking and segment routing. IEEE Access **8**, 93962–93977 (2020)
30. Court, A., Alamleh, H.: Multi-path data transmission to protect data in transit. In: 2023 IEEE International Conference on Consumer Electronics (ICCE), pp. 1–6, January 2023. ISSN 2158-4001
31. Bas, A., Fingerhut, A., Sivaraman, A.: The behavioral model (BMv2), January 2024. Original-date: 2015-01-26T21:43:23Z

Author Index

L. Barolli (Ed.): AINA 2024, LNDECT 202, pp. 483–485, 2024.
https://doi.org/10.1007/978-3-031-57916-5

Printed in the United States
by Baker & Taylor Publisher Services